Botany and Plant Biology

Botany and Plant Biology

Editor: Austin Balfour

www.callistoreference.com

Callisto Reference,
118-35 Queens Blvd., Suite 400,
Forest Hills, NY 11375, USA

Visit us on the World Wide Web at:
www.callistoreference.com

ISBN: 978-1-63239-975-5 (Hardback)

Cataloging-in-Publication Data

Botany and plant biology / edited by Austin Balfour.
    p. cm.
Includes bibliographical references and index.
ISBN 978-1-63239-975-5
1. Botany. 2. Plants. 3. Agriculture. I. Balfour, Austin.
QK45.2 .B68 2018
580--dc23

Table of Contents

Preface

Botany is a branch of biology which studies plants. Some of the topics included in this subject are reproduction, plant diseases, plant taxonomy, structure and growth of plant and biochemistry. New techniques that have been developed in this field include live cell imaging, electron microscopy, optical microscopy, etc. While understanding the long-term perspectives of the topics, the book makes an effort in highlighting their impact as a modern tool for the growth of the discipline. With state-of-the-art inputs by acclaimed experts of this field, this book targets students and professionals.

After months of intensive research and writing, this book is the end result of all who devoted their time and efforts in the initiation and progress of this book. It will surely be a source of reference in enhancing the required knowledge of the new developments in the area. During the course of developing this book, certain measures such as accuracy, authenticity and research focused analytical studies were given preference in order to produce a comprehensive book in the area of study.

This book would not have been possible without the efforts of the authors and the publisher. I extend my sincere thanks to them. Secondly, I express my gratitude to my family and well-wishers. And most importantly, I thank my students for constantly expressing their willingness and curiosity in enhancing their knowledge in the field, which encourages me to take up further research projects for the advancement of the area.

Editor

Alterations in Auxin Homeostasis Suppress Defects in Cell Wall Function

Blaire J. Steinwand[9], Shouling Xu[¤9], Joanna K. Polko, Stephanie M. Doctor, Mike Westafer, Joseph J. Kieber*

Biology Department, University of North Carolina, Chapel Hill, North Carolina, United States of America

Abstract

The plant cell wall is a highly dynamic structure that changes in response to both environmental and developmental cues. It plays important roles throughout plant growth and development in determining the orientation and extent of cell expansion, providing structural support and acting as a barrier to pathogens. Despite the importance of the cell wall, the signaling pathways regulating its function are not well understood. Two partially redundant leucine-rich-repeat receptor-like kinases (LRR-RLKs), FEI1 and FEI2, regulate cell wall function in *Arabidopsis thaliana* roots; disruption of the FEIs results in short, swollen roots as a result of decreased cellulose synthesis. We screened for suppressors of this swollen root phenotype and identified two mutations in the putative mitochondrial pyruvate dehydrogenase E1α homolog, *IAA-Alanine Resistant 4 (IAR4)*. Mutations in *IAR4* were shown previously to disrupt auxin homeostasis and lead to reduced auxin function. We show that mutations in *IAR4* suppress a subset of the *fei1 fei2* phenotypes. Consistent with the hypothesis that the suppression of *fei1 fei2* by *iar4* is the result of reduced auxin function, disruption of the *WEI8* and *TAR2* genes, which decreases auxin biosynthesis, also suppresses *fei1 fei2*. In addition, *iar4* suppresses the root swelling and accumulation of ectopic lignin phenotypes of other cell wall mutants, including *procuste* and *cobra*. Further, *iar4* mutants display decreased sensitivity to the cellulose biosynthesis inhibitor isoxaben. These results establish a role for *IAR4* in the regulation of cell wall function and provide evidence of crosstalk between the cell wall and auxin during cell expansion in the root.

Editor: Ive De Smet, University of Nottingham, United Kingdom

Funding: This project was supported by grants IOS 0624377, MCB-1021704 from the National Science Foundation to JJK and Rubicon grant 825.13.018 of the Netherlands Organization for Scientific Research (NWO) to JKP. The funders had no role in study design, data collection and analysis, decision to publish, or preparation of the manuscript.

Competing Interests: The authors have declared that no competing interests exist.

* E-mail: jkieber@unc.edu

9 These authors contributed equally to this work.

¤ Current address: Department of Plant Biology, Carnegie Institution for Science, Stanford, California, United States of America

Introduction

Cell expansion plays a critical role in plant growth and development. The direction and extent to which cells expand is controlled by the rigid, yet highly dynamic cell wall. The cell wall is a major determinant of cell size and shape and consequently overall plant morphology. In roots, the architecture of the cell wall permits longitudinal cell elongation while restricting radial expansion, which leads to highly asymmetric, anisotropic growth [1–4].

Plant cell walls are composed primarily of load-bearing cellulose microfibrils, cross-linking hemicelluloses, and pectins. Together with a relatively small number of structural proteins, this matrix of polysaccharides lends the wall the strength and rigidity that is required for structural support and plant defense, while simultaneously allowing cells to expand as plants grow and develop [5]. During cell expansion, wall polymers are actively remodeled and rearranged and their synthesis is altered in response to both developmental and environmental cues [6]. The ability of cell walls to maintain structural integrity and function properly as changes in the architecture of the cell wall occur suggests that there is a sensing and feedback system in place to perceive and respond to changes in the wall. Despite a crucial role in the maintenance of plant cell wall function, our current understanding of the components and mechanisms involved in the perception of and response to regulatory input from the wall remains poorly understood.

Several members of the receptor-like kinase (RLK) family have been implicated as sensors of signals from the cell wall. In Arabidopsis, the RLK family is comprised of approximately 600 members, many of which have been shown to act in a variety of different signaling pathways that function throughout plant development [7]. Of those, members of three different subfamilies have been implicated in regulating cell wall function. The wall-associated kinases (WAKs) are tightly bound to the cell wall and are required for normal cell expansion [8–10]. In addition to the WAKs, four members of the *Catharanthus roseus* RLK1-Like (*Cr*RLK1L) subfamily (*HERCULES1, HERCULES2, FERONIA,* and *THESEUS1*) and two members of the leucine-rich repeat (LRR) subfamily (*FEI1* and *FEI2*) have been implicated in cell wall signaling. Although members from each of the three RLK subfamilies are required for cell expansion, only *THESEUS1 (THE1)*, its close homologs, and the FEIs have been linked to cell wall synthesis [11–13].

Mutations in *THE1* suppress ectopic lignin deposition and restore hypocotyl elongation in cellulose-deficient mutants, but do not restore cellulose biosynthesis in the *procuste* mutant [11]. These data suggest that *THE1* plays a role in sensing and actively responding to changes in the cell wall. Disruption of both *FEI1* and *FEI2* leads to a loss of anisotropic growth in rapidly expanding cells of the root elongation zone, but also affects cell expansion in the stamen filament and the hypocotyl of dark-grown seedlings [12]. In addition, the roots of double *fei1 fei2* mutants display ectopic lignin deposition, are hypersensitive to the cellulose synthesis inhibitor isoxaben, and synthesize less cellulose as compared to wild-type roots when seedlings are grown under non-permissive conditions of elevated salt or sucrose [12]. Further, disruption of *FEI2* leads to a reduction in the rays of cellulose observed in the mucilage of wild-type seeds [14,15]. These data suggest that FEI1 and FEI2 positively regulate cell wall function by promoting cellulose synthesis.

The fasciclin-like GPI-anchored extracellular protein SOS5 acts in the FEI pathway to regulate cell wall synthesis [12]. Like *fei1 fei2*, *sos5* mutants display short, swollen roots when grown under the restrictive conditions of elevated salt or sucrose, and this phenotype is reversed in both mutants by blocking ethylene biosynthesis, but not ethylene perception. SOS5 also regulates the synthesis of cellulose during the production of seed coat mucilage [14]. Introduction of *sos5* into the *fei1 fei2* mutant does not cause an additive phenotype, in contrast to other mutants affecting cellulose biosynthesis such as *cobra* [12]. The non-additive phenotype of *fei1 fei2* and *sos5* mutations suggests that the FEI RLKs act in a linear pathway with SOS5 to regulate cellulose synthesis. Taken together, these data suggest an important role for the FEI RLKs/SOS5 pathway in positively regulating cellulose synthesis.

In order to better understand the FEI signaling pathway, we sought to uncover additional components involved in regulating cell wall synthesis in the root. Here we describe the identification and characterization of a suppressor of the *fei1 fei2* mutant. We show that mutations in the previously characterized *IAA-Alanine Resistant 4* (*IAR4*) gene, encoding a putative mitochondrial E1α pyruvate dehydrogenase subunit, suppress the defects in root anisotropic cell expansion exhibited by *fei1 fei2*. *IAR4* was originally identified in a genetic screen for IAA conjugate-resistant mutants [16] and was subsequently identified as an enhancer of *tir1* auxin resistance [17]. Although the precise role of *IAR4* in the auxin biosynthesis pathway remains unclear, *iar4* mutants display phenotypes consistent with reduced endogenous auxin, accumulate IAA-amino acid conjugates, and are rescued by increasing endogenous IAA levels in the plant [16,17]. Thus, IAR4 is predicted to play an important role in maintaining auxin homeostasis. Here we show that reduced auxin function, via either an *iar4* single or a *wei8/tar2* double mutant, suppresses growth isotropy of cell wall mutants, including *fei1 fei2*. Our results shed light on the role of auxin in regulating cell wall function in the Arabidopsis root.

Results

Isolation and characterization of *shou2*

In order to identify additional elements regulating cell wall function, we screened for suppressors of the swollen root phenotype of *fei1 fei2* mutants. An M_2 population of ethyl methanesulfonate mutagenized *fei1 fei2* was screened for suppressors of the conditional short, swollen root phenotype of *fei1 fei2* seedlings. Eight independent suppressor lines that retested as robust *fei2 fei2* suppressors were identified from screening

approximately 200,000 M_2 seedlings representing 30,000 M_1 seeds. We designated these suppressors *shou* mutations (the Chinese word for thin). These suppressors represented seven distinct loci, two of which were allelic and were designated *shou2-1 and shou2-2*. The *fei1 fei2 shou2-1* and *fei1 fei2 shou2-2* lines both had significantly fewer and shorter root hairs. The F_1 of a backcross to the parental *fei1 fei2* line displayed a non-suppressed phenotype, and the suppressor phenotype segregated consistent with 3 non-suppressed: 1 suppressed ratio in the F_2 progeny of this backcross, consistent with *shou2* acting as a single locus, recessive mutation. In addition to the suppression of root length (Fig. 1A, B), the *shou2* mutations also suppress the radial swelling (Fig. 1A) and the radial expansion of cells in the elongation zone (Fig. 1C) in *fei1 fei2* roots. We isolated the *shou2-1* mutation by backcrossing to the wild type. This *shou2-1* single mutant line displayed fewer and shorter root hairs, similar to the *fei1 fei2 shou2-1*. Intriguingly, under the non-permissive conditions used to assess the *fei* phenotype (grown on MS+4.5% sucrose), both the *fei1 fei2* and *shou2-1* parental seedlings displayed roots that were significantly shorter that their wild-type counterparts, despite the fact that the *fei1 fei2 shou2-1* triple mutants displayed nearly wild-type root elongation in the growth conditions used for the suppressor screen.

SHOU2 is allelic to IAR4

We used a map-based positional cloning approach to isolate the *SHOU2* gene. The *fei1* and *fei2* mutations (isolated in the Columbia (Col) ecotype) were introgressed six times into the *Landsberg erecta* (L*er*) ecotype to generate a *fei1 fei2* plant that was largely L*er* except for small regions of DNA near the *fei1* and *fei2* mutations (see Materials and Methods). This line was crossed to *fei1 fei2 shou2* to generate a mapping population for *shou2-1*. Mapping with Col/L*er* SSLPs indicated that *SHOU2* was linked to the top of chromosome 1. Analysis of 350 *fei1 fei2* F_2 progeny with additional molecular markers further delimited *SHOU2* to a 47 kb interval between the F3I6.D and F3I6.F markers (Fig. 2A; Table S1 in File S1). Sequencing of candidate genes within this region identified missense mutations in the first and seventh exon of *IAR4* (AT1G24180) in *fei1 fei2 shou2-1* and *fei1 fei2 shou2-2* respectively. The *shou2-1* allele contains a C→T transition in the fifth codon of the coding region of *IAR4*, which converts an arginine residue to a stop codon. The *shou2-2* mutation is the result of a G→A transition that is predicted to change a glutamate at position 366 to a stop codon (Fig. 2B). To confirm that *shou2* mutations correspond to AT1G24180, we examined the ability of an independent T-DNA insertional allele that contains a T-DNA insertion in the first exon of *IAR4* (SALK_091909) to suppress *fei1 fei2*. This *shou2-3* allele, which has no full-length *IAR4* transcript (Fig. S1 in File S1), was introduced into a *fei1 fei2* mutant line by crossing and the phenotype of the roots was examined in non-permissive conditions (Fig. 2B). Similar to the other alleles, *shou2-3* suppressed the root swelling phenotype of *fei1 fei2*, which confirms that mutations in *IAR4* correspond to *shou2*. Additionally, we performed crosses between *fei1 fei2 shou2-1*, *fei1 fei2 shou2-2* and *fei1 fei2 shou2-3*. The F_1 progeny of all three of these crosses displayed a wild-type root phenotype with no visible swelling of the tip when grown on 4.5% sucrose (Fig. S2 in File S1), which suggests that *shou2-1*, *shou2-2*, *shou2-3* are indeed allelic. To avoid confusion and to be consistent with prior studies, we re-named *shou2-1*, *shou2-2*, and *shou2-3*, to *iar4-5*, *iar4-6*, and *iar4-7* respectively.

A role for auxin in regulating cell wall function

As *IAR4* is involved in the maintenance of auxin homeostasis and mutations in *IAR4* restore anisotropic root growth in *fei1 fei2*, we hypothesized that a reduction in the level of endogenous IAA

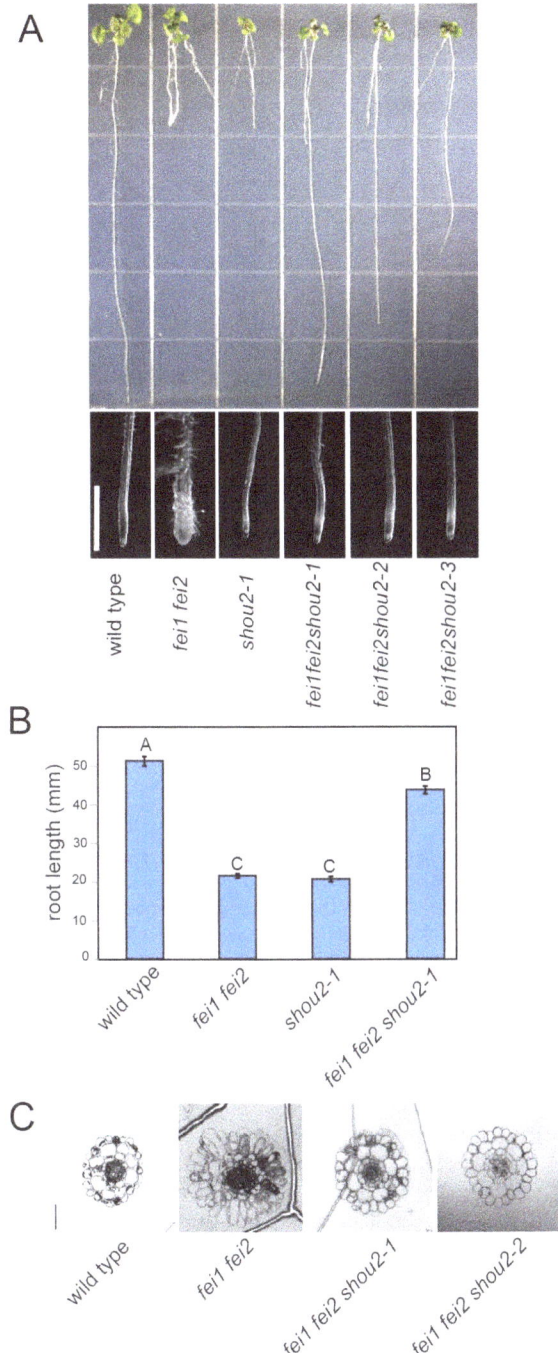

Figure 1. Isolation of the *shou2* suppressor. (A) Top: Phenotypes of indicated seedlings grown on MS medium containing 4.5% sucrose for three weeks. The bottom panels show a close-up of the root tips. Scale bar = 1 mm. **(B)** Quantification of total root length from (A). Values are the means (n = 150) ± SE. Different letters indicate significant differences between groups. Data were analyzed with one-way ANOVA and Tukey's post-hoc comparisons; $P < 0.05$. **(C)** Transverse sections through the root elongation zone of wild type, *fei1 fei2*, *fei1 fei2 shou2-1*, and *fei1 fei2 shou2-2*. Scale bar = 50 μm.

would also suppress the loss of growth anisotropy in *fei1 fei2*. To test this hypothesis, we examined the effect of mutations in the auxin biosynthetic genes *WEI8* and *TAR2* on the *fei1 fei2* root swelling phenotype. *WEI8* and *TAR2* are partially redundant

genes that encode two of the five tryptophan aminotransferases (TAA1) essential for the major auxin biosynthesis pathway in plants. The level of IAA in the roots of double *wei8 tar2* mutants is reduced by 50% relative to the wild type [18], suggesting *WEI8* and *TAR2* are involved in auxin biosynthesis in roots. We generated a *wei8 tar2 fei1 fei2* quadruple mutant; when grown under restrictive conditions, the swelling of the root tip was suppressed in this quadruple *wei8 tar2 fei1 fei2* mutant (Fig. 3). The suppression of the *fei1 fei2* phenotype by *wei8* and *tar2* is similar to the suppression of *fei1 fei2* by iar4, which suggests that auxin is required for the radial cell expansion that occurs in response to decreases in cellulose synthesis in the absence of the FEI proteins. Consistent with this notion, growth of *fei1 fei2 iar4-5* seedlings at higher temperature (28°C), which has been shown to elevate endogenous auxin levels [19] and to suppress *iar4* phenotypes [17], blocks the suppression of root swelling the *iar4-5* mutant (Fig. S3A in File S1). Furthermore, the *fei1 fei2* mutant is slightly hypersensitive to auxin in a root elongation assay (Fig. S3B and C in File S1).

To further explore this hypothesis, we investigated the sensitivity of the *iar4-5* mutant to the cellulose synthesis inhibitor isoxaben. Previous work has shown that loss of growth anisotropy is exacerbated in cell wall mutants treated with isoxaben [20,21]. Consistent with these results, *fei1 fei2* is hypersensitive to isoxaben [12]. However, in contrast to *fei1 fei2*, both the triple *fei1 fei2 iar4-5* and single *iar4-5* are partially resistant to the effects of isoxaben on root swelling (Fig. 4). The suppression of aberrant cell expansion by *iar4-5* suggests that the effect of the loss of cell wall integrity on root morphogenesis can be attenuated by a reduction in auxin function.

The effect of *iar4* on other *fei1 fei2* phenotypes

We have previously shown that the FEI RLKs are required for proper hypocotyl cell expansion in etiolated seedlings and in anchoring pectin in seed coat mucilage to the seed surface [12,14]. The hypocotyls of dark-grown *fei1 fei2* seedlings are significantly wider than those of the wild type [12]. In addition, mutations in *FEI2* lead to disruption of seed coat mucilage [14]. We examined whether mutations in *IAR4* could suppress these additional *fei1 fei2* phenotypes. In contrast to its role in the root, iar4 did not suppress the increased hypocotyl width phenotype of *fei1 fei2* (Fig. 5A and 5B). In fact, the *iar4-5* mutant also had slightly wider hypocotyls and this effect was additive with that of *fei1 fei2*. The additive nature of *iar4-5* and *fei1 fei2* on hypocotyl width suggests that these genes may act in parallel to regulate cell wall function. Unlike in the hypocotyl, mutations in *iar4-5* did not affect the seed coat mucilage of *fei1 fei2*. The seed coat mucilage of the triple *fei1 fei2 iar4-5* mutant resembled that of *fei1 fei2* indicating that mutations in *IAR4* do not suppress this phenotype (Fig. 6).

An additional role for the FEI RLKs is to act additively with COBRA (COB) in stamen filament elongation in the flower. *COBRA* encodes a GPI-anchored protein that associates with the cell wall and is required for the oriented deposition of cellulose in rapidly expanding cells [22]. Like *fei1 fei2*, *cob-1* mutants are deficient in cellulose and as a result display a short, swollen root phenotype that is enhanced by elevated sucrose. Although neither the *fei1 fei2* nor *cob-1* mutants themselves display an obvious floral phenotype, a triple *fei1 fei2 cob-1* mutant has short stamen filaments and as a result is partially infertile [12]. Similar to root cells in the elongation zone, cells of the stamen filament also undergo primarily longitudinal expansion. Therefore, we assessed the ability of *iar4-5* to suppress the short stamen phenotype of *fei1 fei2 cob-1* mutant. Analysis of a quadruple *fei1 fei2 cob-1 iar4-5* mutant indicated that the *iar4-5* allele restores fertility in *fei1 fei2 cob-1*

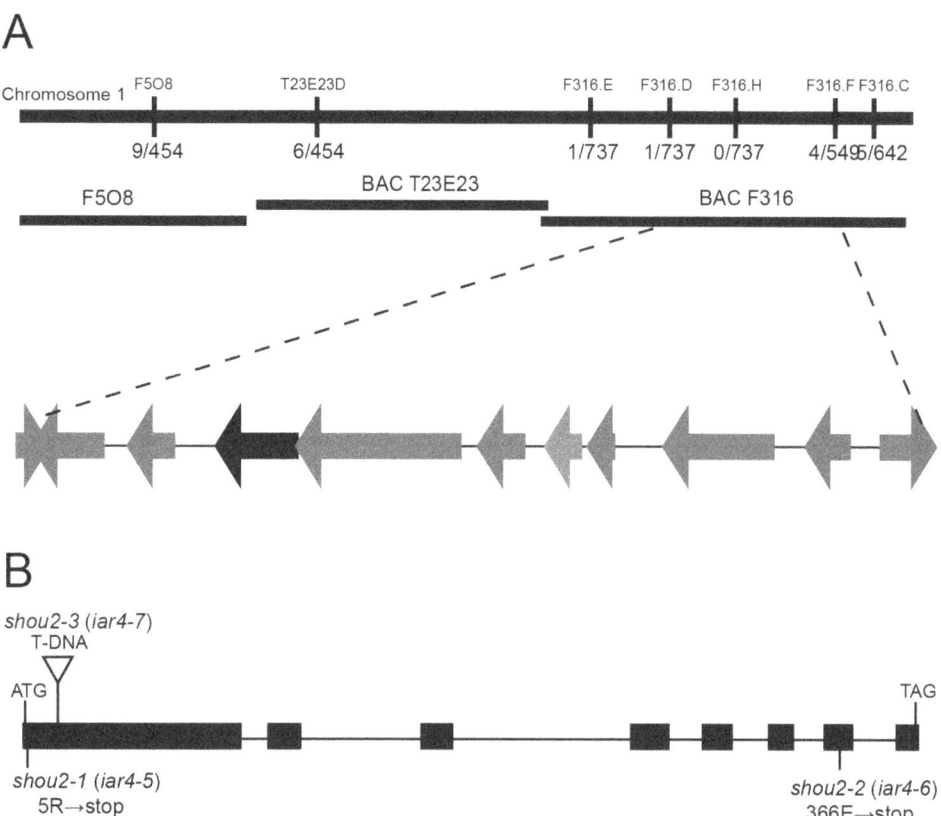

Figure 2. Positional cloning of *SHOU2*. (A) *shou2* was mapped to a region on chromosome 1 between markers F3I6.D and F3I6.F as described in the methods. The name of each DNA marker is shown above and the number of recombinants is indicated below the line. Open reading frames located between markers F3I6.D and F3I6.F are shown below BAC F316. **(B)** Structure of *SHOU2* gene. Boxes represent exons and lines represent the introns. The positions and changes of the three *shou* alleles are indicated. The triangle indicates the position of the DNA insertion in *shou2-3*. The corresponding allele numbers for the *iar4* designations are shown in parentheses.

(Fig. 7A; S4 in File S1). We hypothesized that the restoration of fertility is a result of increased stamen filament length in the quadruple mutant. However, analysis of flower parts showed that in the *fei1 fei2 cob-1 iar4-5* stamen length is further reduced (Fig. 7B), indicating that *iar4-5* does not in fact restore stamen filament length to the *fei1 fei2 cob-1* mutant. The ratio of carpel length to stamen length is increased in the quadruple mutant (Fig. 7C) relative to the infertile *fei1 fei2 cob-1 iar4-5* parent, suggesting that the fertility is most likely rescued due to further decrease in carpel length in *fei1 fei2 cob-1 iar4-5* flowers (Fig. S4 in File S1) rather than increased stamen length.

iar4 is a general suppressor of defects in cell wall synthesis

To ascertain whether loss-of-function mutations in *IAR4* suppress defects in cell expansion exhibited by other cell wall mutants or whether they are specific to the FEI pathway, we crossed *iar4-5* to *sos5*, *procuste (prc;* a null allele of *CESA6),* and a weak allele of *cobra, (cob-1).* When grown in the presence of 4.5% sucrose, each of these mutants displayed a substantial reduction in root length accompanied by radial expansion of cells in the root tip as a result of reduced cellulose biosynthesis. As expected, *iar4* suppressed swelling of the cells in the root tip of *sos5*, which acts in

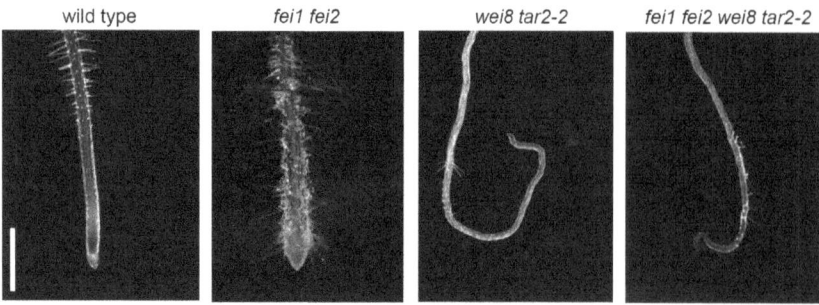

Figure 3. Mutations in the auxin biosynthetic genes *WEI8* and *TAR2* suppress *fei1 fei2*. Phenotypes of the root tips of the indicated seedlings four days after transfer from MS medium containing 0% sucrose to MS medium containing 4.5% sucrose. Scale bar = 1 mm.

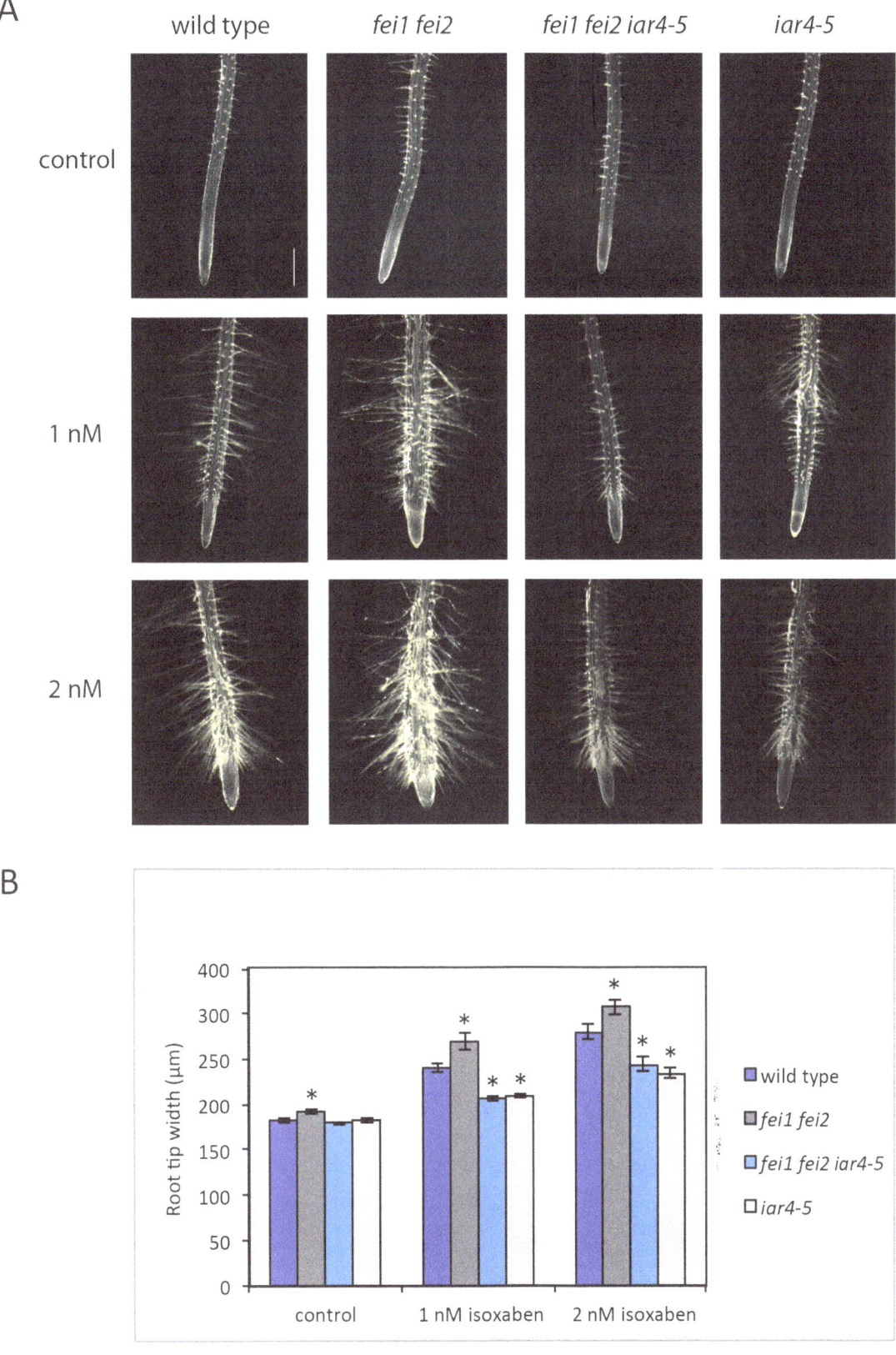

Figure 4. Mutations in *IAR4* confer resistance to isoxaben. (A) Root phenotypes of indicated seedlings germinated and grown for five days on MS medium with 0% sucrose and then transferred for 48 hours to MS medium with 0% sucrose containing either 0 nM (DMSO), 1 nM or 2 nM isoxaben. Representative root tips were imaged using dark field microscopy. Scale bar = 0.5 mm. (B) Quantification of the root tip swelling from (A). The width of the roots was measured at the level of the youngest root hair using ImageJ software [48]. Values are the means (n>8) $\pm$ SE. Data were analyzed by unpaired Student's t-test. Asterisks indicate significant differences relative to the wild type; $P < 0.05$.

A

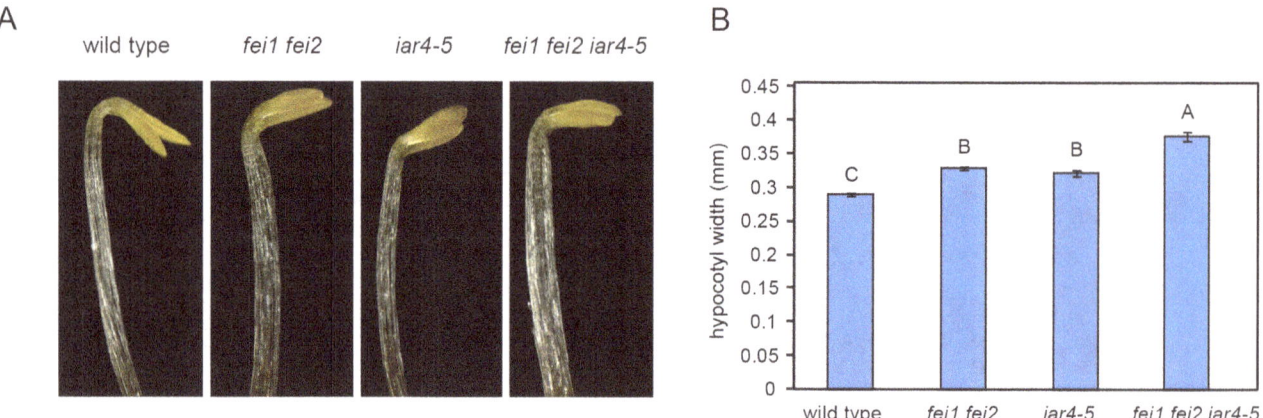

Figure 5. Effect of *iar4-5* mutation on the hypocotyl phenotype of *fei1 fei2*. (**A**) Seedlings of the indicated genotypes were grown for four days in the dark on MS medium with 0% sucrose. Representative seedlings were imaged using dark field microscopy. (**B**) Quantification of hypocotyl widths from seedlings grown as in (A). Values represent the mean (n = 20) ± SE. Different letters indicate significant differences between groups. Data were analyzed with one-way ANOVA and Tukey's post-hoc comparisons; $P < 0.05$.

the FEI pathway (Fig. 8). However, in contrast to the *fei1 fei2 iar4* triple mutant, which displayed a substantial suppression of the root elongation defect observed in both parental lines, the *sos5 iar4* double mutant displays even shorter roots than *sos5* single mutant. Thus, the root of the *sos5 iar4* double mutant is short, but not swollen and thus resembles the *iar4* parental root phenotype. *iar4-5* also suppresses the swollen root phenotypes of both the *cob-1* and *prc* mutants, both of which affect cellulose synthesis independent of the FEI pathway (Fig. 8A), and in both cases it moderately restores the root elongation in these mutants (Fig. 8B).

We next tested whether mutations in *IAR4* could suppress the accumulation of ectopic lignin in these mutants. Lignin is deposited ectopically into the cell wall in response to decreased cellulose synthesis that occurs in cellulose deficient mutants. Previous studies have shown that the roots of *fei1 fei2*, *cob-1*, and *prc* all accumulate ectopic lignin [12,20,23,24]. Interestingly, when we assessed the roots of these cell wall mutants in an *iar4-5*

background using a colorimetric stain, no ectopic lignin deposition was observed (Fig. 9). This result is consistent with the suppression of the root swelling defect in these mutants by *iar4* and suggests that IAR4 is required for the ectopic deposition of lignin that occurs in response to decreased cellulose biosynthesis. Taken together, these observations suggest that *iar4* is not specific to the FEI pathway, but rather acts as a more general suppressor of defects in cellulose biosynthesis.

Discussion

We demonstrate that reducing auxin function, either through loss-of-function mutations in *IAR4* or in the auxin biosynthetic genes *WEI8* and *TAR2*, suppresses the root swelling that occurs in the *fei1 fei2* mutant. Several lines of evidence suggest that *iar4*, and by inference, auxin, acts not in the FEI pathway directly, but rather independently to regulate cell wall function. First, *iar4* acts additively with *fei1 fei2* to increase hypocotyl width and affect floral

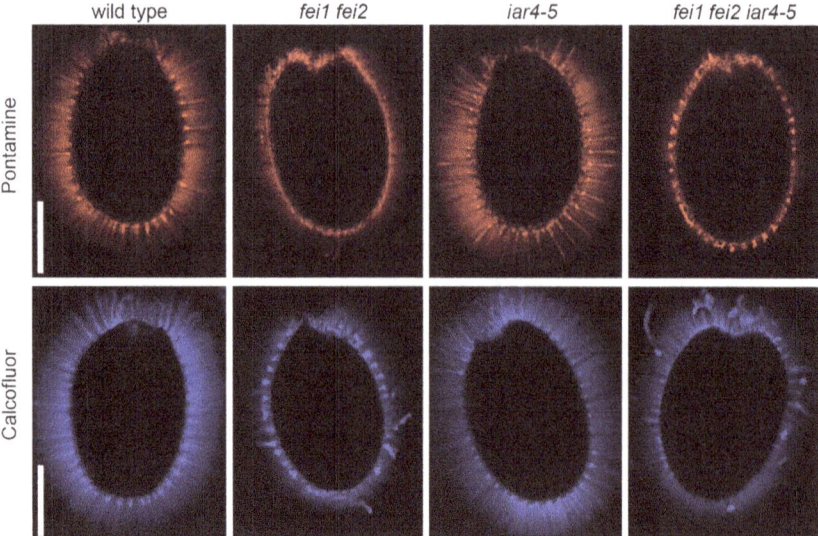

Figure 6. *iar4* does not suppress defects in the seed coat mucilage of *fei1 fei2*. Seeds of the indicated genotype were stained as described in the Methods with either Pontamine fast scarlet S4B or calcofluor as indicated and visualized by confocal microscopy. Scale bar = 0.2 mm.

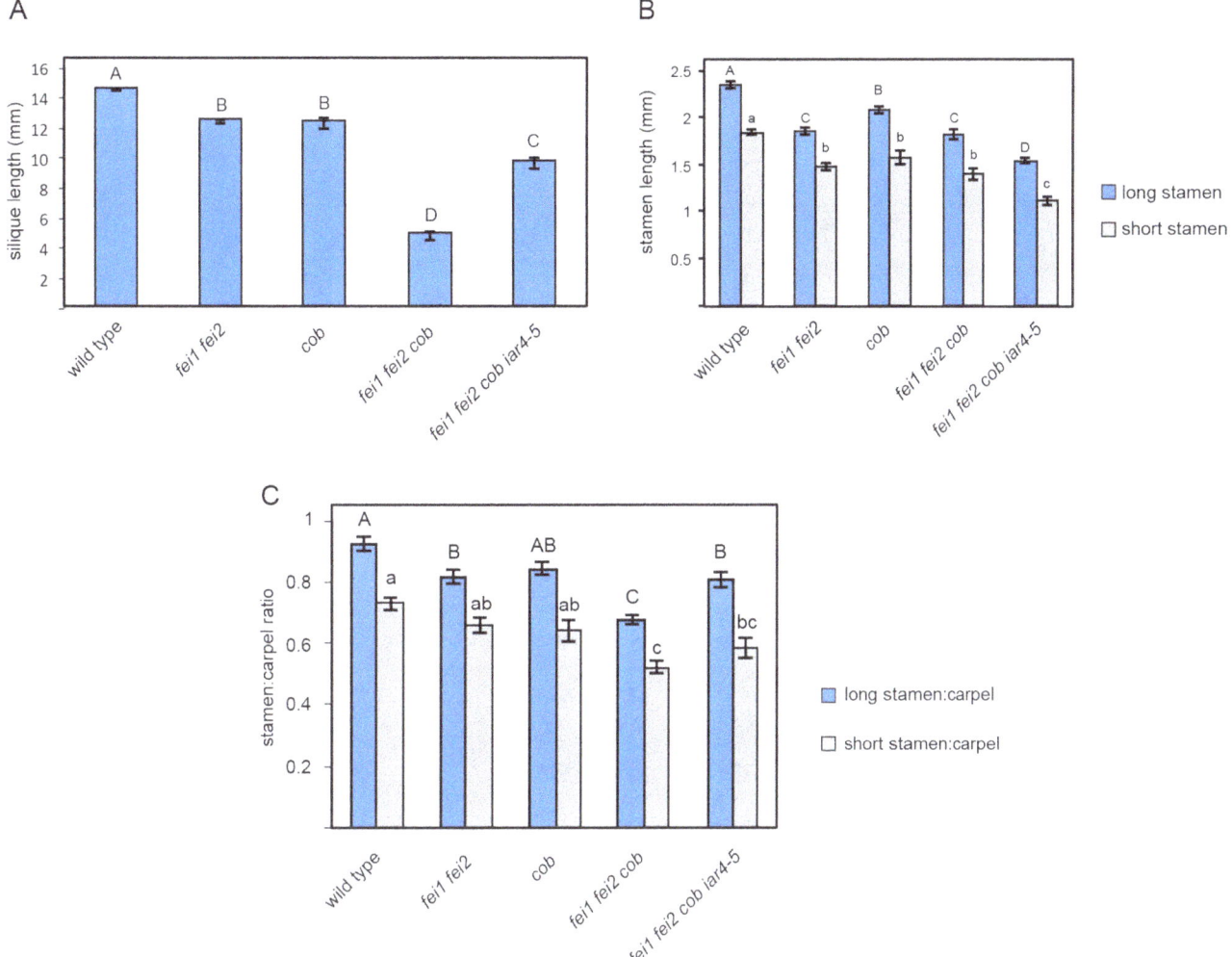

Figure 7. Effect of *iar4-5* on floral phenotypes of *fei1 fei2*. (A) Silique length from indicated genotypes. Plants were grown on soil under 16h light regime for four weeks at 22°C. The first five fully developed siliques from the primary inflorescences were used for the analysis. Values are the mean ± SE (n = 25). **(B)** Stamen length and **(C)** Carpel:stamen ratio of indicated genotypes. Stamens and carpels were removed from individual flowers and measured. Values are the mean ± SE (n>11). Different letters indicate significant differences between groups. Data were analyzed with one-way ANOVA and Tukey's post-hoc comparisons; P<0.05.

development. Second, *iar4* reverts the swollen root phenotype and suppresses the accumulation of ectopic lignin in other cellulose synthesis mutants such as *cob-1* and *prc*, which act in parallel with the FEIs. Finally, the *iar4* mutation confers resistance to isoxaben, which inhibits cellulose synthase function rather than the FEI pathway. The data support a model in which reduced auxin function acts to modulate cell wall function in the root in some way to counteract the effects of reduced cellulose synthesis in the *fei1 fei2* mutant root, as well in other cellulose deficient mutants.

The two aspects of cell expansion are its orientation and extent. The amount and orientation of cellulose microfibrils in the cell wall determines to a large extent the orientation of expansion. In roots, cellulose microfibrils are orientated transversely around the cells, restricting radial expansion leading to principally longitudinal cell expansion [25]. Reduction of cellulose levels, in mutants such as *fei1 fei2* or *prc*, results in a loss of the restriction of radial expansion, and therefore swollen roots. Less is known regarding the regulation of extent of cell expansion. The driving force for cell expansion is the turgor pressure of the cell, which provides the force for increasing the volume of the cell that is opposed by the

rigidity of cell wall, which is controlled by the breaking and reforming of the bonds holding the polymers of the cell wall together. Mutants have been isolated that are defective in root elongation that affect the orientation of cell expansion, the extent of cell expansion, or both processes [26].

Previous studies have linked auxin to the regulation of cell wall function in aerial tissues primarily by affecting the extent of cell expansion. The acid-growth hypothesis attributes auxin-induced cell expansion to the acidification of the cell wall, which results in an increase in the activity of the wall loosening enzymes expansins [27]. Expansins disrupt the non-covalent bonds that form between cellulose and hemicelluloses in the wall and thus promote cell expansion in hypocotyls and modulate the growth of leaves, petioles, and roots [27–29]. Auxin Binding Protein (ABP1) may play an important role in this response; ABP1 activates H$^+$ ATPases and K$^+$ channels at the plasma membrane upon the perception of auxin and is required for cell elongation [30]. In addition to expansins, other wall-loosening enzymes such as xyloglucan hydrolases (XGH) and endotransglycosylases (XET), which cleave and re-graft a major form of hemicellulose,

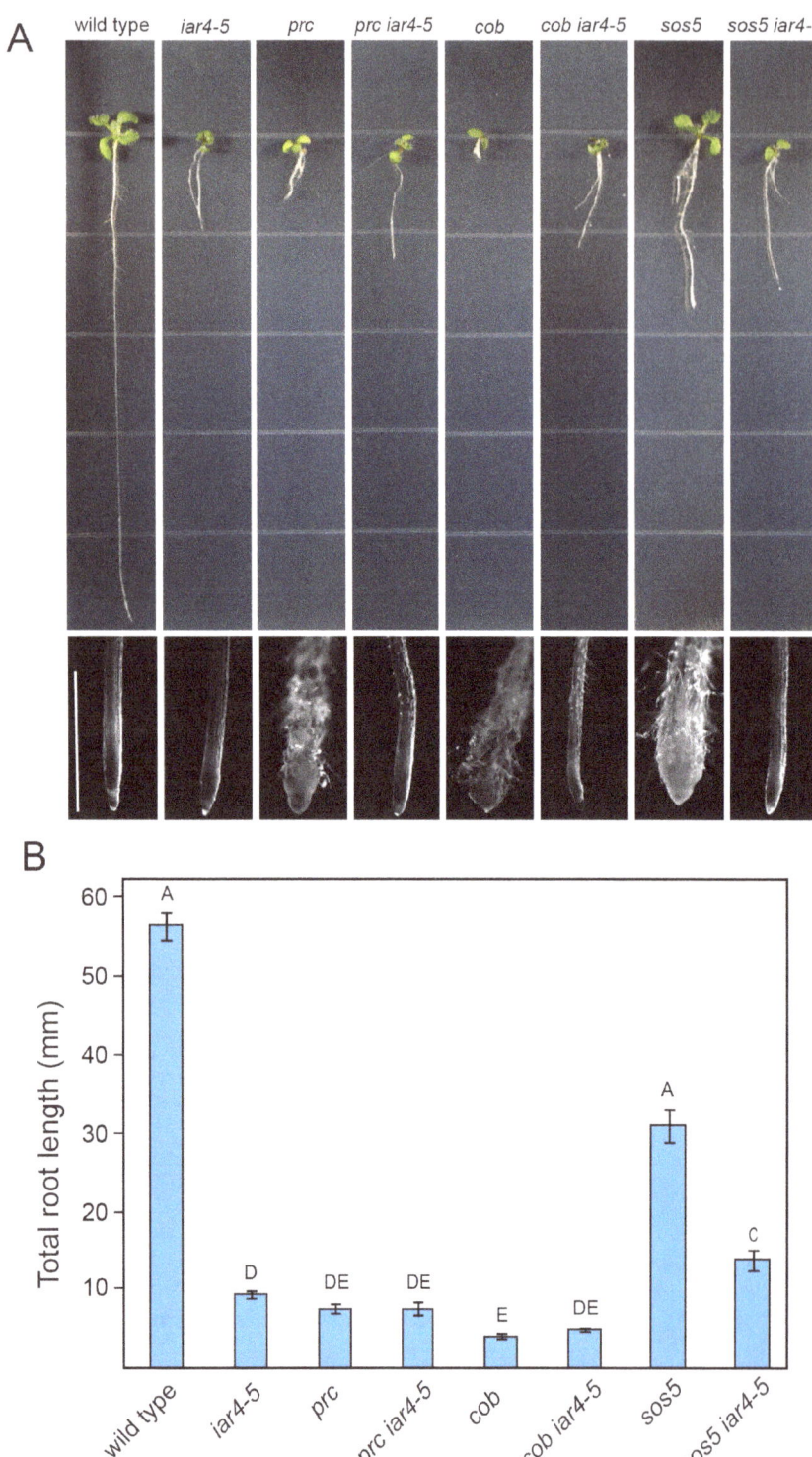

Figure 8. Mutations in *iar4-5* suppress other cell wall mutants. (**A**) Phenotypes of indicated seedlings grown for 14 days on MS medium supplemented with 4.5% sucrose. The bottom is a close-up of the root tips of the seedlings shown above. Scale bar = 1.5 mm (**B**) Quantification of total root elongation. Plants were grown on MS medium supplemented with 4.5% sucrose for 10 days and total root lengths were measured. Values represent means ± SE (n>17). Different letters indicate significant differences between groups. Data were analyzed with one-way ANOVA and Tukey's post-hoc comparisons; $P < 0.05$.

xyloglucan, are also activated upon acidification of the cell wall and in response to auxin [31]. Consistent with these findings, the mechanical extensibility of epidermal cells isolated from azuki bean epicotyls increases dramatically following incubation with XGH [32]. These are among many studies that support a role for auxin in increasing the extensibility of the cell wall in the shoot.

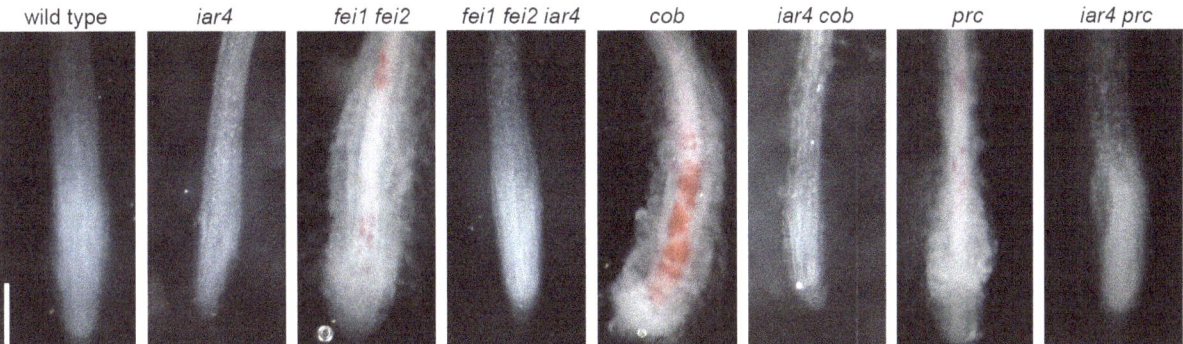

Figure 9. *iar4* suppresses lignin accumulation in cell wall mutants. Phloroglucinol stain for lignin (red) accumulation in root tips of seedlings of the indicated genotype grown on 4.5% sucrose for 2 weeks. Scale bar = 1 mm.

In contrast to the shoot, auxin appears to inhibit root growth and exogenous auxin causes a rapid alkalization of the apoplast in the elongation zone of Arabidopsis roots [33,34]. However, the auxin-insensitive, gain-of-function Aux/IAA mutant, *axr3-1* has a shorter root as compared to the wild type, similar to the short root phenotype of the *iar4* single mutants. Further, auxin represses numerous genes involved in cell wall synthesis and remodeling. Among the genes that are de-regulated in *axr3-1* seedlings treated with IAA are those that encode arabinogalactan proteins (AGPs), expansins (EXP), extensins, proline rich proteins (PRP), xyloglucan endotransglucosylase-hydrolases (XTHs), and pectin methyl-esterases (PMEs) [35]. Although extensins rigidify the cell wall, a disproportionate number of genes repressed in *axr3-1* encoded proteins that loosen the cell wall matrix and thus promote cell elongation [36]. Similarly, mutations in the auxin influx carrier, *lax3*, prevent the induction of expansin expression in the in developing lateral roots of Arabidopsis seedlings. *LAX3* is required for lateral root initiation and its expression precedes the necessary changes in cell wall architecture that are predicted to play a critical role in the emergence of lateral root primordium [37]. The lack of wall plasticity coupled with alterations in the expression of genes that encode cell wall remodeling proteins in the *axr3-1* and *lax3* mutants suggest that auxin may also promote wall loosening in roots. Recently, a detailed study of gene expression kinetics in response to auxin in Arabidopsis roots revealed that a large number of genes involved in cell wall remodeling are regulated in response to exogenous auxin application [38]. Notably, one cluster of auxin down-regulated genes included 57 genes annotated as playing a role in cell wall synthesis/remodeling, including CESA2, suggesting that auxin modulates cell wall synthesis in roots.

The growth of seedlings in the presence of auxin has been shown to lead to root swelling in a manner independent of ethylene biosynthesis [39,40]. This suggests that exogenous auxin decreases the integrity of the cell wall, leading to a loss of growth anisotropy. This is consistent with the suppression of swelling in mutants defective in cellulose biosynthesis by reduction of endogenous auxin that is described here. An important question is by what mechanism does exogenous auxin increase root swelling in wild-type roots, and conversely, how does reduced endogenous auxin suppress swelling in cellulose-deficient roots. One possibility is that auxin negatively regulates cellulose synthesis. Consistent with this, the transcript level of the *CESA2* gene has recently been shown to be decreased in response to auxin [38]. However, the suppression of the *procuste* mutant by *iar4* in our study makes this somewhat unlikely as *procuste* is a null allele of CESA6. However, it is possible that reduced auxin levels may elevate cellulose synthesis

via alternative CESA complexes as CESA6 acts redundantly with CESA2, CESA5, and CESA9 in some Arabidopsis tissues [41]. This scenario is unlikely, at least with respect to CESA5 because mutations in *IAR4* do not suppress the defects in seed coat mucilage production in *fei1 fei2* where CESA5 is required for cellulose biosynthesis in the seed coat [14]. A somewhat more plausible model is that in the long term, auxin modulates the rigidity of the wall not by regulating cellulose synthesis, but by altering other properties of the wall, such as the crosslinking of cellulose microfibrils or the activity of extensins or other cell wall modifying enzymes as described above. Intriguingly, while both *iar4* and *fei1 fei2* mutant roots are short in the presence of elevated sucrose, the triple *iar4 fei1 fei2* mutant displays both a non-swollen root, as well as root elongation comparable to the wild type. In contrast, although mutations in *iar4* suppress the swollen root phenotypes of the other cell wall mutants, they do not fully restore root elongation in *prc*, *cob*, or *sos5*. This suggests that *iar4* does not simply restore cellulose biosynthesis in these mutants, as this would rescue both the swollen root and root length phenotypes. Additionally, it is conceivable the short root phenotype of *iar4* may be partially attributed to decreased cell division in the root apical meristem as auxin has been shown to regulate both the activity and the size of the meristem in the Arabidopsis root [42]. While we isolated *iar4* as a suppressor of *fei1 fei2*, the short root phenotype of both *fei1 fei2* and the *iar4* mutant but not the *fei1 fei2 iar4* triple mutant raises an interesting question. How does *fei1 fei2* suppress *iar4*? It is unlikely that it does so by suppressing any potential effects of *iar4* on cell division rates, and thus likely does so through modulation of cell wall properties.

We propose a model that combines the previously characterized role of the FEI receptor-like kinases in regulating cellulose synthesis with a role for auxin in regulating cell wall rigidification in the root (Fig. 10). We have previously shown that ACC may act as a signal in the FEI pathway to regulate cellulose biosynthesis. The *fei1 fei2* mutations lead to radial cell expansion in the root as a result of decreased cellulose synthesis, which alters wall function such that there is not sufficient force to constrict radial expansion. One model consistent with the data is that decreased auxin results in an increase in the rigidity of the cell wall, which allows the reduced levels of cellulose in the *fei1 fei2* mutant to provide sufficient force to inhibit radial expansion. In most cases this increase in rigidity would also cause a decrease in the overall expansion of the cells, and hence a decrease in the length of the root, such as in observed in the *cob-1 iar4* and *prc iar4* lines. We hypothesize that in the case of the *fei1 fei2 iar4* line, the increased rigidity caused by decreased auxin is precisely balanced by the reduced cellulose levels cause by the *fei1 fei2* mutations, leading to both a lack of swelling and near wild-type elongation. Auxin increases the expression of multiple *ACS* genes [43], which may also have an effect on the FEI signaling pathway. Alternatively, auxin could be involved in the signaling cascade linking perception of perturbation of the cell wall to changes in cell wall synthesis. Interestingly, a recent study has demonstrated that the inhibition of root cell elongation that occurs in response to isoxaben is attenuated by mutations in the *tir1-1* auxin receptor and growth in the presence of the synthetic antagonist of TIR1, PEO-IAA. Furthermore, results from this study indicated that inhibitors of the precursor to ethylene, ACC, fully restore growth anisotropy in the presence of isoxaben and this effect was shown to act independent of ethylene [44]. Consistent with this data, inhibitors of ACC, but not ethylene suppress the swollen root phenotype of *fei1 fei2* [12].

While the data suggest a role of ACC independent of ethylene in regulating cell wall function, ethylene itself clearly also plays a role in cell expansion. Growth of Arabidopsis seedling in the presence of ethylene causes a reduction in cell expansion in many tissues, including the root. In pea epicotyls, ethylene causes a reorientation of the deposition of the cellulose microfibrils in the cortical cells from primarily transverse to primarily longitudinally [45]. In Arabidopsis, ethylene shifts the orientation of cell expansion in the hypocotyl of etiolated seedlings, but not the extent of cell expansion the hypocotyls from etiolated seedlings grown in the presence of ethylene are substantially shorter, which is balanced by an increase in hypocotyl width. In contrast, growth of etiolated seedlings in ethylene causes a strong reduction in root elongation that is likely the results of effects on both the extent and orientation of expansion as the reduction in root elongation is only partly matched by an increase in root width.

The characterization of *iar4* in this study as a suppressor of defects in cell wall synthesis provides evidence that auxin plays a key role in the regulation of primary cell wall function and suggests that wall extensibility may be a major determinant of cell expansion in the root. Whether auxin acts as a general regulator of cell wall function throughout development or participates in the active signaling processes that occur in response to perturbations in the cell wall remains an interesting question for future studies.

Materials and Methods

Plant Material and Growth Conditions

All lines used in this study are in the Columbia (Col-O) ecotype of *Arabidopsis thaliana*, except where noted. The *shou2-3* (SALK_091909) allele was obtained from the SALK T-DNA insertional collection [46]. The *prc-1* [24] and *cob-1* [47] mutants were obtained from the Arabidopsis Biological Resource Center. The *wei8* and *tar2-1* mutants have been previously described (Stepanova *et al.*, 2008). For *in vitro* studies, seeds were surface sterilized, cold treated for 4 days at 4°C, germinated on vertical

plates containing 1 x Murashige and Skoog (MS) salts, 0.6% phytagel (Sigma, St Louis, MO, USA) and either 0% or 4.5% sucrose and grown at 22°C under constant light. For the analysis of root elongation, seeds were germinated on 4.5% sucrose and total root elongation was quantified using ImageJ [48]. For the analysis of auxin sensitivity, seedlings were grown for 4 days on MS media, then transferred to new MS medium supplemented with various auxin levels and root elongation quantified four days after transfer. For the hypocotyl elongation assay, seedlings were exposed to light for 3 hours and grown for 4 days in the dark on MS agar supplemented with 1% sucrose. The width of each hypocotyl was measured 1 mm from the hook of an etiolated seedling. For growth on soil, plants were grown either under constant light or long day conditions at 23°C. For growth in the presence of isoxaben, seedlings were germinated and grown in the absence of sucrose for 5 days then transferred to MS agar supplemented with 0 nM (DMSO control), 1 nM or 2 nM isoxaben for 48 hours. The root width was measured at the level of youngest root hair using ImageJ [48].

Positional Cloning of *shou2*

The *fei1* and *fei2* mutations (Columbia, Col ecotype) were introgressed into Landsberg erecta (Ler) through back crossing with Ler six times and a line homozygous for *fei1* and *fei2* was obtained. Theoretically, after six backcrosses, approximately 98.4% of the genome is Ler, with the exception of regions around the *fei1* and *fei2* mutations, which remain Col. We tested 42 molecular markers across all 5 chromosomes and found only the molecular markers F6NI8 and TIOP12 (close to *FEI1*), and TIJ8 (close to *FEI2*) remained Col. All other 39 markers were homozygous for the Ler SNPs. A mapping population was generated by crossing *fei1 fei2 shou2-1* (Col) to *fei1 fei2* (Ler). Bulk segregant analysis was performed using a total of 42 markers that span the Arabidopsis genome on a pool of DNA obtained from 40 F$_2$ seedlings showing suppression of the *fei1fei2* phenotype. The mutation was initially mapped to an interval spanning markers FI2K8 (7.954Mbp) and FI3K9 (9.744Mbp) on chromosome 1. Fine mapping was facilitated by the root hair phenotype of *shou2* mutants using restriction fragment length polymorphisms and cleaved amplified polymorphic sequence markers. The *shou2-1* mutation was mapped to a ~47-kilobase (kb) region delimited by recombination events between marker F3I6-D (8.552 Mbp) and F3I6-F (8.599Mbp) of chromosome 1. Sequencing of 12 genes within this region identified mutations in the first and seventh exons of At1g24180 in *fei1 fei2 shou2-1* and *fei1 fei2 shou2-2* respectively.

Phloroglucinol Staining

Seedlings were fixed in a solution of three parts ethanol: one part acetic acid for fifteen minutes and transferred to 70% ethanol for 10 minutes. Seedlings were then cleared in chlorohydrate:gly-

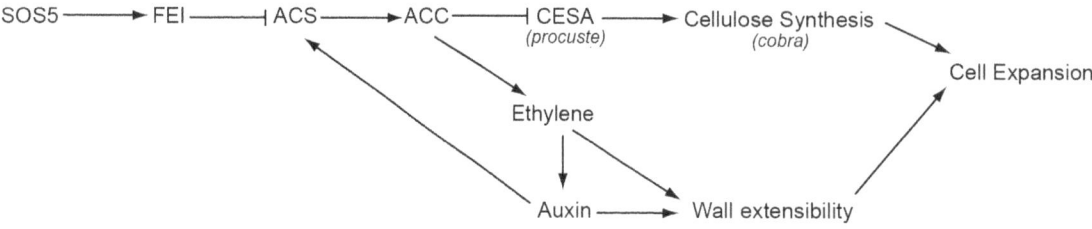

Figure 10. Model of the FEI pathway. Hypothetical model depicting the role of both auxin and the FEI pathway in regulating cell expansion. See text for additional details.

cerol:water (8:1:2) for 5 minutes and stained for a total of 5 minutes in a 2% phloroglucinol-HCl solution.

Microscopy and seed staining

The calcofluor stain was done as described by Willats *et al.* [49]. Seeds were pre-treated with 50 mM EDTA, stained for 20 min in 25 µg/ml fluorescent brightener 28 (Sigma), washed overnight in water and then visualized using a Zeiss LSM710 confocal microscope equipped with a 405 nm laser diode. Pontamine staining was done as described by Anderson *et al.* [50]. Seeds were stained for 30 min in 0.01% Pontamine fast scarlet S4B (Sigma) following a 90 min pre-hydration, washed for 4 hours in water and then visualized using a Zeiss LSM710 confocal microscope equipped with a 561 laser. Flowers and root tips were imaged using bright field microscopy and hypocotyls using dark field microscopy on the compound Leica microscope. Cross sections of the root elongation zone were prepared and imaged as described by Xu *et al.* [12].

Flower and silique analyses

Plants were grown for 4 weeks under the long day conditions. For the silique measurements we used first five siliques from primary inflorescences. For the analysis of flower parts we used stamens and carpels dissected from fully matured flowers. Siliques, stamen filaments and carpel lengths were measured using ImageJ software [48]. The significance of the data was tested by one-way ANOVA with Tukey's post-hoc comparison.

Gene expression analysis

100 mg of root tissue from 8d-old seedlings were harvested and snap frozen in liquid nitrogen. RNA was isolated using the RNeasy extraction kit (Qiagen) and genomic DNA was removed using on-column DNase digestion (Qiagen). 1 µg total RNA was used for cDNA synthesis conducted with random hexamers using the iScript cDNA synthesis kit (Bio-rad). Real-time reverse transcription PCR was performed using the Applied Biosystems ViiA RT-PCR system, the SYBR *Premix Ex Taq* (Takara) and *IAR4* (AT1G24180) specific primers. Relative mRNA values were calculated using the $2^{-\Delta\Delta Ct}$ method [51] with *TUB4* (AT5G44340) as an internal reference gene. Primer sequences are listed in Table S2 in File S1.

Supporting Information

File S1 Supplemental material. Figure S1, T-DNA insertion in *IAR4* results in reduced transcript levels. (A) Cartoon of the *IAR4* gene: boxes represent exons, line introns. The site of the T-DNA insertion in *iar4-7* is indicated by the red triangle and

primers used for the analysis of transcript levels indicated below. (B) *IAR4* transcript levels of different regions (as indicated in (A)); n = 3 (± SE). Transcript levels were calculated using the $2^{(-\Delta\Delta Ct)}$ method as described [51]. Asterisks indicate significant differences between *iar4-7* and the wild type or *fei1 fei2 iar4-7* and *fei1 fei2* ($P<$ 0.05); nd, not detectable. Figure S2, Complementation test of *iar4* alleles. F_1 progeny from indicated crossed were grown on MS media containing 4.5% sucrose for 14 days. Note that the roots of the F_1 seedlings display a non-swollen (i.e. suppressed) phenotype, suggesting that the *iar4-5*, *iar4-6* and *iar4-7* mutations are allelic. Scale bar = 0.5 mm. Figure S3, Effect of altered auxin levels on *fei1 fei2* root growth and on the suppression of *fei1 fei2* by *iar4*. (A) Effect of elevated temperature on the suppression of root swelling of the *fei1 fei2* mutant by *iar4-5*. Four-day old seedlings grown on MS with no sucrose at 22°C were transferred to MS media containing 4.5% sucrose and grown an additional five days at 28°C. (B) Quantification of root elongation of wild-type and *fei1fei2* in response to auxin. Four-day-old seedlings were transferred to media containing the indicated level of auxin and the amount the roots grew after transfer was measured four days later. Values represent the mean of ± SE (n>15). (C) Quantification of relative root elongation of wild-type and *fei1fei2* as in described in A. Values were normalized to the no auxin control. Data were analyzed by Student's t-test; *, $P<0.05$. **, $P<0.01$. The experiment was repeated three times and showed very similar results. Figure S4, *iar4-5* restores fertility of *fei1 fei2 cob-1*. (A) Inflorescence, (B) flower and (C) silique phenotypes of indicated genotypes. Note that *iar4-5* partially restores the silique length of *fei1 fei2 cob*. Plants were grown on soil under long day conditions for four weeks. To visualize stamen length, some petals and sepals were removed from each flower. Bar = 1 mm. (D) Carpel length of indicated genotypes. Different letters indicate significant differences between groups. Data were tested with one-way ANOVA and Tukey's post-hoc analysis; n>13; $P<0.05$. Table S1, Markers used to map *iar4-5*. Table S2, Primers used for gene expression analysis.

Acknowledgments

We thank Jose Alonso at North Carolina State University for *wei8 tar2-1* seeds and Yujin Sun for help with the photography.

Author Contributions

Conceived and designed the experiments: BJS SX JJK. Performed the experiments: BJS SX JKP SMD MW. Analyzed the data: BJS SX JKP SMD JJK. Contributed reagents/materials/analysis tools: BJS SX JKP SMD MW. Wrote the paper: BJS SX JKP JJK.

References

1. Green PB (1980) Organogenesis-a biophysical view. Annu Rev Plant Physiol 31: 51–82.
2. Taiz L (1984) Plant cell expansion: regulation of cell wall mechanical properties. Annu Rev Plant Physiol 35: 585–657.
3. Baskin TI (2005) Anisotropic expansion of the plant cell wall. Annu Rev Cell Dev Biol 21: 203–222.
4. Steinwand BJ, Kieber JJ (2010) The role of receptor-like kinases in regulating cell wall function. Plant Physiol 153: 479–484.
5. Somerville C, Bauer S, Brininstool G, Facette M, Hamann T, et al. (2004) Toward a systems approach to understanding plant cell walls. Science 306: 2206–2211.
6. Pilling E, Höfte H (2003) Feedback from the wall. Curr Opin Plant Biol 6: 611–616.
7. Gish L, Clark S (2011) The RLK/Pelle family of kinases. Plant J 66: 117–127.
8. He Z-H, Fujiki M, Kohorn BD (1996) A cell wall-associated, receptor-like protein kinase. J Biol Chem 271: 19789–19793.
9. Lally D, Ingmire P, Tong HY, He ZH (2001) Antisense expression of a cell wall-associated protein kinase, WAK4, inhibits cell elongation and alters morphology. Plant Cell 13: 1317–1331.
10. Wagner TA, Kohorn BD (2001) Wall-associated kinases are expressed throughout plant development and are required for cell expansion. Plant Cell 13: 303–318.
11. Hématy K, Höfte H (2008) Novel receptor kinases involved in growth regulation. Curr Opin Plant Biol 11: 321–328.
12. Xu SL, Rahman A, Baskin TI, Kieber JJ (2008) Two leucine-rich repeat receptor kinases mediate signaling linking cell wall biosynthesis and ACC synthase in Arabidopsis. Plant Cell 20: 3065–3079.
13. Guo H, Li L, Ye H, Yu X, Algreen A, et al. (2009) Three related receptor-like kinases are required for optimal cell elongation in *Arabidopsis thaliana*. Proc Natl Acad Sci USA 106: 7648–7653.
14. Harpaz-Saad S, McFarlane HE, Xu S, Divi UK, Forward B, et al. (2011) Cellulose synthesis via the FEI2 RLK/SOS5 pathway and cellulose synthase 5 is

required for the structure of seed coat mucilage in Arabidopsis. Plant J 68: 941–953.

15. Harpaz-Saad S, Western TL, Kieber JJ (2012) The FEI2-SOS5 pathway and CELLULOSE SYNTHASE 5 are required for cellulose biosynthesis in the Arabidopsis seed coat and affect pectin mucilage structure. Plant Signal Behav 7: 285–288.

16. LeClere S, Rampey RA, Bartel B (2004) IAR4, a gene required for auxin conjugate sensitivity in Arabidopsis, encodes a pyruvate dehydrogenase E1alpha homolog. Plant Physiol 135: 989–999.

17. Quint M, Barkawi LS, Fan K-T, Cohen JD, Gray WM (2009) Arabidopsis IAR4 modulates auxin response by regulating auxin homeostasis. Plant Physiol 150: 748–758.

18. Stepanova AN, Robertson-Hoyt J, Yun J, Benavente LM, Xie D-Y, et al. (2008) TAA1-mediated auxin biosynthesis is essential for hormone crosstalk and plant development. Cell 133: 177–191.

19. Gray WM, Ostin A, Sandberg G, Romano CP, Estelle M (1998) High temperature promotes auxin-mediated hypocotyl elongation in Arabidopsis. Proc Natl Acad Sci USA 95: 7197–7202.

20. Desprez T, Vernhettes S, Fagard M, Refrégier G, Desnos T, et al. (2002) Resistance against herbicide isoxaben and cellulose deficiency caused by distinct mutations in same cellulose synthase isoform CESA6. Plant Physiol 128: 482–490.

21. Scheible WR, Eshed R, Richmond T, Delmer D, Somerville C (2001) Modifications of cellulose synthase confer resistance to isoxaben and thiazolidinone herbicides in Arabidopsis ixr1 mutants. Proc Natl Acad Sci USA 98: 10079–10084.

22. Roudier F, Fernandez AG, Fujita M, Himmelspach R, Borner GHH, et al. (2005) COBRA, an Arabidopsis extracellular glycosyl-phosphatidyl inositol-anchored protein, specifically controls highly anisotropic expansion through its involvement in cellulose microfibril orientation. Plant Cell 17: 1749–1763.

23. Caño-Delgado A, Penfield S, Smith C, Catley M, Bevan M (2003) Reduced cellulose synthesis invokes lignification and defense responses in Arabidopsis thaliana. Plant J 34: 351–362.

24. Fagard M, Desnos T, Desprez T, Goubet F, Refregier G, et al. (2000) PROCUSTE1 encodes a cellulose synthase required for normal cell elongation specifically in roots and dark-grown hypocotyls of Arabidopsis. Plant Cell 12: 2409–2424.

25. Sugimoto K, Williamson RE, Wasteneys GO (2000) New techniques enable comparative analysis of microtubule orientation, wall texture, and growth rate in intact roots of Arabidopsis. Plant Physiol 124: 1493–1506.

26. Hauser M, Morikami A, Benfey P (1995) Conditional root expansion mutants of Arabidopsis. Development 121: 1237–1252.

27. Hager A (2003) Role of the plasma membrane H+-ATPase in auxin-induced elongation growth: historical and new aspects. Plant Res 116: 483–505.

28. Cosgrove DJ, Gilroy S, Kao T-h, Ma H, Schultz JC (2000) Plant Signaling 2000. Cross Talk Among Geneticists, Physiologists, and Ecologists. Plant Physiol 124: 499–506.

29. Cosgrove DJ, Li LC, Cho H-T, Hoffmann-Benning S, Moore RC, et al. (2002) The growing world of expansins. Plant Cell Physiol 43: 1436–1444.

30. Sauer M, Kleine-Vehn J (2011) AUXIN BINDING PROTEIN1: the outsider. Plant Cell 23: 2033–2043.

31. Lorences E, Zarra J (1987) Auxin-induced growth in hypocotyl seqments of Pinus pinaster Aiton - changes in molecular weight distribution of hemicellulosic polysaccharides. J Exp Bot 38: 960–996.

32. Kaku T, Tabuchi A, Wakabayashi K, Kamisaka S, Hoson T (2002) Action of xyloglucan hydrolase within the native cell wall architecture and its effect on cell wall extensibility in azuki bean epicotyls. Plant Cell Physiol 43: 21–26.

33. Monshausen GB, Miller ND, Murphy AS, Gilroy S (2011) Dynamics of auxin-dependent Ca2+ and pH signaling in root growth revealed by integrating high-resolution imaging with automated computer vision-based analysis. Plant J 65: 309–318.

34. Gjetting SK, Ytting CK, Schulz A, Fuglsang AT (2012) Live imaging of intra- and extracellular pH in plants using pHusion, a novel genetically encoded biosensor. J Exp Bot 63: 3207–3218.

35. Overvoorde P, Okushima Y, Alonso JM, Chan A, Chang C, et al. (2005) Functional genomic analysis of the AUXIN/INDOLE-3-ACETIC ACID gene family members in Arabidopsis thaliana. Plant Cell 17: 3282–3300.

36. Cosgrove DJ (2005) Growth of the plant cell wall. Nat Rev Mol Cell Biol: 850–861.

37. Swarup K, Benkova E, Swarup R, Casimiro I, Peret B, et al. (2008) The auxin influx carrier LAX3 promotes lateral root emergence. Nat Cell Biol 10: 946–954.

38. Lewis DR, Olex AL, Lundy SR, Turkett WH, Fetrow JS, et al. (2013) A kinetic analysis of the auxin transcriptome reveals cell wall remodeling proteins that modulate lateral root development in Arabidopsis. Plant Cell 25: 3329–3346.

39. Alarcon M, Lloret P, Iglesias D, Talon M, Salguero J (2012) Comparison of growth responses to auxin 1-naphthaleneacetic acid and the ethylene precursor 1-aminocyclopropane-1-carboxilic acid in maize seedling root. Acta Biol Cracov Bot 54: 16–23.

40. Eliasson L, Bertell G, Bolander E (1989) Inhibitory action of auxin on root elongation not mediated by ethylene. Plant Physiol 91: 310–314.

41. Persson S, Paredez A, Carroll A, Palsdottir H, Doblin M, et al. (2007) Genetic evidence for three unique components in primary cell-wall cellulose synthase complexes in Arabidopsis. Proc Natl Acad Sci USA 104: 15566–15571.

42. Blilou I, Xu J, Wildwater M, Willemsen V, Paponov I, et al. (2005) The PIN auxin efflux facilitator network controls growth and patterning in Arabidopsis roots. Nature 433: 39–44.

43. Abel S, Nguyen MD, Chow W, Theologis A (1995) ACS4, a primary indoleacetic acid-responsive gene encoding 1-aminocyclopropane-1-carboxylate synthase in Arabidopsis thaliana. J Biol Chem 270: 19093–19099.

44. Tsang DL, Edmond C, Harrington JL, Nühse TS (2011) Cell wall integrity controls root elongation via a general 1-aminocyclopropane-1-carboxylic acid-dependent, ethylene-independent pathway. Plant Physiol 156: 596–604.

45. Lang J, Eisinger W, Green P (1982) Effects of ethylene on the orientation of microtubules and cellulose microfibrils of pea epicotyl cells with polyamellate cell walls. Protoplasma 110: 5–14.

46. Alonso JM, Stepanova AN, Leisse TJ, Kim CJ, Chen H, et al. (2003) Genome-wide insertional mutagenesis of Arabidopsis thaliana. Science 301: 653–657.

47. Schindelman G, Morikami A, Jung J, Baskin TI, Carpita NC, et al. (2001) COBRA encodes a putative GPI-anchored protein, which is polarly localized and necessary for oriented cell expansion in Arabidopsis. Genes Dev 15: 1115–1127.

48. Abramoff MD, Magelhaes PJ, Ram SJ (2004) Image processing with ImageJ. Biophotonics International 11: 36–42.

49. Willats WGT, McCartney L, Knox JP (2001) In-situ analysis of pectic polysaccharides in seed mucilage and at the root surface of Arabidopsis thaliana. Planta 213: 37–44.

50. Anderson CT, Carroll A, Akhmetova L, Somerville C (2010) Real-time imaging of cellulose reorientation during cell wall expansion in Arabidopsis roots. Plant Physiol 152: 787–796.

51. Livak KJ, Schmittgen TD (2001) Analysis of relative gene expression data using real-time quantitative PCR and the 2(-delta delta C(T)) method. Methods 25: 402–408.

2

Comparison of Five Major Trichome Regulatory Genes in *Brassica villosa* with Orthologues within the *Brassicaceae*

Naghabushana K. Nayidu[1,2]*, **Sateesh Kagale**[1,3], **Ali Taheri**[1], **Thushan S. Withana-Gamage**[4],
Isobel A. P. Parkin[1], **Andrew G. Sharpe**[3], **Margaret Y. Gruber**[1]*

1 Agriculture and Agri-Food Canada, Saskatoon Research Centre, Saskatoon, SK, Canada, **2** Department of Biology, University of Saskatchewan, Saskatoon SK, Canada, **3** National Research Council (NRC), Saskatoon SK, Canada, **4** POS Bio-Sciences, Saskatoon, SK, Canada

Abstract

Coding sequences for major trichome regulatory genes, including the positive regulators *GLABRA 1(GL1)*, *GLABRA 2 (GL2)*, *ENHANCER OF GLABRA 3 (EGL3)*, and *TRANSPARENT TESTA GLABRA 1 (TTG1)* and the negative regulator *TRIPTYCHON (TRY)*, were cloned from wild *Brassica villosa*, which is characterized by dense trichome coverage over most of the plant. Transcript (FPKM) levels from RNA sequencing indicated much higher expression of the *GL2* and *TTG1* regulatory genes in *B. villosa* leaves compared with expression levels of *GL1* and *EGL3* genes in either *B. villosa* or the reference genome species, glabrous *B. oleracea*; however, cotyledon *TTG1* expression was high in both species. RNA sequencing and Q-PCR also revealed an unusual expression pattern for the negative regulators *TRY* and CPC, which were much more highly expressed in trichome-rich *B. villosa* leaves than in glabrous *B. oleracea* leaves and in glabrous cotyledons from both species. The *B. villosa* TRY expression pattern also contrasted with TRY expression patterns in two diploid Brassica species, and with the Arabidopsis model for expression of negative regulators of trichome development. Further unique sequence polymorphisms, protein characteristics, and gene evolution studies highlighted specific amino acids in *GL1* and *GL2* coding sequences that distinguished glabrous species from hairy species and several variants that were specific for each *B. villosa* gene. Positive selection was observed for GL1 between hairy and non-hairy plants, and as expected the origin of the four expressed positive trichome regulatory genes in *B. villosa* was predicted to be from *B. oleracea*. In particular the unpredicted expression patterns for *TRY* and CPC in *B. villosa* suggest additional characterization is needed to determine the function of the expanded families of trichome regulatory genes in more complex polyploid species within the *Brassicaceae*.

Editor: John Schiefelbein, University of Michigan, United States of America

Funding: 1. Agriculture and Agri-Food Canada, 2. Canola Council of Canada, 3. The Saskatchewan Agricultural Development Fund, and 4. The Saskatchewan Canola Development Commission. POS Bio-Sciences provided support in the form of a salary for author TSW-G, but did not have any additional role in the study design, data collection and analysis, decision to publish, or preparation of the manuscript.

* E-mail: nayidun@agr.gc.ca (NKN); gruberm@agr.gc.ca (MG)

Introduction

Trichomes are protruding structures which protect plant surfaces against dehydration and insect pests, and provide a storage organ to handle metal toxicity [1,2]. Trichome development has been studied extensively at the molecular level in non-glandular trichomes of the model plant, *Arabidopsis thaliana*, in which trichome regulation is controlled by multiple transcription factors. Among these, a number of major positive regulatory genes for trichome initiation have been studied in detail, including *GLABRA 1(GL1)*, *GLABRA 2 (GL2)*, *GLABRA 3 (GL3)*, *ENHANCER OF GLABRA 3 (EGL3)*, and *TRANSPARENT TESTA GLABRA 1 (TTG1)* [3]. *GL1* was isolated by gene tagging [4] and is a member of the R2R3 activator MYB gene family in *A. thaliana* [5]. *GL3* and *EGL3* encode members of an IIIf subfamily of basic Helix-Loop-Helix (bHLH) proteins [6]. *TTG1* encodes a WD40 domain protein [7]. GL1, TTG1 and GL3/EGL3 are positive patterning proteins which form an activator MYB-bHLH-WD40 (MBW) tri-protein complex that induces the expression of an immediate downstream target gene, *GLABRA 2 (GL2)*. GL2 encodes a homeobox transcription factor and its induction is required both for trichome cell specification and for subsequent phases of

trichome morphogenesis such as cell expansion, branching, and cell wall maturation [8–10]. Mutations in any of these four above-mentioned genes reduce trichome initiation and the density of trichome patterning in *A. thaliana* [11–14] A yeast two hybrid assay demonstrated a direct physical interaction between *GL3* and *GL1*, *TTG1*, and itself; *GL1* and *GL3* co-overexpression confirmed their interaction, but *GL1* and *TTG1* do not physically interact [15]. Mutations in *GL3* modestly reduce trichome density, branching, DNA endoduplication, and trichoblast size [11,15]. *Egl3* mutant plants have no obvious trichome defects, but *gl3egl3* double mutants show a complete glabrous phenotype [14] due to some functional overlap between the two genes [16].

TRIPTYCHON (TRY) is the first identified negative regulator of Arabidopsis trichome initiation [11] and encodes a small single R3-MYB repeat protein which lacks the R2 activation domain [6,17]. The activator proteins of the Arabidopsis MBW tri-protein complex are assumed to locally activate their own expression and that of *TRY* [18]. TRY moves through plasmodesmata into adjacent cells where it and several other potent inhibitor proteins such as CAPRICE (CPC), ENHANCER OF TRY and CPC1 and 2 (ETC1 and ETC3) can competitively bind to the MBW complex to release GL1 [19–23]. This release of GL1 renders the MYB complex inactive at stimulating GL2 expression and prevents trichome initiation in neighbouring cells [2,18]. At the same time, the GL3 protein "traps" and reduces the mobility of TTG1 proteins (which can move from neighboring cells into trichome-initiating cells) by binding to them in trichome-initiating cells [3,24,25].

Trichomes are found on other species within the *Brassicaceae*, but very few details are known about *Brassica* trichome genes and their regulation compared with the model Arabidopsis. *Brassica napus* is an amphidiploid species originating from a cross between *Brassica rapa* (of the A genome) and *Brassica oleracea* (of the C genome) [26]. Leaf trichomes are found in substantial numbers on many lines of *B. rapa*, but not on *B. oleracea*, and the distribution on *B. rapa* is patchy rather than even, the patterning is light, and occurs mainly on only a few tissues (eg. leaves and stems) [27]. Leaf trichomes are mainly found on the adaxial side of the *A. thaliana* and *B. rapa* leaves. In contrast, *B. napus* is practically devoid of trichomes.

Trichome genes that have been characterized from the *Brassicas* only include *B. rapa TTG1*, *GL1*, and *GL2*, but analysis of a full set of regulatory genes has never been conducted within a single species. An orthologue, *BrTTG1*, isolated from a brown-seeded hairy *B. rapa* genotype, was found to functionally complement (rescue) an *A. thaliana ttg1* mutant [28], while the orthologue isolated from *B. rapa* yellow-seeded glabrous germplasm was not functional. *B. rapa* leaf hairiness was associated with nucleotide polymorphisms in the DNA-binding domain of the *BrGL1* gene [29], and several *BrGL1* alleles with varying impact on phenotype have been found. *B. rapa* harbouring the B-allele of *BrGL1* produces hairy plants when transformed with the A-allele and hairless plants with the C-allele [30]. *GL2* promoter regions in *A. thaliana* and *B. napus* have low homology and confer differential expression patterns, such that the *BnGL2* promoter introduced into *A. thaliana* is only expressed at an early stage of trichome development, whereas the native *AtGL2* promoter is expressed throughout the whole process of trichome development [31]. Moreover, glabrous *B. napus* plants expressing the *A. thaliana GL3* from a constitutive promoter develop an extremely dense coverage of trichomes on stems and young leaves [32] compared with overexpression in Arabidopsis. These findings indicate differences between the *Brassica* and *A. thaliana* genes that can impact trichome function and trichome patterning.

Brassica villosa is a wild C genome relative of glabrous *B. oleracea*, yet *B. villosa* is even more densely covered, with evenly distributed trichomes over most surfaces of the plant, than *B. napus* transformed with the *AtGL3* gene [32,33]. Like other *Brassica* species, *B. villosa* trichomes are without branches, whereas trichomes on the model species, *A. thaliana*, are multi-branched. The limited information available for *Brassica* trichome genes and the dense trichome patterning on *B. villosa* organs prompted us to dissect the trichome regulatory gene network in this species. In this regard, we successfully amplified *B. villosa* orthologues to the *GL1*, *GL2*, *EGL3*, *TTG1* and *TRY* coding sequences and compared their expression patterns and sequence structure with corresponding genes available from *B. rapa*, *B. oleracea*, *B. napus*, and *A. thaliana*.

Materials and Methods

Plant Material

A. thaliana (Columbia), *Brassica oleracea* (TO1000 DH3), and *Brassica napus* (cv. Westar) seeds were obtained from germplasm collections at the Saskatoon Research Centre, Saskatoon, Canada. *Brassica villosa* Biv. subsp. *drepanensis* seeds were obtained from the Centre for Genetic Resources, Wageningen, The Netherlands.

Seeds of *B. rapa* (cv. Echo) were obtained from Plant Gene Resources Canada, Saskatoon, Canada. All seeds were grown in soil-less potting mixture in a controlled greenhouse environment (16 h light/8 h dark, 20/17°C) supplemented with halogen lights at Agriculture and Agri-Food Canada, Saskatoon, Canada.

Amplification and Sequence Analysis of Five Major Trichome Genes from B. villosa

Specific primers to amplify *Brassica villosa* (*Bv*) orthologues to *GL1*, *GL2*, *EGL3*, *TTG1* and *TRY* were designed from coding sequences from *B. napus* and *B. rapa* (Table 1). The orthologues were amplified from cDNA developed from *B. villosa* seedling leaf RNA (isolated as described under gene expression profiling). At the time of amplification, the *B. rapa EGL3* (accession HM208589.1) was mis-annotated as *GL3* in NCBI, but the nomenclature was recently corrected. PCR was conducted using 32 cycles of 94°C for 30 sec, 56°C for 30 sec, and 72°C for up to 3 minutes (based on the length of the template cDNA), a final extension time of 5 minutes at 72°C, and PFU Ultra II fusion HotStart DNA polymerase enzyme (Stratagene, USA). Primers were designed starting at the translational start and ending with the 3′ ends of the *B. napus* and *B. rapa* cDNA sequences (including restriction sites) except for *TRY*, for which the forward primer was designed from the 72nd residue upstream from the translational start site (Table 1). At the time of cloning, only an unannotated EST was available for *TRY* rather than a cDNA with a verified translational start site. Amplified sequences were cloned into the pGEM-Teasy vector (Promega, Madison, WI, USA), and then several amplicons (clones) per gene were sequenced in the DNA Technologies Laboratory of National Research Council, Saskatoon, and used to search the NCBI database using BLASTN to determine gene identity.

Construction of Transgenic Expression Vectors and Transformation into the Arabidopsis Wild Type

Binary vectors pMP79-103 and pBI121 were initially modified by inserting a 1.7 kb Hydroperoxide Lyase (HPL) [34] promoter between *Hind*III/*Xba*I sites, and then a 2.2 kb *BvGL2* and 506 bp *BvTRY* cDNA sequence were cloned, respectively, downstream of the promoter using *Bam*HI/*Kpn*I restriction sites. In a separate construct, pMP79-103 vector was modified by inserting a 2.5 kb Elongation factor 1(EF1) promoter between *Hind*III/*Xba*I sites, and then a 1.8 kb *BvEGL3* cDNA sequence was cloned using *Bam*HI/*Pst*I restriction sites. pCAMBIA1305.2 vector was modified by cloning a cassette carrying 570 bp NOS promoter, 1.1 kb cDNA clone of *TTG1* and 400 bp NOS terminator between *Eco*RI and *Hind*III restriction sites. All of these binary constructs were individually transformed into *Agrobacterium tumefaciens* strain GV3101 by electroporation. Arabidopsis plants were transformed with respective plasmids using the floral dip method [34]. Transformed plants were selected on the basis of respective vector based resistance markers. Pictures were taken from T1 generation plants three weeks after germination.

RNA Sequencing for Verification and Expression Profiling of Individual Gene Copies

Fresh-frozen cotyledons and first true-leaves (before the second true leaf emergence) of *B. villosa* and *B. oleracea* were collected for RNA sequencing from seedlings (5–7 plants per replicate) for 3 biological replicates. RNA was extracted using standard methods, then developed into libraries using a TruSeq RNA sample preparation kit (Illumina, San Diego, CA, USA). Library sequencing (100 cycles) was conducted from both ends on an

Table 1. Primers used to isolate/analyze *GL1*, *GL2*, *GL3*, *TTG1* and *TRY* genes from *B. villosa*.

Primers	NCBI accession	Sequence (5′ to 3′)
Coding Sequences from B. napus[1]		
GL1-F	HQ162473.1	GTC<u>AGGATCC</u>ATGAGAACGAGGAGAAGAACAGA
GL1-R		GTC<u>ACTGCAG</u>CTAGAGACAGTAGCCAGTATCA
GL2-F	EU826520.1	GTC<u>AGGATCC</u>ATGTCAATGGCCGTCGAGATGTCA
GL2-R		ATAT<u>CTGCAG</u>TGTTGTGCAGCGTGACAGAGACGA
TRY-F	EE451172.1	AGC<u>TGGATCC</u>GCTTGCATTCTCCAACT
TRY-R		CGC<u>ACTGCAG</u>GCAATTTCGTTATGCTATATG
Coding Sequences from B. rapa[1]		
EGL3-F	HM208589.1	GTC<u>AGGATCC</u>ATGGCTGCTGTAGAAAACAG
EGL3-R		GCAG<u>CTGCAG</u>AGTGCATCTTGAATCATTCCT
TTG1-F	HM208590.1	TAG<u>AGGATCC</u>ATGGACAACGCAGCTCCGGACT
TTG1-R		AGTCGG<u>TACCTC</u>AAACTCTAAGGAGCTGCA
Conserved Regions (for Q-PCR)[2]		
GL1-F		ACTGGGCTGAAGAGGTGTGGA
GL1-R		GAGATGAGTGTTCCAGTGA
GL2-F		CGCTGGCCGGGAGAAAGAGC
GL2-R		GGAGGTTTTTTCTGGATGAA
EGL3-F		ACATTCAATGGAGTTACGGA
EGL3-R		AGAGATTCGTAAAGCTCTCT
TTG1-F		CTCTGGGAGGTCAACGAA
TTG1-R		ATGCTGCACGTGCCTAAC
TRY-F		CATCACTCCTCTTCTCACA
TRY-R		TGTGGTGGGGAAGAAAACAGA
BnEF1F (*B. napus*)		CCCATTCGTCCCCATCTCTGGA
BnEF1R (*B. napus*)		ACGGAGGGGCTTGTCCGAGG

[1]Forward (F) primers were designed at the 5′ coding end and reverse (R) primers to the 3′ end of cDNA sequences from *B. napus* for GL1 and GL2 and from *B. rapa* for EGL3 and TTG1. Forward primers for TRY were designed 72 nucleotides before coding region and reverse primer at 3′ end of the cDNA from *B. rapa*. Underlined sequences indicate incorporated restriction enzyme sites; *Bam*HI and *Pst*I sites in the GL1, GL2, EGL3 and TRY forward and reverse primers, respectively; *Bam*HI and *Kpn*I sites in the TTG1 forward and reverse primers, respectively.
[2]Primers for Q-PCR were designed to conserved regions based on alignments between the *A. thaliana* homologue and the homologues from four *Brassica* species. B nEF1F/BnEF1R are endogenous reference gene primers.

Illumina HiSeq 2000. A total of 51 Gb of *B.oleracea* reads (Cotyledons; 26 Gb and true leaves; 25 Gb) and 60 Gb of *B.villosa* reads (Cotyledons; 33 Gb and true leaves; 27 Gb) were obtained. The RNAseq data was trimmed using trimmomatic ver.0.30 with minimum quality score of 15, removing the first 12 bp, and a minimum length of 20, which was then aligned to the genome of *B. oleracea* with TopHat ver. 2.0.7 (with parameters, including minimum intron length (i) of 20, maximum intron length (I) of 11000, and distance between pair ends (r) of 30) [35]. The aligned RNAseq reads were assembled into transcripts and their relative abundance was estimated as fragments per kilobase of exon per million fragments mapped (FPKM) using Cufflinks software [35]. Differential expression and statistical analysis of individual gene copies was conducted using Cummerbund package in R; http://www.r-project.org/. Raw data from the RNAseq experiment was deposited to NCBI with SRA accession number, SRP035213.

Q-PCR for Total Gene Expression Profiling Relative to B. napus

First true leaves were collected separately from three individual plants at the same stage as mentioned above for *B. villosa* for each of the four other species mentioned under plant material. Total RNA was extracted from the seedling leaves (three independent preparations per species) with a commercial RNA-Easy mini kit (Qiagen, Valencia, CA, USA) and cDNA synthesized using Superscript II^{TM} (Invitrogen, Carlsbad, CA) according to manufacturer's instructions. Quantitative real time-PCR (Q-PCR) was conducted using a Platinum SYBR Green Super Mix-UDG kit (Invitrogen), a CFX96 Real-Time PCR system (BioRad, Hercules, CA, USA), and primers to conserved regions for *GL1*, *GL2*, *EGL3*, *TTG1* and *TRY* within all five species (Table 1). Even though the *B. napus EGL3* coding sequence was not available, the primers to *EGL3*-specific conserved regions in other species amplified the *B. napus* transcripts in Q-PCR reactions. For each pair of gene-specific primers, melting curve analysis was conducted to determine the melting temperature and to ensure a single PCR amplicon of the expected length. Three independent RNA preparations were assayed per species and data was analyzed using CFX Manager Software (BioRad). The expression level of each mRNA was determined using the mean cycle threshold (ΔCT) value normalized to an endogenous reference gene, *BnEF1* (Table 1). Mean values with corresponding standard errors were expressed relative to glabrous *B. napus* leaf tissue and analysed by one-way analysis of variance (ANOVA) using PROC GLM in SAS ver. 9.2 (SAS, 2008) and a completely randomized design (CRD) (with genes and species as the two main factors). Significantly different means were detected by Fisher's protected Least Significant Difference (LSD) tests.

Bioinformatic Analysis of Brassica Trichome Regulatory Genes

A. thaliana nucleotide and protein sequence information for GL1, GL2, EGL3, TTG1, and TRY was collected from http://www.arabidopsis.org. *B. villosa* sequences were obtained as above. *B. napus* sequences for GL1, GL2, TTG-2 and TRY were obtained from EST databases available at http://rapa.agr.gc.ca and http://blast.ncbi.nlm.nih.gov. *B. napus* EST databases were searched at http://napus.agr.gc.ca/aped (*B. napus EGL3* sequence was unavailable). *B. rapa* sequence information was collected from http://brassicadb.org/brad, http://www.plantgdb.org/BrGDB, and http://www.brassica-rapa.org/BRGP/index.jsp. *B. oleracea* sequences were collected from the *B. oleracea* genome database at Agriculture and Agri-Food Canada, Saskatoon (Parkin, unpublished). Nucleotide and translated protein sequences were aligned using Clustal-W in Vector NTI 9 (Invitrogen). Molecular weight (Mr) and isoelectric point [35] were determined on translated protein sequences using http://web.expasy.org/compute_pi/. Amino acid frequencies were determined using http://emboss.bioinformatics.nl/cgi-bin/emboss/fuzzpro. Phylogenetic analysis was carried out using Molecular Evolutionary Genetics Analysis software (MEGA v5.1) and the neighbour joining method of phylogenetic inference with a bootstrap parameter of 1000 replications [36]. Initially, all six copies (ESTs) for *BnTTG1* and the short (<100 bp) *B. rapa* sequences for *BrTRY2* and *BrTRY3* were used in phylogeny testing, but later only the longer overlapping *BnTTG1* consensus sequence and the longer *BrTRY1* were used with the other orthologues to improve the confidence (bootstrap values) of phylogenetic relationships above 50%. Protein conserved domain recognition was determined at http://www.ebi.ac.uk/Tools/pfa/iprscan/.

Evolutionary analysis was performed to obtain more information on similarity and variation between the five major trichome regulatory genes of *B. villosa* and orthologues within three other *Brassicas* and *A. thaliana*. A pairwise comparison of *GL1*, *GL2*, *EGL3*, *TTG1* and *TRY* coding region of the orthologous genes was used to calculate the ratio of non-synonymous amino acid substitution rate (*Ka*) to synonymous substitution rate (*Ks*) using the maximum likelihood algorithm implemented in Phylogenetic Analysis by Maximum Likelihood (PAML) [37]. As with the phylogenetic analysis, all the *B. napus TTG1* coding sequences and the very short *B. rapa TRY2* and *TRY3* sequences were initially aligned with other sequences, but later excluded from the evolutionary analysis due to lack of sufficient overlapping sequence. Generally, a *Ka/Ks* of 1 indicates neutral selection, *Ka/Ks* <1 indicates a functional constraint with purifying selection, and *Ka/Ks* >1 shows accelerated evolution with positive selection [37].

Results

Trichome Regulatory Gene Amplification in B. villosa

Amplification of the coding regions for *B. villosa* orthologues to the *GL1*, *GL2*, *EGL3*, *TTG1* and *TRY* regulatory genes using primers based on available *B. napus* and *B. rapa* sequences gave single PCR bands (in some cases amplifying multiple copies). The bands were cloned, sequenced and used to search the NCBI database to confirm gene identity and to determine copy number. Three independent clones each were sequenced for the *BvGL1*, *BvGL2*, and *BvTTG1* genes from *B. villosa*, and this was expanded to 10 clones for *BvEGL3*, and thirty clones for *BvTRY*. Alignments of the cloned sequences to private (*B. oleracea*) and public (*B. rapa*, *B. napus*) databases, plus RNAseq data, indicated that *B. villosa* contained a single copy each for *GL1* and *GL2* and two unique sequences for *EGL3* (95% amino acid identity) (Table 2; Figure S1, protein sequence shown). Only one *TTG1* copy could be isolated from *B. villosa* using the *B. rapa* primers in spite of a second copy (with only 45% nucleotide identity) in the *B. oleracea* database (Parkin, unpublished). Sequencing of 30 *TRY* clones indicated two copies from *B. villosa* with 100% amino acid identity (Figure S1). RNAseq suggested a 3rd copy for *B. villosa TRY* not seen within the 30 amplicons (Table 2).

Phylogeny of B. villosa Trichome Regulatory Genes

In general, the amplified *B. villosa* translated coding sequences were over 90% identical to other *Brassica* orthologues, especially GL2 and TTG1 (Table S1), with only a few exceptions. *B. villosa* GL1 was only 82% similar to *B. rapa* GL1-2 and 75% similar to AtGL3. *B. villosa* EGL3-1 was 87% similar to *B. oleracea*, *B. rapa*, and Arabidopsis EGL3-2, while *B. villosa* EGL3-2 was 82% similar to the three latter sequences. *B. villosa* TRY-1 and TRY-2 were 86% and 87% similar to *B. oleracea* TRY-3 and *B. rapa* TRY-1, respectively. In contrast, *B. villosa* TTG1-1 and all other orthologues were only 48% similar to *B. oleracea* TTG1-2.

Phylogenetic trees using translated protein sequences indicate that GL1 and GL2 from hairy *B. villosa* are closest to several sequences from the two non-hairy species *B. napus* and *B. oleracea* (Figure 1). The two BvEGL3 sequences both showed strong relationships to EGL3-1 from *B. oleracea* and *B. rapa*, and were less close to *B. rapa* and *B. oleracea EGL3-2* sequences, which were closer to the Arabidopsis EGL3 (*B. napus* EGL3 sequence was not available). BvTTG1 sequence was intermediate between Arabidopsis and *B. napus*, BoTTG1-1, and *B. rapa* TTG1 sequences, and farthest from BoTTG1-2 (which was only 45% similar to BoTTG1-1). BvTRY-1 and BvTRY-2 paired together and fell

Table 2. Database accession numbers for orthologues to the major trichome regulatory genes (*GL1*, *GL2*, *EGL3*, *TTG1* and *TRY*) in *B. villosa*.

Gene	B. villosa (NCBI)	B. oleracea (Unpublished data, Dr. I. Parkin)	B. rapa (BRAD)	B. napus (NCBI)	Arabidopsis (TAIR)	Function in Overexpression And Knockdown studies in B. rapa
GL1	KF188209	Bo7g090950	Bra025311-1 Bra039065-2	HQ162473.1	AT3G27920, Myb-like protein, helps in induction of trichome development	Like Arabidopsis
GL2	KF188210	Bo6g046840	Bra003535	EU826521.1-1 EU826520.1-2	AT1G79840, Homeodomain protein affects trichomes initiation & development	Unknown
EGL3	KF188207-1 KF188208-2	Bo9g029230-1 Bo9g035460-2	Bra027653-1 Bra027796-2	NA	AT1G63650, a bHLH transcription factor 1, mutant has reduced trichomes	Unknown
TTG1	KF188213	Bo7g096780-1 Bo2g159360-2	Bra009770	EF175932.1-1 EF175931.1-2 EU192031.1-3 EF175929.1-4 EF175930.1-5 EU192030.1-6	AT5G24520, WD-40 protein involved in trichome development	Like Arabidopsis
TRY	KF188211-1 KF188212-2	Bo2g046050-1 Bo3g022870-2 Bo9g110930-3	Bra022637-1 Bra026297-2 Bra029089-3	EE451172(EST)	AT5G53200, Myb-like protein, mutation leads to glabrous leaves	Unknown

Genes in the same row are closest orthologues to each other. NA = Sequence not available.

nearest to the pairing between the *B. napus* consensus sequence and the *B. oleracea* TRY-3 sequences. Initially, all six copies (ESTs) for BnTTG1 and the short (<100 bp) *B. rapa* sequences for BrTRY2 and BrTRY3 were used in this phylogeny testing. Later (due to differences in sequence length), only the longer overlapping BnTTG1 consensus sequence and longer BrTRY1 were compared with the other orthologues, and this improved the confidence (bootstrap values) of the phylogenetic relationships.

Total Trichome Regulatory Gene Expression in Leaves of Hairy B. villosa, Diploid Brassica, and A. thaliana, Relative to Glabrous B. napus

Alignment of the *B. villosa* nucleotide sequences with other known *Brassica* and *A. thaliana* trichome regulatory sequences in public and private databases (data not shown) indicated conserved regions from which we designed Q-PCR primers (Table 1) to compare total expression levels of the five genes relative to orthologues of glabrous *B. napus*. In a preliminary Q-PCR expression experiment, all five trichome regulatory genes were highly expressed in trichome-bearing first true leaves of *A. thaliana* (with *GL1*, *GL2*, and *EGL3* being the highest and *TTG1* and *TRY* being the lowest) (Figure S2, insert). In contrast, expression of these orthologues in the first leaf of *B. villosa*, *B. oleracea*, *B. rapa* and *B. napus* (set at 1) (was dramatically lower than in Arabidopsis (Figure S2). *GL1* and *GL2* transcription was highest in *B. villosa* when only the four *Brassica* species were compared. *B. villosa* expression was proportionately more similar to *B. rapa* expression (in leaves bearing a light coverage of leaf trichomes) when total relative expression patterns of each of the four positive regulatory genes were compared across the *Brassica* species. Moreover, the relative expression pattern was highly distinct for the two hairy *Brassica* species compared with the glabrous *B. oleracea* and *B. napus*. Both hairy lines showed proportionately lower total leaf expression for *EGL3* and *TTG1* than *GL1* and *GL2*, but *TRY* expression in *B. rapa* was much lower. In contrast, *B. villosa* with its dense leaf trichome coverage showed high expression for *TRY*.

Copy-specific Trichome Regulatory Gene Expression in B. villosa and its C Genome Relative, B. oleracea

Since total overall transcript levels for *GL1*, *GL2* and unexpectedly for *TRY* were very high in *B. villosa* leaves relative to the other *Brassica* species in the Q-PCR analysis, expression levels were examined for the four trichome positive regulatory genes, and *TRY* by RNA seq. This analyses was expanded to include three other major negative regulatory genes (*CPC*, *ETC1* and *ETC3*) to determine whether any particular copy was more prominently expressed in hairy *B. villosa* first true leaves (and glabrous cotyledons). To do this, we took advantage of a new reference genome in the glabrous C genome relative, *B. oleracea* (Parkin, unpublished). Expression of two copies were detected for *EGL3* and *TTG1* and three copies for *TRY* and *CPC* in *B. villosa* by mapping to the *B. oleracea* genome. RNA seq showed that the single-copy BvGL2 gene and BvTTG1-1 and BoTTG1 have very high transcript levels in hairy *B. villosa* leaves compared with the *GL1* and *EGL3* positive regulatory genes. Transcripts of the single BoGL2 were undetectable in glabrous *B. oleracea* leaves and similarly *TTG1-2* was undetectable in glabrous cotyledons of both species (Figure 2). Expression of the single copy *GL1* gene was very low (<0.5 FPKM) in leaves and equivalent between the two species, and undetectable in cotyledons. Both copies of *EGL3* were expressed at a slightly higher levels in leaves (ranging from 0.3 to 2.5 FPKM) than cotyledons (ranging from 0.05 to 1 FPKM), but the transcript levels were reduced in *B. villosa* leaves compared

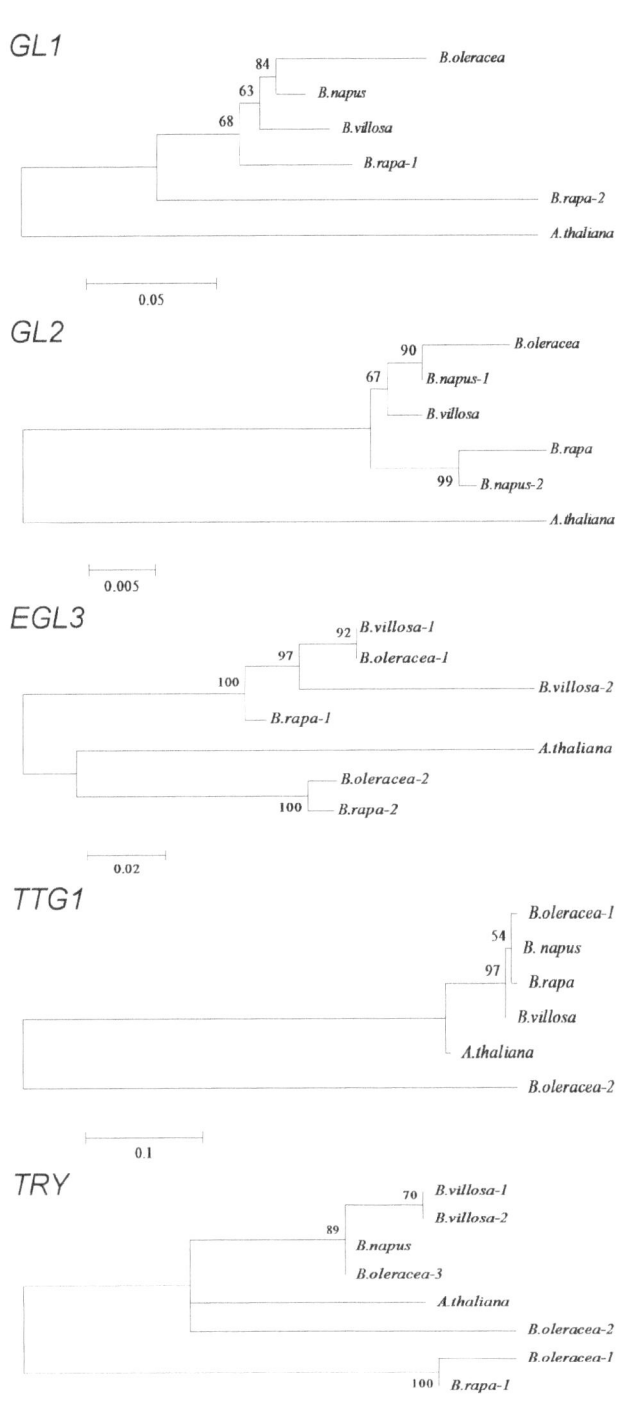

Figure 1. Phylogenetic relationships for the five major trichome regulatory genes present in *Brassica* and *A. thaliana*. Sequences were analysed by the maximum likelihood method with bootstrap values (%) indicated (100% is implicit in vacant branching positions). Scale indicates amino acid substitutions per site. A consensus sequence based on six *B. napus* TTG1 copies in NCBI was used for more robust analysis rather than the individual BnTTG1 copies, which each gave very weak associations due to limited overlapping sequence. *B. rapa* TRY-2 and TRY-3 were also not included since their small size gave spurious weak associations of <50%. Although three copies for *TRY* exist in *B. oleracea* and *B. villosa* (Fig. 2), *TRY-3* could not be cloned from *B. villosa* cDNA due to low expression.

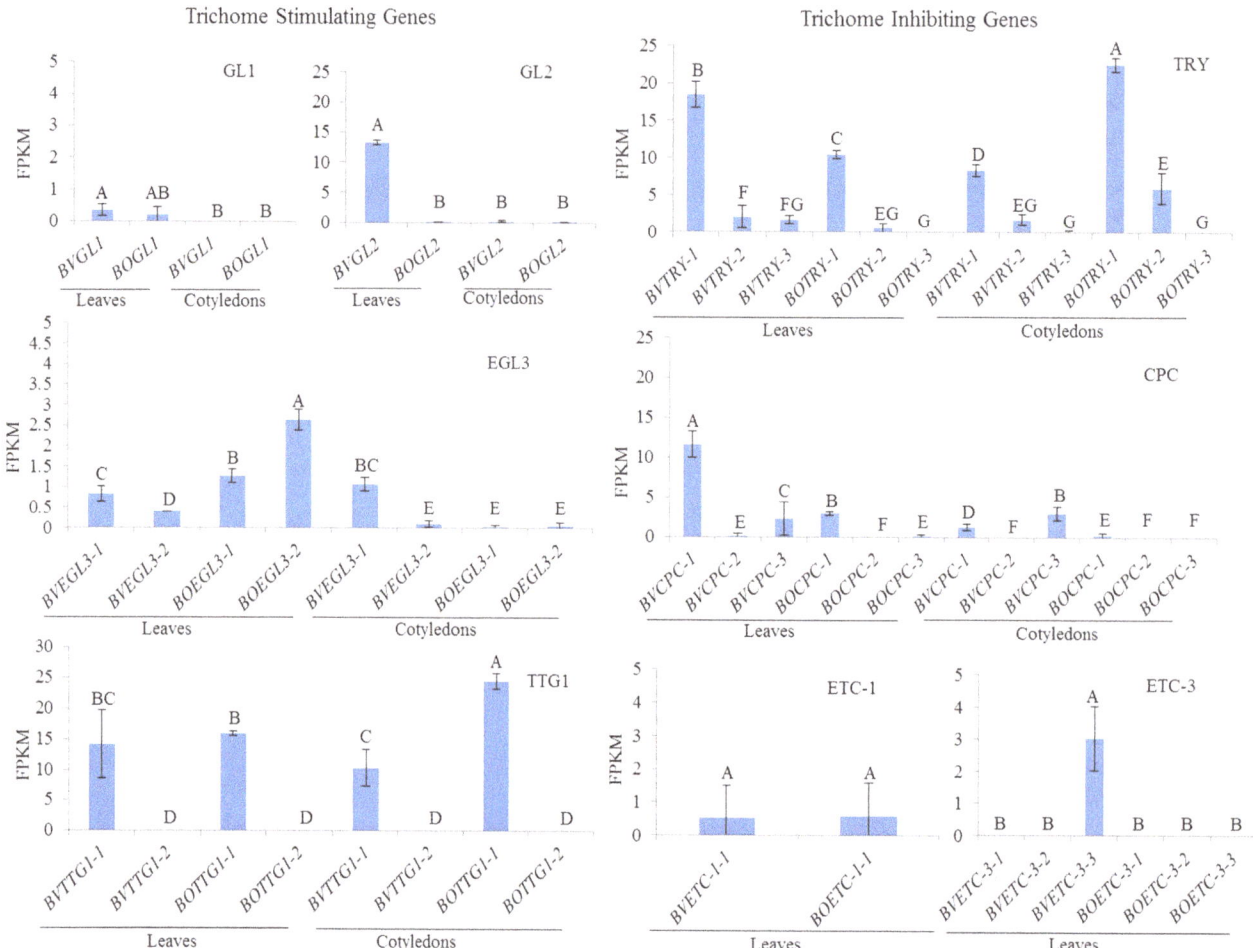

Figure 2. Transcript levels for individual gene copies of the four trichome positive regulatory genes and four negative regulatory genes in *B. villosa* compared with *B. oleracea*. RNAseq data is expressed as fragments per kilobase of exon per million fragments mapped (FPKM). Within each panel, different letters represent significantly different means (± standard error) for 3 independent RNA extractions (1st true leaves or cotyledons from up to 10 plants per extraction) at $p \leq 0.05$.

with *B. oleracea* leaves (Figure 2). Curiously, *BvEGL3-1* had higher leaf transcript levels proportionally to *BvEGL3-2* in *B. villosa*, whereas the converse was true in *B. oleracea*. In cotyledons, *BvEGL3-1* had expression (equivalent to leaf), and both genes were almost undetectable in *B. oleracea* cotyledons (Figure 2). The single copy *B.villosa GL3* has a similar pattern of expression to that of *BvEGL3-1*, with higher expression of *BvGL3* in cotyledons compared to *BoGL3*, whereas higher expression of *BoGL3* in true leaves compared to *BvGL3* (data not shown). The *B. villosa* single copy of *GL2*, copy 1 of *EGL3* and *TTG1* all proved to be functional proteins, since their binary expression constructs enhanced trichome density in Arabidopsis (Figure 3). Preliminary analysis of knockdown lines for these genes in both Arabidopsis and *B. napus* resulted in no observed phenotype changes (Nayidu and Gruber, unpublished).

Transcripts for three different copies of the *TRY* negative regulatory gene and two copies of *BoTRY* were detected (Figure 2). Very high leaf and cotyledon expression was detected for *BvTRY-1* in *B. villosa* (18–20 FPKM) and *BoTRY-1* in *B. oleracea* (9–10 FPKM), while *BvTRY-2/BoTRY-2* and *BvTRY-3/BoTRY-3* transcript levels were much lower in both leaves and cotyledons (Figure 2). The ranking and proportionate expression of individual *TRY* gene copies for *B. oleracea* was similar to *B. villosa*, but 50%

lower in leaves and much higher in cotyledons. The higher combined expression level (FPKM value) for the three *TRY* genes in trichome-rich *B. villosa* leaves correlated with the overall higher total *TRY* expression level measured in the *B. villosa* leaves by Q-PCR. The fact that the relative expression for all three *TRY* genes was quite similar in both species and that *BvTRY-1* and *BoTRY-1* expression was equally high regardless of the extreme differences in leaf trichome density between *B. villosa* and *B. oleracea* was inconsistent with the Arabidopsis model of *TRY* being a negative regulator in *B. villosa*, even though Arabidopsis leaf trichomes were eliminated when Arabidopsis was transformed with a *BvTRY-1* expression construct (Figure 3).

Since the expression pattern of *TRY-1* was higher in trichome-covered *B. villosa* leaves, we took advantage of the new *B. oleracea* genome database to examine individual copies of several other prominent negative regulatory genes, *CPC*, *ETC1*, and *ETC3* ([23]; Table 2). Only one copy existed for *ETC1* in both *B. villosa* and *B. oleracea* and the expression level of both was very low (0.05 FPKM) in leaves (Figure 2) and undetected in cotyledons of both species (data not shown). *ETC3* had three copies in both species, but only *BvETC3-3* was expressed (2–4 FPKM) in *B. villosa* leaves (Figure 2), and the other copies were undetectable in either species and in cotyledons (data not shown). In contrast, all three

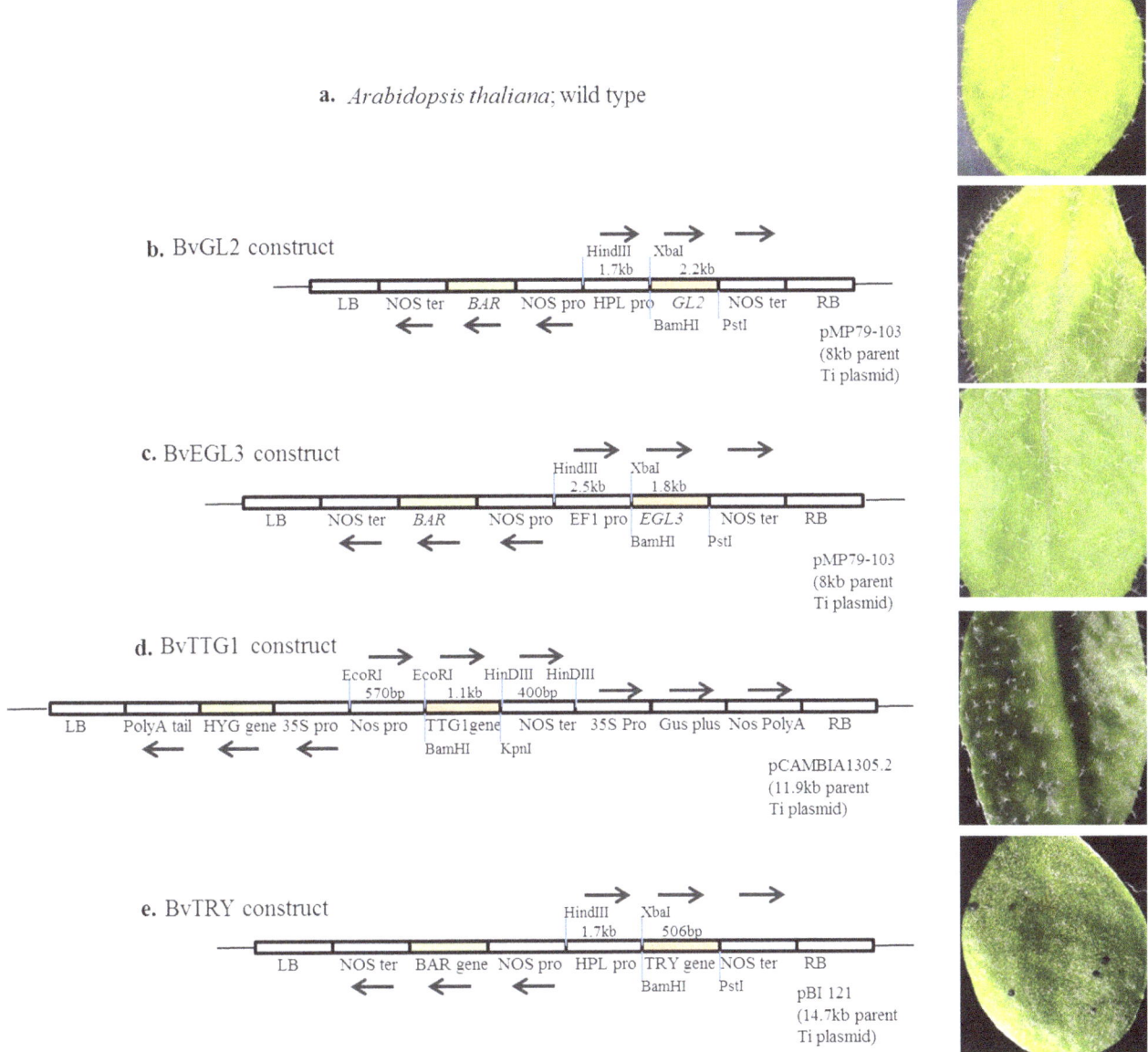

Figure 3. Trichome phenotypes of Arabidopsis transgenic plants overexpressing trichome related genes from *Brassica villosa*. a. Expanded leaf of wild type Arabidopsis (Columbia) showing normal trichome pattern. b. BvGL2 over-expressed transgenic leaf showing increased trichome number. c. BvEGL3 over-expressed transgenic leaf showing increased trichome number. d. BvTTG1 over-expressed transgenic leaf showing increased trichome number. e. BvTRY over-expressed transgenic leaf showing glabrous phenotype. (All photographs were taken with 10X magnification by a compound microscope at 3 weeks after germination).

copies of the *CPC* gene were expressed in leaves of both species, but expression of *CPC-1* was predominant in hairy *B. villosa* leaves and 6-fold higher than *BoTRY-1* in glabrous *B. oleracea* leaves. The other two *CPC* copies had low transcript levels in leaves and cotyledons, and *BvCPC-3* expression (2.5 FPKM) was predominant in cotyledons.

Sequence Comparisons between B. villosa, three other Brassicas and Arabidopsis

Since gene function impacts phenotype from a combination of transcript level and translated protein sequence, amino acid sequence diversity was determined for translated sequences of the five cloned regulatory genes in *B. villosa*, the three other *Brassica* species and *A. thaliana* (Figure S1). Overall, sequences for *B. villosa*

were quite similar to those of the other *Brassicas* and *A. thaliana* in length, Mr and pI, with the following exceptions (Table 3). BvEGL3-2 was 20 kDa shorter than BvEGL3-1 from *B. villosa* and the other EGL3 sequences from *B. oleracea*, *B. rapa*, and Arabidopsis. The Mr of BvTRY-1 was most similar to that of *B. oleracea*, *B. napus*, and Arabidopsis, while BvTRY-2 was smaller. Isoelectric points for proteins from *B. villosa* were similar to those of orthologues in other *Brassica* species and *Arabidopsis* and usually acidic, except in the case of GL1 (Table 3). Here, the *B. villosa* orthologue BvGL1 was neutral and had a pI closest to *B. rapa* BrGL1-1 rather than to alkaline pI of BnGL1-1, BoGL1-1, BrGL1-2, BoTTG1-2 and most of the TRY sequences.

Many amino acid substitutions, additions and deletions were apparent within each of the five trichome genes when comparisons

Table 3. Theoretical protein size (Mr/Number of amino acid) and isoelectric point (*pI*) of trichome regulatory coding sequences in the *Brassicaceae*.

	B. napus (Mr)	B. oleracea (Mr)	B. rapa (Mr)	B. villosa (Mr)	A. thaliana (Mr)	B. napus (pI)	B. oleracea (pI)	B. rapa (pI)	B. villosa (pI)	A. thaliana (pI)
GL1-1 GL1-2	26.0/225	24.7/212	26.1/225 22.7/199	26.1/225	26.3/228	8.80	9.42	7.67 9.22	7.69	6.66
GL2-1 GL2-2	83.7/750 83.7/750	83.6/748	83.8/750	83.5/748	86.5/776	6.58 6.83	6.58	6.33	6.56	6.38
EGL3-1 EGL3-2	ND	64.0/573 67.6/604	66.3/596 67.8/606	66.5/597 46.6/421	66.6/616	ND	5.59 5.15	5.47 5.14	5.52 6.20	5.06
TTG1-1 TTG1-2 TTG1-3 TTG1-4 TTG1-5 TTG1-6	37.3/337 37.2/337 37.2/337 37.3/337 37.4/338 37.2/337	37.2/337 24.5/212	37.3/337	37.3/337	37.9/341	4.66 4.66 4.66 4.66 4.70	4.66 10.77	4.66	4.66	4.71
TRY-1 TRY-2 TRY-3	13.1/107	13.0/107 13.0/106 19.0/160	9.07/75 6.51/54 6.34/51	13.0/107 9.02/75	13.0/106	9.58	9.51 9.24 9.02	9.29 9.51 5.28	9.21 9.39	9.51

ND, not determined (sequence not available). Closest orthologues between the species are positioned within the same row. Data represents all known orthologues and homologues for each species.

were made between *B. villosa* and the four other species. In total, there were 58 amino acid changes (26.2%) out of a total of 237 consensus amino acids (CAA) for GL1, 30 (3.8%) out of 786 CAA for GL2, 92 (15.1%) out of 608 CAA for EGL3-1, 132 (27.2%) out of 485 CAA for EGL3-2, and 111(31.2%) out of 356 CAA for TTG1. TRY genes were the smallest compared to the positive regulatory genes and had the most amino acid differences between the species: 99 (84.6%) out of 117 CAA for TRY-1 and 111 (69.4%) out of 160 CAA for TRY-2 (Figure S1). Out of the total amino acid differences observed, those which were unique to *B. villosa* or distinguished hairy germplasm from glabrous germplasm were more closely examined (Table 4, Figure S1). Several particularly noteworthy variations between hairy and glabrous germplasm involved more extreme charge and hydrophobicity differences that potentially could affect the molecular structure and functional properties of a protein. These included consensus amino acid (CAA) position 223 in the 3′ end of GL1, where the aromatic phenylalanine [38] in *B. villosa* and *B. rapa* and the hydrophobic residue leucine (Leu) in *A. thaliana* and *B. rapa* [hydrophobicity values of 2.8 (F) and 3.8 (L), respectively [39]] were replaced by the slightly hydrophilic serine (Ser, value of − 0.8) in glabrous *B. napus* and *B. oleracea* (Table 4). CAA position 273 in GL2 included a glutamine (Gln) residue in hairy *B. villosa*, *B. napus* (one out of two copies), and the hairy *B. rapa* sequences, and a negatively charged glutamate acid (Glu) residue in *A. thaliana*. These were replaced by a positively charged imidazole ring (histidine, His) in the other copy of glabrous *B. napus* and in glabrous *B. oleracea* sequences. CAA position 439 in GL2-1 included an alanine conserved between hairy *B. villosa*, *A. thaliana*, and *B. rapa* proteins and replaced by a hydrophobic valine in GL2 of glabrous *B. napus* and *B. oleracea*. Finally, three positions in BvGL1, two in BvGL2, two in EGL3, one in BvTTG1, and one in TRY were unique to *B. villosa*. In particularly, the serine in CAA position 154 of all BvTRY sequences was replaced by an arginine in TRY genes from all other species.

Each of the four trichome initiation genes and TRY had domains which were conserved within all the *Brassica* species and *A. thaliana*, but they also showed species distinctions not correlated with a trichome phenotype (Figure S1). GL1 consisted of well known conserved R2 and R3 MYB DNA binding domains, but *B. rapa* GL1 was missing CAAs 1-29 and the *BoGL1* had a nonsense mutation at the end of the sequence and five unique amino acids (CAA position 119-123) in a conserved region immediately downstream of the R3MYB domain. GL2 showed no significant differences in the conserved homeobox domain and had the most consistent amino acid profile across all species compared to the other regulatory genes (Figure S3). However, CAA positions 114 to 118 present in the conserved *A. thaliana* START domain and the unique 29 amino acid (aa) Arabidopsis leader sequence were missing in GL2 from the four *Brassica* species (Figure S1). EGL3 translated sequences were mainly equivalent in their bHLH-DNA binding domains, but variable between CAA 391-3 in all orthologues (Figure S1). Noteworthy was an entirely different sequence post-CAA 426 for *B. villosa* EGL3-2, whereas all other EGL3s were missing aa from 426–455 and 485 to the stop codon, aa 456–484 were completely different, and sequences for *B. oleracea* EGL3s were missing CAAs 164–187. BoTTG1-2 and TRY also showed diversity. TRY was completely conserved in the R3 MYB binding domain but diverse in other areas (Figure S1). BvTRY-2, BrTRY-1 and BrTRY-2 sequences were each missing CAA 53–85, *B. rapa* TRY-3 was missing aa 105–160 within and beyond the R3 binding region, TRY-3 of *B. oleracea* had a unique 52 aa leader sequence, while *B. rapa* TRY-2 and 3 had the smallest R3 MYB binding regions. Except for CAA 1–85 missing in BvTRY-2,

Table 4. Specific amino acids in five trichome regulatory genes within the *Brassicaceae*.

	*Consensus position	A. thaliana hairy	B. rapa AA hairy	B. napus AC glabrous	B.oleracea CC glabrous	B. villosa CC hairy
GL1	137*	Pro	Pro	Pro	Pro	Thr
	154*	Gln	Gln	Gln	Gln	Glu
	202*	Asn	Asn	Asn	Asn	Asp
	223(1) 223(2)	Leu NA	Phe Leu	Ser NA	Ser NA	Phe NA
GL2	273(1) 273(2)	Glu NA	Gln NA	His Gln	His NA	Gln NA
	282*	Tyr	Tyr	Tyr	Tyr	Phe
	287*	Ala	Ala	Ala	Ala	Ser
	439-(1) 439-(2)	Ala	Ala	Val	Val	Ala
EGL3	171(1) * 171(2)	Val NA	Val Val	ND ND	Val Val	Ala Val
	552*	Leu	Leu	ND	Leu	Val
TTG1	4*	Ser	Ser	Ser	Ser	Ala
TRY	154(1) * 154(2) * 154(3) *	Arg NA NA	Arg NIL NIL	Arg NA NA	Arg Arg Arg	Ser Ser NA

*Selected positions in the aligned consensus amino acid sequence (CAA) were selected from Figure S1 if they distinguished hairy from glabrous germplasm (dark arrows in Fig. S1) or were unique to *B. villosa* (*red arrows in Fig. S1). ND, not determined (sequence not available). NIL, missing amino acid. NA, not applicable. Note: Multiple gene copies are only indicated if an amino acid differed between the copies.

BvTRY-1 and BvTRY-2 were identical, but the very low expression of *BvTRY-3* prevented evaluation.

Ka/Ks Amino Acid Substitutions

Leucine (L), serine (S) and arginine (R) were the most abundant amino acids within all five trichome regulatory genes for *B. villosa* as well as the other four species (Figure S3). However, cysteine was completely absent in TRY orthologues (except for *B. oleracea* and *B. rapa* TRY-2, each of which had one cysteine at CAA 95). Moreover, *B. villosa* TRY-2 and *B. rapa* TRY-1 were under-represented in V, F, S, T, D, H, K, and R compared with BvTRY-1 and TRY in other species examined (Figure S3). This led us to determine *Ka/Ks* amino acid substitution values for the five trichome genes to assess whether sufficiently different amino acid substitutions had occurred to potentially change protein function.

Ka/Ks values were much less than one for most pair-wise comparisons between homologues and orthologues, especially for GL2, TTG1, and TRY, indicating mainly synonymous substitutions that would not impact protein structure (Table S1). *Ka/Ks* values for EGL3 sequence comparisons were higher in general than values for GL2, TTG1, and TRY comparisons, but less than one and lower than many of the *Ka/Ks* values determined for GL1 sequences (Table 5; Table S1). Most pairwise comparisons between GL1 in *B. villosa* and each *Brassica* species and Arabidopsis consistently showed synonymous substitutions. However, *Ka/Ks* ratios for BvGL1 and BoGL1 were much closer to one, indicating an evolutionary tendency towards significant change. This was even more extreme between GL1 from *B. villosa* and *B. rapa* (1.04) and from *B. villosa* and *B. napus* (1.87) (Table 5), indicating large substitution differences that may have the potential for functional changes.

Discussion

Brassica villosa is a weedy C genome species of the *Brassicaceae* and is of interest to molecular evolutionists and plant breeders for its dense trichome patterning on most tissues [33]. In the present study, we compared expression patterns and coding sequences for key trichome regulatory genes of *B. villosa* with those of orthologues from *A. thaliana* and several species within *Brassica* A and C genomes, as a means of distinguishing unique *B. villosa* sequence patterns from those of more modest trichome-bearing species and glabrous germplasm. Phylogenetic trees indicated that GL1, GL2, EGL3 and TRY sequences from hairy *B. villosa* are closest to several orthologues from the two non-hairy species *B. napus* and *B. oleracea*. These data confirm the closer relationship of these *B. villosa* positive regulatory trichome genes to those of *B. oleracea* and the C genome of *B. napus* than to Arabidopsis regardless of the trichome density differences. Li and co-workers showed pairing of GL1 sequences from hairy *Brassica incana* (related to *B. villosa*) with those from non-hairy *B. napus* lines (with a 92–94% sequence identity) [29].

Generally, RNA sequencing showed that transcript levels of one of the multiple copies of *EGL3, TTG1, TRY, CPC*, and *ETC3* were predominant compared to levels of the other copies in *B. villosa*. Expression for the trichome stimulating genes, *BvGL2* and *BvTTG1*, and the trichome inhibiting genes, *BvTRY-1* and *BvCPC-1*, was very high in young true leaves of *B. villosa* (where dense trichome coverage is found) and low in cotyledons compared with *GL1, EGL3, ETC1*, and *ETC3* genes. This *B. villosa* RNAseq pattern was distinct and somewhat unexpected since the *BvEGL3* expression was lower than in glabrous *B. oleracea* leaves and the *BvTTG1* expression was quite similar in level to *BoTTG1* expression. Expression of the *BvTRY-1* and *BvCPC-1* was also high in hairy *B. villosa*, and expression of *BoTRY-1* was high in glabrous *B. oleracea* leaves. This was inconsistent with the Arabidopsis model of TRY and CPC as negative regulators of trichome initiation, where enhanced leaf trichome density phenotype occurred when TRY expression was knocked down in Arabidopsis *try* mutants [40]. The data implies that *BvTRY-1, BvCPC-1*, and (potentially) *BvETC3-1* genes may not behave as negative regulators of trichome initiation in *B. villosa*. Protein coding sequences for BvTRY-1 and BvTRY-2 genes were closest to those of non-hairy BoTRY-3 and BnTRY, and all four of these are closer to each other and to *A. thaliana* than to the other *Brassica* TRY genes. Redundant trichome negative regulatory genes exist in the *A. thaliana* model [25] and functional redundancy can speed up gene evolution [41–43]. Hence, *B villosa* may use these R3 regulatory genes with high expression for a different purpose in the densely covered *B. villosa* true leaves. This hypothesis is supported

Table 5. *Ka/Ks** ratios for trichome regulatory gene comparisons between *B. villosa* and three other *Brassica* species and Arabidopsis.

Pairwise Comparison		GL1	GL2	EGL3	TTG1	TRY
B. napus	*B. villosa*	1.87	0.09	NA	0.03	0.23
B. villosa	*B. rapa-1*	1.04	0.06	0.42	0.02	0.19
B. oleracea	*B. villosa*	0.78	0.14	0.51	0.1	0.06
B. villosa	*B. rapa-2*	0.21	–	0.31	–	–+
B.villosa	*A. thaliana*	0.24	0.07	0.2	0.01	0.06
B.napus-2	*B.villosa*	0.09	0.09	0.09	0.09	–
B.villosa-1	*B.villosa-2*	–	–	0.86	–	0.00
B.villosa-1	*B.oleracea-2*	–	–	0.30	0.27	0.15
B.oleracea-1	*B.villosa-2*	–	–	0.86	–	0.06
B.villosa-2	*B.oleracea-2*	–	–	0.40	–	0.15
B.villosa-2	*B.rapa-2*	–	–	0.49	–	–+
B.villosa-2	*A.thaliana*	–	–	0.21	–	0.06
B.napus-3	*B.villosa*	–	–	–	0.05	–
B.napus-4	*B.villosa*	–	–	–	0.03	–
B.napus-5	*B.villosa*	–	–	–	0.03	–
B.napus-6	*B.villosa*	–	–	–	0.03	–
B.villosa-1	*B.oleracea-3*	–	–	–	–	0.19
B.villosa-2	*B.oleracea-3*	–	–	–	–	0.19

**Ka/Ks* (Yang Z, 1997): *Ka*, non-synonymous nucleotide substitution. *Ks*, synonymous nucleotide substitution value.
+*B. rapa*-2 (BRTRY-2) and *B. rapa*-3 (BRTRY-3) amino acid sequences were too short to be included. NA, *B. napus* sequence not available.

by the insertion of *BvTRY-1* into *B. napus*, yielding transgenic plants in which trichome density is not affected even though the same binary construct was used to depress Arabidopsis trichome development (Nayidu and Gruber, unpublished and Figure 3 respectively). In the future, it will be particularly useful to express these *B villosa* genes in a range of other Brassica species and to develop knock-out RNAi lines in *B. villosa* to solve the mystery of their true function. This will depend on the development of a transformation system for *B. villosa*, such as the protocols that now exist for *B. napus*, *B. rapa*, *B. oleracea*, and *B. carinata* [44–47]. Additional analysis of *B. villosa* gene structure is also necessary for a complete understanding of their introns, untranslated regions, and promoters. For example, intron 1 and 30 non-coding nucleotides in *A. thaliana* are important for the expression of the *GL1* gene [5,48,49]. A 620 bp fragment of the *TRY* promoter contains sequences that mediate the repression of its own expression, and deletion of this promoter region can rescue the *A. thaliana try* mutant phenotype [50]. Moreover, the Arabidopsis TRY promoter contains sequences regulated by microRNA-regulated "*SQUAMOSA PROMOTER BINDING PROTEIN LIKE*" (*SPL*) genes [50], and enhanced expression of the SPL-regulator miR156b in Arabidopsis increases trichome density [51].

The high expression pattern for GL2 and TRY-1 in *B. villosa* leaves led us to compare the coding sequences of the five trichome regulatory genes we had isolated from *B. villosa* with those in *B. napus*, two diploid *Brassica* species, and *A. thaliana*. Extreme trichome coverage in *B. villosa* leaves may be due (at least in part) to polymorphism and evolutionary differences that impact on protein function rather than solely to specific gene transcript levels, which are dictated by promoter "strength" and intra-gene regulatory structures. Although analysis showed few differences between most orthologues, overall evolutionary selection was detected between GL1 proteins of hairy (*B. villosa*, *B. rapa*) and

glabrous (*B. napus*) genotypes by their high pairwise *Ka/Ks* values (>1), potentially due to sites involved in adaptive change. Adaptive changes may occur at surprisingly few sites; consequently, the overall *Ka/Ks* ratio for an entire protein may remain dominated by non-adaptive changes and be substantially lower than unity (reviewed in [52]) as seen by the *Ka/Ks* scores for *GL2*, *EGL3*, *TTG1*, and TRY. Moreover, once a protein achieves a new advantageous function, the frequency of non-synonymous substitutions at the adapted sites will be reduced by new functional constraints [52].

A substantial proportion of the differences that distinguished the four *Brassica* trichome regulatory sequences occurred outside of conserved domains. Several of these individual amino acid differences were sufficiently dramatic to potentially change the molecular and functional properties of these proteins. Particularly, three variable positions in GL1 and GL2 translated sequences distinguished hairy *B. villosa* and *B. rapa* from glabrous *B. napus* and *B. oleracea* germplasm, and all the *B. villosa* genes had a minimum of one unique site (and usually more) compared with the other species. Bloomer [53] reported on the effects of natural variation in GL1 from Arabidopsis and suggested that qualitative differences in trichome phenotypes (glabrous or hairy) might have arisen independently several times by three unique protein coding changes and a whole locus deletion. The same authors also suggested that quantitative variation (in trichome density) might have arisen because of completely linked amino acid replacements and mutations in a known (as yet uncharacterized) enhancer region within the *AtGL1* locus. In addition, Li et al. proposed that a 5-bp deletion in exon 3 of the *B. rapa GL1* gene (starting at CAA 110 in the present study) is the basis of a hairless phenotype that arose from a normally hairy double haploid brown-seeded line [29]. Hence, the extremely dense trichome coverage of *B. villosa* could be due to a combination of relatively higher transcription of

GL2, hydrophobic amino acids and evolutionary changes in GL1, as well as the replacement of serine in all BvTRY sequences, and potentially a different function for TRY-1, CPC-1, and ETC3-3 (consistent with their higher *B. villosa* leaf transcript levels). Moreover, the glabrous leaves of the C genome relative *B. oleracea* could result from two non-synonymous substitutions (from asparagine to serine (at CAA 26 and 112), five continuous amino acid replacements (at CAA 119–123), one nonsense mutation (at CAA 224) leading to a shortened GL1 amino acid sequence, and a missing aa stretch in the bHLH DNA binding region of BoEGL3-1 and BoEGL3-2. Our study adds additional sequence variation data to a previous report detailing two 1 bp deletions and 1 bp insertion in exon 3 of the *B. oleracea* GL1 sequence [29]. Changes in protein function with individual amino acid modifications are also seen in other species and genes. Yeam et al. (2007) showed that a G/R polymorphism at aa 107 of the *Capsicum* eukaryotic translation initiation factor 4E protein (eIF4E) is sufficient for the acquisition of resistance against several *Capsicum* and tobacco etch potyvirus strains by expressing the amino acid substituted gene in potato (*Solanum lycopersicum*) [54].

Likely, the genetic variation we have uncovered is the "tip of the iceberg" in terms of variation that affects the function of *Brassica* trichome genes, since the trichome pathway in the simpler *A. thaliana* genome is already considered to be an integrated hierarchy of regulation by complex cell cycle status, transcriptional control and cytoskeletal function [55]. This is confirmed by the ever increasing number of trichome genes being discovered in *A. thaliana* mutant populations (>80 genes known in TAIR) [38] (Taheri et al., 2014 submitted manuscript from the same lab) and by the on-going discovery of *cis*-regulatory sequences that provide greater diversification in gene function, eg. for GL1 and MYB23 [56]. The constant improvement in genomic-scale sequencing technology, SNP analysis, and computation can now be applied to characterize large within-accession and within-species variance in trichome gene patterning in the *Brassicaceae* and identify statistically robust associations hitherto undetectable [57,58].

Conclusion

The present study utilized expression profiling and bioinformatics to compare *B. villosa* trichome regulatory genes with their orthologues in *B. oleracea*, *B. rapa*, *B napus*, and *A. thaliana*. In doing so, we discovered the potential for positive evolutionary selection in the *GL1* gene between *B. villosa*, *B. rapa* and *B. napus*. Several point mutations found within GL1 and GL2 protein sequences correlated with hairy and glabrous phenotypes within these five species and have the potential to be predictive factors. We also discovered high transcript levels for *BvGL2*, *BvTTG1*, *BvTRY-1*, and *BvCPC-1* in *B. villosa* leaves with densely covered trichomes. This *B. villosa* gene expression pattern is contrary to the trichome gene models in Arabidopsis in which cells are glabrous when expression of trichome R3 MYB inhibitor genes is high and trichomes are present in greater density and clustered when these genes are knocked down by mutation [23]. These *B. villosa* genes are now being tested in comparative over-expression studies in

Arabidopsis and *B. napus*. Investigations on these genes should also be expanded to include a much broader range of hairy and glabrous *Brassica* germplasm within the Triangle of U [26].

Supporting Information

Figure S1 Alignment of translated protein sequences for four major trichome positive regulatory genes and one negative regulatory gene in the *Brassicaceae*. GL1: Conserved R2 and R3 MYB DNA binding domains are boxed. A unique 5 aa variable sequence and a unique 14 aa 3′ deleted sequence are designated in the *B. oleracea* sequence by ovals. GL2: The conserved HDZip homeobox domain and a large START domain are boxed. A unique *A. thaliana* 29 aa leader sequence is delineated by an oval. GL3: The conserved bHLH protein binding domain is boxed. TTG1: Two conserved WD40 protein domains are boxed. A unique 44 aa leader sequence in *B. oleracea* is designated by an oval. TRY: The conserved R2 MYB DNA binding domains are boxed. A 33 aa unique deletion in the *B. rapa* gene is designated by an oval. For all 5 genes, unique amino acid modifications are designated for *B. villosa* (red arrows) and for hairy vs. glabrous plants (blue arrows).

Figure S2 Relative expression of four trichome regulatory genes and one negative regulatory gene in leaves of hairy *Brassica* villosa, three other Brassica species, and *A. thaliana*. Main panel shows expressed transcripts from four *Brassica* species. Insert panel shows *A. thaliana* orthologues which are much more highly expressed compared with the *Brassica* species. Expression (Q-PCR) in both panels is relative to glabrous *B. napus* cv. Westar (set at 1). Different letters in both panels indicate significant differences of the means ($\pm$ standard error) at p≤0.05 in an LSD test (SAS, 2008).

Figure S3 Amino acid profiles for five major trichome regulatory sequences from Brassica villosa, three other *Brassica* species, and *A. thaliana*.

Table S1 *Ka/Ks values from pairwise comparisons of all trichome regulatory gene orthologues and homologues within four Brassica species and Arabidopsis.

Acknowledgments

The authors thank Dr. L. Visser at the Center for Genetic Resources, Wageningen, The Netherlands, for the gift of *B. villosa* seeds. Dr. N. Nayidu and Dr. A. Taheri were recipients of Visiting Fellowships to a Canadian Government Laboratory.

Author Contributions

Conceived and designed the experiments: NKN. Performed the experiments: NKN. Analyzed the data: NKN SK AT TSW-G. Contributed reagents/materials/analysis tools: MG IP AS. Wrote the paper: NKN MG.

References

1. Broadhurst CL, Chaney RL, Angle JS, Maugel TK, Erbe EF, et al. (2004) Simultaneous hyperaccumulation of nickel, manganese, and calcium in Alyssum leaf trichomes. Environmental Science & Technology 38: 5797–5802.
2. Larkin JC, Brown ML, Schiefelbein J (2003) How do cells know what they want to be when they grow up? Lessons from epidermal patterning in Arabidopsis. Annual Review of Plant Biology 54: 403–430.
3. Balkunde R, Pesch M, Hülskamp M (2010) Chapter Ten-Trichome Patterning in *Arabidopsis thaliana*: From Genetic to Molecular Models. Current Topics in Developmental Biology 91: 299–321.

4. Herman PL, Marks MD (1989) Trichome development in *Arabidopsis thaliana*. II. Isolation and complementation of the *GLABROUS1* gene. The Plant Cell Online 1: 1051–1055.
5. Oppenheimer DG, Herman PL, Sivakumaran S, Esch J, Marks MD (1991) A myb gene required for leaf trichome differentiation in Arabidopsis is expressed in stipules. Cell 67: 483–493.
6. Zhao H, Wang X, Zhu D, Cui S, Li X, et al. (2012) A Single Amino Acid Substitution in IIIf Subfamily of Basic Helix-Loop-Helix Transcription Factor AtMYC1 Leads to Trichome and Root Hair Patterning Defects by Abolishing

Its Interaction with Partner Proteins in Arabidopsis. Journal of Biological Chemistry 287: 14109–14121.

7. Walker AR, Davison PA, Bolognesi-Winfield AC, James CM, Srinivasan N, et al. (1999) The *TRANSPARENT TESTA GLABRA1* locus, which regulates trichome differentiation and anthocyanin biosynthesis in Arabidopsis, encodes a WD40 repeat protein. The Plant Cell Online 11: 1337–1349.

8. Cristina M, Sessa G, Dolan L, Linstead P, Baima S, et al. (2002) The Arabidopsis Athb-10 (*GLABRA2*) is an HD-Zip protein required for regulation of root hair development. The Plant Journal 10: 393–402.

9. Rerie WG, Feldmann KA, Marks MD (1994) The *GLABRA2* gene encodes a homeo domain protein required for normal trichome development in Arabidopsis. Genes & Development 8: 1388–1399.

10. Szymanski DB, Jilk RA, Pollock SM, Marks MD (1998) Control of GL2 expression in Arabidopsis leaves and trichomes. Development 125: 1161–1171.

11. Hülskamp M, Miséra S, Jürgens G (1994) Genetic dissection of trichome cell development in Arabidopsis. Cell 76: 555–566.

12. Larkin JC, Walker JD, Bolognesi-Winfield AC, Gray JC, Walker AR (1999) Allele-specific interactions between ttg and gl1 during trichome development in *Arabidopsis thaliana*. Genetics 151: 1591–1604.

13. Schellmann S, Hulskamp M (2005) Epidermal differentiation: trichomes in Arabidopsis as a model system. International Journal of Developmental Biology 49: 579.

14. Zhang F, Gonzalez A, Zhao M, Payne CT, Lloyd A (2003) A network of redundant bHLH proteins functions in all TTG1-dependent pathways of Arabidopsis. Development 130: 4859–4869.

15. Payne CT, Zhang F, Lloyd AM (2000) GL3 encodes a bHLH protein that regulates trichome development in Arabidopsis through interaction with GL1 and TTG1. Genetics 156: 1349–1362.

16. Morohashi K, Zhao M, Yang M, Read B, Lloyd A, et al. (2007) Participation of the Arabidopsis bHLH factor GL3 in trichome initiation regulatory events. Plant Physiology 145: 736–746.

17. Schellmann S, Schnittger A, Kirik V, Wada T, Okada K, et al. (2002) *TRIPTYCHON* and *CAPRICE* mediate lateral inhibition during trichome and root hair patterning in Arabidopsis. The EMBO Journal 21: 5036–5046.

18. Ishida T, Kurata T, Okada K, Wada T (2008) A genetic regulatory network in the development of trichomes and root hairs. Annu Rev Plant Biol 59: 365–386.

19. Digiuni S, Schellmann S, Geier F, Greese B, Pesch M, et al. (2008) A competitive complex formation mechanism underlies trichome patterning on Arabidopsis leaves. Molecular Systems Biology 4.

20. Esch JJ, Chen M, Sanders M, Hillestad M, Ndkium S, et al. (2003) A contradictory *GLABRA3* allele helps define gene interactions controlling trichome development in Arabidopsis. Development 130: 5885–5894.

21. Kurata T, Ishida T, Kawabata-Awai C, Noguchi M, Hattori S, et al. (2005) Cell-to-cell movement of the CAPRICE protein in Arabidopsis root epidermal cell differentiation. Development 132: 5387–5398.

22. Wang S, Kwak S-H, Zeng Q, Ellis BE, Chen X-Y, et al. (2007) *TRICHOME-LESS1* regulates trichome patterning by suppressing *GLABRA1* in Arabidopsis. Development 134: 3873–3882.

23. Wester K, Digiuni S, Geier F, Timmer J, Fleck C, et al. (2009) Functional diversity of *R3 single-repeat* genes in trichome development. Development 136: 1487–1496.

24. Bouyer D, Geier F, Kragler F, Schnittger A, Pesch M, et al. (2008) Two-dimensional patterning by a trapping/depletion mechanism: the role of *TTG1* and *GL3* in Arabidopsis trichome formation. PLoS Biology 6: e141.

25. Pesch M, Hülskamp M (2009) One, two, three… models for trichome patterning in Arabidopsis? Current Opinion in Plant Biology 12: 587–592.

26. U N (1935) Genome analysis in *Brassica* with special reference to the experimental formation of *B. napus* and peculiar mode of fertilization. Japanese Journal of Botany 7: 389–452.

27. Agren J, Schemske DW (1994) Evolution of trichome number in a naturalized population of *Brassica rapa*. American Naturalist: 1–13.

28. Zhang J, Lu Y, Yuan Y, Zhang X, Geng J, et al. (2009) Map-based cloning and characterization of a gene controlling hairiness and seed coat color traits in *Brassica rapa*. Plant Molecular Biology 69: 553–563.

29. Li F, Kitashiba H, Nishio T (2011) Association of sequence variation in *BrassicaGLABRA1* orthologs with leaf hairiness. Molecular Breeding 28: 577–584.

30. Li F, Zou Z, Yong HY, Kitashiba H, Nishio T (2013) Nucleotide sequence variation of *GLABRA1* contributing to phenotypic variation of leaf hairiness in *Brassicaceae* vegetables. Theoretical and applied genetics.

31. ZeTao B, GuoHua C, Lei S, LiMing X, ChengAng W, et al. (2010) Isolation and functional characterization of GLABRA2 promoter related to trichome

development in *Brassica napus*. Journal of Agricultural Biotechnology 18: 210–217.

32. Gruber MY, Wang S, Ethier S, Holowachuk J, Bonham-Smith PC, et al. (2006) "HAIRY CANOLA"–Arabidopsis GL3 Induces a Dense Covering of Trichomes on *Brassica napus* Seedlings. Plant Molecular Biology 60: 679–698.

33. Palaniswamy P, Bodnaryk RP (1994) A wild Brassica from Sicily provides trichome-based resistance against flea beetles, *Phyllotreta cruciferae* (Goeze)(Coleoptera: Chrysomelidae). The Canadian Entomologist 126: 1119–1130.

34. Clough SJ, Bent AF (1998) Floral dip: a simplified method for Agrobacterium-mediated transformation ofArabidopsis thaliana. The Plant Journal 16: 735–743.

35. Trapnell C, Roberts A, Goff L, Pertea G, Kim D, et al. (2012) Differential gene and transcript expression analysis of RNA-seq experiments with TopHat and Cufflinks. Nature protocols 7: 562–578.

36. Kumar S, Nei M, Dudley J, Tamura K (2008) MEGA: a biologist-centric software for evolutionary analysis of DNA and protein sequences. Briefings in Bioinformatics 9: 299–306.

37. Yang Z (1997) PAML: a program package for phylogenetic analysis by maximum likelihood. Computer Applications in the Biosciences: CABIOS 13: 555–556.

38. Robinson SJ, Tang LH, Mooney BAG, McKay SJ, Clarke WE, et al. (2009) An archived activation tagged population of *Arabidopsis thaliana* to facilitate forward genetics approaches. BMC Plant Biology 9: 101.

39. Kyte J, Doolittle RF (1982) A simple method for displaying the hydropathic character of a protein. Journal of Molecular Biology 157: 105–132.

40. Kirik V, Simon M, Wester K, Schiefelbein J, Hulskamp M (2004) ENHANCER of TRYand CPC 2 (ETC2) reveals redundancy in the region-specific control of trichome development of Arabidopsis. Plant molecular biology 55: 389–398.

41. Blanc G, Wolfe KH (2004) Functional divergence of duplicated genes formed by polyploidy during Arabidopsis evolution. The Plant Cell Online 16: 1679–1691.

42. Conant GC, Wolfe KH (2008) Turning a hobby into a job: how duplicated genes find new functions. Nature Reviews Genetics 9: 938–950.

43. Cusack BP, Wolfe KH (2007) When gene marriages don't work out: divorce by subfunctionalization. Trends in Genetics 23: 270–272.

44. Babic V, Datla RS, Scoles GJ, Keller WA (1998) Development of an efficient Agrobacterium-mediated transformation system for Brassica carinata. Plant Cell Reports 17: 183–188.

45. Bhalla PL, Singh MB (2008) Agrobacterium-mediated transformation of Brassica napus and Brassica oleracea. Nature protocols 3: 181–189.

46. Poulsen GB (1996) Genetic transformation of Brassica. Plant Breeding 115: 209–225.

47. Radke SE, Turner JC, Facciotti D (1992) Transformation and regeneration of Brassica rapa using Agrobacterium tumefaciens. Plant Cell Reports 11: 499–505.

48. Larkin JC, Oppenheimer DG, Pollock S, Marks MD (1993) Arabidopsis GLABROUS1 gene requires downstream sequences for function. The Plant Cell Online 5: 1739–1748.

49. Wang S, Wang JW, Yu N, Li CH, Luo B, et al. (2004) Control of plant trichome development by a cotton fiber MYB gene. The Plant Cell Online 16: 2323–2334.

50. Pesch M, Hülskamp M (2011) Role of TRIPTYCHON in trichome patterning in Arabidopsis. BMC Plant Biology 11: 130.

51. Wei S, Gruber MY, Yu B, Gao M-J, Khachatourians GG, et al. (2012) Arabidopsis mutant sk156 reveals complex regulation of SPL15 in a miR156-controlled gene network. BMC plant biology 12: 169.

52. Patthy L (2009) Protein evolution: Wiley-Blackwell.

53. Bloomer RH, Juenger TE, Symonds VV (2012) Natural variation in GL1 and its effects on trichome density in *Arabidopsis thaliana*. Molecular Ecology.

54. Yeam I, Cavatorta JR, Ripoll DR, Kang BC, Jahn MM (2007) Functional dissection of naturally occurring amino acid substitutions in eIF4E that confers recessive potyvirus resistance in plants. The Plant Cell Online 19: 2913–2928.

55. Szymanski DB, Lloyd AM, Marks MD (2000) Progress in the molecular genetic analysis of trichome initiation and morphogenesis in Arabidopsis. Trends in Plant Science 5: 214–219.

56. Koornneeff M, Dellaert LWM, Van der Veen JH (1982) EMS-and relation-induced mutation frequencies at individual loci in *Arabidopsis thaliana*(L.) Heynh. Mutation Research/Fundamental and Molecular Mechanisms of Mutagenesis 93: 109–123.

57. Hauser MT, Harr B, Schlötterer C (2001) Trichome distribution in *Arabidopsis thaliana* and its close relative *Arabidopsis lyrata*: molecular analysis of the candidate gene *GLABROUS1*. Molecular Biology and Evolution 18: 1754–1763.

58. Larkin JC, Young N, Prigge M, Marks MD (1996) The control of trichome spacing and number in Arabidopsis. Development 122: 997–1005.

3

Assessing Quantitative Resistance against *Leptosphaeria maculans* (Phoma Stem Canker) in *Brassica napus* (Oilseed Rape) in Young Plants

Yong-Ju Huang[1,2]*, Aiming Qi[1,2], Graham J. King[2,3], Bruce D. L. Fitt[1,2]

1 School of Life and Medical Sciences, University of Hertfordshire, Hatfield, Hertfordshire, United Kingdom, **2** Department of Plant Pathology and Microbiology, Rothamsted Research, Harpenden, Hertfordshire, United Kingdom, **3** Southern Cross Plant Science, Southern Cross University, Lismore, Australia

Abstract

Quantitative resistance against *Leptosphaeria maculans* in *Brassica napus* is difficult to assess in young plants due to the long period of symptomless growth of the pathogen from the appearance of leaf lesions to the appearance of canker symptoms on the stem. By using doubled haploid (DH) lines A30 (susceptible) and C119 (with quantitative resistance), quantitative resistance against *L. maculans* was assessed in young plants in controlled environments at two stages: stage 1, growth of the pathogen along leaf veins/petioles towards the stem by leaf lamina inoculation; stage 2, growth in stem tissues to produce stem canker symptoms by leaf petiole inoculation. Two types of inoculum (ascospores; conidia) and three assessment methods (extent of visible necrosis; symptomless pathogen growth visualised using the GFP reporter gene; amount of pathogen DNA quantified by PCR) were used. In stage 1 assessments, significant differences were observed between lines A30 and C119 in area of leaf lesions, distance grown along veins/petioles assessed by visible necrosis or by viewing GFP and amount of *L. maculans* DNA in leaf petioles. In stage 2 assessments, significant differences were observed between lines A30 and C119 in severity of stem canker and amount of *L. maculans* DNA in stem tissues. GFP-labelled *L. maculans* spread more quickly from the stem cortex to the stem pith in A30 than in C119. Stem canker symptoms were produced more rapidly by using ascospore inoculum than by using conidial inoculum. These results suggest that quantitative resistance against *L. maculans* in *B. napus* can be assessed in young plants in controlled conditions. Development of methods to phenotype quantitative resistance against plant pathogens in young plants in controlled environments will help identification of stable quantitative resistance for control of crop diseases.

Editor: Zhengyi Wang, Zhejiang University, China

Funding: This work was funded by the UK Biotechnology and Biological Sciences Research Council (BBSRC, BB/E001610/1, and BB/I017585/1), the Department for Environment, Food and Rural Affairs (Defra), HGCA, the British Council, DuPont, the Perry Foundation, the Felix Thornley Cobbold Agricultural Trust, and the Chadacre Agricultural Trust. The funders had no role in study design, data collection and analysis, decision to publish, or preparation of the manuscript.

Competing Interests: The authors have the following interests: This study was partly funded by DuPont. There are no patents, products in development or marketed products to declare.

* E-mail: y.huang8@herts.ac.uk

Introduction

With growing concern about world-wide food shortages and climate change [1], protecting food crops against pathogens that cause epidemic diseases is more important than ever. Using resistance against pathogens in crop cultivars is one of the most economical and environmentally friendly methods for control of crop diseases. Two types of cultivar resistance used are quantitative resistance and qualitative resistance [1–4]. Quantitative resistance, which is usually controlled by several genes, is often a 'partial' resistance that does not prevent pathogens from colonisation of plants but decreases symptom severity and/or epidemic progress over time [5–7]. By contrast, qualitative resistance is usually controlled by single, dominant resistance (*R*) genes and often effective in preventing pathogens from colonisation of plants [8–12]. Whilst *R* gene-mediated qualitative resistance is 'complete' resistance, which follows a gene-for-gene interaction, it is generally less durable than quantitative resistance because pathogen populations often rapidly evolve for virulence against the *R* genes [13–16].

Many arable crops rely on quantitative partial resistance that has been selected by breeding 'field' resistance into modern cultivars for control of their diseases. This reliance on resistance is essential for small-holder farmers in many regions of the world where use of fungicides is prohibitively expensive. In regions where crops are grown intensively on larger farms, the reliance on resistance for disease control will increase if use of certain effective fungicides is not permitted by legislation. However, it has been difficult to breed cultivars with quantitative resistance because of the difficulty in assessing it. By contrast, *R* gene-mediated qualitative resistance can be selected reliably by phenotyping in young plants (with distinct resistant and susceptible phenotypes after inoculation). Traditionally, selection for quantitative resistance has relied on field assessments of disease severity made towards the end of the cropping season before harvest [17–19]. Thus, quantitative resistance has also been referred to as 'adult plant resistance' [3] because it has been difficult to assess it in young plants. Furthermore, due to the influence of genotype by environment interactions [16,19], identification of stable quantitative resistance has required large replicated field plot experiments at different locations in different years, which is very costly and time-consuming. There is a need to develop new methods to select for quantitative resistance in arable crops that are rapid, cheap and reliable.

Phoma stem canker, caused by *Leptosphaeria maculans*, is an economically important disease on oilseed rape (*Brassica napus*) in Europe, North America and Australia, causing world-wide losses worth more than £1000 M (at a price of £300 t^{-1}) per cropping season, despite use of fungicides [20–21]. In Europe, these losses will increase if use of the most effective fungicides is no longer permitted by EU legislation (EU directive 91/414, http://ec. europa.eu/food/plant/protection/index_en.htm). Epidemics of phoma stem canker are initiated by air-borne *L. maculans* ascospores (sexual spores) [18,22], which produce germ tubes that infect the leaves of winter oilseed rape in autumn (October/ November), causing leaf lesions [23–24]. Conidia (asexual spores), produced in pycnidia (asexual fruiting bodies) on leaf lesions and spread by rain-splash, can cause secondary leaf infections [25]. From leaf lesions, *L. maculans* grows symptomlessly along the leaf vein/petiole to reach the stem where, in spring/summer (April-July), it causes stem cankers that result in yield losses [9,18].

Both quantitative resistance and *R* gene-mediated qualitative resistance against *L. maculans* have been identified in *B. napus* [3,8,26–27]. *R* gene-mediated resistance against *L. maculans* operates in leaves to prevent leaf lesion development and thus prevent subsequent stem canker development [9,13–14]. Quantitative resistance against *L. maculans* is a partial resistance that does not prevent leaf lesion development but decreases the severity of stem cankers [6,28].

R gene-mediated resistance against *L. maculans* can be selected in young plants by scoring lesion phenotypes after inoculation of cotyledons or leaves [8,14]. However, *R* gene-mediated resistance against *L. maculans* usually loses its effectiveness within three cropping seasons of widespread use in commercial cultivars because of selection for virulence within the variable *L. maculans* populations. For example, the resistance genes *Rlm1* (introduced into several winter oilseed rape cultivars in France) and *LepR3* (introduced into the Australian cultivar Surpass 400 from *B. rapa* var. *sylvestris*) were rendered ineffective after 3 years of commercial use [13–14,29]. This is very costly for breeders since introduction of new resistance into elite commercial cultivars can take 10–15 years [20].

By contrast, quantitative resistance is considered to be more durable than *R* gene-mediated resistance against *L. maculans* [3,30]. However, it has been difficult to select for quantitative resistance against *L. maculans* due to the long period of symptomless growth after initial development of leaf lesions [6]. Furthermore, if effective *R* genes are present, it is difficult to assess quantitative resistance because *R* gene-mediated qualitative resistance operates in leaves to prevent the growth of the pathogen from leaf lesions to stems. Currently, selection of cultivars with quantitative resistance generally relies on winter oilseed rape field experiments in which stem canker severity is assessed just before harvest [18,30–31]. If quantitative resistance can be assessed in young plants, it will not only accelerate the process of breeding for resistance but also save money for the industry. A study of growth of *L. maculans* from the cotyledon to the hypocotyl showed that stem canker symptoms can be produced on hypocotyls after cotyledon inoculation in controlled conditions [28]. However, it was not clear whether resistance to growth of *L. maculans* to cause canker in hypocotyls is the same as that in stems. Recently, a petiole inoculation method

has been used to compare pathogenicity of different *L. maculans* isolates and to screen different cultivars in terms of stem canker development at a young-plant stage in controlled conditions [32–33]. However, there is a need for further work to develop methods to test for quantitative resistance in young plants.

This paper reports work to investigate new methods for assessing quantitative resistance against *L. maculans* in controlled environment experiments with young plants of two oilseed rape doubled haploid (DH) lines differing in quantitative resistance, using different types of inoculum and different inoculation and assessment methods.

Materials and Methods

In this work, assessment of quantitative resistance against growth of *L. maculans* in oilseed rape was considered in two stages; stage 1, resistance against growth along leaf veins/petioles towards the stem; stage 2, resistance against growth in stem tissues to produce stem canker symptoms (Fig. 1). Inoculation methods, type of *L. maculans* inoculum, number of plants inoculated, design of experiments and assessment methods used in each of the controlled environment experiments are described in Table 1.

Preparation of plant material and pathogen inoculum

DH lines A30 and C119, selected from the mapping population developed from a cross between winter oilseed rape Darmor-*bzh* (with quantitative resistance against *L. maculans* and *R* gene *Rlm9*) and the spring oilseed rape Yudal (without quantitative resistance or *R* genes) [30], were used for all the experiments. These two DH lines were derived from microspore cultures of a single F$_1$ hybrid plant from the cross 'Darmor-*bzh* × Yudal'. Use of DH lines can decrease potential variation associated with genetic background. Since no effective major resistance gene segregates in this DH population, 'field' resistance provides an assessment of quantitative resistance against *L. maculans*. When 'field' resistance was assessed by scoring stem canker at the end of the cropping season in winter oilseed rape field experiments in France, A30 developed severe phoma stem cankers whilst C119 developed much less severe stem cankers [27,30]. Similar results were obtained when these two lines

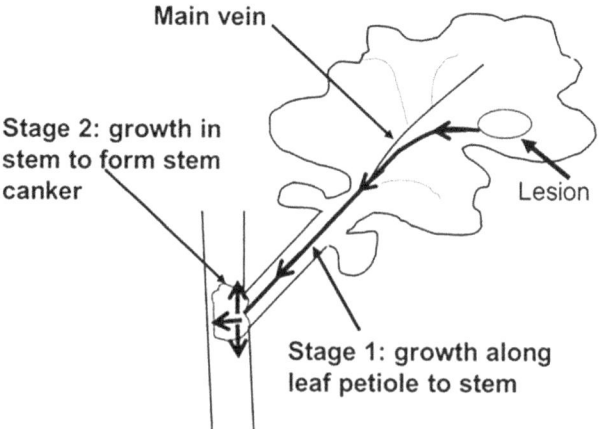

Figure 1. Stages of *Leptosphaeria maculans* **growth in oilseed rape.** Assessment of quantitative resistance against growth of *L. maculans* in oilseed rape was considered in two stages; stage 1, resistance against growth from phoma leaf lesion along the main leaf vein and petiole to the stem; stage 2, resistance against growth in stem to produce stem canker; both stages were investigated in controlled environment experiments.

Table 1. Inoculation methods, type of *Leptosphaeria maculans* inoculum, number of plants inoculated, design of experiments and assessment methods used in each of the controlled environment experiments with the doubled haploid (DH) lines A30 or C119.

Exp.	Inoculum[1]	Design	No. of plants inoculated[2]	Assessment method
L. maculans growth in leaf, leaf lamina inoculation				
LExpt 1	Conidia (GFP)	Randomised block	8×2 plants, 3 leaves per plant	Lesion area[3], distance grown (DG) along vein/petiole viewed by GFP
LExpt 2	Conidia (GFP)	Complete randomised	8 plants, 3 leaves per plant	Lesion area, DG viewed by GFP
LExpt 3	Ascospores	Randomised block	5×2 plants, 3 leaves per plant	DG assessed by extent of necrosis, *L. maculans* DNA in leaf petiole (qPCR)
LExpt 4	Ascospores	Complete randomised	12 plants, 3 leaves per plant	DG assessed by extent of necrosis, qPCR
LExpt 5	Ascospores	Complete randomised	8 plants, 2 leaves per plant	Lesion area, DG assessed by extent of necrosis, qPCR
L. maculans growth in stem, leaf petiole inoculation				
PExpt 1	Conidia (GFP)	Complete randomised	10 plants, 2 petioles per plant	Stem canker score, extent of *L. maculans* growth in stem viewed by GFP
PExpt 2	Ascospores	Complete randomised	5 plants, 3 petioles per plant	Stem canker score, *L. maculans* DNA in stem (qPCR)
PExpt 3	Ascospores	Complete randomised	12 plants, 2 petioles per plant	Stem canker score, qPCR
PExpt 4	Ascospores	Complete randomised	12 plants, 2 petioles per plant	Stem canker score, qPCR

[1]*L. maculans* ascospores obtained from naturally infected oilseed rape stem base debris collected in August 2007; conidia were produced by GFP-transformed isolate ME24/3.13.
[2]Plants were inoculated when they had three fully expanded leaves. For details of leaf lamina or leaf petiole inoculation, see Fig. 2. All experiments were done at 20°C with alternating 12 h light/12 h darkness.
[3]The lesion area was estimated by multiplying the lesion length by lesion width.

were assessed in field experiments in the UK (the mean disease score on a 0–4 scale was 1.9 for A30 and 0.6 for C119 in 2007/2008, and 2.7 for A30 and 1.3 for C119 in the 2008/2009 cropping season). Results of these field experiments show that A30 is susceptible and does not have quantitative resistance whereas C119 has good quantitative resistance against *L. maculans*. Furthermore, these two lines do not have the dwarfing gene (*bzh*) and do not differ greatly in terms of maturity date and plant height. Molecular markers in QTL regions for resistance to *L. maculans* showed that the difference in resistance between these two DH lines was mainly due to resistance inherited from the resistant parent (Personal communication, Dr Regine Delourme). Therefore, these two DH lines were chosen for this work.

Plants of lines A30 or C119 were grown in pots (9 cm diameter) containing a peat-based compost mixed with a soluble fertiliser (WE Hewitt & Sons Ltd, UK). Plants were initially grown in a

Figure 2. Leaf lamina and leaf petiole inoculation. Inoculation of oilseed rape leaf lamina (A) or leaf petiole (B) with conidia or ascospores of *Leptosphaeria maculans*. The arrows indicate the droplets of spore suspension (either conidia or ascospores) placed on the lamina (A) or petiole (B) of leaves of doubled haploid lines A30 (susceptible) or C119 (with quantitative resistance). Plants were inoculated at 21 days after sowing when they had three fully expanded leaves.

glasshouse (20–23°C) and thinned to one plant per pot 10 days after sowing. Three weeks after sowing, when each plant had three expanded leaves, the plants were transferred to a controlled environment cabinet (20°C day/20°C night, 12 h light/12 h darkness, light intensity 210 μmol m^{-2}s^{-1}) for 24 h before inoculation. After inoculation, plants were kept in the cabinet until the end of the experiments.

Both ascospores and conidia were used as inoculum. Although line C119 has resistance gene *Rlm9* (A30 has no *R* genes), the *L. maculans* populations throughout Europe are 100% virulent against *Rlm9* [8,34], so ascospores from natural populations were used. Both the *L. maculans* isolates used as conidial inoculum, ME24 (*AvrLm1, avrLm2, avrLm3, avrLm4, avrLm5, AvrLm6, AvrLm7, avrLm9*) and GFP-expressing isolate ME24/3.13 (GFP-transformed ME24; all work involving the use and storage of genetically modified *L. maculans* was done under Defra licence no. PHL 174E/5443/01/2007), are virulent against *Rlm9*. Conidial suspensions of isolates ME24 and ME24/3.13 were prepared from 12-day-old cultures on V8 agar [9]. The concentration of conidia was adjusted to 10^7 conidia mL^{-1} using a haemocytometer slide. Since phoma stem canker epidemics in crops are initiated by ascospores released from stem debris, *L. maculans* ascospores produced under natural conditions were also used as inoculum. An ascospore suspension of *L. maculans* was prepared from naturally infected UK oilseed rape stem base debris of the cultivar Courage collected after harvest in August 2007 (pieces of stem debris with mature pseudothecia were stored at −20°C until required), using the method described by Huang *et al.* [23]. The concentration of ascospores was adjusted to 10^4 ascospores mL^{-1} using a haemocytometer slide.

Growth of *L. maculans* in leaf petioles

In natural conditions in Europe, phoma stem cankers are usually initiated from phoma lesions on leaves of winter oilseed rape crops in the previous autumn [18,22]. Therefore, to investigate the growth of L. maculans from leaf lesions towards the stems, experiments were done by inoculation of leaf laminas. Before inoculation, each leaf lamina was rubbed gently using a wet tissue so that the droplet of inoculum would not run off. For inoculation with conidia, the rubbed sites were wounded using a sterile pin and a 10 μL droplet of conidial suspension was placed on each wounded site. For inoculation with ascospores, the rubbed sites were not wounded and a 15 μL droplet of ascospore suspension was placed on each site. Each leaf had two inoculation sites, one on each side of the main vein (Fig. 2A). The first two or three fully expanded leaves of each plant were inoculated.

Two experiments were done with conidia of GFP-expressing L. maculans isolate ME24/3.13 and three experiments were done with ascospores (Table 1). After inoculation, plants in trays were covered with tray covers to maintain high relative humidity for 48 h (after inoculation with ascospores) or 72 h (after inoculation with conidia). The time from inoculation to the appearance of phoma leaf lesions was recorded in each experiment. The maximum length and width of each lesion were measured at 17–18 days post inoculation (dpi), then the lesion area (mm^2) was estimated by multiplying the lesion length by lesion width.

For leaf inoculation with conidia of GFP-expressing L. maculans, the distance grown by L. maculans from the inoculation site along leaf the vein/petiole towards the stem was assessed by measuring extent of appearance of GFP fluorescence. At 22–25 dpi, the inoculated leaves were detached at the place where the petiole joins the stem (leaf scar). The leaves were viewed using a Leica MZ FLIII stereo-microscope. For observation of GFP fluorescence, filter GFP2 from Leica Microsystems (Milton Keynes, UK) was used.

For leaf inoculation with ascospores, the growth of L. maculans along the leaf vein/petiole towards the stem was measured by assessing the extent of visible necrosis and by quantification of the pathogen DNA in the tissue using quantitative PCR (qPCR). At 18–20 dpi, the inoculated leaves were detached at the leaf scar. The distance from the inoculation site on the leaf lamina to the furthest visible necrosis on the leaf petiole was measured on each leaf, then a piece of leaf petiole 8 cm long (measured from the inoculation site) was placed in a 15 mL tube to be freeze-dried for DNA extraction and qPCR.

Growth of L. maculans in stem tissues

To ensure that L. maculans reached the stem before the leaf abscission, the petioles of first two or three leaves of each plant were inoculated (Table 1). Before inoculation, the leaf petiole (1 cm from the stem) was gently rubbed using a wet tissue so that the droplet of inoculum would not run off. Then the leaf petiole was wounded using a sterile pin and either a 10 μL droplet of conidial suspension or a 12 μL droplet of ascospore suspension was placed over the wound (Fig. 2B). One experiment was done with conidia of GFP-expressing L. maculans isolate ME24/3.13 and three experiments were done with ascospores (Table 1). After inoculation, plants in trays were covered with tray covers to maintain high relative humidity for 48 h (after inoculation with ascospores) or 72 h (after inoculation with conidia). The time from inoculation to the appearance of first phoma stem canker symptoms at the leaf scar on the stem was recorded for each experiment.

For petiole inoculation with conidia of GFP-expressing isolate ME24/3.13, the growth of L. maculans in the stem was measured both by assessing stem canker severity and by observation of extent of GFP fluorescence within stem tissues. After inoculation, plants were regularly assessed for appearance of visible canker symptoms at the leaf scars of inoculated leaf petioles. At 87 dpi, the inoculated plants were sampled by cutting their stems at the soil surface and these stems were viewed using a Leica MZ FLIII stereo microscope for appearance of GFP fluorescence at the leaf scar of each inoculated leaf. The stems were then cut horizontally at the leaf scar of the inoculated leaf to assess the internal severity of stem canker on a 0–4 scale; where 0 = healthy; 1 = 1–25% stem cross-section necrotic; 2 = 26–50% stem cross-section necrotic; 3 = 51–75% stem cross-section necrotic; 4 = 76–100% stem cross-section necrotic. The scale was modified from that of Zhou et al. [35]. Then stems were cut vertically to assess internal vertical spread of GFP-expressing L. maculans along the stem cortex or pith using a Leica MZ FLIII stereo-microscope.

For leaf petiole inoculation with ascospores, the growth of L. maculans in the stem was measured by assessing stem canker severity and by quantification of the pathogen DNA using qPCR. At 46 dpi, stems of the inoculated plants were sampled by cutting the stem at the soil surface. The stems were then cut horizontally at the leaf scar of the inoculated leaf to assess the internal severity of stem canker on the 0–4 scale. After stem canker assessment, a piece of each stem (5 cm long) was placed in a 50 mL tube to be freeze-dried for DNA extraction and qPCR.

Quantification of L. maculans DNA in leaf petioles and stems

Freeze-dried individual leaf petioles or stems were ground into powder using a mortar and pestle. DNA was extracted from each ground individual leaf petiole or stem using a DNA extraction kit (DNAMITE Plant Kit, Microzone Ltd, West Sussex, UK). The amount of L. maculans DNA in each leaf petiole or stem was quantified using a SYBR green quantitative PCR. The primers LmacF and LmacR, specific for amplification of the internal transcribed spacer (ITS) region of ribosomal DNA of L. maculans, that have been previously used for identification [36] and quantification [6] of L. maculans were used. All qPCR reactions were done in duplicate and each reaction volume was 20 μL, including 0.6 μL of primers at a final concentration of 300 nM, 10 μL of SYBR Green JumpStart Taq ReadyMix (Sigma, UK), 0.08 μL of ROX internal reference dye and the 50 ng DNA sample. Nuclease-free water (Sigma, UK) was used as the no-template control. All reactions were done in 96×0.2 mL PCR plates (ABgene) covered with cap strips, using a Stratagene Mx3000P quantitative PCR machine. The thermocycling profile consisted of an initial cycle of 95°C for 2 min, followed by 40 cycles of 95°C for 15 s, 60°C for 30 s, 72°C for 45 s and a read step at 83°C for 15 s, then a dissociation stage (thermal profile: 95°C 1 min, 60°C 1 min, 95°C 15 s). In each qPCR run, a standard dilution series consisting of 10000, 1000, 100, 10 and 1 pg of DNA of L. maculans isolate ME24 was included to produce a standard curve. A standard curve was generated by plotting the amount of L. maculans DNA against the threshold cycle (Ct) value for the series of dilutions. The amount of L. maculans DNA for each unknown sample was extrapolated from the Ct value and the value obtained from the standard curve using Stratagene MxPro-Mx3000 P v3.2 software.

Statistical analysis

For experiments to assess growth of L. maculans in leaf petioles, the data on leaf lesion area, distance grown by L. maculans along leaf petioles and amount of L. maculans DNA in leaf petioles were analysed by analysis of variance to assess the differences between DH lines A30 and C119 or between different experiments. For experiments to assess growth of L. maculans in stem tissues, the data on stem canker severity and amount of L. maculans DNA in stems were analysed by analysis of variance to assess the differences between DH lines A30 and C119 or between different experiments. To compare different methods used to assess quantitative resistance to *L. maculans* at growth stages 1 or 2, linear regressions were done for individual experiments; these regression lines were compared to assess whether they differed in intercept and/or slope for different experiments. The data on amount of *L. maculans* DNA in petioles or stems were log$_{10}$-transformed before regression analysis. For stage 1 assessments, linear regressions were done for leaf lesion area against distance grown or amount of *L. maculans* DNA and for amount of *L. maculans* DNA against distance grown in the leaf vein/petiole. For stage 2 assessments, linear regressions were done for amount of *L. maculans* DNA in the stem against stem canker severity score. All the analyses were done using GENSTAT statistical software [37].

Results

Growth of *L. maculans* in leaf petioles

There were differences between conidial inoculum and asco-spore inoculum in incubation period (time from inoculation to appearance of lesions) for phoma leaf lesion development, with the incubation period shorter for ascospores than for conidia. Phoma leaf lesions were observed at 7 dpi for plants inoculated with ascospores (Fig. 3B) but phoma leaf lesions were not observed on plants inoculated with conidia until 10 dpi (Fig. 3A). There was no difference between lines A30 and C119 in incubation period or visual appearance of phoma leaf spot symptoms, whether the plants were inoculated with conidia or ascospores.

In experiments with conidia of GFP-expressing isolate ME24/3/.13 (LExpt 1 & 2; Table 1), the leaf lesions on A30 were larger

Figure 3. Phoma leaf spot symptoms produced in controlled conditions. Phoma leaf spot symptoms on leaves of doubled haploid lines A30 (susceptible) or C119 (with quantitative resistance) inoculated with conidia or ascospores of *Leptosphaeria maculans*. Lesions on leaves of A30 at 15 days post inoculation (dpi) with conidia (A) or 10 dpi with ascospores (B); lesions on leaves of A30 (C) and C119 (D) at 22 dpi; symptomless growth of GFP-expressing *L. maculans* (isolate ME24/3.13) along a leaf vein towards the petiole of A30 viewed with brightfield illumination (E-1) or a GFP filter (E-2) at 18 dpi.

and the distances grown by *L. maculans* in petioles of A30 were greater than those on C119 (Fig. 3C, d). At 17 dpi, leaf lesion area (average of LExpt 1 & 2) on A30 (60.7 mm^2) was significantly larger than on C119 (27.4 mm^2) (*P*<0.001, 28 d.f., SED 7.4). GFP-expressing *L. maculans* (Fig. 3E) was observed in petioles of both A30 and C119; the distance grown by *L. maculans* (average of LExpt 1 & 2) was greater in petioles of A30 than in petioles of C119 (Fig. 4A) (*P*<0.001, 28 d.f., SED 1.6). There was a significant difference between LExpt 1 and LExpt 2 in distance grown by GFP *L. maculans* (*P*<0.05, 28 d.f., SED 1.6), since the distance grown by GFP *L. maculans* was assessed at 22 dpi in experiment LExpt 1 and 25 dpi in LExpt 2. However, there was no significant interaction between experiment and DH line in distance grown along the leaf petiole.

In experiments with ascospore inoculum (LExpt 3, 4 & 5; Table 1), the distance grown (average of LExpt 3, 4 & 5) by *L. maculans* (measured by the extent of necrosis) was greater in petioles of A30 than in petioles of C119 (*P*<0.001, 54 d.f., SED 1.3) (Fig. 4A). There was a significant difference between the three experiments in distance grown by *L. maculans* (*P*<0.001, 54 d.f., SED 1.4). However, there was no significant interaction between experiment and DH line. The difference between these three experiments may have occurred because the distance grown in leaf petioles was assessed at 18 dpi in LExpt 3 and at 20 dpi in LExpt 4 and LExpt 5, because there was no significant difference between LExpt 4 and LExpt 5. When symptomless growth of *L. maculans* in petioles was measured by qPCR, there was a significant difference between A30 and C119 in the amount of *L. maculans* DNA (average of LExpt 3, 4 & 5), with the amount of *L. maculans* DNA greater in leaf petioles of A30 than that in leaf petioles of C119 (*P*<0.001, 54 d.f., SED 2.9; Fig. 4B). There was no difference between experiments in amount of *L. maculans* DNA. Leaf lesion area was assessed in LExpt 5 (not assessed in LExpt 3 and LExpt 4); the leaf lesion area on A30 (327.5 mm^2) was larger than on C119 (137.6 mm^2) at 18 dpi (*P*<0.01, 14 d.f., SED 0.6).

For the experiments with conidial inoculum, there were significant linear relationships between distance grown (*d*) by *L. maculans* and leaf lesion area (*f*) in the leaf petiole in LExpt 1 (*d* = 0.32*f*+12.7, R^2 = 0.64, *P*<0.001, 14 d.f.) (Fig. 5A) and LExpt 2 (*d* = 0.31*f*+14.1, R^2 = 0.62, *P*<0.001, 13 d.f.). Comparison of these two regression lines showed that there was no significant difference in intercept (*P*>0.85) or slope (*P*>0.71) between them. A significant linear relationship was observed in experiment LExpt 5 with ascospores (*d* = 3.90*f*+10. 7, R^2 = 0.40, *P*<0.01, 14 d.f.). There was also a significant linear relationship between leaf lesion area (*f*) and *L. maculans* DNA (*n*) in the leaf petiole (*f* = 130.8*n*−163.0, R^2 = 0.59, *P*<0.001, 13 d.f.) (Fig. 5B). For the three experiments with ascospore inoculum, there were significant linear relationships between *L. maculans* DNA (*n*) and distance grown (*d*) by *L. maculans* in experiments LExpt 3 (*n* = 0.5*d*+3.0, R^2 = 0.58, *P*<0.001, 18 d.f.), LExpt 4 (*n* = 0.7*d*+1.6, R^2 = 0.78, *P*<0.001, 22 d.f.; Fig. 5C) and LExpt 5 (*n* = 0.7*d*+1.7, R^2 = 0.54, *P*<0.005, 13 d.f.). Comparison of these regression lines showed that the slope did not differ significantly (*P*>0.36) but the intercepts differed significantly (*P*<0.001) between them.

Growth of *L. maculans* in stem tissues

For leaf petioles inoculated with conidia of GFP-expressing isolate ME24/3.13, although no visible symptoms were observed at the leaf scars (Fig. 6A-1) at the time when the inoculated leaves abscised, growth of *L. maculans* in the leaf scar tissues was observed by using a GFP filter (Fig 6A-2) at 25 dpi. By 36 dpi, visible stem canker symptoms had started to develop at the leaf scars on some plants of line A30 (Fig 6B-1), as confirmed by observing GFP

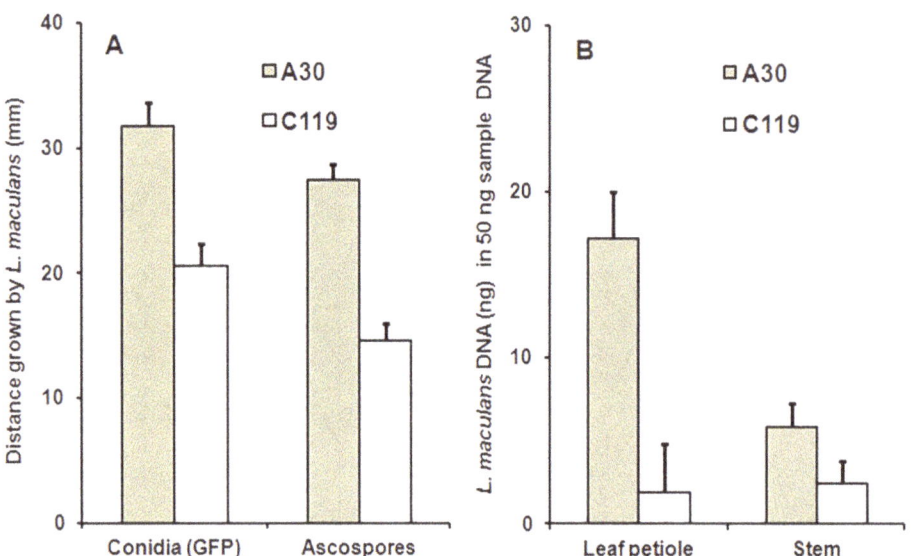

Figure 4. Distance grown along leaf petioles and *Leptosphaeria maculans* DNA in petioles and stems. Leaf laminas or leaf petioles of doubled haploid lines A30 (susceptible) or C119 (with quantitative resistance) were inoculated with conidia of GFP-expressing *L. maculans* isolate ME24/3.13 or ascospores from a natural population. The distance grown by *L. maculans* along the leaf vein/petiole towards the stem (leaf lamina inoculation, Fig. 2A) was measured by assessing extent of GFP fluorescence at 22–25 days post inoculation (dpi) for conidial inoculation (A) or by assessing extent of the visible necrosis (A) and by quantification of amount of *L. maculans* DNA (B) using qPCR at 18–20 dpi for ascospore inoculation. The growth of *L. maculans* in stem tissue (petiole inoculation, Fig. 2B) was assessed by quantification of amount of *L. maculans* DNA using qPCR (B) at 46 dpi inoculation with ascospores. Bars show standard errors (A, 28 d.f. for conidia and 54 d.f. for ascospores; B, 54 d.f. for leaf petiole and 50 d.f. for stem).

fluorescence (Fig 6B-2). Necrotic stem canker lesions were assessed at 87 dpi; there was a significant difference ($P<0.05$, 17 d.f., SED 0.96) between A30 and C119 (Fig. 7), with stem canker more severe on A30 (Fig. 6C) than on C119 (Fig. 6D). After it reached the stem, GFP-expressing *L. maculans* was observed to spread both vertically up/down the stem and horizontally towards the stem

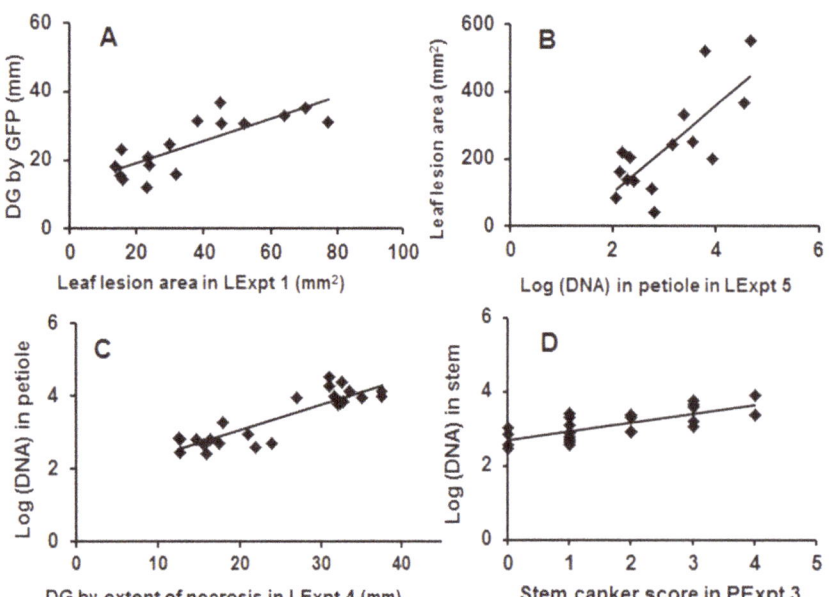

Figure 5. Relationships between different methods for assessment of quantitative resistance. Different methods were used to assess growth of *Leptosphaeria maculans* in leaf petioles or in stems of doubled haploid lines A30 (susceptible) or C119 (with quantitative resistance). Relationships between leaf lesion area and distance grown (DG) by *L. maculans* in the leaf petiole viewed by GFP in experiment LExpt 1 (A; $R^2 = 0.64$) or amount of *L. maculans* DNA in leaf petiole in experiment LExpt 5 (B; $R^2 = 0.59$); or between amount of *L. maculans* DNA and DG in leaf petiole viewed by extent of necrosis in experiment LExpt 4 (C; $R^2 = 0.78$); or between amount of *L. maculans* DNA in the stem and stem canker severity score in experiment PExpt 3 (D; $R^2 = 0.56$). The amount of *L. maculans* DNA in the petiole or stem was log$_{10}$-transformed. Details of these experiments are presented in Table 1.

pith. GFP-expressing *L. maculans* was observed in the pith of more A30 plants (80%) than C119 plants (56%) and spread further both up and down internal stem tissues of A30 (27.7 mm) than those of C119 (14.8 mm). After incubation of the stem cross-sections in darkness for 22 h, *L. maculans* grew more slowly in the cortex of C119 (Fig. 8B) than in that of A30 (Fig. 8A) but there was no difference between C119 (Fig. 8 D) and A30 (Fig. 8 C) in *L. maculans* growth after *L. maculans* had reached the stem pith.

There were differences between conidial inoculum and ascospore inoculum in incubation period for phoma stem canker development, with the incubation period shorter for ascospores than for conidia. Phoma stem cankers were observed at 25 dpi for plants inoculated with ascospores and at 36 dpi for plants inoculated with conidia. Stem canker developed more rapidly and more severely on plants inoculated with ascospores than on plants inoculated with conidia (Fig. 6C-F; Fig. 7).

In experiments with ascospore inoculum (PExpt 2, 3 & 4), 42% of inoculated leaves of A30 and 38% of those of C119 were senescent by 25 dpi. Stem canker developed more rapidly in A30 than in C119. By 32 dpi, 63% of inoculated leaves of A30 and 8% of inoculated leaves of C119 had developed stem canker symptoms at their leaf scars in PExpt 3; 60% of inoculated leaves of A30 and 30% of inoculated leaves of C119 had developed stem canker symptoms at their leaf scars by 33 dpi in PExpt 4. By 46 dpi, the internal stem canker severity score (average of PExpt 2, 3 & 4) was significantly greater in A30 than in C119 ($P<0.001$, 50 d.f., SED 0.27) (Fig. 7); the amount of *L. maculans* DNA was significantly greater in stems of A30 than in stems of C119 ($P<0.05$, 50 d.f., SED 1.48; Fig. 4B). However, there was a significant difference between the three experiments in stem canker severity ($P<0.01$, 50 d.f., SED 0.28) and in the amount of *L. maculans* DNA in stem tissues ($P<0.05$, 50 d.f., SED 1.59). There was no significant

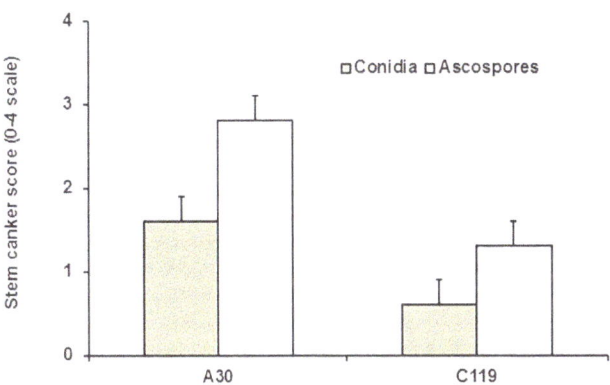

Figure 7. Severity of phoma stem canker symptoms. Leaf petioles of oilseed rape doubled haploid lines A30 (susceptible) or C119 (with quantitative resistance) were inoculated with conidia of isolate ME24/3.13 or ascospores of *Leptosphaeria maculans*; severity of stem canker was assessed on a 0–4 scale at 46 days post inoculation (dpi) (ascospore inoculum) or 87 dpi (conidial inoculum). Bars show standard errors (17 d.f. for conidial inoculation; 50 d.f. for ascospore inoculation).

interaction between experiment and DH line in either stem canker severity or the amount of *L. maculans* DNA.

There were significant linear relationships between amount of *L. maculans* DNA (n) in stem tissues and stem canker severity score (s) in experiments PExpt 2 ($n = 0.87s-1.61$, $R^2 = 0.86$, $P<0.001$, 7 d.f.) and PExpt 3 (s) ($n = 0.24s+2.70$, $R^2 = 0.56$, $P<0.001$, 21 d.f.) (Fig. 5D). Differences between these two regression lines in both the slope ($P<0.001$) and the intercept ($P<0.001$) were significant. There was no significant linear relationship between amount of *L.*

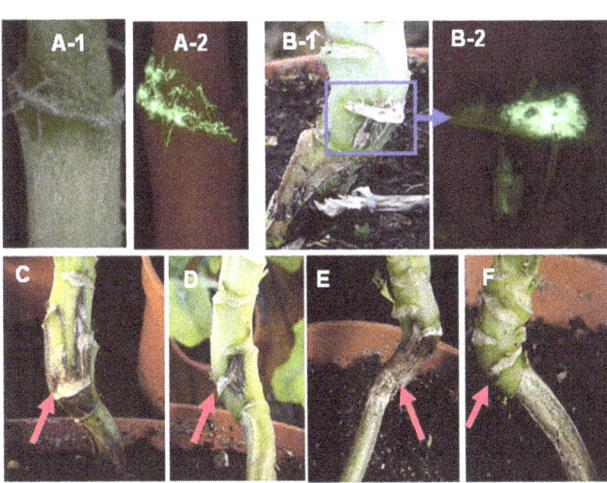

Figure 6. Phoma stem canker symptoms produced in controlled conditions. Leaf petioles of oilseed rape doubled haploid lines A30 (susceptible) or C119 (with quantitative resistance) were inoculated with conidia of GFP-expressing *Leptosphaeria maculans* isolate ME24/3.13 (A, B, C) or ascospores produced on stem debris under natural conditions (E, F). No visible stem canker symptom at the leaf scar (A-1) of A30 after the inoculated leaf had abscised at 25 days post inoculation (dpi) but growth of the pathogen was observed using a GFP2 filter (A-2). Stem canker was visible at the leaf scar of A30 at 36 dpi (B-1) and growth of *L. maculans* was visualised by GFP fluorescence (B-2) (the selected region in B-1 was viewed using a GFP2 filter). Symptoms at leaf scars of A30 (C, E) or C119 (D, F) at 87 dpi with conidia (C; D) or at 31 dpi with ascospores (E, F) (arrows indicate the leaf scars of inoculated leaves).

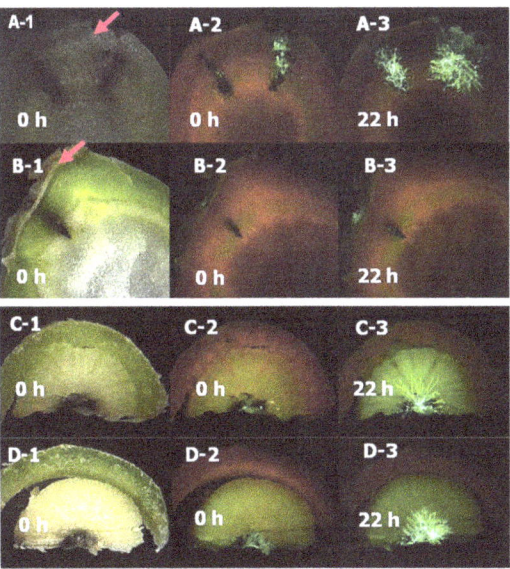

Figure 8. Growth of *Leptosphaeria maculans* in the stem cortex or pith. Leaf petioles of oilseed rape doubled haploid lines A30 (susceptible) or C119 (with quantitative resistance) were inoculated with conidia of GFP-expressing *L. maculans* isolate ME24/3.13, stems of A30 (A, C) and C119 (B, D) were cut horizontally at the leaf scar (A, B) of the inoculated leaf or at the hypocotyl (C, D) 2–3 cm below the leaf scar at 87 days post inoculation (dpi). Stem cross-sections were viewed with brightfield illumination (A-1, B-1, C-1, D-1) or a GFP2 filter (A-2-3, B-2-3, C-2-3, D-2-3) immediately after cutting (0 h) or after incubation for 22 hours in darkness at 20°C (22 h).

maculans DNA in stem tissues and stem canker severity score in experiment PExpt 4.

Discussion

Results of these experiments with DH lines C119 and A30 suggest that *B. napus* quantitative resistance against *L. maculans* can be detected in young plants in controlled conditions by leaf lamina or leaf petiole inoculation. These methods should be further evaluated on a wide range of cultivars or breeding lines with different levels of quantitative resistance to develop methods can be used reliably by breeders to select for quantitative resistance in young plants. This will not only save money but also accelerate the process of breeding for resistance. Traditionally, breeding for quantitative resistance against *L. maculans* in winter oilseed rape has relied on disease assessments before harvest 10–11 months after sowing of field experiments [3,18,30]. By contrast, an assessment of quantitative resistance in young plants in controlled conditions takes only 1–2 months by leaf lamina inoculation or 2–3 months by leaf petiole inoculation (Table 2). Furthermore, controlled environment assessment of young plants can test large numbers of lines. Such controlled environment assessment methods could be used for pre-selection of breeding material to decrease the number of lines required for final field testing and could thus save costs.

Differences between A30 and C119 in growth of *L. maculans* in leaf petioles and stems suggest that quantitative resistance against *L. maculans* decreases the severity of stem canker by impeding the growth of *L. maculans* in both petiole and stem tissues. The shorter distance grown by *L. maculans* in petioles of C119 than in those of A30 suggests that quantitative resistance in C119 impeded the growth of *L. maculans* along the petiole towards the stem. However, in previous work with commercial cultivars Darmor (with quantitative resistance) and Eurol (without quantitative resistance), there was no difference between the two cultivars in growth of *L. maculans* along the petiole although there was a difference between them in severity of stem canker [6]. The observation that growth of GFP-labelled *L. maculans* was slower in the stem cortex of C119 than in the cortex of A30 but did not differ between A30 and C119 after *L. maculans* reached the stem pith suggests that quantitative resistance in C119 mainly operated during growth of *L. maculans* from the stem cortex to the stem pith. This is consistent with results of previous work [6]. The development of more severe necrosis in hypocotyls of A30 than in those of C119 [28] suggests that quantitative resistance against *L. maculans* may also impede the growth of the pathogen from the cotyledon to the hypocotyl. There is a need to test a wide range of cultivars and DH lines under different conditions to investigate the relationship between quantitative resistance against *L. maculans* in leaf petiole and stem tissues identified in controlled environment experiments and quantitative 'field' resistance identified by end of cropping season disease assessments in field conditions.

Results of these experiments with C119 and A30 inoculated with either conidia or ascospores suggest that leaf lesion area might be used to assess quantitative resistance against *L. maculans* at stage 1 (Fig. 1). The good positive relationship between leaf lesion area and distance grown by *L. maculans* along the petiole towards the stem suggests that *L. maculans* may reach the stem more quickly and subsequently cause more severe stem canker before harvest if

Table 2. Comparison of different methods for assessing *Brassica napus* quantitative resistance against *Leptosphaeria maculans*.

Inoculum	Assessment method	Advantages	Disadvantages	Duration	Ref. [1]
Field expt., natural inoculation					
Ascospores	Score canker severity on stem before harvest	Test of field resistance under natural conditions, reliable results, no need to wound leaf	Variation between years and regions, costly to assess large numbers of lines, cannot distinguish quantitative from *R* gene resistance when effective *R* gene is present in the plant	10–11 months[2]	1, 2, 3, 4
CE expt., cotyledon inoculation – growth in hypocotyl					
Conidia	Score canker severity on hypocotyl	Easy to produce genetically homogeneous inoculum, can test large numbers of plants	Tests few isolates, less infective than ascospores, variation between expts, not easy to produce canker symptoms, hypocotyl infection does not occur in field in Europe	60–90 days	5
CE expt., leaf lamina inoculation – growth in leaf petiole					
Conidia	Extent (distance) of visible necrosis or GFP	Easy to produce homogeneous inoculum, rapid, can test large numbers of plants, visualize symptomless growth if GFP used	Tests few isolates, less infective than ascospores, need to wound leaf, may not distinguish lines with moderate resistance, restriction on use of GFP	40–45 days	3, 6
Ascospores	Extent (distance) of visible necrosis, qPCR	Easy to assess, tests natural population, can test large number of plants, no need to wound leaf, can distinguish quantitative from *R* gene resistance	Difficult to produce large amount of inoculum, inoculum not genetically homogeneous, may not distinguish moderate resistance lines	35–40 days	6
CE expt., leaf petiole inoculation – growth in stem tissue					
Conidia	Extent of visible necrosis or GFP	Easy to produce genetically homogeneous inoculum, visualize symptomless growth if GFP used	Tests few isolates, less infective than ascospores, restriction on use of GFP	70–90 days	6
Ascospores	Extent of visible necrosis, qPCR	Easy to assess, test natural population, results close to natural conditions, reliable results	Difficult to produce large amount of inoculum, inoculum not genetically homogeneous, cannot distinguish quantitative from *R* gene resistance when effective *R* gene is present	50–60 days	6

[1] 1,Delourme et al., 2006; 2, Fitt et al., 2006; 3, Huang et al., 2009; 4, Pilet et al., 1998; 5, Travadon et al., 2009; 6, this study.
[2] To obtain a reliable estimate, it is necessary to do several experiments at different sites in different seasons due to G × E interactions.

it is growing from large leaf lesions than if it is growing from small leaf lesions. This suggests that the measurement of areas of phoma leaf lesions might provide a more accurate assessment of quantitative 'field' resistance than assessing the numbers of phoma leaf lesions in field experiments in autumn since there is a poor relationship between the number of phoma leaf lesions in autumn and the subsequent severity of stem canker in summer [38]. Furthermore, it is quicker and more reliable to measure lesion area than to measure distance grown by *L. maculans* along leaf vein/petiole. For example, with conidial inoculation (e.g. LExpt 1 &2), lesion area can be measured at 17–18 dpi and no difference was observed between experiments. However, distance grown along leaf petiole cannot be measured until 22–25 dpi and differences between experiments were observed. Measurement of leaf lesion area is easy and simple. If area of leaf lesions could be used to assess quantitative resistance in field conditions, it would be a valuable method for breeders. Leaf lesion area has been used to assess quantitative resistance against other fungal pathogens [5,39–40]. However, there is a need to test a wide range of cultivars under different conditions to investigate the relationship between the area of phoma leaf lesions in autumn and the severity of phoma stem canker in the following summer.

Comparison of different inoculation methods for assessing *B. napus* quantitative resistance against *L. maculans* in young plants suggests that leaf petiole inoculation is better than other methods for use in controlled environments (Table 2). Although canker symptoms can be produced on hypocotyls after cotyledon inoculation [28], production of canker symptoms requires 4 weeks at 6°C followed by 6 weeks at 20°C and hypocotyl infection is not representative of what happens in field conditions in the UK. Furthermore, the structure of hypocotyl is different from that of stem [41]. Although the leaf lamina inoculation method is quicker than leaf petiole inoculation, the main damage caused by *L. maculans* is canker on stems, and results of leaf lamina inoculation experiments need to be further confirmed by assessing canker severity on stems. The consistency of the results of the four petiole inoculation experiments for stem canker score (i.e. less severe stem canker on C119 than on A30) suggests that quantitative resistance against *L. maculans* can be reliably assessed in young plants during colonisation of stem tissues (stage 2) after petiole inoculation. This conclusion is supported by previous work on cultivars Darmor and Eurol [6].

Results of these experiments with different types of inoculum suggest that ascospore inoculum of *L. maculans* is more effective than conidial inoculum in terms of leaf lesion and stem canker development. The incubation periods for leaf lesion and stem canker development may have been shorter for ascospore inoculum than for conidial inoculum because ascospores germinate more rapidly and contain more nutrients for the pathogen to colonise the host than conidia. Although phoma stem canker epidemics are initiated by ascospores released from crop debris, much experimental work with young plants to study resistance has used conidial inoculum [8,33,42]. Use of ascospores from natural populations as inoculum can simulate what happens in natural conditions and is more likely to identify 'quantitative resistance' expressed in field conditions. Recent work has shown that some components of quantitative resistance against *L. maculans* can be isolate-specific [43]; use of conidia of single *L. maculans* isolates may not accurately assess quantitative resistance while use of ascospore inoculum from natural populations is more likely to do so. Therefore, use of ascospore inoculum is more reliable than use of conidial inoculum for detecting and assessing quantitative resistance.

Comparison of different assessment methods used in controlled environments suggests that leaf lesion area could be a good measurement for use at stage 1 and stem canker severity score could be a good measurement for use at stage 2 (Table 2). At stage 1, measurement of lesion area was better than the other three methods used (GFP, qPCR and visible distance grown) and it might be used in field experiments whereas it is difficult to measure the spread of *L. maculans* from the leaf lesion along the leaf petiole in field conditions. In addition, assessment of leaf lesions (appearance and size) can distinguish between quantitative resistance (typical large grey lesion) and *R* gene-mediated resistance (small dark lesion) against *L. maculans* [6,9]. Although GFP and qPCR are valuable methods for assessing symptomless growth of *L. maculans* in leaf or stem tissues, GFP can be used only in controlled environments and is not practical for use with large numbers of lines. Whilst qPCR can be used with large numbers of lines, it is expensive and not always reliable. Because qPCR is very sensitive and *L. maculans* has a symptomless growth phase, *L. maculans* DNA was detected in stems (Fig. 5D) or hypocotyls [28] with no visible necrosis. By contrast, scoring stem canker severity in controlled environment experiments is more reliable at stage 2. More severe stem cankers developed on A30 than on C119 in all the four experiments with petiole inoculation. Results of stem canker assessment in young plants in controlled conditions (stage 2) could be used to predict how the lines or cultivars will operate in field conditions.

Development of methods to phenotype quantitative resistance against plant pathogens in young plants under controlled conditions will improve the process of breeding cultivars for control of crop diseases. Assessment of quantitative trait loci (QTL) for resistance is often inconsistent due to interactions with environment [4,16,31]. Detection of quantitative resistance in controlled conditions can avoid the influence of fluctuations in natural weather conditions. Furthermore, with the development of methods to assess quantitative resistance in controlled environments, it is possible to test effectiveness of the quantitative resistance under a range of different controlled conditions, which will help to optimise the use of quantitative resistance in different natural environments. Due to the incomplete nature of quantitative resistance and the difficulties in assessing it, mechanisms of operation of quantitative resistance are poorly understood [4]. Quantitative resistance has been thought to be expressed at later stages of plant development and it has been referred to as 'adult plant resistance'. There has been little work to investigate potential operation of quantitative resistance in young plants; results of this work suggest that quantitative resistance can be expressed at early stages of plant development. With the use of next generation sequencing techniques and reduced costs, more genome sequences of both pathogens and their hosts are available [44–47]. Identification of stable QTL in controlled conditions will improve our understanding of mechanisms of operation of quantitative resistance, which will help in the breeding of cultivars with durable resistance to contribute to food security.

Acknowledgments

We thank Regine Delourme and Hortense Brun for providing the seeds of DH lines A30 and C119 and for their advice throughout this work, including comments on this paper, Zhunyan Huang for help with DNA extraction, Jean Devonshire for help with GFP microscopy, Sandra Harvey for help with inoculation and data curating. All work involving the use and storage of genetically modified *L. maculans* was done under Defra licence no. PHL 174E/5443(01/2007).

Author Contributions

Conceived and designed the experiments: YJH GJK BDLF. Performed the experiments: YJH. Analyzed the data: YJH AQ. Contributed reagents/materials/analysis tools: YJH AQ GJK. Wrote the paper: YJH AQ BDLF.

References

1. Beddington J (2010) Food security: contributions from science to a new and greener revolution. Philosophical Transactions of the Royal Society B: Biological Sciences 365: 61–71.

2. Lindhout P (2002) The perspectives of polygenic resistance in breeding for durable disease resistance. Euphytica 124: 217–226.

3. Delourme R, Chèvre AM, Brun H, Rouxel T, Balesdent MH, et al. (2006) Major gene and polygenic resistance to Leptosphaeria maculans in oilseed rape (Brassica napus). European Journal of Plant Pathology 114: 41–52.

4. Poland JA, Balint-Kurti PJ, Wisser RJ, Pratt RC, Nelson RJ (2008) Shades of gray: the world of quantitative disease resistance. Trends in Plant Sciences 14: 21–29.

5. Chartrain L, Brading PA, Widdowson JP, Brown JKM (2004) Partial resistance to septoria tritici blotch (Mycosphaerella graminicola) in wheat cultivars Arina and Riband. Phytopathology 94: 497–504.

6. Huang YJ, Pirie EJ, Evans N, Delourme R, King GJ, et al. (2009) Quantitative resistance to symptomless growth of Leptosphaeria maculans (phoma stem canker) in Brassica napus (oilseed rape). Plant Pathology 58: 314–323.

7. Brun H, Chèvre AM, Fitt BDL, Powers F, Besnard AL, et al. (2010) Quantitative resistance increases the durability of qualitative resistance to Leptosphaeria maculans in Brassica napus. New Phytologist 185: 285–299.

8. Balesdent MH, Attard A, Ansan-Melayah D, Delourme R, Renard M, et al. (2001) Genetic control and host range of avirulence towards Brassica napus cvs Quinta and Jet Neuf in Leptosphaeria maculans. Phytopathology 91: 70–76.

9. Huang YJ, Evans N, Li ZQ, Eckert M, Chèvre AM, et al. (2006) Temperature and leaf wetness duration affect phenotypic expression of Rlm6-mediated resistance to Leptosphaeria maculans in Brassica napus. New Phytologist 170: 129–141.

10. Larkan NJ, Lydiate DJ, Parkin IAP, Nelson MN, Epp DJ, et al. (2013) The Brassica napus blackleg resistance gene LepR3 encodes a receptor-like protein triggered by the Leptosphaeria maculans effector AVRLM1. New Phytologist 197: 595–605.

11. Hammond-Kosack KE, Kanyuka K (2007) Resistance genes (R genes) in plants. Encyclopedia of Life Sciences, John Wiley & Sons, Ltd. doi: 10.1002/9780470015902.a0020119.

12. Bent AF, Mackey D (2007) Elicitors, effectors, and R genes: the new paradigm and a lifetime supply of questions. Annual Review of Phytopathology 45: 399–436.

13. Rouxel T, Penaud A, Pinochet X, Brun H, Gout L, et al. (2003) A 10-year survey of populations of Leptosphaeria maculans in France indicates a rapid adaptation towards the Rlm1 resistance gene of oilseed rape. European Journal of Plant Pathology 109: 871–881.

14. Sprague SJ, Balesdent MH, Brun H, Hayden HL, Marcroft SJ, et al. (2006) Major gene resistance in Brassica napus (oilseed rape) is overcome by changes in virulence population of Leptosphaeria maculans in France and Australia. European Journal of Plant Pathology 114: 33–40.

15. Soanes DM, Talbot NJ (2008) Moving targets: rapid evolution of oomycete effectors. Trends in Microbiology 16: 507–510.

16. McDonald B (2010) How can we achieve durable disease resistance in agricultural ecosystems? New Phytologist 185: 3–5.

17. Czembor PC, Arseniuk E, Czaplicki A, Song Q, Cregan PB, et al. (2003) QTL mapping of partial resistance in winter wheat to Stagonospora nodorum blotch. Genome 46: 546–554.

18. Fitt BDL, Brun H, Barbetti MJ, Rimmer SR (2006) World-wide importance of phoma stem canker (Leptosphaeria maculans and L. biglobosa) on oilseed rape (Brassica napus). European Journal of Plant Pathology 114: 3–15.

19. Oliver RP, Rybak K, Shankar M, Loughman R, Harry N, et al. (2008) Quantitative disease resistance assessment by real-time PCR using the Stagonospora nodorum-wheat pathosystem as a model. Plant Pathology 57: 527–532.

20. Fitt BDL, Fraaije BA, Chandramohan P, Shaw MW (2011) Impacts of changing air composition on severity of arable crop disease epidemics. Plant Pathology 60: 44–53.

21. Stonard JF, Latunde-Dada AO, Huang YJ, West JS, Evans N, et al. (2010) Geographic variation in severity of phoma stem canker and Leptosphaeria maculans/L. biglobosa populations on UK winter oilseed rape (Brassica napus). European Journal of Plant Pathology 126: 97–109.

22. Huang YJ, Fitt BDL, Jedryczka M, Dakowska S, West JS, et al. (2005) Patterns of ascospore release in relation to phoma stem canker epidemiology in England (Leptosphaeria maculans) and Poland (L. biglobosa). European Journal of Plant Pathology 111: 263–277.

23. Huang YJ, Toscano-Underwood C, Fitt BDL, Hu XJ, Hall AM (2003) Effects of temperature on ascospore germination and penetration of oilseed rape (Brassica napus) leaves by A-group or B-group Leptosphaeria maculans (phoma stem canker). Plant Pathology 52: 245–255.

24. Toscano-Underwood C, West JS, Fitt BDL, Todd AD, Jedryckzka M (2001) Development of phoma lesions on oilseed rape leaves inoculated with ascospores of A-group and B-group Leptosphaeria maculans (stem canker) at different temperatures and wetness durations. Plant Pathology 50: 28–41.

25. Travadon R, Bousset L, Saint-Jean S, Sache I (2007) Splash dispersal of Leptosphaeria maculans pycnidiospores and the spread of blackleg on oilseed rape. Plant Pathology 56: 595–603.

26. Rimmer SR (2006) Resistance genes to Leptosphaeria maculans in Brassica napus. Can J Plant Pathol 28: S288–S297.

27. Jestin C, Lode M, Vallee P, Domin C, Falentin C, et al. (2011) Association mapping of quantitative resistance for Leptosphaeria maculans in oilseed rape (Brassica napus L.). Molecular Breeding 27: 271–287.

28. Travadon R, Marquer B, Ribule A, Sache I, Masson JP, et al. (2009) Systemic growth of Leptosphaeria maculans from cotyledons to hypocotyls in oilseed rape: influence of number of infection sites, competitive growth and host polygenic resistance. Plant Pathology 58: 461–469.

29. Marcroft SJ, Van de Wouw, Salisbury PA, Potter TD, Howlett BJ (2012) Effect of rotation of canola (Brassica napus) cultivars with different complements of blackleg resistance genes on disease severity. Plant Pathology 61: 934–944.

30. Pilet ML, Delourme R, Foisset N, Renard M (1998) Identification of loci contributing to quantitative field resistance to blackleg disease, causal agent Leptosphaeria maculans (Desm.) Ces. et de Not., in winter rapeseed (Brassica napus L.). Theoretical and Applied Genetics 96: 23–30.

31. Delourme R, Brun H, Ermel M, Lucas MO, Vallee P, et al. (2008) Expression of resistance to Leptosphaeria maculans in Brassica napus double haploid lines in France and Australia is influenced by location. Annals of Applied Biology 153: 259–269.

32. Brun H, Ermel M, Mabon R, Miteul H, Balesdent MH, et al. (2011) A method to phenotype the aggressiveness of Leptosphaeria maculans isolates on stems of oilseed rape in controlled conditions. Abstract book of the 13th International Rapeseed Congress, Crop Protection Section: 507.

33. Delourme R, Miteul H, Ermel M, Brun H (2011) Assessment of quantitative resistance to Leptosphaeria maculans isolates in controlled conditions. Abstract book of the 13th International Rapeseed Congress, Crop Protection Section: 505.

34. Stachowiak A, Olechnowicz J, Jedryczka M, Rouxel T, Balesdent MH, et al. (2006) Frequency of avirulence alleles in field populations of Leptosphaeria maculans in Europe. European Journal of Plant Pathology 114: 67–75.

35. Zhou Y, Fitt BDL, Welham SJ, Gladders P, Sansford CE, et al. (1999) Effects of severity and timing of stem canker (Leptosphaeria maculans) symptoms on yield of winter oilseed rape (Brassica napus) in the UK. European Journal of Plant Pathology 105: 715–728.

36. Liu SY, Liu Z, Fitt BDL, Evans N, Foster SJ, et al. (2006) Resistance to Leptosphaeria maculans (phoma stem canker) in Brassica napus (oilseed rape) induced by L. biglobosa and chemical defence activators in field and controlled environments. Plant Pathology 55: 401–412.

37. Payne RW, Harding SA, Murray DA, Soutar DM, Baird DB, et al. (2011) The Guide to GenStat Release 14, Part 2: Statistics. 997 pp. Hemel Hempstead, UK: VSN International Ltd.

38. Powers SJ, Pirie EJ, Latunde-Dada AO, Fitt BDL (2010) Analysis of leaf appearance, leaf death and phoma leaf spot, caused by Leptosphaeria maculans, on oilseed rape (Brassica napus) cultivars. Annals of Applied Biology 157: 55–70.

39. Talukder ZI, Tharreau D, Price AH (2004) Quantitative trait loci analysis suggests that partial resistance to rice blast is mostly determined by race-specific interactions. New Phytologist 162: 197–209.

40. Silva MR, Martinelli JA, Federizzi LC, Chaves MS, Pacheco MT (2012) Lesion size as a criterion for screening oat genotypes for resistance to leaf spot. European Journal of Plant Pathology 137: 315–327.

41. Sprague SJ, Watt M, Kirkegaard JA, Howlett BJ (2007) Pathways of infection of Brassica napus roots by Leptosphaeria maculans. New Phytologist 176: 211–222.

42. Raman R, Taylor B, Marcroft S, Stiller J, Eckermann P, et al. (2012) Molecular mapping of qualitative and quantitative loci for resistance to Leptosphaeria maculans causing blackleg disease in canola (Brassica napus L.). Theoretical and Applied Genetics 125: 405–418.

43. Marcroft ST, Elliott VL, Cozijnsen AJ, Salisbury PA, Howlett BJ, et al. (2012) Identifying resistance genes to Leptosphaeria maculans in Australian Brassica napus cultivars based on reactions to isolates with known avirulence genotypes. Crop & Pasture Science 63: 338–350.

44. Rouxel T, Grandaubert J, Hane JK Hoede C, van de Wouw A (2011) Effector diversification within compartments of the Leptosphaeria maculans genome affected by Repeat-Induced Point mutations. Nature Communications 2: 202. (Doi: 10.1038/ncomms1189).

45. Rouxel T, Balesdent MH (2013) From model to crop plant–pathogen interactions: cloning of the first resistance gene to Leptosphaeria maculans in Brassica napus. New Phytologist 197: 356–358.

46. Wang X, Wang H, Wang J, Sun R, Wu J, et al. (2011) The genome of the mesopolyploid crop species Brassica rapa. Nature Genetics 43: 1035–1039.

47. Hayward A, McLanders J, Campbell E, Edwards D, Batley J (2012) Genomic advances will herald new insights into the Brassica: Leptosphaeria maculans pathosystem. Plant Biology 14 (Suppl. 1): 1–10.

4

A *GmRAV* Ortholog Is Involved in Photoperiod and Sucrose Control of Flowering Time in Soybean

Qingyao Lu[1], Lin Zhao[1]*, Dongmei Li[1], Diqiu Hao[1], Yong Zhan[2], Wenbin Li[1]*

1 Key Laboratory of Soybean Biology of Chinese Education Ministry (Key Laboratory of Biology and Genetics & Breeding for Soybean in Northeast China, Ministry of Agriculture), Northeast Agriculture University, Harbin, China, 2 Agricultural Academy of Shi He Zi, Xinjiang Province, China

Abstract

Photoperiod and sucrose levels play a key role in the control of flowering. *GmRAV* reflected a diurnal rhythm with the highest expression at 4 h after the beginning of a dark period in soybean leaves, and was highly up-regulated under short-day (SD) conditions, despite of not following a diurnal pattern under long-day (LD) conditions. *GmRAV-i* (*GmRAV*-inhibition) transgenic soybean exhibited early flowering phenotype. Two of the *FT* Arabidopsis homologs, *GmFT2a* and *GmFT5a*, were highly expressed in the leaves of soybeans with inhibition (-i) of *GmRAV* under SD conditions. Moreover, the transcript levels of the two *FT* homologs in *GmRAV-i* soybeans were more sensitive to SD conditions than LD conditions compared to the WT plant. *GmRAV-i* soybeans and *Arabidopsis rav* mutants showed more sensitive hypocotyl elongation responses when compared with wild-type seedlings, and *GmRAV*-ox overevpressed in tobacco revealed no sensitive changes in hypocotyl length. These indicated that *GmRAV* was a novel negative regulator of SD-mediated flowering and hypocotyl elongation. Although sucrose has been suggested to promote flowering induction in many plant species, high concentration of sucrose (4% [w/v]) applied into media defer flowering time in Arabidopsis wild-type and *rav* mutant. This delayed flowering stage might be caused by reduction of *LEAFY* expression. Furthermore, Arabidopsis *rav* mutants and *GmRAV-i* soybean plants were less sensitive to sucrose by the inhibition assays of hypocotyls and roots growth. In contrast, transgenic *GmRAV* overexpressing (-ox) tobacco plants displayed more sensitivity to sucrose. In conclusion, *GmRAV* was inferred to have a fundamental function in photoperiod, darkness, and sucrose signaling responses to regulate plant development and flowering induction.

Editor: Tianzhen Zhang, Nanjing Agricultural University, China

Funding: This study was conducted in the Key Laboratory of Soybean Biology of Chinese Education Ministry, Soybean Research & Development Center, CARS and the key Laboratory of Northeastern Soybean Biology and Breeding/Genetics of Chinese Agriculture Ministry, financially supported by National Core Soybean Genetic Engineering Project (Contract No. 2013ZX08004-1 003), Chinese National Natural Science Foundation (31271748, 60932008, 31101169, 31201227), National 973 Project (2012CB126311), National 863 Project (2013AA102602, 2012AA101106-1-9), Chinese Key Projects of Soybean Transformation (2013ZX08004-005), Provincial/National Education Ministry project (1252G014, 1252-NCET-005, 20122325120012), and Provincial/National Education Ministry for the team of soybean molecular design. Youth Backbone Research program of Provincial Education Department (1253G010). The funders had no role in study design, data collection and analysis, decision to publish, or preparation of the manuscript.

Competing Interests: The authors have declared that no competing interests exist.

* E-mail: wenbinli@neau.edu.cn (WL); zhaolinneau@126.com (LZ)

Introduction

In many plant species, flowering time is strongly influenced by environmental factors where photoperiod plays a prominent role [1]. Plants perceive light through its phytochromes and crypto-chromes, which transfer the signal to the plant internal circadian clock system. There is an increasing evidence for conservation of flowering pathways between many plant species. Flowering time is regulated by multiple and to some extent redundant pathways that can promote or delay flowering [2]. *Arabidopsis thaliana*, a model organism, is a long-day plant, and there are many pathways, such us photoperiod, vernalization, gibberellic acid and autonomous reactions, involved in control of its floral transition [3–6]. In contrast, soybean (*Glycine max*) is a short-day plant, and photoperiod controls its duration in both pre- and post-flowering phases [7]. Therefore, photoperiod is an important environmental cue that determines flowering time in soybean [8]. The term 'critical photoperiod' is described as the duration of daylight period under which the plant is induced to flower, and determines plant transition from vegetative to reproductive stage. Sensitivity to

photoperiod limits the adaptation of soybean to a wider range of latitude [9]. The identification of the genetic components contributing to the photoperiodic control of flowering time in soybean was recently limited.

Florigen (FT) is a hypothetical leaf-produced signal that moves from phloem to induce flowering at shoot apex. The expression of *FT* gene (the flowering integrator genes, FLOWERING LOCUS T [10], and its orthologs are critical for flowering in plants [11,12]. Two soybean *FT* homologs (*GmFT2a* and *GmFT5a*) have florigen-like functions and their transcript levels are upregulated under SD conditions (SDs) [13].

Not only carbohydrates provide energy and carbon sources for plants, but also act as essential regulators during their growth and development [14,15], as evidenced by the variety of sugar sensing and signaling mechanisms that have been uncovered [16,17]. Carbohydrates seem to regulate many essential processes, including photosynthesis, sucrose synthesis and degradation, flowering, and senescence [18,19].

There has been a certain amount of evidence suggesting that sucrose promotes flowering in most species [20]. In Arabidopsis, the induction of flowering in wild-type plants by LDs causes an early and transient increase in sucrose export from leaves. The efficiency of floral induction by a single LD is reflected by the amplitude of an increase in exported sucrose [21]. Rolda'n et al. reported that *in vitro* culture of plants on medium containing 1% (w/v) sucrose, partially rescued the phenotypes of late-flowering mutants [22]. In contrast, Zhou et al. reported that high levels of glucose in the medium delayed flowering in Arabidopsis [23]. Masa-aki et al. also analyzed the effects of sugar on development and floral transition [24]. In an early flowering mutant *tfl1*, 5% (w/v) sucrose in the medium delayed floral transition. It was concluded that the inhibition was caused by metabolic rather than its osmotic effects. Recently, King et al. reported that *FT* and sucrose may regulate flowering as 'a florigen' in plants [25].

Despite their metabolic role, glucose and fructose play additional signaling functions in plant cells [26]. Oligosaccharides derived from the cell wall also function as signals in the processes of regulation of hypocotyl elongation [27], fruit ripening [28] and defense mechanisms to pathogens [29]. Also, root growth was considerably more sensitive to carbon source than hypocotyl elongation [30]. In this study, the effects of sugars on a range of growth and developmental parameters in *Arabidopsis thaliana*, tobacco and soybean were measured. The effects of sugars on growth and developmental processes in plants at earlier stages of vegetative development and the flowering timing were also investigated.

In Arabidopsis, the RAV subfamily belongs to one of the largest and most diverse family of transcription factors AP2/ERBP. RAV proteins function in the involvement in cold tolerance, dehydration, and circadian rhythm clock. *GmRAV* (DQ147914) [31,32] was one of four *RAV2*-like paralogues in the soybean genome. *GmRAV* may be a complete functional orthologue of any *AtRAV2* family member [33].

In this study, *GmRAV* was overexpressed in transgenic tobacco and inhibited in transgenic soybean, which showed that *GmRAV* was a responder in the photoperiodic control of flowering time and sugar signaling. We found that *GmRAV* transcript exhibited a circadian rhythm under SDs and decreased significantly in leaves by exogenous sucrose application. A detailed phenotypic characterization, along with genetic and physiological analysis, indicated that *GmRAV* was inferred to be a signaling component involved in regulation of plant development and flowering time.

Materials and Methods

Plant Materials and Growth Conditions

Arabidopsis thaliana Columbia (Col-0) ecotype was used in this study as wild-type plant (Lehle Seeds, Round Rock, TX). The Salk T-DNA knockout mutant line of *AtRAV* (At1g25560; SALK_029626c) was obtained from the *Arabidopsis* Biological

Table 1. List of primers for transgenic plants detection used in the present study.

Primer name	Primer sequence
pat-F (soybean)	GCACCATCGTCAACCACTAC
pat-R	TGAAGTCCAGCTGCCAGAAAC
Salk_029626c LP	AATCTCATGTGAACCCCCTTC
Salk_029626c RP	CGCTGATGCTTCTCGTAAATC
Salk_029626c LB	ATTTTGCCGATTTCGGAAC

Table 2. List of primer for real-time PCR analysis used in the present study.

Primer name	Primer sequence
At18SrRNA-F	CGTCCCTGCCCTTTGTACAC
At18SrRNA-R	CGAACACTTCACCGGATCATT
AtLEAFY-F	TGTGAACATCGCTTGTCGTC
AtLEAFY-R	TAATACCGCCAACTAAAGCC
GmACTIN4-F	GTGTCAGCCATACTGTCCCCATTT
GmACTIN4-R	GTTTCAAGCTCTTGCTCGTAATCA
GmRAV-F	GGTTCGGATGGTGTAGGGAAGAGAA
GmRAV-R	TTACAAAGCTCCAATTACTTTTAAC
GmFT2a-F	GGATTGCCAGTTGCTGCTGT
GmFT2a-R	GAGTGTGGGAGATTGCCAAT
GmFT5a-F	GCCTTACTCCAGCTTATACT
GmFT5a-R	GGCATGCTCTAGCATTGCAA

Resource Center (ABRC). *GmRAV-i* soybean, *GmRAV-ox* tobacco and *GmRAV* promoter::GUS transgenic Arabidopsis seeds were provided by our lab (Northeast Agricultural University, Harbin, China) [33]. The primer pairs for transgenic plants detection were listed in Table 1. Arabidopsis seeds were surface sterilized, placed in Petri dishes containing solid Murashige and Skoog (MS) medium, and stratified for 3 days at 4°C. Subsequently, the seedlings were placed in a vertical orientation in the growth chamber at 22°C under LDs (16 h/8 h light/dark). T4 generation *GmRAV-ox* tobacco and T6 generation transgenic *GmRAV-i* soybean [33] were grown in a growth chamber at 25°C, and illuminated with 200 μmol·m^{-2}·s^{-1} fluorescent lights.

Seeds of soybean cultivars 'Dong Nong 42' and 'Dong Nong 47' (provided by Northeast Agricultural University, Harbin, China; photoperiod sensitive) and 'GmRAV-i transgenic soybean' were grown at 25°C under LDs with 250 μmol·m^{-2}·sec^{-1} white light. The plants were transferred to SDs under the same temperature regime after V2 stage. Experiments were conducted under LDs of 16 h/8 h light/dark and SDs of 8 h/16 h light/dark. Seeds of the WT plants and Arabidopsis *rav* mutants were sowed in solid MS medium in Petri dishes, and then were conducted for the same treatments as above.

In diurnal expression analysis, pieces of young fully developed trifoliate leaves were sampled as a bulk of three plants grown under LDs at 15 days after emergence (DAE) every 4 h starting at dawn for a total of 24 h. Also, plant tissues were harvested from root, stem, leaf, trifoliate leaf, flower bud, pod and immature seed at 12 h after dawn under SDs. In time course-dependent expression analysis, the trifoliate leaves from 'Dong Nong 47' soybean plants were sampled at 12 h after dawn by bulk from four individual plants grown in SDs and LDs at 17, 20, 23, 26, 29, 32, 35 and 45 DAE. The dates of the first flower appearance and flower bud formation at each node were recorded individually.

RNA Isolation and Quantitative Real-time RT-PCR (qRT-PCR) Analysis

Total RNA was extracted from soybean and *Arabidopsis* seedlings with RNAiso Plus Kit (TaKaRa, Japan). The total RNA was reverse-transcribed into first-strand cDNA in a 20 μL volume with PrimeScript RT reagent Kit (TaKaRa, Japan). qRT-PCR analysis was carried out using SYBR Premix Ex Taq II

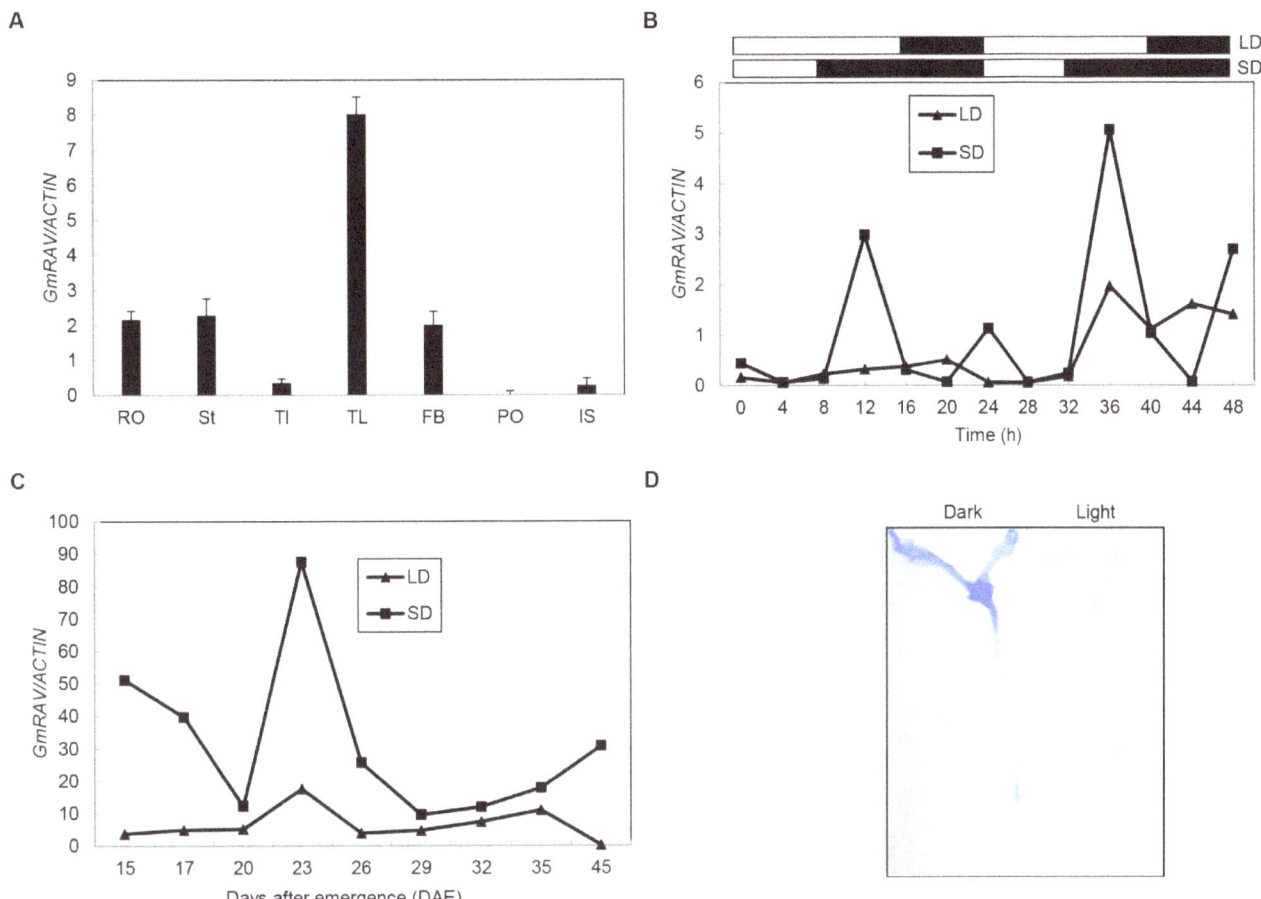

Figure 1. Quantitative real-time RT-PCR analysis of transcript level of *GmRAV* gene under SDs and LDs. A, Tissue-specific expression of soybean *GmRAV* in SDs. Tissues tested are leaf (TI), trifoliate leaf (TL), stem (St), root (RO), pod (PO), flower bud (FB), and immature seed (IS) (plants aged 21 d). B, Relative transcript levels of *GmRAV* mRNA in soybean leaves under SDs and LDs. Soybean leaves were harvested every 4 h for 48 h at 25-day-old under LDs and SDs. *Open* and *closed boxes* indicate days and nights. C, Time course-dependent expression in LDs. Soybean 'Dong Nong 47' plants were grown under LDs for 10 d and were transferred to LDs or SDs before sampling. Relative transcript levels were analyzed by qRT-PCR. D, Histochemical detection of *GmRAV*–GUS promoter activity in transgenic *Arabidopsis* seedlings. 4-day-old seedlings were grown on MS medium.

(TaKaRa, Japan) in a 25 µL reaction, containing 2 µL of cDNA, 12.5 µL SYBR Premix Ex Taq II (2×), 1 µL of 10 µM forward primer, 1 µL of 10 µM reverse primer and 8.5 µL of water. The reaction was performed in the Thermal Cycler Dice Real Time System. The thermal cycle used was as follows: 95°C for 30 s; 40 cycles of 95°C for 5 s, 60°C for 20 s and 72°C for 20 s. Soybean actin 4 (*GmACTIN*; GenBank accession number AF049106) and Arabidopsis 18 s rRNA (GenBank accession number X16077.1) were included as inner references for soybean and Arabidopsis genes. The total RNA was used as templates in qRT-PCR reactions with the primers of *GmRAV*, *GmFT2a*, *GmFT5a* and *AtLEAFY* genes. The primer pairs were listed in Table 2. PCR reactions were performed according to the manufacturer's instructions on the Chromo 4 real time DNA amplification system (BioRad, USA). Data were analyzed using the comparative Ct method. Further qRT-PCR analysis were performed as described above. The analysis were done using the DNA Engine Opticon 2 System (MJ Research, USA). The sequences reported in this paper have been deposited in the GenBank/EMBL/DDBJ database with accession numbers AB550122 (*GmFT2a*), AB550126 (*GmFT5a*) for cDNA sequences of soybean cultivar 'Dong Nong 50' and Genbank accession number AF010190.2 (*AtLEAFY*) for cDNA sequences of Arabidopsis.

GUS Assays

GUS activity was assayed in T3 transgenic Arabidopsis plants. GUS histochemical staining and GUS activity measurements (using about 40–50 seedlings in each sample) were carried out following the procedures described by Jefferson et al. [34].

Sucrose Stress Assay

The sucrose sensitivity assay for hypocotyl elongation was carried out by germinating the wild-type and *rav* transgenic seeds. The change in root and hypocotyl length was used as a measure to check the sensitivity of the plants using the Image-J program (http://rsb.info.nih.gov/ij/docs/menus/file.html) after 15 days.

For flowering assay, the wild-type and transgenic seeds were grown on MS medium supplemented with different concentrations of sucrose during their lifetime. Plants were growing on MS with 2% and 4% (w/v) sucrose, and after 15 days were transferred into soil till flowering. Flowering time was measured by scoring the time from sowing to first flower.

Statistical Analysis

Data was presented as means ± standard error of means. The statistical comparisons were made using Student's t test at p<0.01 or p<0.05.

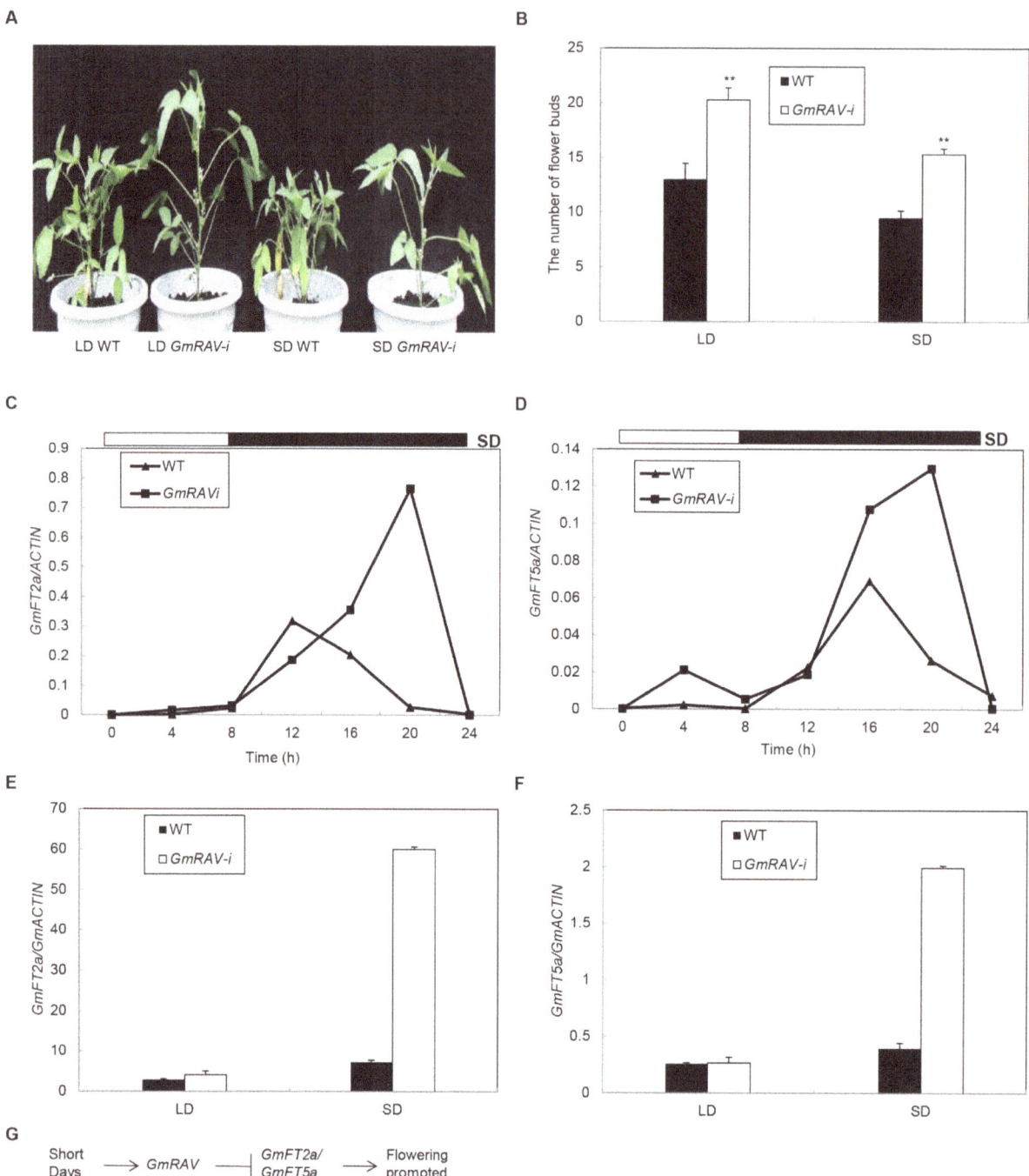

Figure 2. A, Phenotypes of the T6 generation *GmRAV-i* soybean under LDs and SDs. 50-day-old seedlings of WT and *GmRAV-i* transgenic soybean under LDs and SDs at Harbin (planted at May 28). B, The compare of the number of flower buds between WT plants and *GmRAV-i* soybeans under LDs and SDs. The flower buds of 50 plants were measured for each treatment. Error bars represent the SE. **Significant differences in comparison to the non-transgenic lines at P<0.01 (Student's t test). C and D, Diurnal expression of soybean FT homologs: *GmFT2a* and *GmFT5a* in *GmRAV-i* soybeans grown under SDs (8 h/16 h light/dark). Trifoliate leaves were sampled every 4 h at 15 DAE. White and black bars at the top represent light and dark phases, respectively. Samples were processed and analyzed by RT-PCR as described in Experimental procedures. The levels of *GmACTIN* expression were used as a normalization control, respectively. Average and SE values for three replications are given for each data point. E and F, Relative transcript levels of *GmFT2a* and *GmFT5a* mRNA in *GmRAV-i* soybean leaves under SDs and LDs. Soybean leaves were harvested at 4 h before dawn at 25-day-old under LDs and SDs. G, Pathway controlling flowering in response to short days in soybean.

Table 3. Comparison of growth parameters of transgenic T6 *GmRAV-i* soybean plants and wild type (WT) that were under LDs and SDs at Harbin.

		LD WT	LD *GmRAV-i*	SD WT	SD *GmRAV-i*
44 day	Plant height (cm)	25.72	29.62**	20.32	27.63**
	Flower bud number	12.97	20.26**	8.43	16.33**
Maturity stage	Plant height (cm)	65.73	71.13**	53.15	68.23**
	Internode number	19.00	20.33**	15.60	16.00**
	Branch number	4.80	3.67*	4.30	3.33*
	Pod number per plant	85.93	99.33	51.15	80.67
	Seed number per plant	111.47	161.60*	67.34	141.67*
	Seed weight per 100 (g)	6.19	6.47	6.00	6.34

*Differences in comparison to the wild type at $0.01 < P < 0.05$ (Student's t test),
**Significant differences in comparison to the wild type at $P < 0.01$ (Student's t test).

Results

The Accumulation of GmRAV Transcript is Regulated by Photoperiod and Darkness

Transcription profiles of *GmRAV* were analyzed in various tissues of 'Dong Nong 42' soybean grown under inductive SDs at 12 h after dawn by quantitative real-time RT-PCR. *GmRAV* mRNA was present in all organs examined, including leaf (Tl), trifoliate leaf (TL), stem (St), root (RO), pod (PO), flower bud (FB), and immature seed (IS). In SDs, the mRNA abundance of *GmRAV* was the highest in trifoliate leaves and the lowest in pods (Fig. 1A).

The diurnal circadian rhythm of *GmRAV* expression was examined by quantitative real-time RT-PCR in trifoliate leaves sampled at 25 d (transferred at 10 d). *GmRAV* transcript exhibited a diurnal circadian rhythm under SDs, suggesting that their expression was partly regulated by circadian clock genes. The expression level of *GmRAV* increased slightly 4 h after dawn, reaching a peak 4 h after the beginning of the dark period and decreased toward dawn, reaching the lowest 4 h before dawn under SDs (Fig. 1B). The amplitude of *GmRAV* mRNA increased significantly under SDs when compared with LDs.

The abundance of the *GmRAV* mRNA in leaves during the shift from SDs to LDs was investigated (Fig. 1B). The time course-dependent expression patterns of *GmRAV* were also analyzed in 'Dong Nong 47' plants grown under SDs and LDs using RNAs isolated from trifoliate leaves that were sampled at 4 h after dusk. The levels of *GmRAV* transcripts under SDs were relatively low at 20 DAE but increased sharply to their maximum levels at 23 DAE and thereafter decreased until 45 DAE (the time of flower bud formation) (Fig. 1C). In contrast, under LDs, the transcript levels of *GmRAV* increased slightly at 23 DAE, and thereafter decreased showing lower levels than that under SDs at all the times (Fig. 1C). Overall, the gene expression studies indicated that *GmRAV* was SD-inducible gene in soybean leaves.

In addition, promoter activity of *GmRAV* was measured to determine whether continuous darkness could increase *GmRAV* expression like in SDs (Fig. 1B). In 7-day-old seedlings, *GUS* expression of *GmRAV* promoter via histochemical GUS assay was more predominantly detected under continuous darkness than in continuous light conditions in both cotyledons and hypocotyls (Fig. 1D). These results suggested that *GmRAV* was SD and darkness-inducible gene.

Effects of GmRAV on Photoperiod Controlling of Flowering Time FT Homologs in GmRAV-i Transgenic Soybean

The transcript abundance of *GmRAV* was affected by day length in soybean leaves, the diurnal phase of *GmRAV* mRNA expression was regulated by the circadian clock, and higher levels of *GmRAV* mRNA were accumulated under SDs than under LDs. To examine whether *GmRAV* acted in a photoperiod functional context, the flowering time in WT and *GmRAV-i* soybeans under SDs and LDs were analyzed. The early-flowering phenotype soybean mutant *GmRAV-i* was observed despite of the day length in both LDs and SDs compared to the WT (Fig. 2A). Apart from their flowering-time phenotype, *GmRAV-i* soybeans also displayed complex pleiotropic alterations of vegetative development. Earlier emergence, reduced numbers of branches and leaves, longer petioles, larger leaves, increased apical dominance, and earlier flowering and maturity were the major phenotypic effects of *GmRAV-i* in all T6 generation plants in both SDs and LDs (Fig. 2A, Table 3).

In Arabidopsis, *FT* transcript levels oscillated with distinct circadian rhythms (Suárez-López et al., 2001). To check *FT* transcription levels in soybeans, the diurnal circadian rhythm of *FT* gene expression was analyzed by quantitative real-time RT-PCR for *GmFT2a* and *GmFT5a* in trifoliate leaves sampled at 15 d after emergence (DAE) in *GmRAV-i* soybean. The expression level of both *GmFT2a* and *GmFT5a* reached a peak 4 h after the beginning of the dark period and decreased toward dawn in *GmRAV-i* soybean plants under SDs. The amplitude and overall level of *GmFT2a* and *GmFT5a* mRNA were much higher in *GmRAV-i* soybean plants than in WT plants under SDs (Fig. 2C, D). Moreover, the transcript abundance of *GmFT2a* and *GmFT5a* was highly affected in *GmRAV-i* transgenic soybean leaves compared to the wild-type seedlings under SDs than under LDs (Fig. 2E, F). The results indicated that *GmRAV* was a SD-inducible flowering repressor in the flowering response of SD-induced soybeans by repressing positive regulator *GmFT2a* and *GmFT5a* gene expression. This work therefore described the conservation of components and sequence order of a pathway controlling flowering in response to day length. It revealed that the promotion of flowering in short days in *GmRAV-i* soybean resulted from the repression of *GmFT2a* and *GmFT5a* by *GmRAV* (Fig. 2G).

A

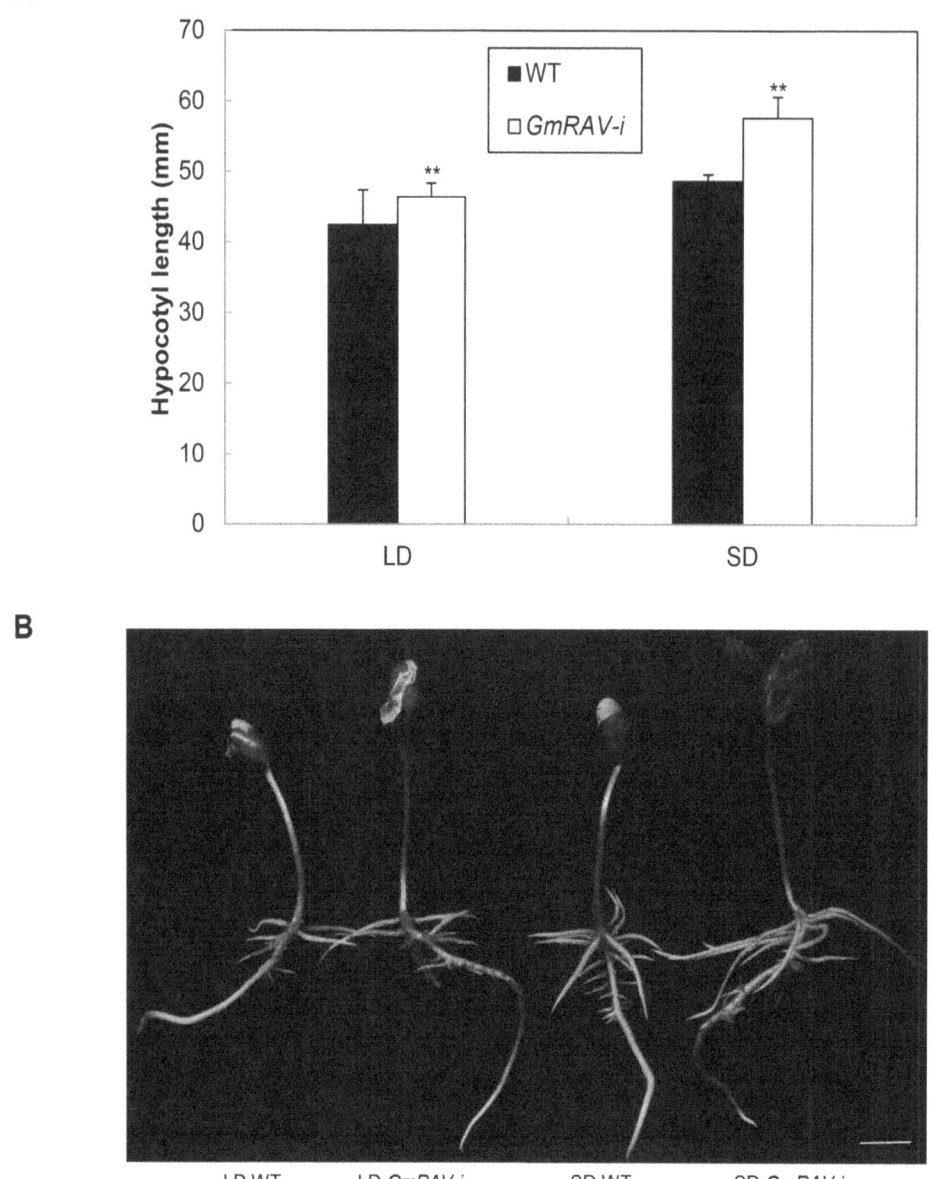

B

LD WT   LD *GmRAV-i*   SD WT   SD *GmRAV-i*

Figure 3. Effects of day length on hypocotyl length in 9-day-old wild-type and *GmRAV-i* soybean seedlings under LDs and SDs. A, Histograms of the mean (*n* = 20) for seedlings grown on medium. All seedlings were transgenic for the soybeans indicated. The seedlings were scored 9 d after sowing. Scale bar = 10 mm. **Significant differences in comparison to the non-transgenic lines at P<0.01 (Student's t test). B, Representative seedlings are shown.

Effects of GmRAV on the Photoperiod Controlling by Hypocotyl Elongation

To further investigate whether *GmRAV* was SD-inducible, the test of hypocotyl elongation was conducted in LD and SD-grown knock-out soybean mutant *GmRAV-i*, and *Arabidopsis rav* mutants and wild-type plants. *GmRAV-i* soybeans showed SD -mediated hypocotyl elongation responses compared to the wild-type seedlings (Fig. 3A, B). In *GmRAV-i* soybeans, *GmRAV* enhanced SD-mediated hypocotyl elongation response. In comparison, *Arabidopsis rav* mutant displayed the same hypocotyl elongation response to SDs (Fig. 4A). This was evident when grown in LDs or SDs (Fig. 4B, C). These results indicated that *GmRAV* also played a negative role in SD-mediated regulation of hypocotyl elongation.

Effects of Sucrose on the Flowering of Arabidopsis Rav Mutant

Masa-aki et al. reported that 2 weeks culture was enough to observe the negative effects of high levels of sucrose on floral transition [24]. Therefore, wild-type plants and *Arabidopsis rav* mutant seedlings were grown in culture medium containing 2% (w/v) and 4% (w/v) sucrose for 2 weeks, respectively and then were transferred to soil. Flowering time of Arabidopsis WT and *rav* mutants were examined on MS medium containing 4% sucrose in LDs. In comparison to the plants that were grown on 2% sucrose plates, *rav* mutants showed a 2-day delay in flowering time, whereas WT plants showed 4-day delay (Fig. 5A). These results supported our hypothesis that Arabidopsis *rav* mutants responded

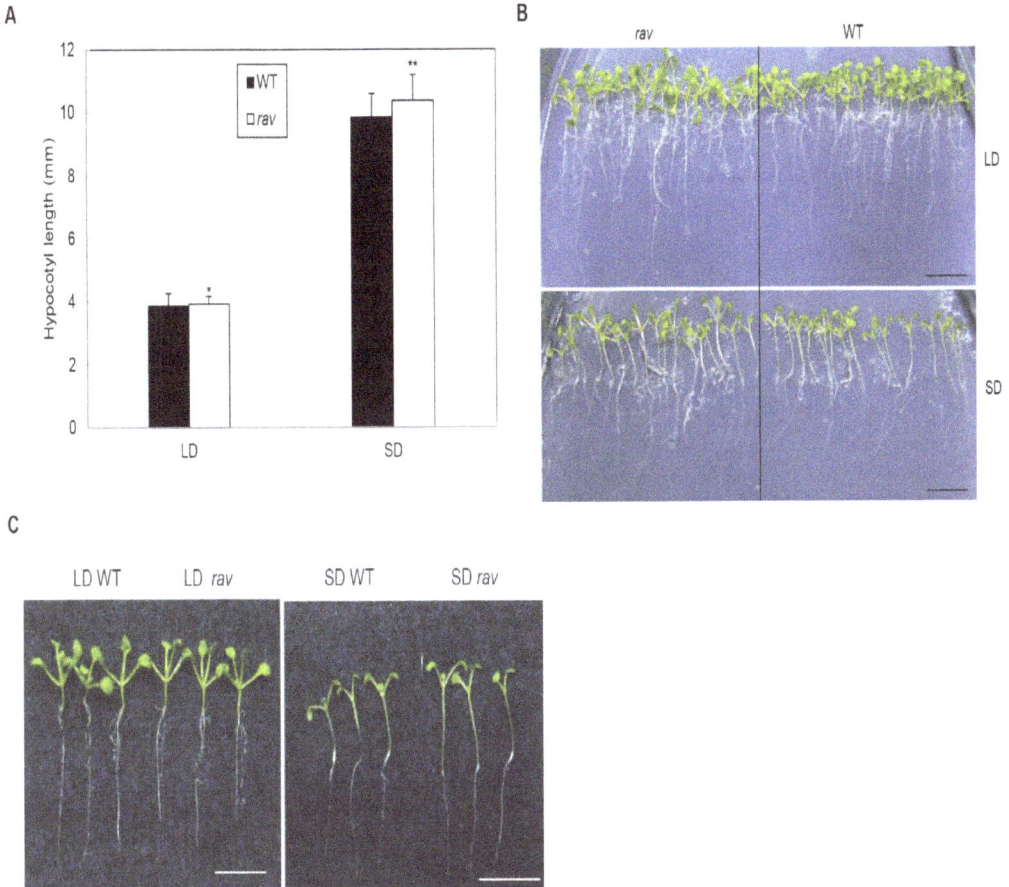

Figure 4. Effects of day length on hypocotyl length in 9-day-old wild-type and Arabidopsis *rav* mutants under LDs and SDs. A, Histograms of the mean (*n* = 20) for seedlings grown on medium. The seedlings were scored 9 d after sowing. Scale bar = 10 mm. *differences in comparison to the non-transgenic lines at 0.01<P<0.05, **Significant differences in comparison to the non-transgenic lines at P<0.01 (Student's t test). B, Phenotype of 9-day-old WT seedlings and Arabidopsis *rav* mutants on MS medium under LDs and SDs. C, Representative seedlings are shown.

to sucrose signaling in plant growth and development. Sucrose affecting flowering time was also observed in the developmental phenotypes (Fig. 5B). Overall, these results indicated that *AtRAV* participated in the regulation of high levels of sucrose-dependent flowering response.

To investigate the reason why high concentration sucrose could delay flowering, different levels of *FT*, *SOC1/AGL20* and *LEAFY* (*LFY*) expressions were analyzed by reverse transcriptase (RT)-PCR in Arabidopsis WT plants and *rav* mutants grown respectively on media with 2% and 4% sucrose for 15 d. Both expression levels of *FT* and *SOC1/AGL20* in Arabidopsis *rav* mutants as well as WT plants on media with 2% (w/v) were identical with on media with 4% (w/v) sucrose under LD conditions (data not shown). *LFY* was expressed in the leaf primordium before the transition to flowering. In this study, *LFY* expression levels were reduced under the supplementary of 4% (w/v) sucrose compared to 2% (w/v) sucrose (Fig. 5C). The results suggested that increased concentration of sucrose could lead to decrease the expression of *LFY* gene. It also showed that high concentration of sucrose in growth media delayed the flowering time in Arabidopsis. Moreover, transcript level of *LFY* was greatly reduced in *rav* mutants than in WT plants on media containing high level of sucrose under LDs. The results displayed that *GmRAV* delayed

flowering in high level of sucrose by regulating the expression of *LFY*.

The Responses of Arabidopsis Rav Mutant, GmRAV-ox Tobacco and GmRAV-i Soybean Seedlings to Exogenous Sucrose

To analyze the possible function of *GmRAV* in response to sucrose stress, we studied the responses to exogenous sucrose application using Arabidopsis *rav*, *GmRAV-ox* tobacco, and *GmRAV-i* soybean plants. Sensitivity of *Arabidopsis rav* mutants and *GmRAV-ox* tobacco seedlings in response to sucrose was tested in root and hypocotyl by growth inhibition assays. The hypocotyl and root lengths of *GmRAV-ox* tobacco and WT plants were inhibited by both concentrations of sucrose (2% and 4%) (Fig. 6A), but *GmRAV-ox* tobacco were more remarkably reduced compared to WT plants (Fig. 6B, C). These results suggested that *GmRAV-ox* tobacco were more sensitive to sucrose than WT plants in the assays of hypocotyls and roots growth inhibition. Fig. 7A showed that the hypocotyl and root lengths of Arabidopsis *rav* mutants and WT plants were inhibited by both concentrations of sucrose. However, inhibited extent of *rav* mutants and WT plants by sucrose in hypocotyl and root inhibition assays was identical (Fig. 7B, C). Likewise, the growth in soybean *GmRAV-i* and WT plants in terms of hypocotyl lengths, main root length and the

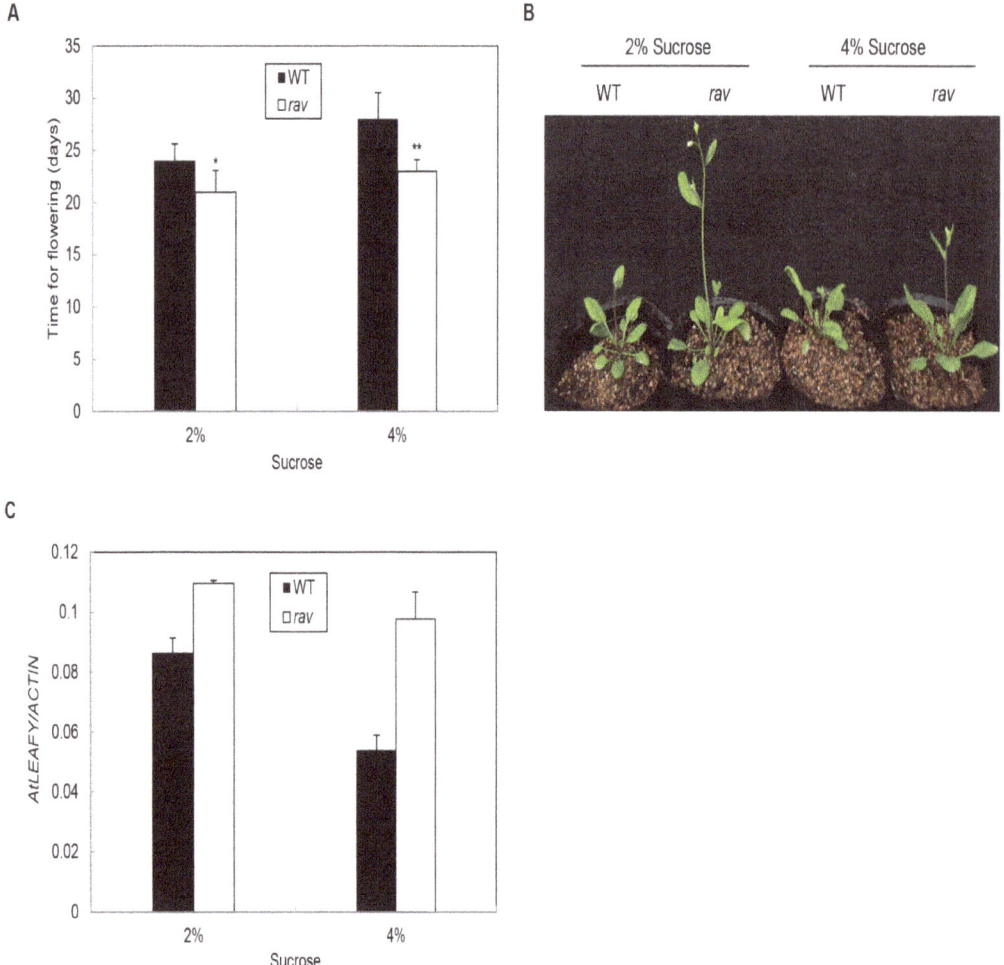

Figure 5. Effects of sucrose on flowering time for Arabidopsis. A, Arabidopsis *rav* mutants and WT seedlings were grown on media with various concentrations of sucrose for 2 weeks, and then transferred to soil under LDs. The flowering time of seedlings with 2% sucrose was WT (24±0.5) and Arabidopsis *rav* mutant (21±0.8). WT plants are the control for Arabidopsis *rav* mutants. Values are the average of 30 to 45 plants. The error bars indicate one SE of the mean. Similar results were obtained in two independent experiments. *Differences in comparison to the non-transgenic lines at 0.01<P<0.05, **Significant differences in comparison to the non-transgenic lines at P<0.01 (Student's t test). B, Phenotypes of the Arabidopsis *rav* mutant. 23-day old seedlings of WT and Arabidopsis *rav* mutant under natural day length (LD) with treated by 2% and 4% sucrose. C, Quantitative real-time RT–PCR analysis of *LFY* expression in Arabidopsis *rav* mutants. Control amplification of *18 s rRNA* transcript indicated equal amounts of cDNA.

number of lateral roots was all inhibited by both concentrations of sucrose (Fig. 8A). Hypocotyl, root length, and the number of lateral roots of *GmRAV-i* soybeans were less inhibited by sucrose than in WT plants (Fig. 8 B–D). Therefore, *GmRAV-i* seedlings exhibited insensitive phenotypes, as compared with the WT plant, during sucrose-mediated root and hypocotyl growth inhibition. Overall, in agreement with the *GmRAV-ox* tobacco results, the Arabidopsis *rav* mutants and the *GmRAV-i* soybean seedlings showed a significantly decreased sensitivity to sucrose by root and hypocotyl growth inhibition assays.

The addition of chemicals to a medium changes both the chemical composition and the osmotic potential of medium. To identify whether the negative effects of sucrose on development and flowering were due to metabolic or osmotic factors, we examined the effects of both mannitol and sorbitol. These two sugar alcohols were widely used as the osmotic controls in plant. Therefore, a combined osmotic control value was calculated from both mannitol- and sorbitol-grown plants. Sugar effects were compared to the combined osmotic control value. Plants grown in

the absence of any supplemental carbon (sugar alcohol or sucrose) were also measured for each parameter in order to observe the overall effect of increasing osmolarity. Increasing osmolarity of the growth media by sugar alcohols inhibited the plant growth when compared to untreated media. The behavior of Arabidopsis *rav* mutant, *GmRAV-ox* tobacco and *GmRAV-i* soybean seedlings with 200 mM sugar alcohols on MS media showed no difference from that of the WT seedlings in hypocotyl, root and the number of lateral roots growth inhibition assay (Fig. 9 A–E). However, *rav* mutants and soybean *GmRAV-i* in terms of hypocotyl lengths, main root length and the number of lateral roots of development that exhibited insensitive phenotypes were less inhibited by 200 mM sugars (6.8%) (Fig. 9A, C–E), which indicated *RAV* gene played a negative role in sucrose-mediated regulation of hypocotyl and root elongation due to metabolic effects rather than osmotic effects.

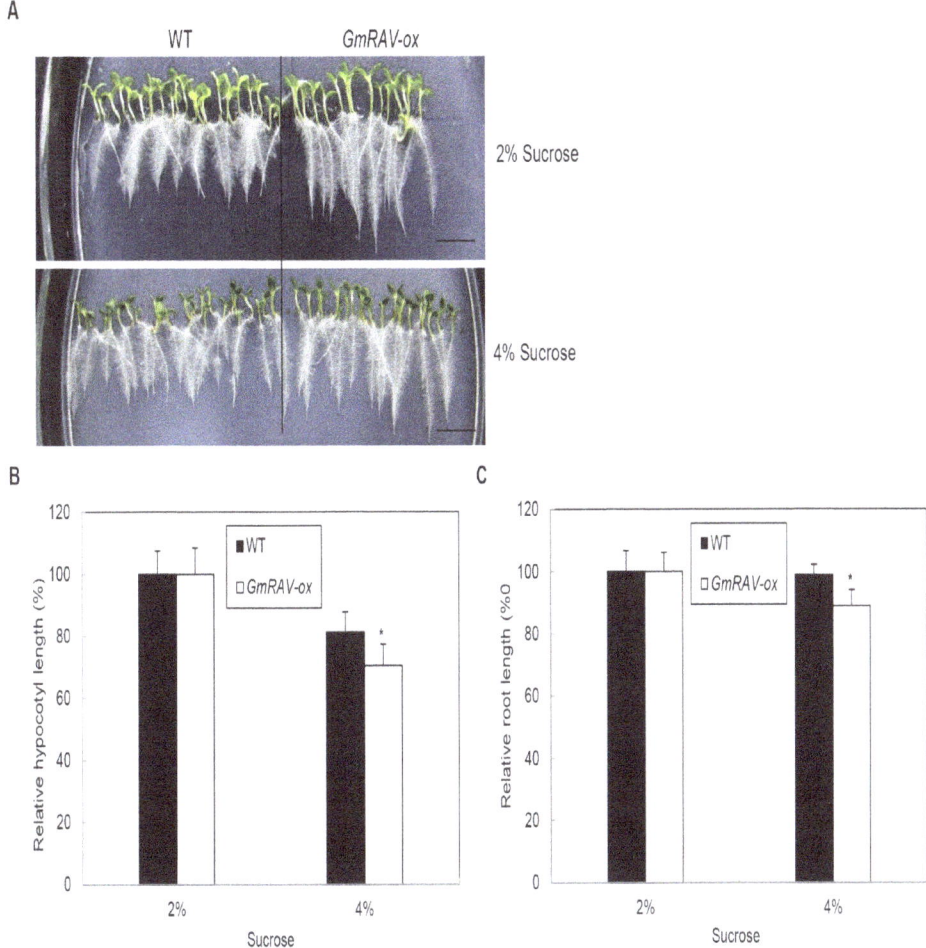

Figure 6. Response of wild-type and *GmRAV-ox* tobacco seedling to sucrose. A and B, Relative hypocotyl and root growth in response to various concentrations of sucrose. The length of hypocotyl of tobacco seedlings grown 2% (w/v) sucrose was 4.91±0.75 mm for the WT and 5.85±0.76 mm for *GmRAV-ox*. Root length, of seedlings grown 2% (w/v) sucrose was 15.8±1.21 mm in the WT and 18.8±1.94 mm in *GmRAV-ox*. The hypocotyl and root length of 20–30 seedlings were measured for each treatment. Error bars represent the SE. C, Phenotype of 7-day-old WT seedlings and T3 generation *GmRAV-ox* tobaccos on MS medium containing 2% and 4% sucrose. *Differences in comparison to the non-transgenic lines at 0.01<P<0.05 (Student's t test).

Discussion

GmRAV is a Novel Negative Regulator of SD-mediated Flowering and Hypocotyl Elongation

The time of flowering induction determines to a large extent the reproductive success of plants. Plants integrate diverse environmental and endogenous signals to ensure the timely transition from vegetative to flowering period. In many plant species, floral transition is strongly controlled by the circadian clock. The clock with a period close to 24 h serves to coordinate diurnal rhythms with physiology and behavior. GmRAV belongs to the RAV protein family containing two domains: the AP2 and the B3 DNA-binding domain. Given that long-day plant Arabidopsis *TEM1*, *TEM2* and chestnut *CsRAV1* genes were circadian regulated [35,36], we examined the possibility that soybean gene was rhythmically expressed in short-day plant soybean leaves. Higher level of *GmRAV* transcripts was accumulated in soybean leaves in SDs than in LDs which was almost suppressed, and it was also regulated by the circadian clock. The *GmRAV* mRNA reached a peak 4 h after the beginning of the dark period in SDs, whereas *CsRAV1* mRNA peaked at noon, *TEM1* and *TEM2* peaked at dusk

in LDs. This different time of expression in day suggests that although *RAV* gene shows high homology among different day-length plants, it may play different roles in different day lengths. Furthermore, the time course-dependent expression pattern of *GmRAV* was analyzed in 'Dong Nong 47' plants grown under SDs and LDs, and showed that the level of *GmRAV* transcript in SDs was higher than in LDs till flowering, confirming that *GmRAV* was SD-inducible gene.

We further analyzed the function of repressing flowering of *GmRAV* gene based on the earlier flowering phenotypes of *GmRAV-i* soybeans than WT under both SDs and LDs. In Arabidopsis, *TEM1 and TEM2* also acted as direct *FT* repressors and repressed flowering under LD and SD conditions [35,37]. Likewisely, the mRNA levels of *GmFT2a* and *GmFT5a* were also examined in WT and *GmRAV-i* soybeans under SDs, and they were evidently enhanced in earlier flowering *GmRAV-i* soybeans, which indicated that *GmRAV* played a negative role by repressing *FT* genes in the SD-mediated photoperiod control of flowering in soybean, whereas in Arabidopsis, *TEM1* repressed flowering by repressing *FT* genes in LDs [35]. We speculated that a genetic pathway

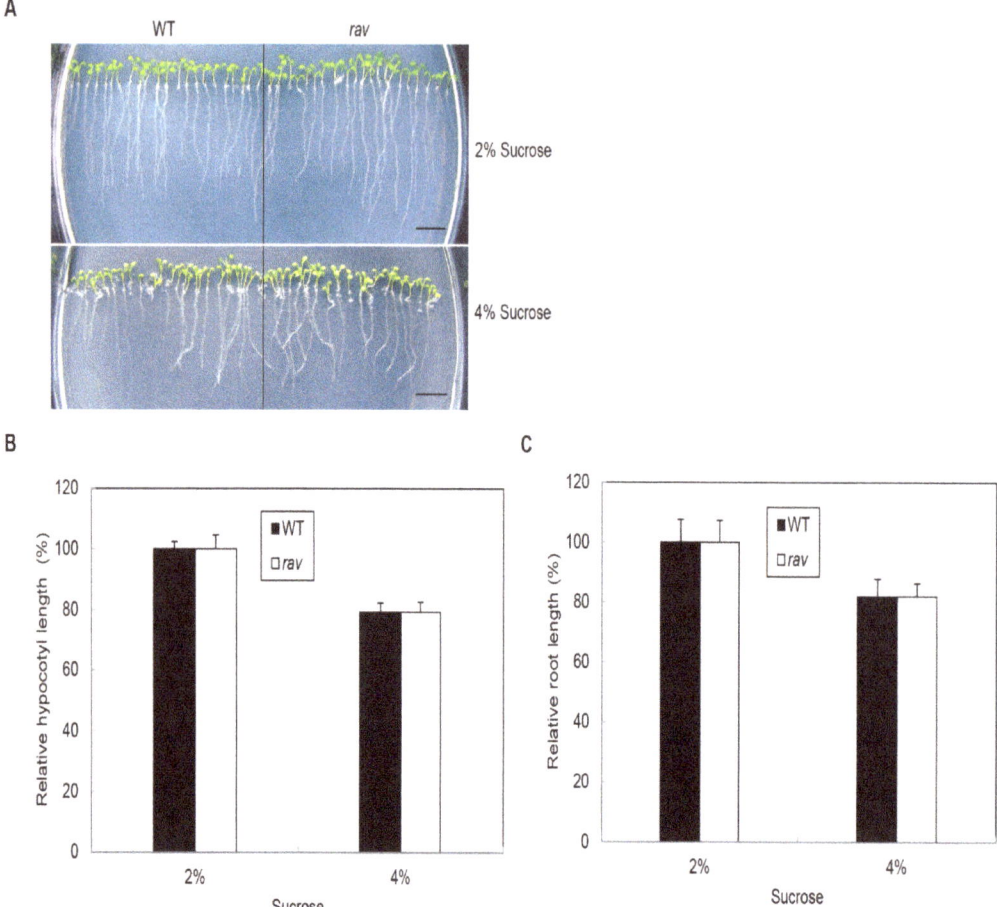

Figure 7. Response of WT and Arabidopsis *rav* mutants to sucrose on development. A and B, Relative hypocotyl and root growth in response to various concentrations of sucrose. Hypocotyl length, as a percentage of the untreated control, of seedlings grown on sucrose. The length of hypocotyl of Arabidopsis seedlings grown 2% (w/v) sucrose was 4.26±0.24 mm for the WT and 4.06±0.46 mm for Arabidopsis *rav* mutant. Root length, as a percentage of the untreated control, of seedlings grown on sucrose. Root length, of seedlings grown without sucrose was 26.70±1.16 mm in the WT and 28.24±1.33 mm in Arabidopsis *rav* mutant. The hypocotyl and root length of 20–30 seedlings were measured for each treatment. Error bars represent the SE. C, Phenotype of 7-day-old WT seedlings and Arabidopsis *rav* mutants on MS medium containing 2% and 4% sucrose.

similar to that in Arabidopsis was conserved in the photoperiod control of flowering in soybean, a SD plant.

Furthermore, in photoperiod control of hypocotyl elongation assays, both *GmRAV-i* soybeans and Arabidopsis *rav* mutants were hypersensitive in SD-mediated promotion of hypocotyl elongation. Therefore, we concluded that the *RAV* role was also conserved in the photoperiod control of hypocotyl elongation in soybean and Arabidopsis.

GmRAV Affected Development and Flowering Time in the Presence of Exogenous Sucrose

Carbohydrates are thought to play a crucial role in the regulation of flowering. The relation between sugar metabolism/signaling and floral transition received extensive attention lately [25]. Sugar signaling was of great importance in flowering time control, which directly affected yield [38,39]. The work of Heyer et al. already provided clear evidence that flowering time control is strongly influenced by modifying sugar balances in the apex [40]. Several sucrose signaling insensitive Arabidopsis mutants have been identified based on the effect of high levels of external sugars on seedling growth and development such as *cai* (carbohydrate

insensitive) [41], *isi* (impaired sugar induction) [42], *lba* (low levels of β-amylase) [43], *rsr* (reduced sugar response) [44], *sis* (sugar insensitive) [45], *sun* (sucrose uncoupled) mutants [46]. For example, *sig* (sucrose insensitive growth) mutant was selected on media containing 350 mM sucrose [47]. Similarly, Arabidopsis *rav* mutants and soybean *GmRAV-i* in terms of hypocotyl lengths, main root length and the number of lateral roots of development which exhibited insensitive phenotypes were less inhibited by 200 mM sugars.

To ensure uniform sugar responses, the initial mutant isolation screens were mostly performed on media containing high sucrose concentrations, raising concerns about the physiological relevance of sugar regulation. Moreover, the phenotypes could also be influenced by osmotic stress in the media. In this study, we treated Arabidopsis *rav* mutants, *GmRAV-ox* tobaccos and *GmRAV-i* soybeans with 200 mM supplemental carbon sources, and observed that transgenic plants and WT were almost uniformly inhibited by only supplemental sugar alcohols. A high concentration of sucrose added into basal medium also inhibited development both WT and transgenic plants. But we found the behavior of Arabidopsis *rav* mutant, *GmRAV-ox* tobacco and *GmRAV-i* soybean seedlings with 200 mM sugar alcohols on MS media

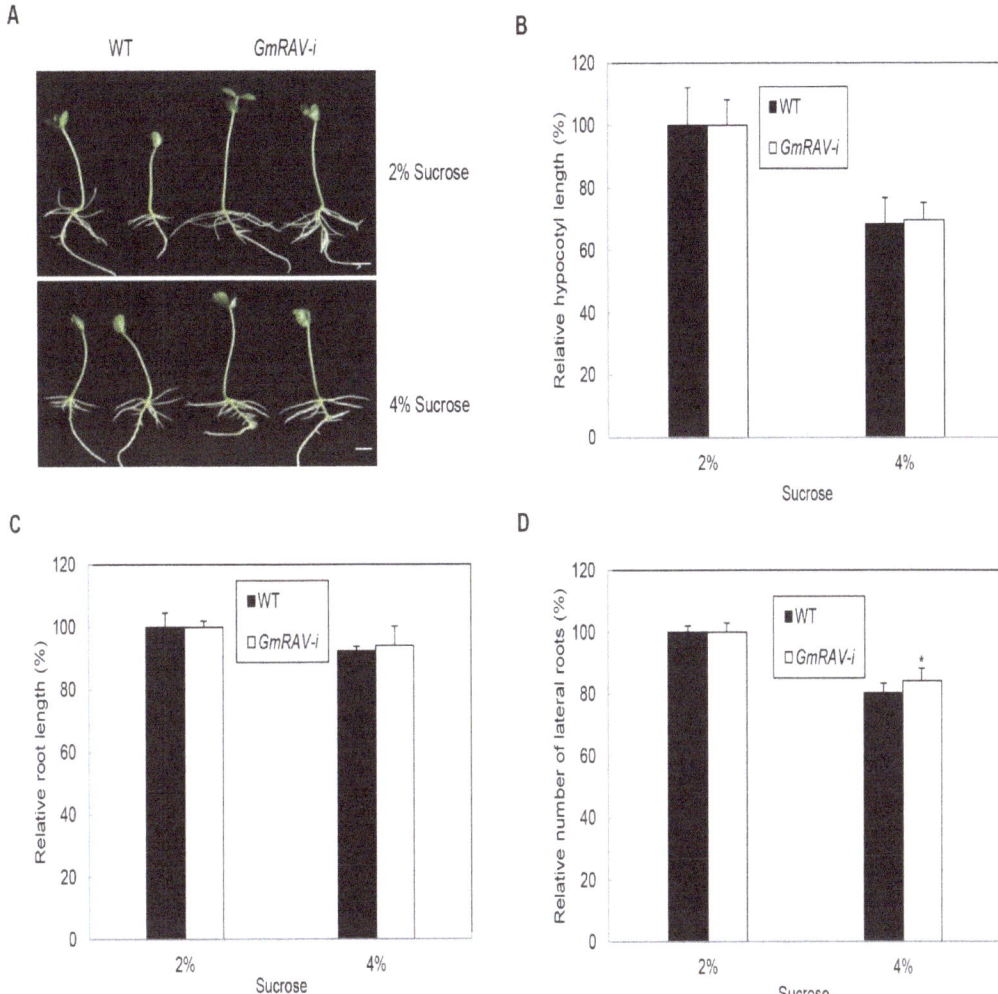

Figure 8. Response of WT and *GmRAV-i* soybean seedlings to sucrose. A and B, Relative hypocotyl and root growth in response to various concentrations of sucrose. Hypocotyl length, as a percentage of the untreated control, of seedlings grown on sucrose. The length of hypocotyl of soybean seedlings grown without sucrose or glucose was 57.17±1.30 mm for the WT and 61.17±1.97 mm for *GmRAV-i* soybean. Root length, as a percentage of the untreated control, of seedlings grown on sucrose. Root length, of seedlings grown without sucrose or glucose was 60.33±1.30 mm in the WT and 65.67±1.97 mm in *GmRAV-i*. Error bars represent the SE. C, Relative number of lateral roots in response to various concentrations of sucrose. The number of lateral roots, as a percentage of the untreated control, of seedlings grown on sucrose. Error bars represent the SE. The number of lateral roots of seedlings grown without sucrose and glucose was 17.83±0.75 for the WT and 21.17±0.98 for *GmRAV-i* soybean. The number of lateral roots of 20–30 seedlings was scored for each treatment. *Differences in comparison to the non-transgenic lines at 0.01<P<0.05. D, Phenotypes of 7-day-old WT and *GmRAV-i* soybean seedlings on MS medium containing 2% and 4% sucrose. *Differences in comparison to the non-transgenic lines at 0.01<P<0.05 (Student's t test).

showed no difference from that of the WT seedlings in hypocotyl, root and the number of lateral roots growth. Thus, the specific effects of sucrose treatment could be attributed to the chemistry of the sucrose itself rather than to osmotic effects of the sucrose, which indicated *RAV* gene played a negative role in sucrose signaling due to metabolic effects rather than osmotic effects.

In Arabidopsis, the delay in flowering time caused by high concentrations of glucose in media was previously reported [23]. Sugar seemed to affect a specific part of the vegetative phase, rather than all phases [24]. Flowering time of Arabidopsis *rav* mutants were less reduced than WT under high concentration of 4% sucrose condition, which showed Arabidopsis *rav* mutants were also insensitive to sucrose in flowering time. Moreover, the expression of *LFY* was much more down-regulated in *Arabidopsis* WT plants compared to *rav* cultivated in high concentrations of

sucrose, indicating that *RAV* might be also a positive regulated factor in flowering time inhibition assays.

In summary, as an important process in plant reproduction, flowering time is finely controlled by complex network. *GmRAV* was both involved in negative regulation of the photoperiodic control of flowering time responses and positive regulation of sucrose control of flowering time.

Acknowledgments

We are grateful to Dr. Xinan Zhou (Oilcrops research institute, Chinese academy of agricultural sciences) for his contributions for invaluable experimental technical assistance.

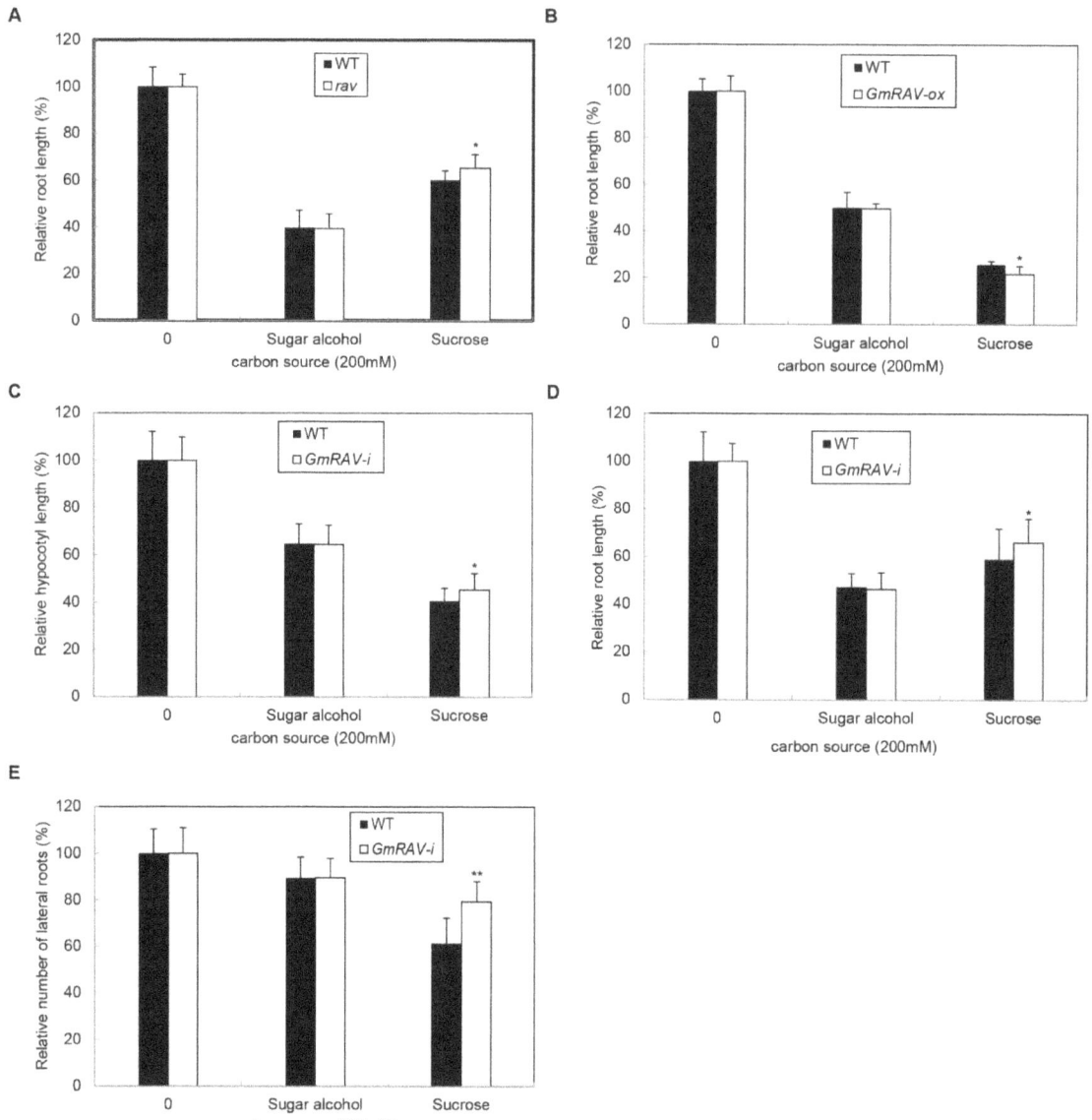

Figure 9. Response of the WT plants, Arabidopsis *rav* mutants and *GmRAV-ox* tobaccos or *GmRAV-i* soybean seedlings to full strength MS media supplemented with the indicated sugar concentration. A and B, Relative root length in response to 200 mM carbon source in 12-day-old Arabidopsis and tobaccos. Root length, as a percentage of the untreated control, of seedlings grown on carbon source. Error bars represent the SE. Root length, of seedlings grown without carbon sucrose was WT (15.96±1.2 mm) and Arabidopsis *rav* mutant (19.25±0.8 mm), and for tobaccos without carbon sucrose was WT (15.29±0.9 mm) and *GmRAV-ox* (18.4±1.2 mm). Root length of 20–30 seedlings was measured for each treatment. C and D, Relative hypocotyl and root length in response to 200 mM carbon source in 8-day-old soybeans. Hypocotyl and root length, as a percentage of the untreated control, of seedlings grown on carbon source. Error bars represent the SE. Hypocotyl length without carbon source was WT (50.67±1.2 mm) and *GmRAV-i* (54.83±0.9 mm), and main root length was WT (67±1.2 mm) and *GmRAV-i* (79.83±0.7 mm). E, Relative number of lateral roots in response to 200 mM carbon source. Number of lateral roots, as a percentage of the untreated control, of seedlings grown on carbon source. Error bars represent the SE. The number of lateral roots, of seedlings grown without carbon source was WT (16.33±0.5) and *GmRAV-i* (21±1.0). Number of lateral roots of 20–30 seedlings was scored for each treatment. *differences in comparison to the non-transgenic lines at 0.01<P<0.05, **Significant differences in comparison to the non-transgenic lines at P<0.01 (Student's t test).

Author Contributions

Conceived and designed the experiments: WL LZ. Performed the experiments: QL DH. Analyzed the data: QL DL. Contributed reagents/materials/analysis tools: DL YZ. Wrote the paper: WL LZ.

References

1. Coupland G (1995) Genetic and environmental control of flowering time in Arabidopsis. Trends Genet 11: 393–397.

2. Amasino RM (1996) Control of flowering time in plants. Curr Opin Genet Devel 6: 480–487.

3. Koornneef M, Alonso-Blanco C, Vries HB, Hanhart CJ, Peeters AJM (1998) Genetic interactions among late-flowering mutants of Arabidopsis. Genetics 148: 885–892.
4. Koornneef M, Alonso-Blanco C, Peeters AJM, Soppe W (1998) Genetic control of flowering time in Arabidopsis. Annu. Rev. Plant Physiol. Plant Mol Biol 49: 345–370.
5. Piñeiro M, Coupland G (1998) The control of flowering time and floral identity in Arabidopsis. Plant Physiol 117: 1–8.
6. Simpson GG, Dean C (2002) Arabidopsis, the Rosetta stone of flowering time. Science 296: 285–289.
7. Zhang L, Wang R, Hesketh JD (2001) Effects of photoperiod on growth and development of soybean floral bud in different maturity. Agron J 63: 944–948.
8. Hadley P, Roberts EH, Summerfield RJ, Minchin FR (1984) Effects of temperature and photoperiod on flowering in soybean Glycine max (L.) Merrill.: a quantitative model. Ann Bot 53: 669–681.
9. Destro D, Carpentieri-Pipolo V, Kiihl RAS, Almeida LA (2001) Photoperiodism and Genetic Control of the Long Juvenile Period in Soybean: A Review. Crop Breed Appl Biotech 1: 72–79.
10. Komeda Y (2004) Genetic regulation of time to flower in Arabidopsis thaliana. Annu Rev Plant Biol 55: 521–535.
11. Hiraoka K, Daimon Y, Araki T (2008) FT protein: a universal long distance mobile signal in seed plants? Plant Morphol 19: 3–13. (in Japanese with English abstract).
12. Hiroyuki T, Shojiro T, Reina K, Shimamoto K (2008) Florigen and the Photoperiodic Control of Flowering in Rice. RICE 1: 25–35.
13. Kong F, Liu B, Xia Z, Sato S, Kim BM, et al. (2010) Two coordinately regulated homologs of FLOWERINGLOCUS T are involved in the control of photoperiodic flowering in soybean. Plant Physiol 154: 1220–1231.
14. Sheen J (1994) Feedback control of gene expression. Photosynth Res 39: 427–438.
15. Dangl JL, Preuss D, Schroeder JL (1995) Talking through walls: signaling in plant development. Cell 83: 1071–1077.
16. Sheen J, Zhou L, Jang JC (1999) Sugars as signaling molecules. Curr Opin Plant Biol 2: 410–418.
17. Rollard F, Windeerikx J, Thevelein JM (2001) Glucose-sensing mechanisms in eukaryotic cells. Trends Biochem Sci 26: 310–317.
18. Sheen J (1990) Metabolic repression of transcription in higher plants. Plant Cell 2: 1027–1038.
19. Chen M, Liu L, Chen Y, Wu H, Yu S (1994) Expression of alphamylases, carbohydrate metabolism, and autophagy in cultured rice cells is coordinately regulated by sugar nutrient. Plant J 6: 625–636.
20. Bernier G, Havelange A, Houssa C, Petitjean A, Lejeune P (1993) Physiological signals that induce flowering. Plant Cell 5: 1147–1155.
21. Corbesier L, Lejeune P, Bernier G (1998) The role of carbohydrates in the induction of flowering in Arabidopsis thaliana: comparison between the wild-type and a starchless mutant. Planta 206: 131–137.
22. Rolda'n M, Go'mez-Mena C, Ruiz-Garc'a L, Salinas J, Martı'nez-Zapater JM (1999) Sucrose availability on the aerial part of the plant promotes dark-morphogenesis and flowering in Arabidopsis. Plant J 20: 581–590.
23. Zhou L, Jang JC, Jones TL, Sheen J (1998) Glucose and ethylene signal transduction crosstalk revealed by an Arabidopsis glucose-insensitive mutant. Proc Natl Acad Sci USA 95: 10294–10299.
24. Ohto M, Onai K, Furukawa Y, Aoki E, Araki T, et al. (2001) Effects of Sugar on Vegetative Development and Floral Transition in Arabidopsis. Plant Physiol 127: 252–261.
25. King RW (2012) Mobile signals in daylength-regulated flowering: gibberellins, flowering locus T, and sucrose. Russ. J. Plant Physiol. 59: 479–490.
26. Rolland F, Moore B, Sheen J (2002) Plant sugar sensing and signaling. Plant Cell 14: S185–S205.
27. York WS, Darvill AG, Albersheim P (1984) Inhibition of 2, 4- dichlorophenox-yacetic acid-stimulated elongation of pea stem segments by a xyloglucan oligosaccharide. Plant Physiol 75: 295–297.
28. Priem B, Gross KC (1992) Mannosyl- and xylosyl-containing glycans promote tomato (Lycopersicon esculentum Mill.) fruit ripening. Plant Physiol 98: 399–401.
29. Shibuya N, Minami E (2001) Oligosaccharide signaling for defense responses in plants. Physiol Mol Plant Pathol 59: 223–233.
30. Stevenson CC, Harrington GN (2009) The impact of supplemental carbon sources on Arabidopsis thaliana growth, chlorophyll content and anthocyanin accumulation. Plant Growth Regul 59: 255–271.
31. Zhao L, Luo Q, Yang C, Han Y, Li W (2008) A RAV-like transcription factor controls photosynthesis and senescence in soybean. Planta 227: 1389–1399.
32. Schmutz J, Cannon SB, Schlueter J, Ma J, Mitros T, et al. (2010) Genome sequence of the palaeopolyploid soybean. Nature 463: 178–183.
33. Zhao L, Hao D, Chen L, Lu Q, Zhang Y, et al. (2012) Roles for a soybean RAV-like orthologue in shoot regeneration and photoperiodicity inferred from transgenic plants. J. Exp.Bot. 63: 3257–3270.
34. Jefferson RA, Kavanagh TA, Bevan MW (1987) GUS fusions: β-glucuronidase as a sensitive and versatile gene fusion marker in higher plants. EMBO Journal 20: 3901–3907.
35. Castillejo C, Pelaz S (2008) The balance between CONSTANS and TEMPRANILLO activities determines FT expression to trigger flowering. Curr Biol 18: 1338–1343.
36. Moreno-Cortés A, Hernández-Verdeja T, Sánchez-Jiménez P, González-Melendi P, Aragoncillo C, et al. (2012) CsRAV1 induces sylleptic branching in hybrid poplar. New Phytologist, 194: 83–90.
37. Osnato M, Castillejo C, Matías-Hernández L, Pelaz S (2012) TEMPRANILLO genes link photoperiod and gibberellin pathways to control flowering in Arabidopsis. Nature Commun. DOI: 10.1038/ncomms1810.
38. Amasino RM (2010) Seasonal and developmental timing of flowering. Plant J. 61: 1001–1013.
39. Huang H, Yan P, Lascoux M, Ge X (2012) Flowering time and transcriptome variation in Capsella bursa-pastoris (Brassicaceae). New Phytol 194: 676–689.
40. Heyer AG, Raap M, Schroeer B, Marty B, Willmitzer L (2004) Cell wall invertase expression at the apical meristem alters floral, architectural, and reproductive traits in Arabidopsis thaliana. Plant J. 39: 161–169.
41. Boxall SF, Gissot L, Graham IA (1997) Arabidopsis thaliana mutants that are carbohydrate insensitive. Plant Physiol 114: S247.
42. Rook F, Bevan MW (2003) Genetic approaches to understanding sugar-response pathways. J. Exp Bot 54: 495–501.
43. Mita S, Murano N, Akaike M, Nakamura K (1997) Mutants of Arabidopsis thaliana with pleiotropic effects on the expression of the gene for beta-amylase and on the accumulation of anthocyanin that is inducible by sugars. Plant J. 11: 841–851.
44. Martin T, Hellmann H, Schmidt R, Willmitzer L, Frommer WB (1997) Identification of mutants in metabolically regulated gene expression. Plant J. 11: 53–62.
45. Laby RJ, Kincaid MS, Kim D, Gibson SI (2000) The Arabidopsis sugar-insensitive mutants sis4 and sis5 are defective in abscisic acid synthesis and response. Plant J. 23: 587–596.
46. Dijkwel PP, Huijser C, Weisbeek PJ, Chua NH, Smeekens SC (1997) Sucrose control of phytochrome a signaling in Arabidopsis. Plant Cell 9: 583–595.
47. Pego JV, Kortstee AJ, Huijser C, Smeekens SC (2000) Photosynthesis, sugars and the regulation of gene expression. J Exp Bot 51: 407–416.

5

Pirin1 (PRN1) Is a Multifunctional Protein that Regulates Quercetin, and Impacts Specific Light and UV Responses in the Seed-to-Seedling Transition of *Arabidopsis thaliana*

Danielle A. Orozco-Nunnelly[1], DurreShahwar Muhammad[1], Raquel Mezzich[1], Bao-Shiang Lee[2], Lasanthi Jayathilaka[2], Lon S. Kaufman[1], Katherine M. Warpeha[1]*

1 Molecular, Cell and Developmental Group, Department of Biological Sciences, Department of Biological Sciences, University of Illinois at Chicago (UIC), Chicago, Illinois, United States of America, **2** Protein Research Laboratory, University of Illinois at Chicago (UIC), Chicago, Illinois, United States of America

Abstract

Pirins are cupin-fold proteins, implicated in apoptosis and cellular stress in eukaryotic organisms. Pirin1 (PRN1) plays a role in seed germination and transcription of a light- and ABA-regulated gene under specific conditions in the model plant system *Arabidopsis thaliana*. Herein, we describe that PRN1 possesses previously unreported functions that can profoundly affect early growth, development, and stress responses. *In vitro*-translated PRN1 possesses quercetinase activity. When PRN1 was incubated with G-protein-α subunit (GPA1) in the inactive conformation (GDP-bound), quercetinase activity was observed. Quercetinase activity was not observed when PRN1 was incubated with GPA1 in the active form (GTP-bound). Dark-grown *prn1* mutant seedlings produced more quercetin after UV (317 nm) induction, compared to levels observed in wild type (WT) seedlings. *prn1* mutant seedlings survived a dose of high-energy UV (254 nm) radiation that killed WT seedlings. *prn1* mutant seedlings grown for 3 days in continuous white light display disoriented hypocotyl growth compared to WT, but hypocotyls of dark-grown *prn1* seedlings appeared like WT. *prn1* mutant seedlings transformed with GFP constructs containing the native *PRN1* promoter and full ORF (*PRN1*::PRN1-GFP) were restored to WT responses, in that they did not survive UV (254 nm), and there was no significant hypocotyl disorientation in response to white light. *prn1* mutants transformed with *PRN1*::PRN1-GFP were observed by confocal microscopy, where expression in the cotyledon epidermis was largely localized to the nucleus, adjacent to the nucleus, and diffuse and punctate expression occurred within some cells. WT seedlings transformed with the *35S*::PRN1-GFP construct exhibited widespread expression in the epidermis of the cotyledon, also with localization in the nucleus. PRN1 may play a critical role in cellular quercetin levels and influence light- or hormonal-directed early development.

Editor: Gloria Muday, Wake Forest University, United States of America

Funding: This work has been supported by National Science Foundation grant MCB-0848113 (NSF: http://www.nsf.gov/div/index.jsp?div = MCB) to Katherine M. Warpeha and Lon S. Kaufman. The funder had no role in study design, data collection and analysis, decision to publish, or preparation of the manuscript.

Competing Interests: The authors have declared that no competing interests exist.

* E-mail: kwarpeha@uic.edu

Introduction

Flavonoids are a class of phenylpropanoids, that are induced in germinating seedlings by UV and blue-light (B) [1], and are reported to play many diverse, but not fully understood, roles in plant physiology [2,3]. Quercetin (a flavonol, a sub-class of the flavonoids) accumulates in young Arabidopsis seedlings of 4–7 days (d) old, particularly in the cotyledonary node, hypocotyl/root transition zone and root tip, with glycosylated flavonoids in the cotyledon [4]. *In vivo* evidence indicates that flavonols regulate auxin accumulation in the *transparent testa 4* (*tt4*; makes no flavonols, including kaempferol and quercetin) mutants [5,6]. It has also been reported that auxin accumulates in *rol1-2* (*repressor of lrx1*) mutant seedlings due to flavonol-induced changes to auxin transport [7]. Comparing auxin transport in *tt4* and *tt7* (makes kaempferol but not quercetin) mutants, it was shown that derivatives of quercetin can inhibit basipetal auxin transport,

elongation, and gravitropism [8]. Quercetin due to its distinct structure can function as an antioxidant and UV screening compound [3,9]. The specific functions of flavonols in the seed-to-seedling developmental transition are still poorly understood, and may involve protein(s) that are still undescribed. Quercetin may also play important roles within the plant cell. Saslowsky et al. demonstrated that flavonoids and specific biosynthetic enzymes of flavonoids were present both in the nucleus and cytosol [10], and Peer et al., found quercetin in the nuclear region, endomembrane system and plasma membrane [4].

Quercetin is cleaved by quercetinase proteins, resulting in carbon monoxide and 2-protocatechuoylphloroglucinol carboxylic acid [11]. Adams and Jia demonstrated that the protein Pirin (PRN or PIR) of both human and bacteria have quercetinase activity *in vitro* [12]. Pirins have rapidly become a focus of interest given reported roles in metabolism [13], apoptosis [14,15], cellular stress [16] and malignancy [17,18]. Pirins are highly conserved members of the cupin super-family [19] found in prokaryotes, fungi, plants, and expressed at low levels in all examined cell types in mammals [20]. Pirin was originally identified as a transcription co-factor, interacting with the heterotrimeric nuclear factor I/ CCAAT box transcription factor NFY (aka NFI/CTF1; HAP) to drive adenovirus DNA replication and polymerase II transcription [20].

In plants, pirin investigations have been limited to specific periods of the life cycle (i.e. germination, and transition from vegetative to reproductive growth) [21]. Pirin was first identified in plants by Orzaez et al. in tomato, as potentially being involved in programmed cell death [15]. Arabidopsis Pirin1 (PRN1) was found via yeast-2-hybrid screening using the G-protein-α subunit (GPA1) as bait, and was shown to have a specific role in ABA regulation of germination in *Arabidopsis thaliana* [21]. PRN1 has been reported to play roles in several specific contexts, such as the B-induction of *LhcB* expression [22] and defense against seedling infection by *Cryptococcus* fungi as inferred by T-DNA insertion mutant analysis [23]. *PRN1* is also reported in high-throughput data derived from carbon status changes [24], meta-analysis of microarrays of plant hormone regulation [25], and in expression analysis where *PRN1* is induced by drought [26]. Although PRN1 is implicated in aspects of abiotic and biotic physiology, its different potential activities of transcription cofactor and possible quercetinase activity have yet to be reconciled.

G proteins are well known to be involved in a number of plant responses to various stimuli [27]. To date in Arabidopsis, GPA1 and PRN1 are reported in regulating germination [21]; and GPA1, PRN1 and NFY form a signal transduction chain, responsible for B-transcription of *LhcB* in 6-d-old dark-grown (etiolated) seedlings [22]. In earlier studies on PRN1, we had focused on a simple developmental period, from 0 (seed) to 6-7-d-old, utilizing only completely etiolated Arabidopsis seedlings [22]. However, study of *prn1* mutants and infection with fungi in dim and bright sunlight conditions has indicated that PRN1 may play a role in photomorphogenic transition and development, as well as in defense [23].

Given PRN1's reported effects on seed-to-seedling development [21,22], and its potential action in stress/defense [23], the activities of this protein are still largely unknown. Based on what is reported for human and bacterial pirins, we hypothesized that PRN1 may have a similar capability to cleave quercetin. Secondly, we hypothesized if PRN1 can regulate quercetin levels, it may affect growth of the young seedling, considering past reports of quercetin's influence on auxin accumulation and transport. To explore PRN1's specific effects, we conducted quercetinase enzyme assays with *in vitro* translated PRN1, we observed phenotypes in seedling development (in light and darkness) and stress responses by using a T-DNA insertion mutant of *PRN1*, and explored native promoter *PRN1*- and *35S*-driven expression of PRN1 in transgenic seedlings. We explored responses of seedlings grown in complete darkness, and under white light growth conditions (continuous white light for 3 d; and 16:8 light:dark for 6 d). We focused on the cotyledon and hypocotyl due to the most prominent phenotypes observed. The data considered support the hypotheses that PRN1 may regulate quercetin levels in at least the epidermal layer cells of the cotyledon, and PRN1 may play a role in light regulation of hypocotyl growth and orientation.

Materials and Methods

Chemicals

All chemicals, unless otherwise noted, were obtained from Sigma (St. Louis, MO).

Plant Materials, Seed Stocks and Accessions

Seeds of wild type (WT) Columbia *Arabidopsis thaliana* and mutants carrying a T-DNA insertion within the coding region of *PRN1* (SALK_006939), and the coding region of *ADT3* (published also as *PREPHENATE DEHYDRATASE* [*PD1*]), but referred to as *AROGENATE DEHYDRATASE* [*ADT3*] after description of the ADT family, henceforth is referred to as *ADT3*; SALK_029949) were obtained from the Arabidopsis Biological Resource Center (Columbus, OH) [28]. The mutant lines are homozygous null for the reported insertions. Plants intended for seed stocks were grown in Scott Metromix 200 (Scotts; Marysville, OH) in continuous white light [21]. Sequence data from this article can be found in the EMBL/GeneBank data libraries under accession numbers At2g27820 (ADT3), At3g59220 (PRN1).

Plant Growth Conditions for Experiments

Seedlings of *Arabidopsis thaliana* WT or T-DNA insertion mutants were grown on 0.5× Murashige and Skoog medium, 0.8% agarose phytatrays as described [21]. The growth medium contained no added sugars, hormones, vitamins or other nutrients. For all experiments, seedlings were sterilized in a bleach solution, all sterilization, planting and manipulations were carried out under dim green light [29]. Seeds were then washed in sterile water, mixed with low melt agarose then sown on phytatrays, and subsequently sealed in light-proof black plastic boxes. Planted seeds were stratified for 48 h in complete darkness at 4°C as described [29] without a light treatment. Cold-vernalized seeds were then moved to appropriate dark (dark-grown = continuous darkness) and/or light conditions (phytatrays or vertical trays moved from dark boxes to open boxes), detailed below. Seedlings were grown between 3 d and 7 d for most experiments in either complete darkness or in white light (continuous or 16:8, 10^2 μmol m^{-2} s^{-1}) chambers as described, at 20°C. Vertical growth experiments were as described here. Seeds were prepared as described for phytatrays, except no top agarose was used, with sterilized and washed seeds set directly onto the 0.5× medium (same as phytatray medium). Plates were sealed with parafilm then taped in vertical position in black plastic boxes, were stratified for 48 h in complete darkness at 4°C, then moved to the appropriate growth condition for the experiment.

Determination of Quercetin Levels

6-d-old dark-grown seedlings were given a 10^4 μmol m^{-2} total dose of 317 nm UV, as quercetins are known to be induced by UV-B [9]. Six h later the aerial portions of sets of seedlings (1 full phytatray of seedlings per sample; 100 μL of dry seed sown/ phytatray) were harvested directly into liquid nitrogen under dim green light, then finely ground, with similar sample preparation as described [29], except the final dried samples were weighed, then further purified by dissolving them in water and extracting with ethyl acetate. The ethyl acetate layer was dried and re-dissolved in a water/methanol mixture for analysis. Liquid chromatography-mass spectrometry analysis of hydrolyzed quercetin and kaempferol was carried out using an Agilent 6410 Triple Quad mass spectrometer (Agilent Technologies; Santa Clara, CA) coupled with an Agilent HPLC 1200 series chromatographic system. The system was operated using Agilent Mass-Hunter workstation software. Separation was achieved on a column HPLC chip (Agilent G4240, C18, 300A°, 43 mm chip) using water and methanol as the mobile phase. The mobile phase was pumped at a 500 nL/min flow rate. The initial mobile phase consisted of 50% methanol for 1 min, and the amount of methanol was increased linearly to 90% over 2.5 min. A multiple-reaction motoring method in the negative ion mode was used for the analysis of quercetin (m/z 301.1−>121.1) and kaempferol (m/z 285.1−> 117.1). Values were determined as ng/g liquid nitrogen ground weight. Resultant data obtained from each data pair (WT vs. *prn1*) were analyzed by a two-tailed T test, where data was entered into GraphPad to determine SD of the ratios for 4 replicates.

Synthesis of PRN1 and GPA1 Protein

Full-length PRN1 and GPA1 templates were prepared, amplified and purified as described previously [21]. PRN1 and GPA1 protein were separately produced in coupled *in vitro* transcription/translation reactions using the TNT T7 Coupled Wheat Germ Extract System (Promega; Madison, WI) as directed, with a 90 min translation time at 30°C, and as previously described, with 50% concentration of extract by using a microcentrifugation concentrating filter to maximize retention of protein 30 kD and above (Millipore, Billerica, MA) [29].

Quercetinase Assay

For a final volume of 500 µL, freshly *in vitro*-translated PRN1 extract (16 µg) was incubated with a final concentration of 10 µM quercetin (diluted from 5.0 mM DMSO stock immediately before assay) in a total reaction volume including 1 µL of 0.5× MS medium, and utilizing the buffer and method of quercetinase determination of activity described in the assay of Adams and Jia [12]. For G-protein assays, 1 µM $MgCl_2$ and 150 mM NaCl were additionally included in the assay buffer. Reactions were carried out in darkness, in a 26°C circulating water bath for 15 min; then the quercetin 2,3-dioxygenase activity of PRN1 was determined by measuring absorbance as described [12], using a Perkin Elmer scanning spectrophotometer (Perkin Elmer; Waltham, MA).

In vitro GPA1-PRN1 Activity Assay

GPA1 was locked in either the active or inactive confirmation by using non-hydrolysable analogs, in a pre-incubation of 16 µg of the *in vitro* translation GPA1 product extract with a final concentration of 100 µM GTPγS (non-hydrolysable GTP analog), or GDPβS (non-hydrolysable GDP analog) in HEPES, pH 7.5, in 100 µL total volume, overnight in darkness at 10°C under rotation [29]. The "activated" and "inactivated" GPA1 mixtures were then separately incubated with PRN1 in a 1:1 volume ratio (26°C for 15 min, circulating water bath) and then tested for quercetinase activity as described above. The 'buffer only' scan was subtracted from all spectra.

UV-C Kill Assay

Seeds were planted and seedlings grown as described above. After the cold-stratification, the seedlings were transferred to complete darkness. UV-C treatments were administered on d 6 as previously described [30] with 4 or 8 min of overhead (254 nm) radiation administered. Seedlings were returned to complete darkness for 24 h. UV-C killing doses for *adt3* or WT were determined by dose-response 'titration' experimentation. Seedlings were photographed from the side.

Light-grown Hypocotyl Orientation Assay

Seeds were planted and grown on phytatrays as described above. After cold-stratification, the seedlings were transferred to continuous white light. After 72 h, they were photographed from the side using a dissection microscope, and hypocotyl angle (in reference to the horizontal plane) was measured. Two sets of 30–35 seedlings were planted, then, due to highly similar means at time of observations, data were pooled for analysis. For the complementation experiments, the seedlings grown on phytatrays were photographed from overhead at 72 h after stratification. Different sets of seeds were planted and grown on vertical plates in both darkness and 16:8 light conditions for 6 d, then were photographed to observe the seedling phenotype(s). Dissection microscope images from vertical plate seedlings were reassembled using a stitching plugin for ImageJ [31].

Cloning

Standard molecular biology techniques and the Gateway system (Invitrogen) were used for all cloning procedures. PCR fragments were created using the primers, shown in Table S1. WT Arabidopsis genomic DNA was used to generate the *PRN1* promoter fragment, and WT Arabidopsis cDNA was used to generate the PRN1 ORF fragment. The *PRN1* full promoter fragment was then cloned into the Invitrogen pDONR P4-P1R vector, and the *PRN1* ORF fragment was cloned into the Invitrogen pENTR/D-TOPO vector. LR reactions (Invitrogen) were then performed with verified entry clones to obtain expression clones. The pGreen [32] binary vector derivative containing a NOS terminator with a C-terminal GFP fusion and spectinomycin and BASTA resistance genes [33] was used for the native promoter construct (*PRN1*::PRN1-GFP). The pEarleyGate 103 [34] vector containing an OCS terminator with a *35S* promoter, a C-terminal GFP-His fusion and kanamycin and BASTA resistance genes was used for the overexpression construct (*35S*::PRN1-GFP). All constructs were confirmed via restriction enzyme digest, PCR, and sequencing. Verified expression clones were transformed in *prn1* mutant or WT backgrounds via floral dip [35]. Transformed seeds/seedlings were grown under BASTA selection. Homozygous transformed seedlings were selected in the third generation (T3) for the native promoter construct (*PRN1*::PRN1-GFP) in *prn1* background, and at T2 for the overexpression construct (*35S*::PRN1-GFP) in WT background.

Visualization of Transgene Expression

Transgenic and untransformed 6-d-old dark-grown seedlings were fixed using paraformaldyhyde (2.5%), then were mounted in Prolong Gold antifade reagent with DAPI (Life Technologies: Grand Island, NY) on glass slides. Seedlings were harvested in dim green light into fixative, then all other manipulations were done in lab lighting. Images were obtained using an Andor WD Spinning Disk confocal system (Yokagawa CSU-W1 with standard 50 µm pinholes) using iQ2 software. For each replicate 12–20 whole seedlings were screened and areas of fluorescence examined in 1.0 µm thick horizontal sections, where z slices were collected through the complete seedling. Exposures were identical among samples, and contrast uniform on each channel using 3 solid-state lasers. All wavelengths were detected sequentially using narrow band pass filters, basically eliminating any potential for crosstalk. DAPI was stimulated using a 405 nm diode laser (200 ms exposure, laser intensity 64%), and detected with Semrock Brightline, Single Band Fluorescence Filter, 447 Centre; GFP was stimulated with a 488 nm diode laser (400 ms exposure, laser intensity 48.4%), and detected with Semrock Brightline, Single Band Fluorescence Filter, 525 Centre; red fluorescing molecules (largely protochloryllides so etioplasts were identifiable in samples) were stimulated with a 561 nm diode laser (350 ms exposure, laser intensity 67.7%), and detected with Semrock Brightline, Single Band Fluorescence Filter, 607 Centre. At least three separate transgenic lines were used to confirm all reported expression data. Images were prepared with ImageJ software (NIH, Bethesda, MD).

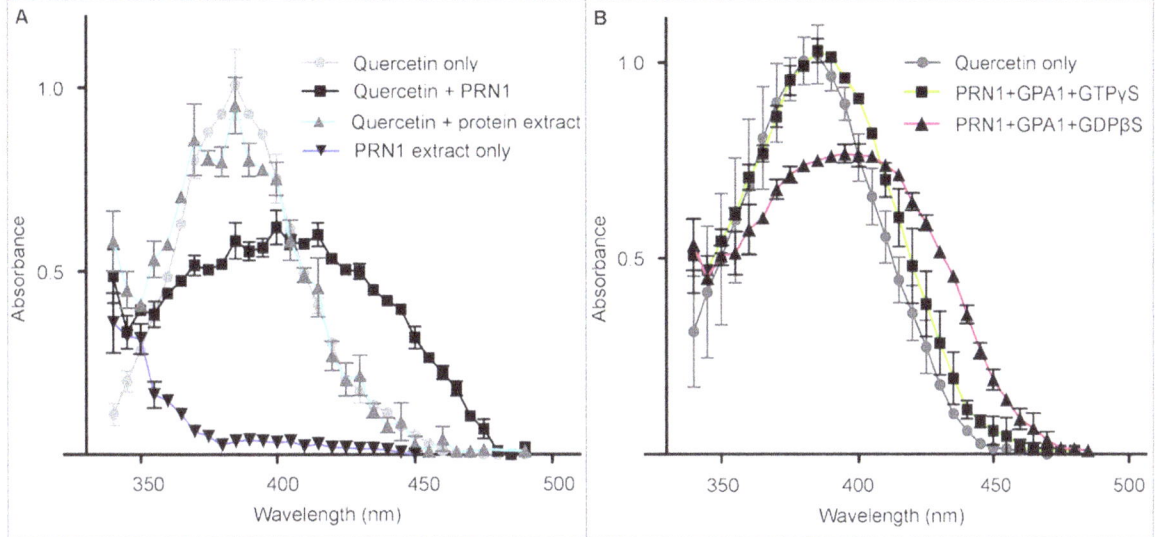

Figure 1. *In vitro*-translated PRN1 has quercetinase activity. A. *in-vitro*-translated PRN1 possesses quercetinase activity. *In vitro*-translated PRN1 was incubated with a final concentration of 10 μM quercetin in a 500 μL total reaction volume, or as a control, 10 μM quercetin alone, in buffer as described [12]. Reactions were carried out in darkness in a 26°C water bath for 15 min. After incubation at 26°C, absorption spectra of quercetin alone (Quercetin only), Quercetin+PRN1, Quercetin+protein extract, or PRN1 extract only were determined using a Perkin Elmer scanning spectrophotometer (Perkin Elmer; Waltham, MA). n = 3, SD shown. **B. The conformational status of GPA1 affects the quercetinase ability of PRN1.** *In vitro*-translated GPA1 was pre-incubated with non-hydrolyzable analogs (either GTPγS or GDPβS) overnight at 10°C, then was incubated with PRN1, then (PRN1+GPA1+GTPγS or PRN1+GPA1+GDPβS) was added to buffer+quercetin to assess the effects on quercetinase activity as shown. Absorption spectra of quercetin alone (Quercetin only) also shown. n = 3, SD shown.

Epidermal Peel

Cotyledon epidermal peels were conducted as described [36], with specific modifications. Hollister 7730 medical adhesive (Hollister, Libertyville, IL) was sprayed across a glass slide, then waved in air to remove bubbles. After 5 min, live 6-day seedling cotyledons were set adaxial side down, then pressed into adhesive by parafilm to ensure evenly adhered surface. Cells were scraped away with a curved microspatula 3 min later. Prolong Gold antifade reagent with DAPI stain was set on top of the cells for 5 min, then cover slides were mounted for analysis via spinning disk confocal microscopy. *prn1* mutants or *prn1*-transformed with *PRN1::PRN1-GFP* or WT-transformed with *35S::PRN1-GFP* were viewed on spinning disk confocal after DAPI-stain was applied. 10–15 cotyledons were viewed per replicate of 3 independent replicates, utilizing the same lasers and imaging conditions described above for visualization of transgene expression.

Statistics

Data shown in figures were entered into Prism v. 5.0 (GraphPad Software, Inc., GraphPad.com), where mean and SD or SEM are shown for most graphic depictions. For enzymatic activity curves SD is shown on the figures. For bar graphs, quercetin quantification (ng/g liq nitrogen ground mass) is shown as ng/g. The WT and *prn1* samples where planted, harvested and extracted as pairs where they were compared, respectively. A two-tailed T test was used with significance set at p<.05. SD are shown on the relevant Figure. For hypocotyl orientation data, a non-parametric statistical analysis was performed for all data. The Kolmogorov-Smirnov normality test indicated that the WT and *prn1* data were non-normally distributed. Subsequently, for the angle of hypocotyl orientation, a Mann-Whitney test was used to compare the *prn1* hypocotyl angle to the WT angle, where on the figure SEM is shown.

Results

PRN1 Possesses Quercetinase Activity

Based on reports that *in vitro*-translated human and bacterial Pirin could act as quercetinases [12], *in vitro*-translated PRN1 was analyzed for the ability to cleave quercetin (Figure 1A). Utilizing the buffer of Adams and Jia [12], quercetin has a reported absorbance maximum of 384 nm, which we observed herein for quercetin (Figure 1A). Upon incubation with *in vitro*-translated PRN1, the absorbance maximum shifted to ~405–410 nm by 15 min at 26°C, indicating cleavage of quercetin [12], and confirming that nascent PRN1 could potentially function as a quercetinase (Figure 1A). Incubation of quercetin with the *in vitro* translation components alone did not result in absorbance changes (Figure 1A).

Quercetinase Activity is Observed When GPA1 is in the Inactive Conformation

To address potential regulation of the quercetinase activity of the nascent PRN1 protein obtained by *in vitro* translation, we explored the interactions between GPA1 and PRN1 (original interaction reported by [21], further explored by [22], and observed in high-throughput yeast-2-hybrid assays [37]). We pre-incubated *in vitro*-translated GPA1 with GTPγS (non-hydrolyzable GTP analog) or GDPβS (non-hydrolyzable GDP analog), in order to produce GPA1 maintained in an activated (GTPγS) or inactivated (GDPβS) conformation, respectively. GPA1 in the active or inactive confirmation was then incubated with *in vitro*-translated PRN1 and quercetin. The resulting absorbance profiles (Figure 1B) indicated that GPA1 in its inactive confirmation permits quercetinase activity of PRN1. When GPA1 was maintained in its active confirmation no quercetinase activity was measurable.

prn1 Mutants can Produce Stable Elevated Amounts of Quercetin Induced by Brief UV-B Treatment

To further explore the possible role of PRN1 as a quercetinase in plants, etiolated, 6-d-old Arabidopsis WT and *prn1* seedlings were both irradiated by UV-B (317 nm), then 6 h post-irradiation, aerial portions of seedlings were harvested. Levels of extractable hydrolyzed quercetin, as well as the closely related compound kaempferol were quantified (Figure 2A). Quercetin levels in *prn1* mutants were approximately twice as high as those detected in WT, while kaempferol levels remained effectively unchanged (Figure 2A). This finding supports the results of the *in vitro* quercetinase assays, suggesting that if PRN1 can act as a quercetinase, lack of PRN1 may result in elevated quercetins in mutant seedlings, since there may be no mechanism for cleavage of quercetin to maintain a WT level.

prn1 Mutants can Survive UV-C Radiation That Kills WT

If young, dark-grown *prn1* seedlings potentially have more or accumulate more quercetin, then one would expect the *prn1* mutants to exhibit improved resistance to cellular stress caused by high energy UV (UV-C; 254 nm). We confirmed this hypothesis by using a published assay [30], where we exposed WT and *prn1* insertion mutants to a brief dose of UV-C, then returned the seedlings to darkness (Figures 2B & 2C). 24 h after the UV-C treatment, WT seedlings survived, while radiation-sensitive *adt3* (T-DNA insertion of AROGENATE DEHYDRATASE 3, sensitive to radiation as prior reported [30]) seedlings died, and *prn1* mutant seedlings survived (Figure 2B). When the UV-C dose was doubled, WT seedlings died, and *prn1* seedlings not only survived, but also exhibited increased hypocotyl shortening (Figure 2C compared to 2B). *prn1* seedlings (T3, homozygous)

transformed with a native *PRN1* promoter and PRN1 ORF construct were killed by the 8 min of UV-C, complementing the *prn1* mutation (Figure 2D).

prn1 Mutants Exhibit an Altered Angle of Hypocotyl Growth When Grown in White Light

Since flavonoids are induced in seedlings by light [1], we also observed the growth of *prn1* and WT seedlings grown under continuous white light for 3 d on phytatrays. *prn1* mutants exhibited an average hypocotyl growth angle (from the horizontal) of 56.6 (+/−3.55 SEM) degrees, while the WT seedlings exhibited an average of 78.0 (+/−2.06 SEM) degrees (Figure 3). Examples of the phenotype are shown (Figure 3A–E). Since the Kolmogorov-Smirnov normality test indicated that the WT and *prn1* data were not normally distributed (WT: KS = 0.222 and p<0.010, *prn1*: KS = 0.146 and p<0.010; Figure S1A and S1B, respectively), a non-parametric statistical analysis was performed. We found that the *prn1* shoot growth angle was significantly different than the WT angle (Mann-Whitney test; p<0.001) (Figure 3F). The disoriented hypocotyl phenotype was complemented (oriented like WT) in *prn1* seedlings transformed by the native construct (*PRN1*::PRN1-GFP; Figure S1C). No gross phenotypic differences were observed for *prn1* seedlings compared to WT seedlings from 6-d-old dark-grown conditions on vertical plates (Figure S2). For 6-d-old light-grown (16:8) seedlings on vertical plates (Figure S3), we observed a hypocotyl phenotype similar to that reported for 3-d continuous white light *prn1* seedlings (Figure 3), and also observed shorter roots, which was complemented by transformation with the native promoter construct (*PRN1*::PRN1-GFP) (Figure S3). However, when WT were transformed with *35S*::PRN1-GFP, the hypocotyl phenotype was reminiscent of the deficiency (*prn1*

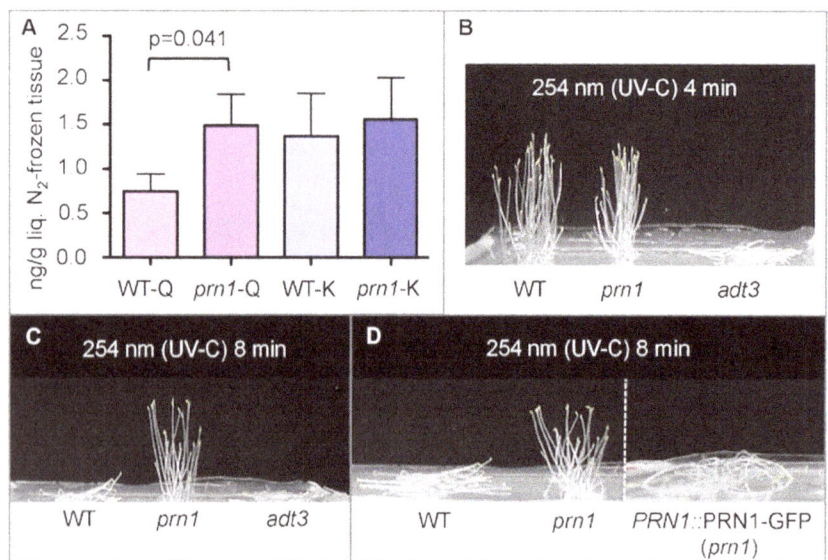

Figure 2. *prn1* mutants accumulate excess quercetin, and can survive UV-C irradiation that kills WT. A. *prn1* mutants accumulate more quercetin compared to WT seedlings. 6-d-old dark-grown seedlings were treated with a brief pulse of UV (317 nm), returned to darkness, then harvested 6 h later to determine the total extractable quercetin (Q) or kaempferol (K) in WT or *prn1* seedlings (WT-Q; *prn1*-Q; WT-K; *prn1*-K). The levels of Q and K are indicated for *prn1* and WT and SD are shown. n = 4 **B & C. *prn1* mutants survive a UV-C radiation treatment that kills WT seedlings.** Seedlings grown in complete darkness as described in methods. UV-C treatments were administered on d 6 as described [30], with 4 min (B) or 8 min (C) of 254 nm (UV-C) radiation. Seedlings were returned to complete darkness for 24 h. UV-C killing doses for *adt3* or WT were determined by prior 'titration' experimentation [30]. Seedlings were photographed from the side. n = 4. **D. *prn1* seedlings transformed with *PRN1*::PRN1-GFP construct are killed by 8 min of UV-C, similar to WT.** Seeds (30) of WT, *prn1* or *prn1* transformed with *PRN1*::PRN1-GFP (*PRN1*::PRN1-GFP *prn1*) were sown, then seedlings grown in complete darkness. UV-C treatments were administered on d 6 as described [30], with an 8 min dose of 254 nm (UV-C) radiation. Seedlings were returned to complete darkness for 24 h. Seedlings were photographed from the side. n = 4.

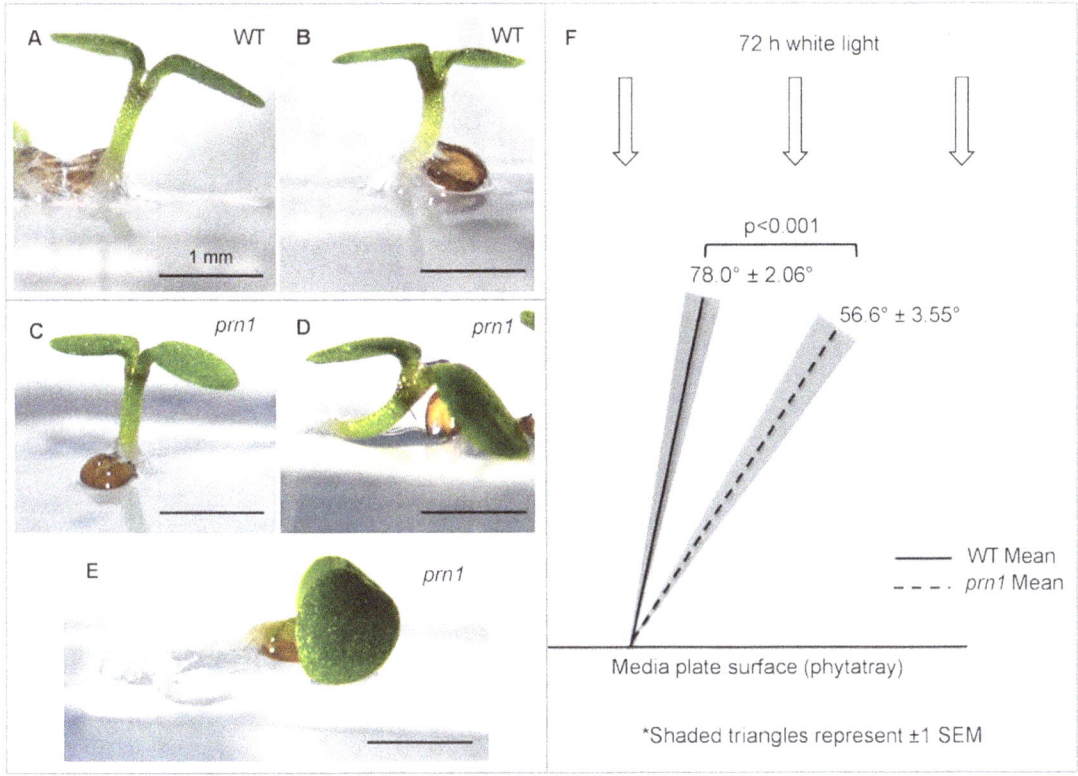

Figure 3. 3-d-old light-grown *prn1* mutants exhibit abnormal light-grown shoot orientation phenotype. Three (72 h)-d-old white light-grown seedlings were photographed from the side using a dissecting microscope to compare WT (**A–B**) and *prn1* mutant (**C–E**) seedlings. Images are representative of the range of orientations observed. Scale bars each represent 1.0 mm. Hypocotyl angles were measured for individual seedlings in reference to the horizontal phytatray surface (where vertical = 90°) and mean angles were calculated (**F**). Means are shown by lines (solid for WT; dotted for *prn1*) and SEMs are represented by shaded triangles. n = 65. SEM shown.

mutant seedlings), but with noticeably smaller cotyledons, hypo-cotyl and root (Figure S3).

PRN1-GFP Fluorescence is Observed in Cells of the Epidermis, Associated with the Nucleus in 6-d-old Dark-grown Seedlings

Subcellular localization of PRN1 was explored by utilizing PRN1-GFP flourescence. *prn1* mutant seedlings were transformed with *PRN1*::PRN1-GFP (T3), or WT seedlings were transformed with *35S*::PRN1-GFP (T2). Homozygous seedlings were grown under selection in complete darkness for 6 d, then images captured on a spinning disk confocal microscope. Flourescence was visible in some cells of the epidermis of the cotyledon, associated with nuclei (both diffuse and punctate), as well as diffuse within the cell cytoplasm of some cells in *prn1* transformed with *PRN1*::PRN1-GFP (Figure 4). Flourescence was greater in the transgenic plants overexpressing PRN1 (WT transformed with *35S*::PRN1-GFP) with localization to the nucleus, and diffuse and variable flourescence in the rest of the cell, visible in most of the cells of the epidermis (Figure 4). Areas of interest from Figure 4 images were enlarged to better observe the details over a series of 5 z-slices of 1 μm thickness (Figures 5 and 6). Enlargement of transformed (*PRN1*::PRN1-GFP) *prn1* (Figure 5) indicated fluorescence coinciding with DAPI stain throughout the nucleus, both diffuse and punctate. Flourescence was also observed immediately outside of the nucleus, possibly reflecting PRN1 trafficking to the nucleus or surrounding ER; diffuse and some punctate fluorescence was visible elsewhere in the cells shown (Figure 5). Some of the

punctate fluorescence could correspond to developing plastids, as there are examples of fluorescence of GFP overlapping with 561 nm, which would be expected to cause etioplasts to fluoresce due to protochlorophyllides. Enlargement of the results for transformed WT (*35S*::PRN1-GFP) seedling (Figure 6) similarly revealed GFP fluorescence throughout the nucleus, with punctate structures within the nucleus, and fluorescence throughout the epidermal cells, diffuse and punctate. Like Figure 5, there is overlapping fluorescence with 561 nm in some examples. We also performed epidermal peel samples from living, non-fixed cotyledons, from seedlings grown 6 d in darkness or grown under 16:8 (Figures S4 & S5). GFP-flourescence was largely found in the nuclei, and was diffuse in extranuclear locations.

PRN1 Promoter Possesses Regulatory Elements of Abiotic and Hormonal Signaling, and Some Specific Life Cycle Elements

The non-coding region upstream of the *PRN1* START codon (2,651 bp, inclusive of the 231 bp 3' UTR end of At3g59210) to the gene upstream was cloned as the PRN1 native promoter, and was analyzed for regulatory elements and motifs (Figure S6; [38,39]). The motifs indicate potential roles in carbon metabolism, hormone-signaling, light-signaling, and abiotic stress-responses, much of which is consistent with reported high-throughput data [24–26]. Lapik and Kaufman [21] also described aspects of the promoter 10 years ago, showing that there were indications for light and hormone regulation, specifically ABA, confirmed by exogenous-ABA regulation of the transcript and complementation

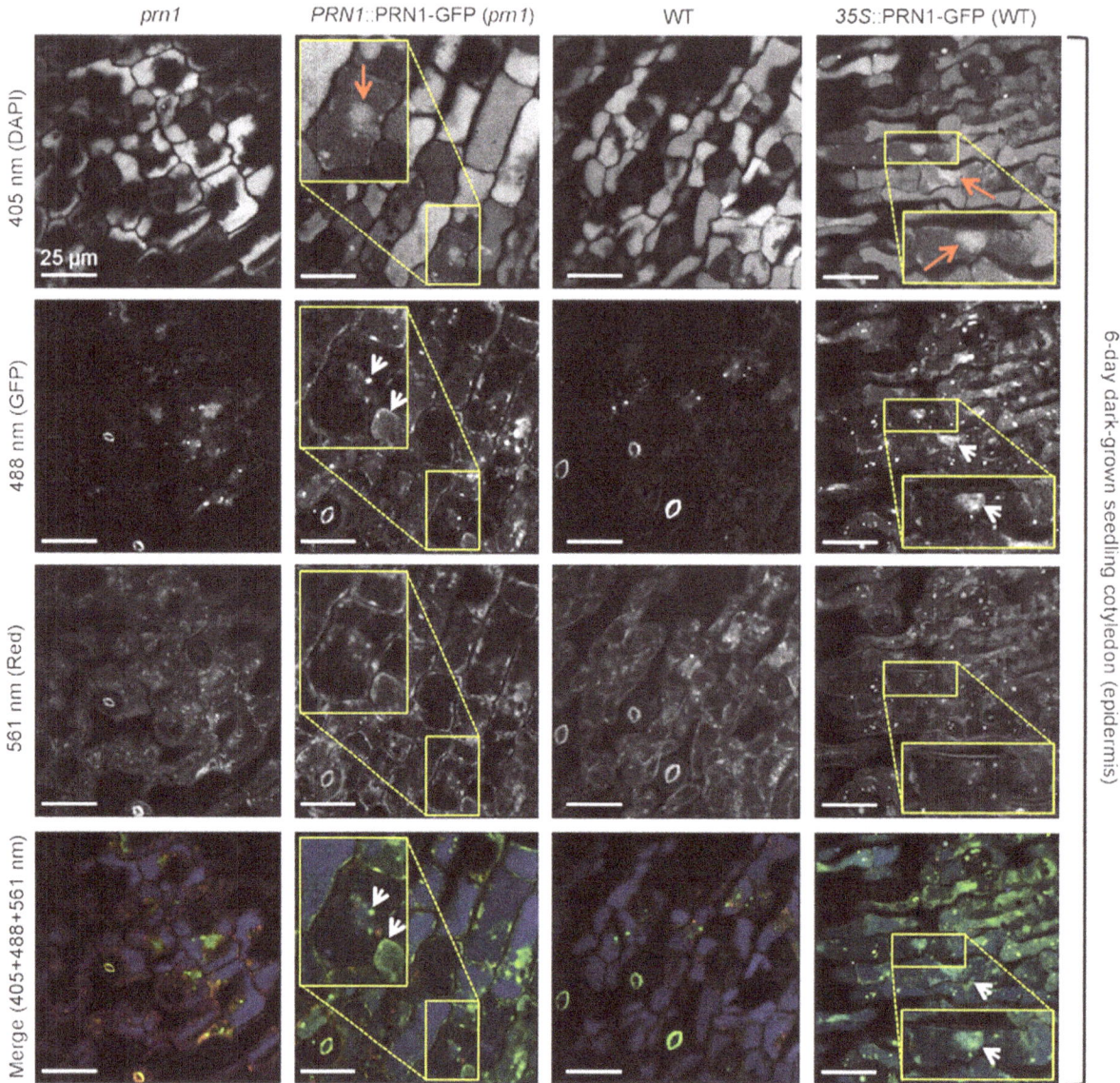

Figure 4. Subcellular localization of PRN1 in dark-grown seedlings. Constructs and transgenic plants are described in methods. 6-d-old dark-grown *prn1* whole seedlings transformed with *PRN1::PRN1-GFP* (T3) or WT seedlings transformed with *35S::PRN1-GFP* were fixed, stained with DAPI, mounted on slides, then photographed on a spinning disk confocal using steady state lasers 405 nm, 488 nm, and 561 nm. All images shown are from the cotyledon epidermis layer. Panel rows are indicated by wavelength, with individual channels in black and white. In the merge, 405 nm (DAPI) is false-colored blue, 488 nm (GFP) is false-colored green, and 561 nm is false-colored red (Red). Images are representative with no alteration of the fluorescence within the image field, and images represent an optical section of 1 μm thickness. The column heading indicates the different seedling lineages. Yellow boxes on the figure indicate areas of interest and show an enlargement. Red arrows indicate DAPI stain in nucleus. White arrows indicate GFP accumulation in 488 nm. Untransformed WT and *prn1* seedlings are also shown on the figure. Scale bars each represent 25 μm. n = 4 biological replicates, with at least 12–20 individual seedlings viewed per replicate; images are representative.

[21]. The eFP browser [40] indicates that PRN1 is predicted to localize to the nucleus based on published experimental data and the WOLFPSORT prediction method, and possibly to the chloroplasts and cytosol, with a low-to-medium confidence value. In addition, PRN1 expression is shown to increase in response to ABA (guard cells), ACC, and abiotic signals, although conditions of seedling growth (age, sucrose in medium etc.) varied widely in the database [40].

Discussion

Integration of a number of external environmental signals occurring simultaneously is necessary for seedling establishment. This transitional period from seed to seedling is complex, as limited materials including quercetins, potent antioxidants and other metabolites are stored in seeds to support seedlings until they become photosynthetically competent [41]. G-proteins and their effectors may be critical mechanisms to regulate this transition. Most higher plants only code for 1–2 copies of a G-protein-α subunit in the genome, yet there are numerous activities known to be associated with G-protein signaling in plants [27,42]. This

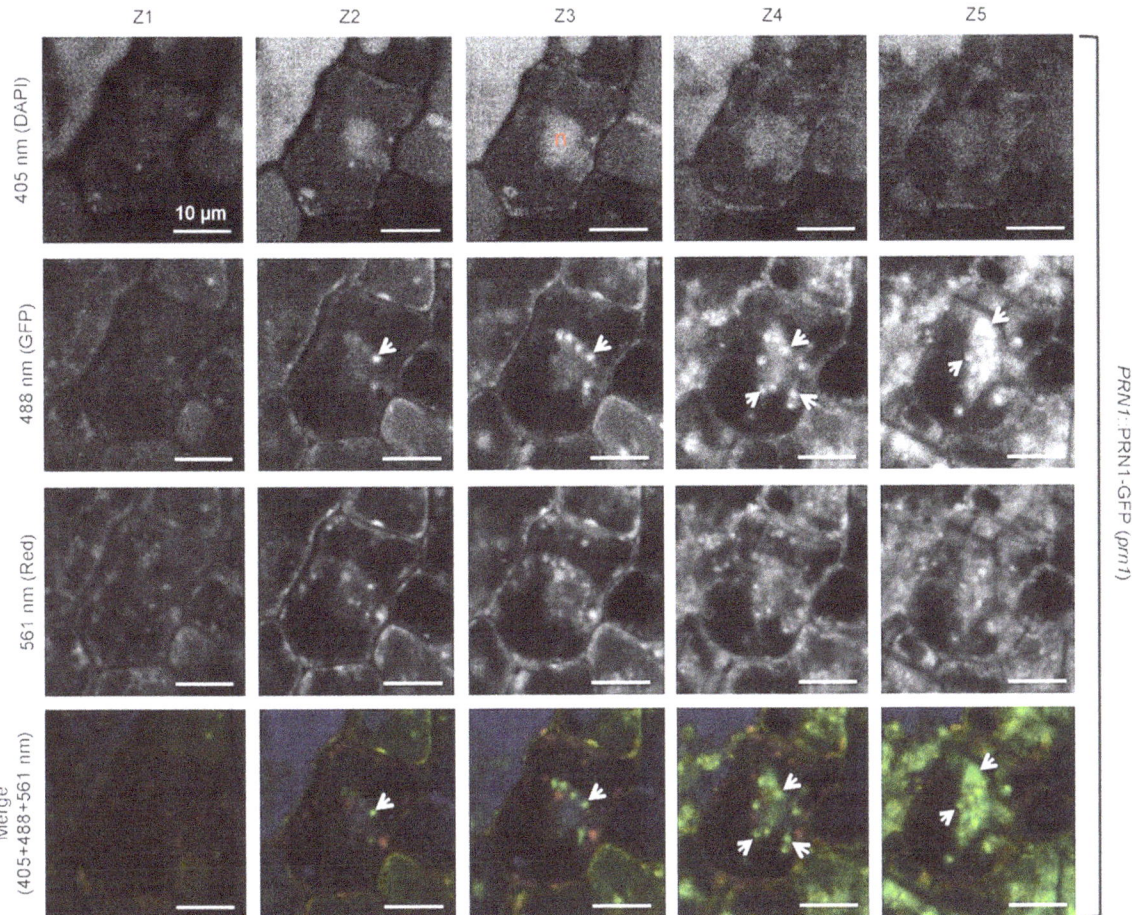

Figure 5. Enlargement and z-slices of region of interest in seedling cotyledon epidermis: *prn1* transformed with *PRN1*::PRN1-GFP. Microscopy and techniques are the same as described for Figure 4. Each row shows 5 consecutive z-slices. The red "n" indicates a nucleus (DAPI stained). GFP expression of interest is indicated by white arrows. Scale bar represents 10 μm.

conundrum is often explained by pointing to the relatively large number of proposed effectors identified. The data presented herein indicate distinct and separable activities for a single effector, which may represent yet another dimension by which plant G-Protein signaling functions.

GPA1 is a Possible Molecular Switch Regulating PRN1 Activity in the Cell

GPA1 is involved in many early processes in the seed-to-seedling transition ([43,44], and reviewed in [27]). The data presented herein, coupled with previously published results [21,22] indicate that the single-copy GPA1 may modulate PRN1 activity in Arabidopsis in multiple ways during seedling development. Given that GPA1 may spend a majority of time in the GTP-bound state [42,45] we would speculate that PRN1 spends a reduced time or has less opportunity to act as a quercetinase (GPA1-GDP/inactive form). PRN1 may function primarily in the nucleus or perhaps in association with the nucleus (potentially the ER) shown herein, and at least in Arabidopsis, act as a transcription co-factor [22]. The quercetinase activity in the GPA1-PRN1 *in vitro* experiment appeared less than the PRN1-alone experiment (Compare Fig. 1B to Fig. 1A). Due to the importance of quercetin as an antioxidant and UV-screening molecule [3,9,46], and due to quercetin's reported function as a transcriptional repressor [47–

49], we would speculate that PRN1 may have reduced quercetinase activity in an organelle or cell, unless quercetin levels were attaining high or toxic levels. There are other proteins, as yet unknown, that may also interact with PRN1 or GPA1, altering PRN1 function(s). High-throughput yeast-2-hybrid assay data indicate that PRN1 interacts with a number of proteins [37]. However, separation of actual activities i.e. that of transcriptional regulator [22] versus quercetinase may be difficult to achieve, as Adams et al. have suggested in a review that pirins may have a role in protecting transcriptional machinery from the inhibitory effects of quercetin [50]. When considered with data of other organisms that show quercetin can affect transcription [47–49,51,52], it is possible that the transcriptional co-factor and quercetinase actions of PRN1 may be linked. Further studies are in progress by our lab group to better understand the functional relationships of PRN1 activities in plant cells.

Regulation of Cellular Quercetin Levels and UV-induced Screening

PRN1 clearly affects quercetin levels in aerial portions of Arabidopsis (Figure 2). The changes that occur in the *prn1* mutant, increased quercetin and perhaps other flavonols or quercetin-derived structures, proved to be important in screening from high energy UV (Figure 2B,2C,2D). Several functions had been

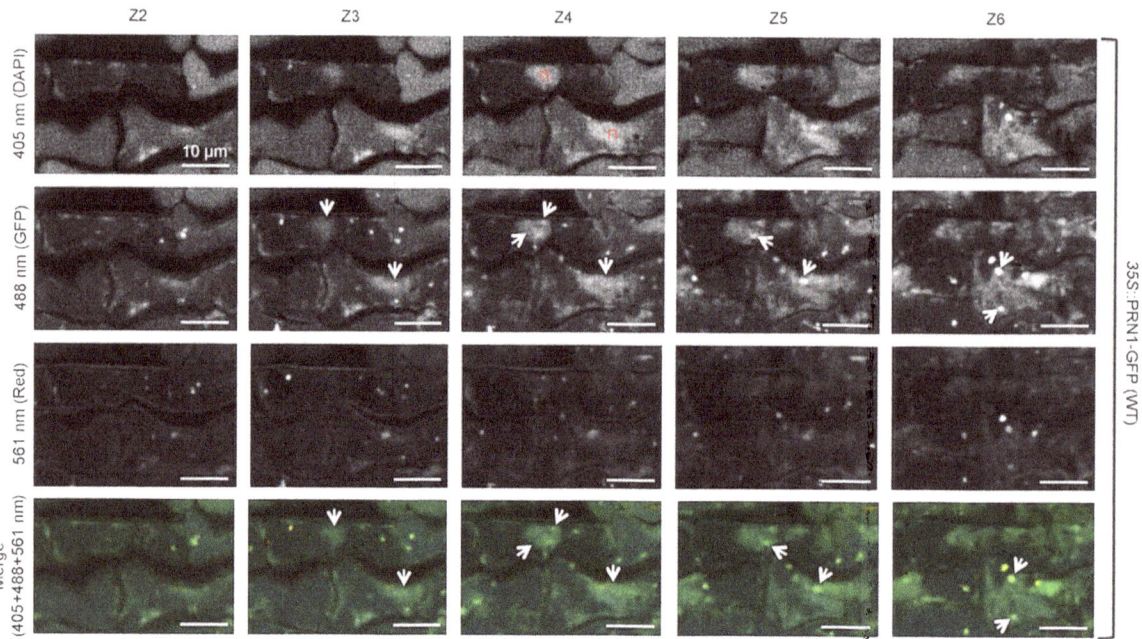

Figure 6. Enlargement and z-slices of region of interest in seedling cotyledon epidermis: WT transformed with *35S*::PRN1-GFP.
Microscopy and techniques are the same as described for Figure 4. Each row shows 5 consecutive z-slices. The red "n" indicates a nucleus (DAPI stained). GFP expression of interest is indicated by white arrows. There is expression in the rest of the cell, largely diffuse. Scale bar represents 10 μm.

proposed for quercetin, but it is still not fully understood what chemical, or structural roles quercetin may play in the young seedling. Flavonoids play many diverse roles in plant responses to the external environment, and in protecting young seedlings from stress [46]. Quercetin, specifically, is known to be a potent antioxidant [3], a UV-screening compound [9], may be a regulator of auxin accumulation or transport [5–8], and even a modulator of transcriptional activity [47,48].

prn1 Mutant and Transgenic Whole Seedling Responses to White Light

An additional phenotype of *prn1* mutants was revealed in this study. When *prn1* seedlings were grown under white light, hypocotyls deviated from WT gravitropic orientation (Fig. 3), a phenotype not observed under growth in continuous darkness. This disorientation was not a curvature, rather, the hypocotyls were straight, whether grown in 3-d continuous white light or 6-d under the 18:6 growth protocol. Under the 16:8 protocol at 6 d on vertical plates, it was also noticed that roots were shorter. WT seedlings transformed with *35S*::PRN1-GFP produced hypocotyls that were disoriented similar to *prn1* mutants, and seedling organs were all smaller than *prn1* mutants or WT–leaves, hypocotyls and roots. Both hypocotyl orientation and size of the seedling may be directly related to the changes that PRN1 may have on quercetin(s), where an alteration in quercetin levels may in turn affect auxin synthesis, metabolism or auxin transport.

Auxin influences cell division, cell elongation and cell size, and organ growth and development, and these auxin impacts are affected by flavonoids, reviewed in [53]. Poppe et al. described disorientation of the hypocotyl in far red light for WT Col, and a similar disorientation range for specific *phyB* mutants, in Landsberg and Nossen ecotypes [54]. The underlying cause of the disoriented hypocotyl phenotype of the *prn1* mutants is yet unknown, but this light-specific response may be due to a

phytochrome effect on high quercetin levels in the *prn1* hypocotyl or the hypocotyl/root junction. Correct auxin transport is reported to be necessary for orientation and hypocotyl elongation in light-grown, but not dark-grown seedlings [55]. It was subsequently shown that there was a difference in auxin (IAA) transport in dark-grown versus light grown seedlings (lower in dark) [56]. In a detailed study of light affects on auxin transport and biosynthesis, Liu et al. showed that there is a difference in transport and metabolism of IAA between the upper (meristem, cotyledons and hook) and lower hypocotyl in tomato seedlings due to phytochrome action, and white light altered auxin transport of Arabidopsis seedlings, comparing basipetal to acropetal transport [57]. While we showed increased quercetin levels in *prn1* mutant shoots, in the overexpressor transformed seed lines (WT transformed with *35S*::PRN1-GFP), it would be expected that quercetins and perhaps flavonoids in general would be reduced from more PRN1 being present and potentially active as a quercetinase. It is envisioned that either way, increased or decreased, quercetin may affect directly or indirectly auxin transport. However, reduction or changes in ratios of flavonoid species (presumably in PRN1-overexpressor lines) on leaf size or development are difficult to predict without direct quantitation. How quercetin levels may affect scavenging of ROS due to IAA catabolism and resultant impacts on cell expansion would also be hard to predict [53], and hence the effects of overexpression of PRN1 will require additional experimentation to fully understand. Further studies are underway to quantitate PRN1's regulation of quercetin levels in the shoot apex, cotyledon, hypocotyl, root-shoot junction and root itself. Subsequent effects on auxin synthesis, conjugation with molecular species of interest, and localization in the young seedling are also under investigation.

Localization and Reconciliation of Activities of PRN1

PRN1 can act as a transcription co-factor [22] and quercetinase (herein), and has also been reported to interact with different proteins, including NF-Y [22] and GPA1 [21,22,37] where data indicate different subcellular locations. PRN1 was identified at multiple subcellular locations herein (Figures 4,5), where transcription cofactor activity would be expected to occur in the nucleus, but quercetinase activity could also occur. Quercetinase activity may be occurring at multiple intracellular sites, as Saslowsky et al. demonstrated that flavonoids and specific biosynthetic enzymes of flavonoids were present both in the nucleus and cytosol [10]. GPA1 is often discussed as a plasma-membrane localized protein, but it has also been found localized to the golgi and ER [58], indicating that there is opportunity for PRN1 to interact with GPA1 in multiple locations. The localization of PRN1 in the transformed *pm1* mutant indicates clusters of cells expressing PRN1-GFP. The fluorescence data possibly indicate trafficking of PRN1 in response to a local environmental signal, or perhaps a plasmodesmata-communicated signal, reminiscent of that observed for intercellularly-trafficked proteins in the leaf, for example, KNOTTED1 [59]. With a number of roles in the cell possible, the microenvironment may greatly affect expression or post-translational modification, linked to specific internal cues, or even associated with the cell cycle status, as has been reported in human cells [14,60]. Quercetin is reported to be localized to many different parts of the cell [4,10], perhaps to selectively carry out its many diverse functions.

This G-protein-regulated Pathway may be Involved with Early Acclimation/Regulation of Cellular Stress Responses, both Abiotic and Biotic

GPA1 (and the Gα subunit of rice [61]) is reported to be involved in responses related to stress [62,63] [64]. PRN1 interacts with GPA1 and as a result may play roles in responses to environmental stimuli. Quercetin is among the most abundant of the phenylpropanoids, and may assist in the prevention of damaging effects of many different types of abiotic and biotic stresses [46]. Enhanced levels of quercetin correlate with survival of *pm1* mutant seedlings after treatment with levels of UV-C normally lethal for WT, and the surviving seedlings, albeit shorter, do not appear to be damaged (Figure 2). This suggests that a major function of this signaling mechanism is to moderate protection of the young seedling from situations such as exposure to full sunlight, or the harmful radicals that can be generated via any cellular stress. While herein we have used UV radiation to study PRN1, it is possible that it may be activated by a host of stressors, biotic and abiotic. ABA is known to be a hormone involved in signaling related to different stress scenarios (reviewed in [68]), and *pm1* mutants have been reported to possess an enhanced sensitivity to ABA-inhibition of germination [21]. ABA involvement in the responses would be consistent with findings describing ABA activation of plant G-protein mediated pathways [42,65–67] and our own data regarding B and ABA signaling [21,22,29]. Several recent studies have linked Pirins with the mechanisms of how plants are infected by various organisms, including Haustorium development in the in the parasitic plant *Triphysaria versicolor* [69] and survival of the human fungal pathogen *Cryptococcus neoformans* [23], the latter of which may be due to the LAC1 gene, where the *C. neoformans* laccase catalyzes the oxidation of quercetin [70]. Therefore PRN1's role as a quercetinase, in a biotic context, may be influencing its biological functions. Consistent with a role in stress-mediation, quercetin was observed to bind the ER stress-induced kinase-endonuclease IRE1, and to function alongside stress signals from the ER lumen in modulating IRE1 activity in a yeast model [71]. The roles of PRN1 as a quercetinase and transcriptional co-factor, and its multiple sites of localization, may

represent a plant-specific adaptation of multi-function, subcellular multitasking in higher plants.

Supporting Information

Figure S1 Analysis of the hypocotyl orientation phenotype of 3-d-old white light grown seedlings. Seedlings were grown for 3 d in white light, and hypocotyl angle was measured for individual seedlings in reference to the horizontal phytatray surface (where vertical = 90°), n = 65. A non-parametric statistical analysis was performed, where the Kolmogorov-Smirnov normality test indicated that the WT and *pm1* data were not normally distributed (**1A**. WT: KS = 0.222, p<0.010, *pm1* **1B**. KS = 0.146 and p<0.010. **1C**. When *pm1* mutants were transformed with *PRN1*::PRN1-GFP (*PRN1*::PRN1-GFP (*pm1*)) and grown in white light for 3 d, the seedlings exhibited a restored WT hypocotyl orientation. WT and *pm1* seedlings are also shown on the figure. Seedlings on the phytatray were imaged from above. Scale bar = 1 mm.

Figure S2 Hypocotyl orientation responses of seedlings in 6-d complete darkness. Seeds of WT, *pm1*, transformed line *PRN1*::PRN1-GFP (*pm1*) and transformed line 35S::PRN1-GFP (WT) were sown on vertical plates, grown for 6 d in complete darkness, then photographed to view full seedling. Representative images are shown. Scale bar = 2 mm.

Figure S3 Hypocotyl orientation responses of seedlings in 6-d white light (16:8). Seeds of WT, *pm1*, transformed line *PRN1*::PRN1-GFP (*pm1*) and transformed line 35S::PRN1-GFP (WT) were sown on vertical plates, grown for 6 d in white light (16:8) then photographed to view full seedling. Representative images are shown. Scale bar = 2 mm.

Figure S4 Epidermal cotyledon peel of 6-d dark-grown seedlings indicates mainly nuclear localization. Epidermal peels of live cotyledons of 6-d dark-grown seedlings of *pm1* mutants or *pm1*-transformed with *PRN1*::PRN1-GFP were viewed on spinning disk confocal after DAPI-stain. 10–15 cotyledons were viewed per replicate of 3 independent replicates. Images are representative. n = nucleus; scale bar = 25 μm.

Figure S5 Epidermal cotyledon peel of 6-d 16:8-grown seedlings indicates mainly nuclear localization. Epidermal peels of live cotyledons of 16:8 dark-grown seedlings of *pm1* mutants or *pm1*-transformed with *PRN1*::PRN1-GFP or WT-transformed with 35S::PRN1-GFP were viewed on spinning disk confocal after DAPI-stain. 10–15 cotyledons were viewed per replicate of 3 independent replicates. Images are representative. n = nucleus; scale bar = 25 μm.

Figure S6 Possible cis-regulatory elements of *PRN1*. Frequently-repeated (≥20) cis motifs in the *PRN1* (At3g59220) promoter region (+ & − strand; 2,651 bp), determined from the "Database of plant cis-acting regulatory DNA elements" (http://www.dna.affrc.go.jp/PLACE/) [38,39]. Y = T/C; N = G/A/C/T; R = A/G; W = A/T; nd = not determined. Each motif is represented by a symbol, and the approximate location of each repeat is displayed along the positive (+) and negative (−) strands of the *PRN1* promoter (from 5′ to 3′ direction). The 5′-UTR region of PRN1 is represented with an asterisk "*" (in the black box), and the beginning of the *PRN1* open reading frame is designated by its start codon start codon "ATG" (in gray box). The white part of the 5′ to 3′ bar represents the 3′UTR (223 bp) of At3g59210, a gene that putatively codes for a protein with homology to F-box/RNI-like superfamily, cyclin-like, and LRR2 proteins.

Table S1 Primers used for cloning.

Acknowledgments

We gratefully acknowledge our undergraduate researchers Ms. Jennifer Baek and Ms. Ashley Williams at UIC for excellent technical assistance with experiments, and Dr. Jeremy Lynch (UIC) for assistance with confocal microscopy, and Dr. Nava Segev lab (UIC) for assistance with spectrophotometer. We also extend our gratitude to Dr. Terri Long (North Carolina State University) for advice on cloning.

Author Contributions

Conceived and designed the experiments: DAO-N KMW LSK. Performed the experiments: DAO-N DM RM B-SL LJ KMW. Analyzed the data: DAO-N DM B-SL LJ KMW. Contributed reagents/materials/analysis tools: DAO-N DM KMW. Wrote the paper: DAO-N KMW.

References

1. Kubasek WL, Shirley BW, Mckillop A, Goodman HM, Briggs W, et al. (1992) Regulation of Flavonoid Biosynthetic Genes in Germinating Arabidopsis Seedlings. Plant Cell 4: 1229–1236.
2. Winkel-Shirley B (2001) It takes a garden. How work on diverse plant species has contributed to an understanding of flavonoid metabolism. Plant Physiol 127: 1399–1404.
3. Agati G, Azzarello E, Pollastri S, Tattini M (2012) Flavonoids as antioxidants in plants: Location and functional significance. Plant Science 196: 67–76.
4. Peer WA, Brown DE, Tague BW, Muday GK, Taiz L, et al. (2001) Flavonoid Accumulation Patterns of Transparent Testa Mutants of Arabidopsis. Plant Physiology 126: 536–548.
5. Brown DE, Rashotte AM, Murphy AS, Normanly J, Tague BW, et al. (2001) Flavonoids Act as Negative Regulators of Auxin Transport in Vivo in Arabidopsis. Plant Physiology 126: 524–535.
6. Murphy A, Peer WA, Taiz L (2000) Regulation of auxin transport by aminopeptidases and endogenous flavonoids. Planta 211: 315–324.
7. Kuhn BM, Geisler M, Bigler L, Ringli C (2011) Flavonols Accumulate Asymmetrically and Affect Auxin Transport in Arabidopsis. Plant Physiology 156: 585–595.
8. Lewis DR, Ramirez MV, Miller ND, Vallabhaneni P, Ray WK, et al. (2011) Auxin and Ethylene Induce Flavonol Accumulation through Distinct Transcriptional Networks. Plant Physiology 156: 144–164.
9. Rozema J, Björn LO, Bornman JF, Gaberščik A, Häder DP, et al. (2002) The role of UV-B radiation in aquatic and terrestrial ecosystems–an experimental and functional analysis of the evolution of UV-absorbing compounds. Journal of Photochemistry and Photobiology B: Biology 66: 2–12.
10. Saslowsky DE, Warek U, Winkel BSJ (2005) Nuclear Localization of Flavonoid Enzymes in Arabidopsis. Journal of Biological Chemistry 280: 23735–23740.
11. Oka T, Simpson FJ (1971) Quercetinase, a Dioxygenase Containing Copper. Biochemical and Biophysical Research Communications 43: 1-&.
12. Adams M, Jia ZC (2005) Structural and biochemical analysis reveal pirins to possess quercetinase activity. Journal of Biological Chemistry 280: 28675–28682.
13. Soo PC, Horng YT, Lai MJ, Wei JR, Hsieh SC, et al. (2007) Pirin regulates pyruvate catabolism by interacting with the pyruvate dehydrogenase E1 subunit and modulating pyruvate dehydrogenase activity. Journal of Bacteriology 189: 109–118.
14. Gelbman BD, Heguy A, O'Connor TP, Zabner J, Crystal RG (2007) Upregulation of pirin expression by chronic cigarette smoking is associated with bronchial epithelial cell apoptosis. Respiratory Research 8.
15. Orzaez D, de Jong AJ, Woltering EJ (2001) A tomato homologue of the human protein PIRIN is induced during programmed cell death. Plant Molecular Biology 46: 459–468.
16. Brzoska K, Stepkowski TM, Kruszewski M (2011) Putative proto-oncogene Pir expression is significantly up-regulated in the spleen and kidney of cytosolic superoxide dismutase-deficient mice. Redox Report 16: 129–133.
17. Licciulli S, Cambiaghi V, Scafetta G, Gruszka AM, Alcalay M (2010) Pirin downregulation is a feature of AML and leads to impairment of terminal myeloid differentiation. Leukemia 24: 429–437.
18. Bergman AC, Alaiya AA, Wendler W, Binetruy B, Shoshan M, et al. (1999) Protein kinase-dependent overexpression of the nuclear protein pirin in c-JUN and RAS transformed fibroblasts. Cellular and Molecular Life Sciences 55: 467–471.
19. Dunwell JM, Culham A, Carter CE, Sosa-Aguirre CR, Goodenough PW (2001) Evolution of functional diversity in the cupin superfamily. Trends in Biochemical Sciences 26: 740–746.
20. Wendler WMF, Kremmer E, Forster R, Winnacker EL (1997) Identification of Pirin, a novel highly conserved nuclear protein. Journal of Biological Chemistry 272: 8482–8489.
21. Lapik YR, Kaufman LS (2003) The Arabidopsis cupin domain protein AtPirin1 interacts with the G protein alpha-subunit GPA1 and regulates seed germination and early seedling development. Plant Cell 15: 1578–1590.
22. Warpeha KM, Upadhyay S, Yeh J, Adamiak J, Hawkins SI, et al. (2007) The GCR1, GPA1, PRN1, NF-Y signal chain mediates both blue light and abscisic acid responses in Arabidopsis. Plant Physiology 143: 1590–1600.
23. Warpeha KM, Park Y-d, Williamson PR (2013) Susceptibility of Intact Germinating Arabidopsis thaliana to the Human Fungal Pathogen Cryptococcus. Applied and Environmental Microbiology.
24. Usadel B, Bläsing OE, Gibon Y, Retzlaff K, Höhne M, et al. (2008) Global Transcript Levels Respond to Small Changes of the Carbon Status during Progressive Exhaustion of Carbohydrates in Arabidopsis Rosettes. Plant Physiology 146: 1834–1861.
25. Nemhauser JL, Hong F, Chory J (2006) Different Plant Hormones Regulate Similar Processes through Largely Nonoverlapping Transcriptional Responses. Cell 126: 467–475.
26. Catala R, Ouyang J, Abreu IA, Hu Y, Seo H, et al. (2007) The Arabidopsis E3 SUMO Ligase SIZ1 Regulates Plant Growth and Drought Responses. The Plant Cell Online 19: 2952–2966.
27. Perfus-Barbeoch L, Jones AM, Assmann SM (2004) Plant heterotrimeric G protein function: insights from Arabidopsis and rice mutants. Current Opinion in Plant Biology 7: 719–731.
28. Alonso JM, Stepanova AN, Leisse TJ, Kim CJ, Chen H, et al. (2003) Genome-Wide Insertional Mutagenesis of Arabidopsis thaliana. Science 301: 653–657.
29. Warpeha KM, Lateef SS, Lapik Y, Anderson M, Lee BS, et al. (2006) G-protein-coupled receptor 1, G-protein G alpha-subunit 1, and prephenate dehydratase 1 are required for blue light-induced production of phenylalanine in etiolated Arabidopsis. Plant Physiology 140: 844–855.
30. Warpeha KM, Gibbons J, Carol A, Slusser J, Tree R, et al. (2008) Adequate phenylalanine synthesis mediated by G protein is critical for protection from UV radiation damage in young etiolated Arabidopsis thaliana seedlings. Plant Cell Environ 31: 1756–1770.
31. Preibisch S, Saalfeld S, Tomancak P (2009) Globally optimal stitching of tiled 3D microscopic image acquisitions. Bioinformatics 25: 1463–1465.
32. Hellens R, Edwards EA, Leyland N, Bean S, Mullineaux P (2000) pGreen: a versatile and flexible binary Ti vector for Agrobacterium-mediated plant transformation. Plant Molecular Biology 42: 819–832.
33. Lee J-Y, Colinas J, Wang JY, Mace D, Ohler U, et al. (2006) Transcriptional and posttranscriptional regulation of transcription factor expression in Arabidopsis roots. Proceedings of the National Academy of Sciences 103: 6055–6060.
34. Earley KW, Haag JR, Pontes O, Opper K, Juehne T, et al. (2006) Gateway-compatible vectors for plant functional genomics and proteomics. The Plant Journal 45: 616–629.
35. Clough SJ, Bent AF (1998) Floral dip: a simplified method for Agrobacterium-mediated transformation of Arabidopsis thaliana. The Plant Journal 16: 735–743.
36. Young JJ, Mehta S, Israelsson M, Godoski J, Grill E, et al. (2006) CO2 signaling in guard cells: Calcium sensitivity response modulation, a Ca2+-independent phase, and CO2 insensitivity of the gca2 mutant. Proceedings of the National Academy of Sciences 103: 7506–7511.
37. Klopffleisch K, Phan N, Augustin K, Bayne RS, Booker KS, et al. (2011) Arabidopsis G-protein interactome reveals connections to cell wall carbohydrates and morphogenesis. Molecular Systems Biology 7.
38. Prestridge DS (1991) SIGNAL SCAN: a computer program that scans DNA sequences for eukaryotic transcriptional elements. Computer applications in the biosciences : CABIOS 7: 203–206.
39. Higo K, Ugawa Y, Iwamoto M, Korenaga T (1999) Plant cis-acting regulatory DNA elements (PLACE) database: 1999. Nucleic Acids Research 27: 297–300.
40. Winter D, Vinegar B, Nahal H, Ammar R, Wilson GV, et al. (2007) An "Electronic Fluorescent Pictograph" Browser for Exploring and Analyzing Large-Scale Biological Data Sets. PLoS ONE 2: e718.
41. Routaboul JM, Kerhoas L, Debeaujon I, Pourcel L, Caboche M, et al. (2006) Flavonoid diversity and biosynthesis in seed of Arabidopsis thaliana. Planta 224: 96–107.
42. Temple BRS, Jones AM (2007) The plant heterotrimeric G-protein complex. Annual Review of Plant Biology 58: 249–266.
43. Alvarez S, Hicks LM, Pandey S (2011) ABA-Dependent and -Independent G-Protein Signaling in Arabidopsis Roots Revealed through an iTRAQ Proteomics Approach. Journal of Proteome Research 10: 3107–3122.

44. Ullah H, Chen JG, Young JC, Im KH, Sussman MR, et al. (2001) Modulation of cell proliferation by heterotrimeric G protein in Arabidopsis. Science 292: 2066–2069.

45. Johnston CA, Taylor JP, Gao Y, Kimple AJ, Grigston JC, et al. (2007) GTPase acceleration as the rate-limiting step in Arabidopsis G protein-coupled sugar signaling. Proc Natl Acad Sci U S A 104: 17317–17322.

46. Korkina LG (2007) Phenylpropanoids as naturally occurring antioxidants: from plant defense to human health. Cell Mol Biol (Noisy-le-grand) 53: 15–25.

47. Boege F, Straub T, Kehr A, Boesenberg C, Christiansen K, et al. (1996) Selected Novel Flavones Inhibit the DNA Binding or the DNA Religation Step of Eukaryotic Topoisomerase I. Journal of Biological Chemistry 271: 2262–2270.

48. Ruiz PA, Braune A, Hölzlwimmer G, Quintanilla-Fend L, Haller D (2007) Quercetin Inhibits TNF-Induced NF-κB Transcription Factor Recruitment to Proinflammatory Gene Promoters in Murine Intestinal Epithelial Cells. The Journal of Nutrition 137: 1208–1215.

49. Spencer JPE, Rice-Evans C, Williams RJ (2003) Modulation of Pro-survival Akt/Protein Kinase B and ERK1/2 Signaling Cascades by Quercetin and Its in Vivo Metabolites Underlie Their Action on Neuronal Viability. Journal of Biological Chemistry 278: 34783–34793.

50. Adams MA, Suits MDL, Zheng J, Jia Z (2007) Piecing together the structure–function puzzle: Experiences in structure-based functional annotation of hypothetical proteins. PROTEOMICS 7: 2920–2932.

51. Ciolino HP, Daschner PJ, Yeh GC (1999) Dietary flavonols quercetin and kaempferol are ligands of the aryl hydrocarbon receptor that affect CYP1A1 transcription differentially. Biochem J 340: 715–722.

52. Xing N, Chen Y, Mitchell SH, Young CYF (2001) Quercetin inhibits the expression and function of the androgen receptor in LNCaP prostate cancer cells. Carcinogenesis 22: 409–414.

53. Peer WA, Blakeslee JJ, Yang H, Murphy AS (2011) Seven Things We Think We Know about Auxin Transport. Molecular Plant 4: 487–504.

54. Poppe C, Hangarter RP, Sharrock RA, Nagy F, Schäfer E (1996) The light-induced reduction of the gravitropic growth-orientation of seedlings of Arabidopsis thaliana (L.) Heynh. is a photomorphogenic response mediated synergistically by the far-red-absorbing forms of phytochromes A and B. Planta 199: 511–514.

55. Jensen PJ, Hangarter RP, Estelle M (1998) Auxin Transport Is Required for Hypocotyl Elongation in Light-Grown but Not Dark-Grown Arabidopsis. Plant Physiology 116: 455–462.

56. Rashotte AM, Poupart J, Waddell CS, Muday GK (2003) Transport of the Two Natural Auxins, Indole-3-Butyric Acid and Indole-3-Acetic Acid, in Arabidopsis. Plant Physiology 133: 761–772.

57. Liu X, Cohen JD, Gardner G (2011) Low-Fluence Red Light Increases the Transport and Biosynthesis of Auxin. Plant Physiology 157: 891–904.

58. Weiss CA, White E, Huang H, Ma H (1997) The G protein α subunit (GPα1) is associated with the ER and the plasma membrane in meristematic cells of Arabidopsis and cauliflower. FEBS Letters 407: 361–367.

59. Kim JY, Yuan Z, Cilia M, Khalfan-Jagani Z, Jackson D (2002) Intercellular trafficking of a KNOTTED1 green fluorescent protein fusion in the leaf and shoot meristem of Arabidopsis. Proceedings of the National Academy of Sciences 99: 4103–4108.

60. Licciulli S, Luise C, Scafetta G, Capra M, Giardina G, et al. (2011) Pirin Inhibits Cellular Senescence in Melanocytic Cells. American Journal of Pathology 178: 2397–2406.

61. Komatsu S, Yang G, Hayashi N, Kaku H, Umemura K, et al. (2004) Alterations by a defect in a rice G protein α subunit in probenazole and pathogen-induced responses. Plant, Cell & Environment 27: 947–957.

62. Nilson SE, Assmann SM (2010) The α-Subunit of the Arabidopsis Heterotrimeric G Protein, GPA1, Is a Regulator of Transpiration Efficiency. Plant Physiology 152: 2067–2077.

63. Zhang W, Jeon BW, Assmann SM (2011) Heterotrimeric G-protein regulation of ROS signalling and calcium currents in Arabidopsis guard cells. Journal of Experimental Botany 62: 2371–2379.

64. Joo JH, Wang SY, Chen JG, Jones AM, Fedoroff NV (2005) Different signaling and cell death roles of heterotrimeric G protein alpha and beta subunits in the arabidopsis oxidative stress response to ozone. Plant Cell 17: 957–970.

65. Pandey S, Assmann SM (2004) The Arabidopsis putative G protein-coupled receptor GCR1 interacts with the G protein alpha subunit GPA1 and regulates abscisic acid signaling. Plant Cell 16: 1616–1632.

66. Zhao ZX, Stanley BA, Zhang W, Assmann SM (2010) ABA-Regulated G Protein Signaling in Arabidopsis Guard Cells: A Proteomic Perspective. J Proteome Res 9: 1637–1647.

67. Pandey S, Chen JG, Jones AM, Assmann SM (2006) G-protein complex mutants are hypersensitive to abscisic acid regulation of germination and postgermination development. Plant Physiol 141: 243–256.

68. Cutler SR, Rodriguez PL, Finkelstein RR, Abrams SR (2010) Abscisic Acid: emergence of a core signaling network. Annual Review of Plant Biology 61: 651–679.

69. Bandaranayake PCG, Tomilov A, Tomilova NB, Ngo QA, Wickett N, et al. (2012) The TvPirin Gene Is Necessary for Haustorium Development in the Parasitic Plant Triphysaria versicolor. Plant Physiol 158: 1046–1053.

70. Pereira L, Bastos C, Tzanov T, Cavaco-Paulo A, Guebitz GM (2005) Environmentally friendly bleaching of cotton using laccases. Environmental Chemistry Letters 3: 66–69.

71. Wiseman RL, Zhang YH, Lee KPK, Harding HP, Haynes CM, et al. (2010) Flavonol Activation Defines an Unanticipated Ligand-Binding Site in the Kinase-RNase Domain of IRE1. Molecular Cell 38: 291–304.

6

Expression and Functional Characterization of the *Agrobacterium* VirB2 Amino Acid Substitution Variants in T-pilus Biogenesis, Virulence, and Transient Transformation Efficiency

Hung-Yi Wu[1,2], Chao-Ying Chen[2], Erh-Min Lai[1,2]*

1 Institute of Plant and Microbial Biology, Academia Sinica, Taipei, Taiwan, **2** Department of Plant Pathology and Microbiology, National Taiwan University, Taipei, Taiwan

Abstract

Agrobacterium tumefaciens is a phytopathogenic bacterium that causes crown gall disease by transferring transferred DNA (T-DNA) into the plant genome. The translocation process is mediated by the type IV secretion system (T4SS) consisting of the VirD4 coupling protein and 11 VirB proteins (VirB1 to VirB11). All VirB proteins are required for the production of T-pilus, which consists of processed VirB2 (T-pilin) and VirB5 as major and minor subunits, respectively. VirB2 is an essential component of T4SS, but the roles of VirB2 and the assembled T-pilus in *Agrobacterium* virulence and the T-DNA transfer process remain unknown. Here, we generated 34 VirB2 amino acid substitution variants to study the functions of VirB2 involved in VirB2 stability, extracellular VirB2/T-pilus production and virulence of *A. tumefaciens*. From the capacity for extracellular VirB2 production (ExB2$^+$ or ExB2$^-$) and tumorigenesis on tomato stems (Vir$^+$ or Vir$^-$), the mutants could be classified into three groups: ExB2$^-$/Vir$^-$, ExB2$^-$/Vir$^+$, and ExB2$^+$/Vir$^+$. We also confirmed by electron microscopy that five ExB2$^-$/Vir$^+$ mutants exhibited a wild-type level of virulence with their deficiency in T-pilus formation. Interestingly, although the five T-pilus$^-$/Vir$^+$ uncoupling mutants retained a wild-type level of tumorigenesis efficiency on tomato stems and/or potato tuber discs, their transient transformation efficiency in *Arabidopsis* seedlings was highly attenuated. In conclusion, we have provided evidence for a role of T-pilus in *Agrobacterium* transformation process and have identified the domains and amino acid residues critical for VirB2 stability, T-pilus biogenesis, tumorigenesis, and transient transformation efficiency.

Editor: Ching-Hong Yang, University of Wisconsin-Milwaukee, United States of America

Funding: This work was supported by a research grant from the National Science Council (NSC 101-2321-B-001 -033 -) and a grant from Academia Sinica to EML. The funders had no role in study design, data collection and analysis, decision to publish, or preparation of the manuscript.

Competing Interests: The authors have declared that no competing interests exist.

* Email: emlai@gate.sinica.edu.tw

Introduction

Agrobacterium tumefaciens is a Gram-negative plant pathogenic bacterium that causes crown gall disease in a wide range of plants [1]. *A. tumefaciens* can sense plant-released phenolic compounds (e.g., acetosyringone; AS) to activate the expression of virulence factors for infection. The VirA/VirG two-component system is responsible for the phenolics-induced virulence (*vir*) gene expression [2]. Transfer DNA (T-DNA) located on the tumor-inducing (Ti) plasmid is recognized and processed by the VirD1/VirD2 relaxosome-like protein complex and covalently linked with the VirD2 protein to form the T-strand. The T-DNA and effector protein substrates are transferred through the VirB/VirD4 assembled type IV secretion system (T4SS) into host plant cells [2–4].

The *A. tumefaciens* VirB/VirD4 T4SS consists of the envelope-spanning translocation channel and the extracellular T-pilus structure [5–8]. Accumulating biochemical and genetic data suggest a possible VirB/D4 T4SS assembly and T-DNA translocation pathway [9,10]. The T-DNA immunoprecipitation (TrIP) technique revealed that T-DNA was first recruited by the VirD4 coupling protein, then the substrate passed to the inner-membrane–associated ATPase VirB11. The VirD4/VirB4 ATPases provide energy for T-DNA transfer to the inner-membrane proteins VirB6/VirB8 followed by passage to VirB2/VirB9 presumably localized at the distal end of the T4SS transmembrane complex [10]. Recent cryo-electron microscopy (cryo-EM) and crystallographic structure studies of the *Escherichia coli* conjugative plasmid pKM101 revealed that the T4SS core complex consisted of 14 copies of each of the VirB7-like TraN, VirB9-like TraO, and VirB10-like TraF subunits forming two layers of a double-walled ring structure inserted in the inner and outer membranes [11,12]. Remarkably, eight T4SS proteins (VirB3–VirB11) encoded by the R388 conjugative plasmid can assemble into an approximately 3-MDa nanomachine spanning the double membranes, which was visualized and reconstructed by electron microscopy [13]. VirB10 may function dynamically to couple cytoplasmic-membrane ATPases with ATP energy to gate the outer-membrane translocation channel via a conformational switch. Interestingly, VirD4 coupling protein not only functions as a receptor for protein substrates [14]; a recent study revealed that DNA but not protein binding to VirD4 and VirB11 activates the VirB10 structural transition and enables DNA transfer [14]. This study also suggested that translocation of DNA and protein substrates through T4SS may be mechanistically distinct processes [14].

Agrobacterium T-pilus is composed of the major subunit VirB2 and the minor component VirB5 [7,15,16]. All VirB proteins (VirB1 to VirB11) but not the VirD4 coupling protein is essential for T-pilus biogenesis [6]. Pilin subunits typically undergo an additional post-translational modification reaction after removal of its N-terminal signal peptide, including acetylation of F-like pilin or cyclization of P-like pilin and T-pilin [17–20]. The 12.3-kDa VirB2 precursor is processed into a 7.2-kDa product by removal of a long N-terminal signal peptide in both *E. coli* and *A. tumefaciens*, but the cyclization occurs only in *A. tumefaciens* in a Ti-plasmid–independent manner [21]. Consistent with predicted topology [19], experimental evidence revealed that processed VirB2 consists of two hydrophobic trans-membrane domains linked by an intervening hydrophilic loop (residues 90–94) inside the cytoplasm and by a periplasmic loop formed by linkage between hydrophilic N- and C-termini at residues 48 and 121 [22].

In contrast to much better-defined functions for the T4SS translocation channel, the roles of T4SS extracellular pilus remain obscure. T4SS substrate translocation from bacteria into host cells likely requires close contact with target cells [23] and pili may play a role during this process. A recent study revealed that the *A. tumefaciens* VirB/D4 T4SS forms helically arranged foci around the bacterial cell that may help maximize effective contact and transfer of substrate to host cells [24–26]. Therefore, T-pili may help *A. tumefaciens* bind to plant cells for close contact [25]. Alternatively, pili may also provide a channel for substrate translocation through their hollow lumen [27,28]. The transfer of DNA via the T4SS pili could be observed by direct visualization of low-efficiency conjugal transfer events via *E. coli* F-pili when cells were separated up to 1.2 μm [29]. Furthermore, DNA could be detected in the F-pilus channels during conjugation [30] and the *Helicobacter pylori* T4SS substrate protein CagA could be detected at the tip of the pilus [31]. However, mutants that block biogenesis of the T-pilus but not substrate transfer could be isolated by amino acid substitution in several T4SS components such as VirB6, VirB9, VirB10 and VirB11 of *A. tumefaciens* [14,32–36]. Because VirB2 protein is a T4SS component required for substrate translocation, isolation of these "uncoupling" mutants suggested that intracellular VirB2 but not its assembled T-pilus is required for T4SS-mediated T-DNA/effector translocation. Thus, the role of T-pilus remains unknown.

In this study, we used site-directed mutagenesis to generate various VirB2 single amino acid substitution variants to identify the amino acid residues critical for VirB2 stability, extracellular VirB2/T-pilus production, and virulence. Notably, we isolated five T-pilus⁻/Vir⁺ uncoupling mutants that retain a wild-type level of tumorigenesis efficiency on tomato stems and/or potato tuber discs but are highly attenuated in transient transformation efficiency in *Arabidopsis* seedlings. These data suggest a role of T-pilus in the *Agrobacterium* transformation process.

Results

VirB2 family proteins comprise variable N-terminal signal peptides and conserved C-terminal processing products

To determine the amino acid residues critical for the function of VirB2 in *Agrobacterium* virulence and T-pilus production, we first compared the amino acid sequences of VirB2 homologs encoded by various agrobacteria and rhizobia (Figure 1). The N-terminal

signal peptide of VirB2 is variable, but the mature processed T-pilin region (including periplasmic, trans-membrane and cytoplasmic domains) is highly conserved (Figure 1 and Table S1). A conserved motif PAxAQ at the processing site is critical for precise signal peptide removal and cyclization of processed T-pilin [17,21].

Identification of domains and amino acid residues critical for VirB2 stability, processing, and extracellular VirB2 production

For full complementation of *Agrobacterium* virulence in a *virB2* in-frame deletion mutant, *virB2* must be co-expressed with adjacent genes and driven by its native promoter [37]. Therefore, we cloned the DNA fragment containing the *virB* promoter and *virB1*, *virB2* and *virB3* genes (*virB*p-*B1*-*B2*-*B3*) into a broad host-range plasmid, pRL662, for expression of wild-type VirB2 and all VirB2 variants in the *virB2* in-frame deletion mutant (Δ*virB2*) derived from the *A. tumefaciens* wild-type C58 strain. The conserved or non-conserved amino acid residues near the processing site or different domains within the T-pilin region were randomly chosen for substitution with Alanine (A). *A. tumefaciens* cells induced for T-pilus production were scraped off the agar plate, resuspended in acidic phosphate buffer (pH 5.3), and centrifuged to obtain the S1 fraction. Cell pellets were resuspended again and subjected to shearing to obtain the S2 fraction enriched for T-pilus. Both intracellular and extracellular VirB2 levels were restored to wild-type levels in Δ*virB2* complemented with wild-type VirB2 (pVirB2), which was consistent with its full complementation by tumorigenesis analysis on tomato stems (Figure 2). Consistent with our previous study [21], we detected both pro-pilin (12.3-kDa unprocessed VirB2 precursor, named VirB2p) and T-pilin (7.2-kDa processed mature VirB2, named VirB2m) inside cells, but only processed T-pilin was detected extracellularly (Figure 2, also see Figure S1 for lower intensity of western blot signals). Extracellular VirB2 was more abundant in the S2 than S1 fraction, and all variants with no detectable VirB2 in the S2 fraction also did not produce any VirB2 signals in the S1 fraction (Figure 2 and S2). Because the detection of VirB2 in the sheared S2 fraction by western blot analysis agrees with the observation of T-pilus by electron microscopy and vice versa [6], we used western blot analysis of the S2 fraction as a first step to screen the mutant phenotype for T-pilus production.

For all five VirB2 variants with amino acid substitutions near the processing site (from P44 to G51), we detected both VirB2p and wild-type levels of VirB2m within the cells (Figure 2). Notably, extracellular VirB2 levels were reduced from the P44A and S49A variants and not detected from the A47V variant. Increased intracellular VirB2p from the P44A variant suggested a role of P44 for processing efficiency. The absence of extracellular VirB2 and slower migration of the intracellular VirB2m from the A47V variant relative to wild-type VirB2m implied that the A47V variant may undergo incorrect processing or cyclization, thus leading to defects in production of extracellular VirB2 or T-pilus. Strikingly, all variants except the R91A variant accumulated comparable intracellular VirB2m levels, in which only the unprocessed but not processed R91A variant could be detected (Figure 2). Because R91 is the sole positively charged residue within the cytoplasmic domain, substitution of Arginine 91 with Alanine may break the "positive inside rule" [38] and result in the instability of the R91A variant after processing.

Little or no extracellular VirB2 could be detected in all the mutants with amino acid substitutions in trans-membrane domain

1 (TM1) (Figure 3). In contrast, substitutions located in the N-terminal periplasmic domain (N-PP), trans-membrane domain 2 (TM2), and C-terminal periplasmic domain (C-PP) had differential effects on the accumulation of extracellular VirB2m. Therefore, these domains of processed VirB2 are all indispensable, but the integrity of the TM1 and its adjacent regions are most critical for production of extracellular VirB2m.

Identification of VirB2 variants uncoupling virulence and T-pilus biogenesis phenotypes

Tumorigenicity of each VirB2 variant was first determined by infection on tomato stems, and their virulence was evaluated by the occurrence or size of the tumors formed. From the capacity for extracellular VirB2 production (ExB2$^+$ or ExB2$^-$) and tumorigenesis on tomato stems (Vir$^+$ or Vir$^-$), the mutants were classified into three groups: ExB2$^-$/Vir$^-$, ExB2$^-$/Vir$^+$, and ExB2$^+$/Vir$^+$ (Figure 3). Although the combination of extracellular VirB2 production and virulence phenotypes theoretically can result in four groups of mutant phenotypes, we identified only three groups from our mutant pools. Group I mutants with the ExB2$^-$/Vir$^-$ phenotype led us to identify the amino acid residues crucial for both extracellular VirB2 production and tumorigenesis. These amino acid residues are mostly dispersed across all domains within the T-pilin region, but three (M88A, F89A, R91A) are located at the junction of TM1 and cytoplasmic domain (CP). Group II mutants with the ExB2$^-$/Vir$^+$ phenotype showed the amino acids residues critical for extracellular VirB2 production but dispensable for tumorigenesis. Strikingly, all mutants defective in virulence also lost extracellular VirB2 production (group I mutants) and all mutants capable of extracellular VirB2 production were virulent (group III mutants). These VirB2 amino acid residues required for A. tumefaciens to form a functional T4SS for substrate transfer may also be essential for production of extracellular VirB2.

Among the three groups of mutants, we were particularly interested in the five variants classified in the ExB2$^-$/Vir$^+$ group because these VirB2 variants (D55A, I85A, L94A, M107A, A110G) can incite the wild-type size of tumors on tomato stem but did not produce detectable extracellular VirB2. Although the loss of extracellular VirB2 production agrees with the loss of T-pilus biogenesis [6], we could not exclude that the failure to detect extracellular VirB2 may be due to the formation of short or structurally distinct T-pilus recalcitrant to isolation by shearing from the mutants. Thus, we negatively stained the A. tumefaciens cells expressing wild-type VirB2 and these five ExB2$^-$/Vir$^+$ variants by uranyl acetate and examined them by transmission electron microscopy (TEM) to observe the bacterial cells and associated surface structures. Similar to previous reports [6,7,17], we observed T-pilus as a rigid or semi-rigid long filament (500 nm to 2 μm) ~10-nm wide in the wild-type complemented strain ΔvirB2(pVirB2) in an AS-induction-dependent manner (Figure 4A). In contrast, no T-pilus-like filament could be detected from these ExB2$^-$/Vir$^+$ mutants (Figure 4B–F). These results agree with the lack of extracellular VirB2 detected by western blot analysis and suggest that these five mutants represent an uncoupling phenotype of virulence and T-pilus biogenesis (Figure 2). Thus, we defined the A. tumefaciens strains expressing these five VirB2 variants (D55A, I85A, L94A, M107A, A110G) as T-pilus$^-$/Vir$^+$ uncoupling mutants.

T-pilus$^-$/Vir$^+$ uncoupling mutants show highly attenuated transient transformation efficiency in Arabidopsis seedlings

The evidence that the T-pilus$^-$/Vir$^+$ uncoupling mutant retains the ability to incite tumors on tomato stems suggested that the T-pilus may not be essential for virulence. To test whether these T-pilus$^-$/Vir$^+$ uncoupling mutants may quantitatively affect Agrobacterium virulence or transformation efficiency, we first used the quantitative tumor assay on potato tuber discs to determine their virulence. Two T-pilus$^-$/Vir$^+$ uncoupling mutants (L94A and A110G) and a randomly selected ExB2$^+$/Vir$^+$ mutant (G121A) all retained the wild-type tumorigenesis efficiency when infected with 10^6 or 10^4 cells (Figure S3). Because tumorigenesis is a complex and long process that may not be sensitive enough to detect quantitative differences in T-DNA transfer efficiency, we adapted a transient transformation assay recently developed in our laboratory [39] for further analysis. This system used the T-DNA–encoded β-glucuronidase gusA (GUS) gene as a reporter to quantitatively monitor transient transformation efficiency in Arabidopsis seedlings. The T-DNA vector pBISN1 harboring the gusA-intron [40] was transformed into an A. tumefaciens strain with virulence gene expression induced by acetosyringone (AS) before infection of 4-day-old seedlings. The seedlings were co-cultured with pre-induced A. tumefaciens in the presence of AS, and GUS activity was determined to monitor transient transformation efficiency at 3 days post-infection (dpi). Remarkably, all five T-pilus$^-$/Vir$^+$ uncoupling mutants showed highly reduced GUS stains in cotyledons with 3- to 4-fold lower GUS activity as compared to the wild type (Figure 5A and 5C). In contrast, five ExB2$^+$/Vir$^+$ mutants exhibited higher transient transformation efficiency than all T-pilus$^-$/Vir$^+$ uncoupling mutants, four showing efficiency comparable to that of the wild type (Figure 5B and C). These results suggest the importance of T-pilus in the Agrobacterium transformation process.

Transient transformation assays of Arabidopsis seedlings with wounded cotyledons

In contrast to tumor assays on tomato stems and potato tuber discs, with A. tumefaciens cells infected on wounded tissues, no intentional wounding was included during A. tumefaciens infection in Arabidopsis seedlings. Thus, we tested whether the T-pilus$^-$/Vir$^+$ uncoupling mutants could efficiently infect wounded tissue of Arabidopsis seedlings. Because cotyledons are highly transformed in the Arabidopsis seedling transient transformation assay, we wounded cotyledons of 7-day-old Arabidopsis seedlings by using a needle. Similar to the method used to infect wounded tomato stems and potato tuber discs, overnight-grown A. tumefaciens cells without AS induction were used to infect wounded cotyledons of Arabidopsis seedlings. As shown in Figure 6A, we detected large GUS stains extending from the wound site of cotyledons on co-culture with cells containing wild-type VirB2 or the ExB2$^+$/Vir$^+$ G121A variant, which also revealed wild-type tumorigenesis efficiency in potato tuber discs (Figure S3) and transient transformation efficiency in intact Arabidopsis seedlings (Figure 5A and C). In contrast, GUS signals were detected only at a focused wound site of cotyledons and not expanded to unwounded regions on co-culture with the two T-pilus$^-$/Vir$^+$ uncoupling mutants tested (L94A and A110G). As a control, 7-day-old seedlings were infected with pre-induced A. tumefaciens cells under unwounded conditions (Figure 6B). We detected 3- to 4- fold lower GUS activity in the uncoupling mutants as compared with A. tumefaciens cells producing wild-type VirB2 or the ExB2$^+$/Vir$^+$ G121A variant, which

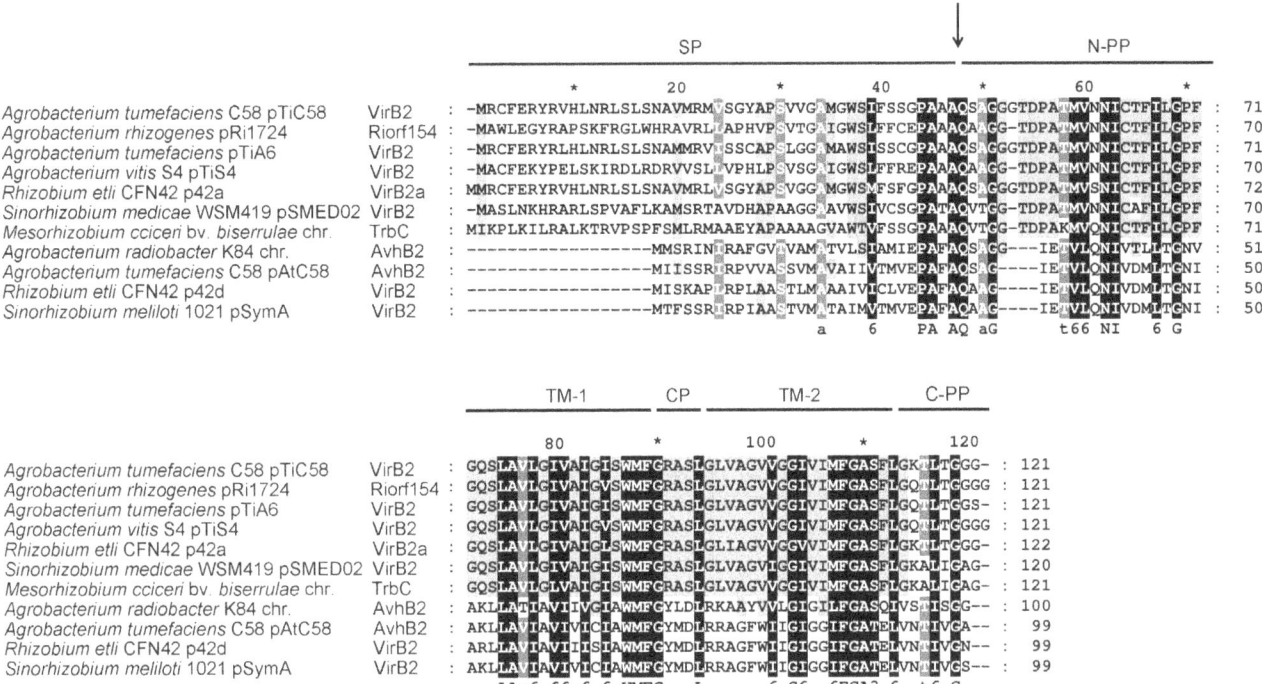

Figure 1. Amino acid sequence alignment of VirB2 family proteins. Multiple amino acid sequence alignment of VirB2 homologues with ClustalW2 [50]. The organism/plasmid name for each homolog is indicated on the left of the aligned sequence. UniProt accession numbers: *Agrobacterium tumefaciens* pTiC58, P17792; *Agrobacterium rhizogenes* pRi1724, Q9F5A1; *Agrobacterium tumefaciens* pTiA6, P05351; *Agrobacterium vitis* pTiS4, B9K417; *Rhizobium etli* p42a, Q2K2L1; *Sinorhizobium medicae* pSMED02; A6UMA7, *Mesorhizobium ciceri* chromosome (chr.), E8TGI0; *Agrobacterium radiobacter* K84 chromosome (chr.), B9JE70; *Agrobacterium tumefaciens* pAtC58, Q7D3S1; *Rhizobium etli* p42d, Q8KIM6; *Sinorhizobium meliloti* pSymA, Q92YZ4. The amino acid residues identical in all proteins are in black and those conserved in most but not all of the proteins are in gray. The arrow indicates the processing site of VirB2 encoded by pTiC58. Each region/domain of VirB2 is indicated: SP, signal peptide; TM-1, trans-membrane domain 1; CP, cytoplasmic domain; TM-2, trans-membrane domain 2; N-PP, N-terminal periplasmic domain; C-PP, C-terminal periplasmic domain.

efficiently infects both cotyledons and newly emerged true leaves not restricted to specific regions (Figure 6A and B). Because T-DNA transient transformation activity does not require the step(s) of T-DNA integration into the plant genome [40], T-pilus may play a role in the T-DNA transfer process at steps before T-DNA integration when infecting wounded and unwounded cotyledons of *Arabidopsis* seedlings.

Discussion

In this study, we identified the VirB2 key amino acid residues involved in VirB2 stability, extracellular VirB2/T-pilus production, and virulence of *A. tumefaciens* and discovered a role for the T-pilus in enhancing transient transformation efficiency. By screening 34 VirB2 variants for their ability to promote T-pilus production and tumorigenesis, we found that all mutants that are capable of producing extracellular VirB2 are virulent, and all mutants with loss of virulence (no tumor) also do not produce extracellular VirB2. Because no mutants with the ExB2⁺/Vir⁻ phenotype could be isolated in our screen, the VirB2 amino acid residue essential for forming a functional T4SS for substrate translocation may also be required for production of extracellular VirB2 and/or T-pilus.

Interestingly, the G119A variant generated in this study showed low amounts of extracellular VirB2 and full virulence phenotypes (Figure 2), whereas the substitution of Glycine 119 with Cysteine in pTiA6 VirB2 caused loss of virulence but retained function in

producing wild-type levels of extracellular VirB2 and T-pilus [22]. To clarify the phenotype discrepancy between pTiC58 G119A and pTiA6 G119C, we generated the pTiC58 VirB2 G119C amino acid substitution as for 34 other VirB2 variants and expressed it in C58 Δ*virB2*. By examining the levels of extracellular VirB2 in strains producing wild-type VirB2 and the G119A and G119C variants in parallel, all three independent G119C variants and the G119A variant produced low levels of extracellular VirB2 as compared with wild-type VirB2 (Figure S4). However, by examining all three independent pTiC58 G119C variants in three independent experiments, we did not observe any rigid and long T-pilus-like filament with ~10 nm diameter. In contrast, we detected the presence of T-pilus produced with the pTiC58 G119A variant, although the number of T-pilus observed was less frequent than with the wild-type VirB2 (Figure 4).

Transient transformation efficiency was comparable with the G119A variant and the wild type VirB2, but no transient GUS activity was detected in *Arabidopsis* seedlings infected with the G119C variant (Figure S5). Consistently, the G119A variant also retained the wild-type tumorigenicity in tomato stems and potato tuber discs, whereas the G119C variant showed highly attenuated induction of tumors on potato tuber discs (Figure S5A). Since G119A and G119C variants both caused reduced or abolished extracellular VirB2/T-pilus production, Glycine 119 may be critical for the efficient assembly of T-pilin into extracellular T-pilus. The nonpolar amino acid at position 119 seems critical for

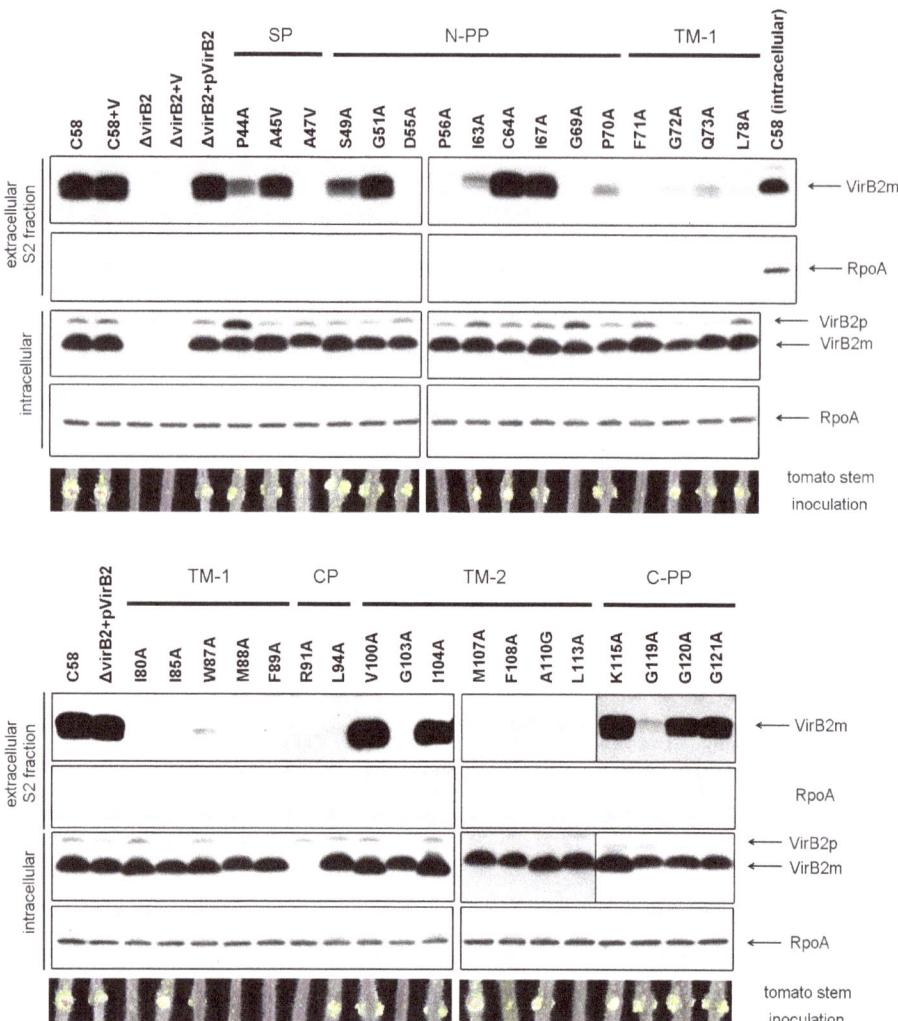

Figure 2. Western blot analysis of the intracellular and extracellular S2 fractions, and tumor assays on tomato stem of VirB2 variants. *A. tumefaciens* cells grown on acetosyringone (AS)-induced AB-MES (pH 5.5) agar at 19°C for 3 days [7] were collected to isolate intracellular proteins and extracellular S2 fractions. C58, *A. tumefaciens* wild type strain C58; V, empty vector pRL662; ΔvirB2, virB2 deletion mutant; ΔvirB2(pVirB2), expression of wild type *virBp-B1-B2-B3* in ΔvirB2. Western blot analysis with antisera against VirB2 B24 peptide or B23 peptide (for variants in C-PP) or RNA polymerase RpoA, as an internal control. Unprocessed VirB2 precursor is indicated as VirB2p and processed mature VirB2 as VirB2m. Shows representative tumor assay results on tomato stems. Similar results were obtained from at least three independent experiments (3–5 plants for each mutant in each independent experiment). Each region/domain of VirB2 is indicated as described in Figure 1.

efficient T-DNA and/or effector translocation because replacing nonpolar Glycine 119 with polar Cysteine but not nonpolar Alanine strongly suppressed the tumorigenesis/transformation efficiency. Although the loss of extracellular VirB2 in both S1 and S2 fractions is consistent with the lack of any observable long and rigid T-pilus-like filament in our five T-pilus⁻/Vir⁺ uncoupling mutants, these methods may have limitations to detect short T-pilus produced on bacterial cell surfaces. It is also possible that certain VirB2 variants may assemble T-pilus only during infection *in planta*. To this end, our data revealed the positive correlation of long and rigid T-pilus with the ability to induce high transient transformation efficiency in *Arabidopsis*, although we observed no dose-dependent correlation. This result highlighted a critical role of T-pilus involved in transient transformation efficiency in *Arabidopsis* seedlings, but T-pilus seems to be dispensable for inciting tumors in wounded plant tissues. Consistently, the T-pilus⁻/Vir⁺ un-

coupling mutants were isolated from amino acid substitution in other T4SS components including VirB6, VirB9, VirB10, and VirB11 [14,32–36].

The lack of T-pilus with the pTiC58 G119C variant is in contrast to the abundant T-pilus production from the pTiA6 G119C variant, which is also defective in substrate transfer based on the deficiency in IncQ plasmid transfer and tumorigeneisis on the wound site of *Kalanchoe daigremontiana* leaves [22]. However, all the Cysteine-substitution VirB2 variants created by Kerr and Christie [22] were generated in a VirB2 C64S (Cysteine 64 substitution by Serine) background to perform Cysteine labeling for mapping the VirB2 membrane topology. Although pTiA6 C64S variant exhibited near-wild-type substrate translocation frequency, whether the T-pilus⁺/Vir⁻ phenotype observed by the pTiA6 VirB2 C64S-G119C variant is solely caused by G119C substitution remains to be determined. Thus, the difference in T-pilus production phenotypes between pTiC58 and pTiA6 VirB2

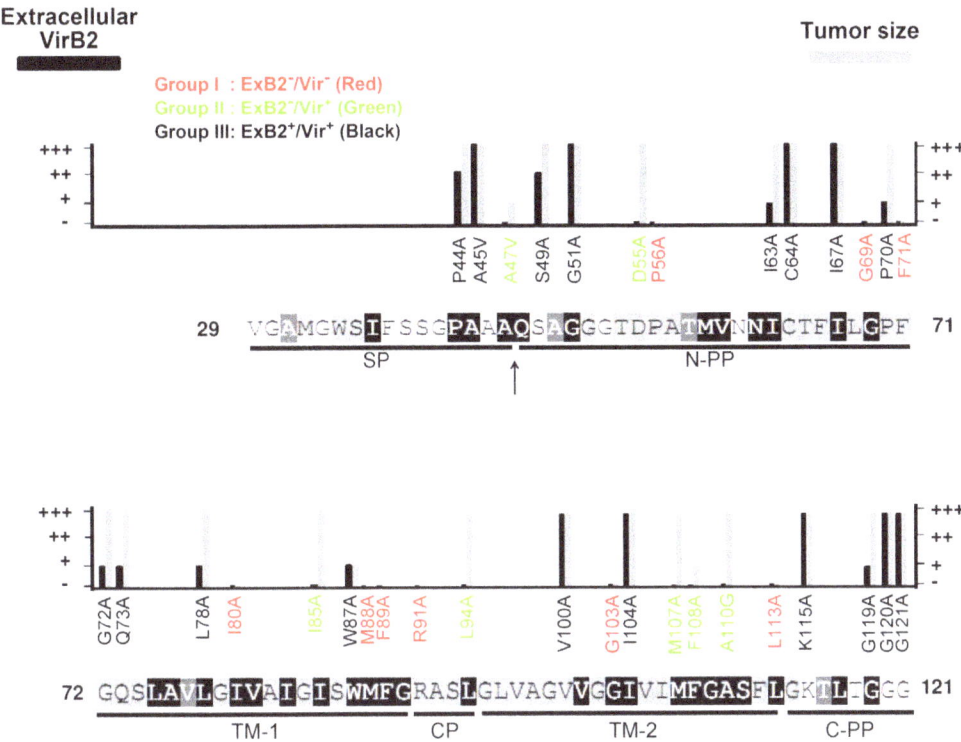

Figure 3. Phenotype summary of VirB2 variants. VirB2 amino acid substitutions are indicated with the levels of extracellular VirB2 (ExB2) production and occurrence or size of tumor formation on tomato stems (Vir) with wild type (+++), modest reduction (++), highly attenuation (+), or loss (−). VirB2 protein sequences with indicated conserved amino acid residues, regions/domains, and the processing site (indicated by an arrow) are presented as described in Figure 1. The 34 VirB2 variants are classified into three groups: ExB2⁻/Vir⁻, ExB2⁻/Vir⁺, and ExB2⁺/Vir⁺ shown in red, green and black, respectively.

G119C variants could be caused by this additional C64S mutation created in pTiA6. In addition, another four pTiC58 VirB2 Alanine substitution variants showed contrasting phenotypes in virulence or extracellular VirB2/T-pilus production as compared with pTiA6 VirB2 Cysteine substitution variants (Table S4). Thus, future work by creating the VirB2 single amino acid substitution with identical amino acid is required to confirm the phenotype discrepancy observed between two Ti plasmids. Replacing a specific amino acid residue by an amino acid with both similar and opposite biochemical features may be critical to unambiguously identify the role of each specific amino acid in contributing the observed phenotype.

Of note, pre-induction of *A. tumefaciens vir* gene expression before infection is critical for successful transient transformation in intact *Arabidopsis* seedlings without intentional wounding (data not shown). In contrast, this *vir* pre-induction is not required when infecting wounded cotyledons of *Arabidopsis* seedlings (Figure 6A). Since T-pilus⁻/Vir⁺ uncoupling mutants caused reduced transient transformation efficiency when infecting *Arabidopsis* seedlings with wounded or unwounded cotyledons, T-pilus may contribute to enhance *Agrobacterium* transient transformation efficiency at both wounded and unwounded infection conditions. However, in contrast to the more pronounced GUS activity detected further from wound sites of cotyledons infected with T-pilus⁺-producing strains, GUS signals were detected only at a focused wound site of cotyledons infected with the T-pilus⁻/Vir⁺ uncoupling mutants. Thus, T-pilus may be dispensable for infection at a wound site but critical for infecting unwounded tissues/cells. This possibility may also explain why these T-

pilus⁻/Vir⁺ uncoupling mutants retained their ability to incite tumor formation when infecting wounded tomato stems and/ or potato tuber discs in this study and in wounded *K. daigremontiana* leaves in other studies [14,32–36].

In summary, this study provides compelling evidence for a role of T-pilus in the *Agrobacterium* transformation process, although the mechanisms involved remain unclear. Recent microscopy studies revealed the formation of T4SS helical array around the bacterial cell and observed a pilus-like structure connecting bacterial cells to the plant cell [24,25]. However, the Ti plasmid and T-pilus were found not required for *A. tumefaciens* attachment to the plant cell [41]. Thus, T-pilus may modulate plant responses to achieve optimal transformation efficiency. VirB5 was found localized in the T-pilus tip [42] and extracellular VirB5 can accelerate *Agrobacterium* transformation efficiency [43]. T-pilus may provide a vehicle to localize VirB5 or other substrates to the correct location and exert their functions on/in plants or the T-pilus/VirB2 may play a direct role. The exact role and molecular mechanisms underlying how T-pilus contributes to *Agrobacterium* transformation process await future investigation.

Materials and Methods

Bacterial strains, growth and T-pilus induction conditions

Bacterial strains and plasmids used are in Table S2. *A. tumefaciens* and *E. coli* were grown in 523 medium [44] at 28°C and LB at 37°C, respectively, with appropriate antibiotics. The plasmids were maintained by the addition of 50 µg/ml gentamycin (Gm)

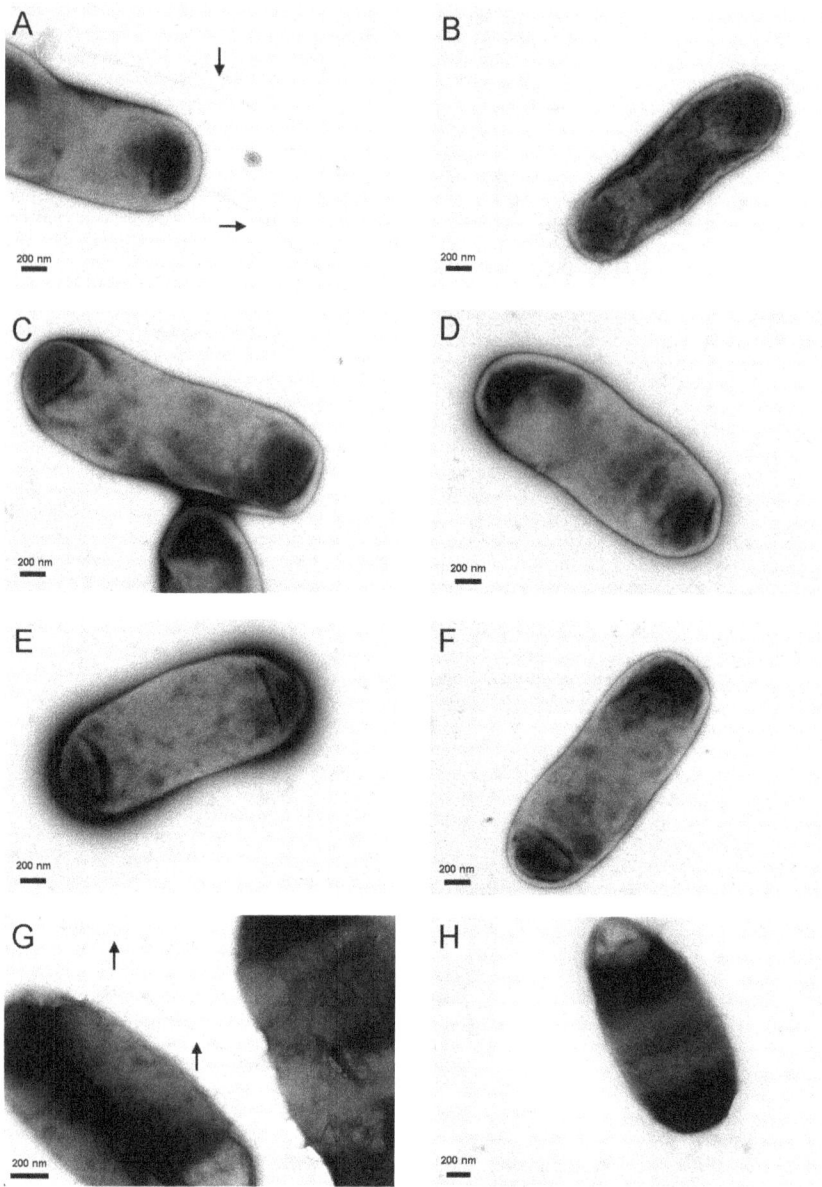

Figure 4. T-pilus observation by transmission electron microscopy (TEM). *A. tumefaciens* cells grown on T-pilus induction condition were collected and stained with 2% uranyl acetate to visualize T-pilus by TEM. Shows representative TEM image of *A. tumefaciens* strains producing wild-type VirB2 (A) or VirB2 variants D55A (B), I85A (C), L94A (D), M107A (E), A110G (F), G119A (G) and G119C (H). The rigid, long T-pilus is indicated by an arrow (A, G). Scale bar: 200 nm. All samples were examined for T-pilus formation by examining hundreds of bacterial cells from at least two independent experiments.

and 20 µg/ml kanamycin (Km) for *A. tumefaciens* and 20 µg/ml Km and 50 µg/ml Gm for *E. coli*. The T-pilus induction condition was as described [7]. Briefly, overnight culture of *A. tumefaciens* cells grown in 523 medium with appropriate antibiotics [44] were harvested and resuspended in liquid AB-MES minimal medium, pH 5.5 [7], with an adjusted OD_{600} of 0.1 for 4 hr without antibiotics. In total, 500 µl bacterial suspension was spreaded onto solid AB-MES medium with 200 µg/ml acetosyringone (AS) in a 150-mm Petri dish and incubated at 19°C for 3 days without antibiotics.

Construction of mutant strains and complementing plasmids

Gene replacement with the suicide plasmid pJQ200KS to generate the *virB2* in-frame deletion mutant (deletion of amino acid residues 4 to 113) followed a previous study [45]. For construction of pJQ-*virB2* for gene replacement, *virB2*-up (550-bp) and *virB2*-down (720-bp) DNA fragments were amplified by *pfu* polymerase and digested with SacI/SpeI and SpeI/XhoI separately and ligated into pJQ200KS at SacI/XhoI sites. For generating a plasmid to express *virB2* for complementation (pVirB2), a 2.1-kb fragment the *virB* promoter and *virB1*, *virB2* and *virB3* genes (*virB*p-*B1-B2-B3*) was PCR-amplified from the *A.*

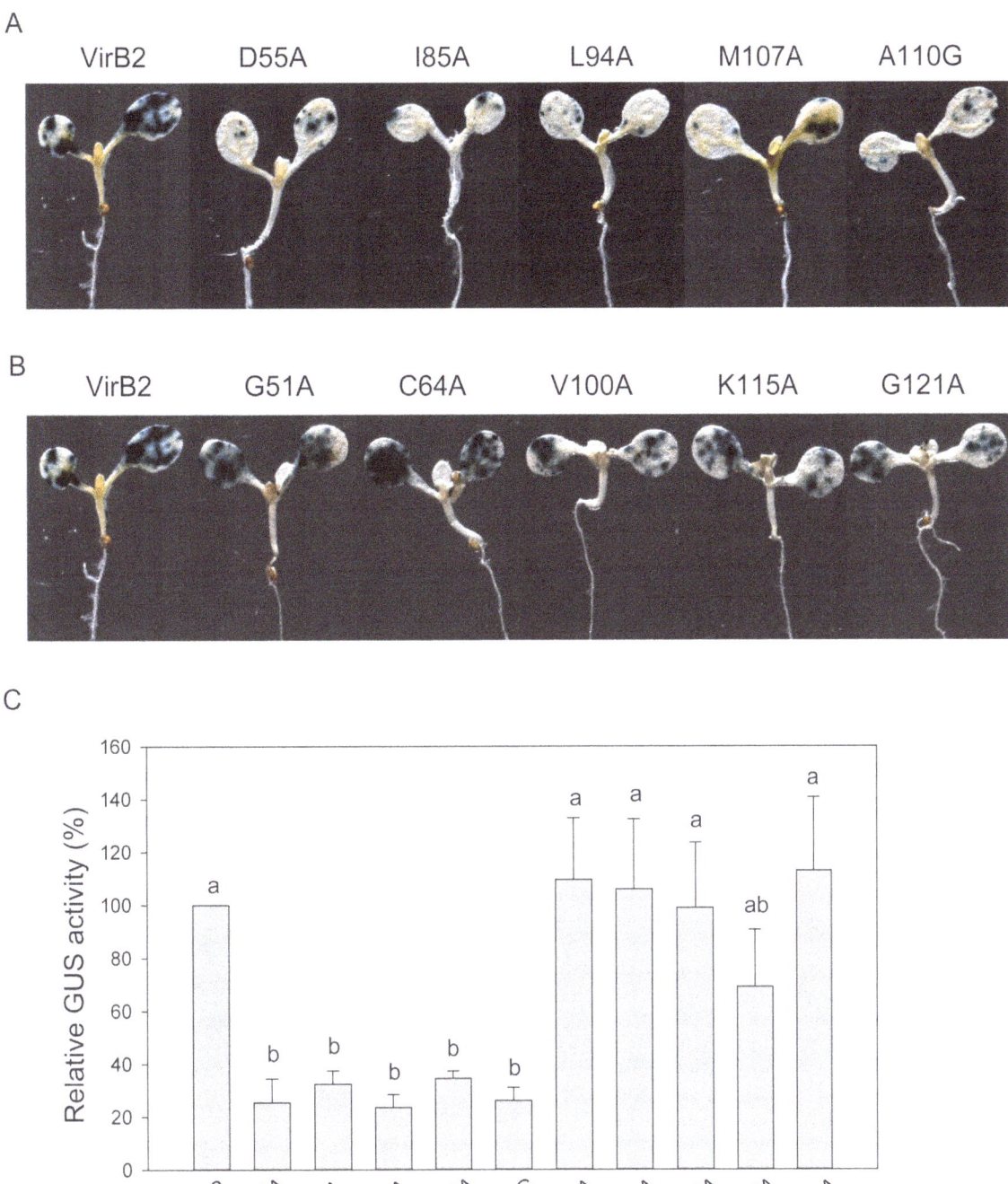

Figure 5. T-pilus⁻/Vir⁺ uncoupling mutants show highly attenuated transient transformation efficiency in *Arabidopsis* seedlings. *A. tumefaciens* strains expressing the wild type or variants of VirB2 harboring the T-DNA vector pBISN1 were used to infect 4-day-old *Arabidopsis* seedlings. GUS activity as a reporter of transient transformation efficiency was determined by GUS staining (A and B) or quantitative activity assay (C) at 3 dpi. (A) GUS staining of T-pilus⁻/Vir⁺ uncoupling mutants and (B) T-pilus⁺/Vir⁺ mutants. (C) Quantitative GUS activity of all mutants. Data are mean±SD of 4 biological repeats from 2 independent experiments (10 seedlings in each biological repeat). The data were analyzed by ANOVA for statistical classification, which revealed two groups (groups a and b) of strains differing in transient GUS activity.

tumefaciens C58 genome and ligated into pRL662 at SpeI/XhoI sites. For generating *virB2* mutants with single amino acid substitutions, corresponding primers in Table S3 were used to amplify mutated sequence templates, followed by further ampli-

fication by the universal primer pair VirBp-B1-SpeI-F and VirB3-XhoI-R to generate the 2.1-kb fragment for ligation into pRL662 at SpeI/XhoI sites. Sequences of the entire 2.1-kb fragment (*virB*p-

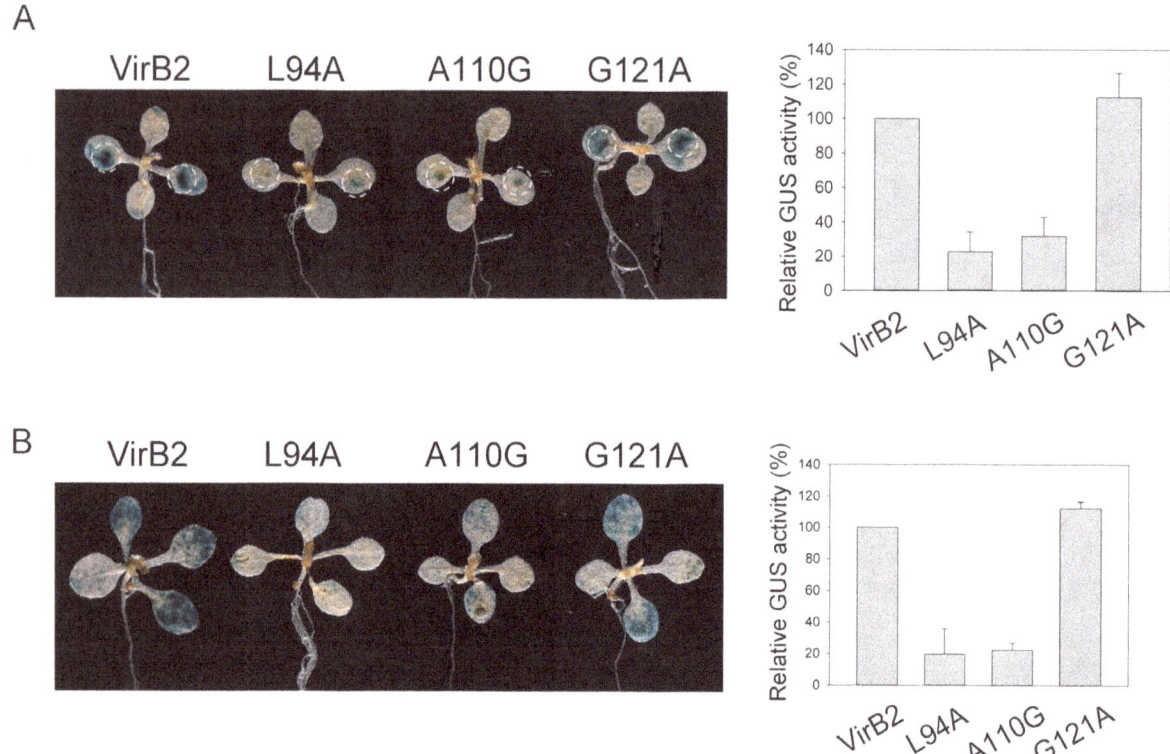

Figure 6. Transient transformation assay in *Arabidopsis* seedlings with or without wounded cotyledons. *A. tumefaciens* strains expressing the wild type or variants of VirB2, L94A (T-pilus⁻/Vir⁺), A110G (T-pilus⁻/Vir⁺) and G121A (T-pilus⁺/Vir⁺), harboring the T-DNA vector pBISN1 were used to infect 7-day-old *Arabidopsis* seedlings. GUS activity as a reporter of transient transformation efficiency was determined by GUS staining or quantitative activity assay at 3 dpi. (A) Infection of *Arabidopsis* seedlings with cotyledons wounded by a needle before infection. *A. tumefaciens* cells grown in 523 overnight culture without AS induction were used to infect the wounded *Arabidopsis* seedlings in the absence of AS. (B) Infection of *Arabidopsis* seedlings without intentional wounding, in which *A. tumefaciens* cells were pre-induced by AS for infection and co-cultured in the presence of AS. Data for quantitative GUS activity are mean±SD of four biological repeats from two independent experiments (10 seedlings in each biological repeat).

B1-B2-B3) were confirmed by DNA sequencing to ensure that no additional mutations occurred via PCR.

Isolation of intracellular and extracellular fractions

The procedure to isolate extracellular fractions was as described previously [7], with minor modifications. *A. tumefaciens* cells were scraped off a 150-mm AB-MES, pH 5.5, agar plate by adding 2 ml buffer A (10 mM phosphate buffer, pH 5.3); the resulting cell suspension was centrifuged (13000×g, 4°C, 10 min) to collect the supernatant, named the S1 fraction. The resulting pellet was resuspended again in buffer A to OD_{600} 10 and divided into 1-ml aliquots. The cells were sheared through a 26-g needle syringe five times and harvested by centrifugation (13000×g, 4°C, 10 min) to collect the supernatant containing sheared T-pilus, named the S2 fraction. Both S1 and S2 fractions were collected and filtrated through a 0.22-μm low-protein-binding membrane (Minisart RC 15, Sartorius stedim biotech) to remove contaminating bacterial cells and precipitated by trichloroacetic acid (TCA) as described previously [45]. The sheared pellet was normalized to OD_{600} 10 and designated as the intracellular fraction.

Tumor assay on tomato stems

Tomato cultivar FARMERS 301 from KNOWN-YOU SEED CO. (Kaohsiung, Taiwan) was grown at 23°C with a 16-/8-hr light/dark cycle. Two- to 3-week-old seedlings were wounded by use of a needle and inoculated with 5 μl *A. tumefaciens* cell suspension prepared as follows. *A. tumefaciens* cells were grown on a 523 agar plate at 28°C for 48 hr. Freshly grown colonies were re-suspended in 0.9% sodium chloride and adjusted to 10^8 and 10^6 CFU/ml for inoculation. Tumors were observed about 4 to 5 weeks after inoculation.

SDS-PAGE and western blot analysis

Proteins were separated by 12% or 16% tricine SDS-PAGE [46] followed by western blot analysis as described [45]. For each protein sample, an equivalent number of cells was mixed with an equal volume of 2x SDS-PAGE loading buffer (0.1 M Tris-Cl, pH 6.8, 4% SDS, 0.1% bromophenol blue, 20% glycerol, 200 mM dithiothreitol) and incubated at 100°C for 10 min before loading. Polyclonal antisera VirB2-B23 (against the N-terminal region of processed VirB2 T-pilin encoded by pTiC58), VirB2-B24 (against the C-terminal region of processed VirB2 T-pilin encoded by pTiC58) [47] and RNA polymerase α-subunit RpoA [48] were used as primary antibodies. Horseradish peroxidase-conjugated goat anti-rabbit immunoglobulin G (Chemichem) was the secondary antibody and chemiluminescence was detected by the Western Lightning system (Perkin Elmer, Boston, MA) with X-ray film (Amersham).

Electron microscopy

Procedures for negative staining were as described previously [6] with minor modifications. Briefly, *A. tumefaciens* cells grown under T-pilus induction conditions were collected, washed with pure water and re-suspend in 10 mM Tris buffer at pH 7.5. The bacterial suspension was deposited on a copper grid with carbon-Formvar film support for 1 min, rinsed with pure water for a few seconds, and stained with 2% uranyl acetate for 1 min. The samples were examined under a PHILIPS-CM100 transmission electron microscope (TEM) at 80 kV.

Transient transformation assay in *Arabidopsis* seedlings

The method for transient transformation assay in *Arabidopsis* seedlings was as described [39]. In brief, *Arabidopsis thaliana* mutant *efr-1* lacking the elongation factor Tu (EF-Tu) receptor (SALK_044334) was used. Seeds were sterilized in 50% bleach and 0.05% Trition X-100 for 10 min and washed with sterile water five times and incubated in a 4°C refrigerator for 3 days before germination. Seeds were germinated in 1-ml MS liquid medium (1/2 MS salts, 0.5% sucrose, pH 5.7) in a 6-well plate (10 seedlings in each well) at 22°C, 16-/8-hr light/dark cycle for 4 or 7 days (indicated as 4- or 7-day-old seedlings). For *A. tumefaciens vir* gene pre-induction and infection, overnight cultured *A. tumefaciens* cells were re-suspended to OD_{600} 0.2 in AB-MES, pH 5.5, with 200 μM AS for growth at 28°C for 14 to 16 hr. Cells were re-suspended in infection medium (1/2 MS salts, 0.5% sucrose, 50 μM AS, pH 5.7) and adjusted to OD_{600} 0.02 and co-cultured with *Arabidopsis* seedlings at 22°C for 3 days. After co-cultivation, seedlings were stained with X-Gluc staining solution for 6 to 12 hr at 37°C or GUS activity was measured according to the *Arabidopsis* protocol [49].

Tumor assay on potato tuber discs

Potato tumor assay was performed as described [45]. Briefly, overnight cultured *A. tumefaciens* cells were sub-cultured by 10X dilution and grown at 28°C to OD_{600} 1.0. Cells were washed with 0.9% sodium chloride and re-suspended in 0.9% sodium chloride at 10^8 and 10^6 CFU/ml. A total of 40 to 60 potato tuber disks were placed on water agar with each potato tuber disk infected with 10 μl bacterial culture and incubated at 22°C for 2 days. Disks were then placed on water agar supplemented with 100 μg/ml Timentin and tumors were scored with number of tumors/disc after incubation at 22°C for 3 to 4 weeks.

Supporting Information

Figure S1 Western blot analysis of the extracellular S2 fraction showing both high and low intensity. *A. tumefaciens* cells grown on AS-induced AB-MES (pH 5.5) agar at 19°C for 3 days [7] were collected to isolate the extracellular S2 fractions. C58, *A. tumefaciens* wild type strain; V, empty vector pRL662; Δ*virB2*, *virB2* deletion mutant; Δ*virB2*(pVirB2), expression of wild type *virBp-B1-B2-B3* in Δ*virB2*. Western blot analysis with antisera against VirB2 B24 peptide or B23 peptide (for variants in C-PP) or RNA polymerase RpoA, as an internal control. Unprocessed VirB2 precursor is indicated as VirB2p and processed mature VirB2 as VirB2m. Each region/domain of VirB2 is indicated as described in Figure 1. Western blot images with high (longer exposure time) and low intensity (shorter exposure time) are shown.

Figure S2 Western blot analysis of S1 fraction. *A. tumefaciens* cells grown on AS-induced AB-MES (pH 5.5) agar at 19°C for 3 days [7] were collected to isolate intracellular proteins and extracellular S1 fraction. C58, *A. tumefaciens* wild type strain; V, empty vector pRL662; Δ*virB2*, *virB2* deletion mutant; Δ*virB2*(pVirB2), expression of wild type *virBp-B1-B2-B3* in Δ*virB2*. Western blot analysis with antisera against VirB2 B24 peptide or RNA polymerase RpoA, as an internal control. Processed mature VirB2 as VirB2m.

Figure S3 Potato tumor assay of *A. tumefaciens* strains expressing wild-type VirB2 or variants of L94A (T-pilus⁻/Vir⁺), A110G (T-pilus⁻/Vir⁺) and G121A (T-pilus⁺/Vir⁺). *A. tumefaciens* cells at 10^8 and 10^6 CFU/ml were used for infection. The potato tuber disks were placed on water agar, infected with 10 μl of bacterial cultures, and incubated at 22°C for 2 days. Disks were placed on water agar supplemented with 100 μg/ml Timentin and incubated at 22°C. Tumors were scored after 3 weeks. Data are mean±SEM of number of tumors averaged from 40–60 disks. Similar results were obtained from at least two independent experiments.

Figure S4 Western blot analysis of the intracellular and extracellular S1 and S2 fractions of *A. tumefaciens* strains expressing wild-type VirB2, G119A, or G119C variants. *A. tumefaciens* cells grown on AS-induced AB-MES (pH 5.5) agar at 19°C for 3 days [7] were collected to isolate the intracellular and extracellular S1 and S2 fractions. *A. tumefaciens* strain producing wild-type VirB2, G119A variant, and three independent colonies of G119C variant (G119C-1,-2 or -3) were analyzed. Western blot analysis with antisera against VirB2 B23 peptide or RNA polymerase RpoA, as an internal control. Processed mature VirB2 is indicated as VirB2m.

Figure S5 Tumorigenesis and transient transformation assays of *A. tumefaciens* strains expressing wild-type VirB2, G119A, or G119C variants. (A) Potato tumor assay. *A. tumefaciens* cells at 10^8 and 10^6 CFU/ml were used for infection. The potato tuber disks were placed on water agar, infected with 10 μgl of bacterial cultures, and incubated at 22°C for 2 days. Disks were placed on water agar supplemented with 100 μg/ml Timentin and incubated at 22°C. Tumors were scored after 3 weeks. Data are mean±SEM of number of tumors averaged from 40–60 disks. Similar results were obtained from at least two independent experiments. (B) Transient transformation assay in *Arabidopsis* seedlings. *A. tumefaciens* strains expressing wild-type or variants of VirB2 harboring T-DNA vector pBISN1 were used to infect 4-day-old *Arabidopsis* seedlings. GUS activity as a reporter for transient transformation efficiency was determined by GUS staining or quantitative activity assay at 3 dpi. Data for quantitative GUS activity are mean±SD of four biological repeats from two independent experiments (10 seedlings in each biological repeat).

Table S1 Amino acid sequence homology analysis of selected VirB2 family proteins.

Table S2 Bacterial strains and plasmids.

Table S3   Primer list.

Table S4   Phenotype comparisons of pTiC58 and pTiA6 VirB2 variants.

Acknowledgments

We thank the members of the Lai laboratory for stimulating discussion and Drs. Stanton Gelvin and Lan-Ying Lee for pBISN1. We also acknowledge the technical support of the Plant Cell Biology Core Laboratory and DNA Sequencing Laboratory at the Institute of Plant and Microbial Biology, Academia Sinica, for electron microscopy and DNA sequencing, respectively.

Author Contributions

Conceived and designed the experiments: HYW EML. Performed the experiments: HYW. Analyzed the data: HYW CYC EML. Contributed reagents/materials/analysis tools: HYW EML. Wrote the paper: HYW EML.

References

1. Smith EF, Townsend CO (1907) A plant-pumor of bacterial origin. Science 25: 671–673.
2. McCullen CA, Binns AN (2006) *Agrobacterium tumefaciens* and plant cell interactions and activities required for interkingdom macromolecular transfer. Annu Rev Cell Dev Biol 22: 101–127.
3. Alvarez-Martinez CE, Christie PJ (2009) Biological diversity of prokaryotic type IV secretion systems. Microbiol Mol Biol Rev 73: 775–808.
4. Gelvin SB (2010) Plant proteins involved in *Agrobacterium*-mediated genetic transformation. Annu Rev Phytopathol 48: 45–68.
5. Christie PJ (2001) Type IV secretion: intercellular transfer of macromolecules by systems ancestrally related to conjugation machines. Mol Microbiol 40: 294–305.
6. Lai EM, Chesnokova O, Banta LM, Kado CI (2000) Genetic and environmental factors affecting T-pilin export and T-pilus biogenesis in relation to flagellation of *Agrobacterium tumefaciens*. J Bacteriol 182: 3705–3716.
7. Lai EM, Kado CI (1998) Processed VirB2 is the major subunit of the promiscuous pilus of *Agrobacterium tumefaciens*. J Bacteriol 180: 2711–2717.
8. Thanassi DG, Bliska JB, Christie PJ (2012) Surface organelles assembled by secretion systems of Gram-negative bacteria: diversity in structure and function. FEMS Microbiol Rev 36: 1046–1082.
9. Baron C (2006) VirB8: a conserved type IV secretion system assembly factor and drug target. Biochem Cell Biol 84: 890–899.
10. Cascales E, Christie PJ (2004) Definition of a bacterial type IV secretion pathway for a DNA substrate. Science 304: 1170–1173.
11. Chandran V, Fronzes R, Duquerroy S, Cronin N, Navaza J, et al. (2009) Structure of the outer membrane complex of a type IV secretion system. Nature 462: 1011–1015.
12. Fronzes R, Schafer E, Wang L, Saibil HR, Orlova EV, et al. (2009) Structure of a type IV secretion system core complex. Science 323: 266–268.
13. Low HH, Gubellini F, Rivera-Calzada A, Braun N, Connery S, et al. (2014) Structure of a type IV secretion system. Nature 508: 550–553.
14. Cascales E, Atmakuri K, Sarkar MK, Christie PJ (2013) DNA substrate-induced activation of the *Agrobacterium* VirB/VirD4 type IV secretion system. J Bacteriol 195: 2691–2704.
15. Backert S, Fronzes R, Waksman G (2008) VirB2 and VirB5 proteins: specialized adhesins in bacterial type-IV secretion systems? Trends Microbiol 16: 409–413.
16. Schmidt-Eisenlohr H, Domke N, Angerer C, Wanner G, Zambryski PC, et al. (1999) Vir proteins stabilize VirB5 and mediate its association with the T pilus of *Agrobacterium tumefaciens*. J Bacteriol 181: 7485–7492.
17. Eisenbrandt R, Kalkum M, Lai EM, Lurz R, Kado CI, et al. (1999) Conjugative pili of IncP plasmids, and the Ti plasmid T pilus are composed of cyclic subunits. J Biol Chem 274: 22548–22555.
18. Kalkum M, Eisenbrandt R, Lanka E (2004) Protein circlets as sex pilus subunits. Curr Protein Pept Sci 5: 417–424.
19. Lai EM, Kado CI (2000) The T-pilus of *Agrobacterium tumefaciens*. Trends Microbiol 8: 361–369.
20. Lawley TD, Klimke WA, Gubbins MJ, Frost LS (2003) F factor conjugation is a true type IV secretion system. FEMS Microbiol Lett 224: 1–15.
21. Lai EM, Eisenbrandt R, Kalkum M, Lanka E, Kado CI (2002) Biogenesis of T pili in *Agrobacterium tumefaciens* requires precise VirB2 propilin cleavage and cyclization. J Bacteriol 184: 327–330.
22. Kerr JE, Christie PJ (2010) Evidence for VirB4-mediated dislocation of membrane-integrated VirB2 pilin during biogenesis of the *Agrobacterium* VirB/VirD4 type IV secretion system. J Bacteriol. 192: 4923–34.
23. Hayes CS, Aoki SK, Low DA (2010) Bacterial contact-dependent delivery systems. Annu Rev Genet 44: 71–90.
24. Aguilar J, Cameron TA, Zupan J, Zambryski P (2011) Membrane and core periplasmic *Agrobacterium tumefaciens* virulence Type IV secretion system components localize to multiple sites around the bacterial perimeter during lateral attachment to plant cells. MBio 2: e00218–00211.
25. Aguilar J, Zupan J, Cameron TA, Zambryski PC (2010) *Agrobacterium* type IV secretion system and its substrates form helical arrays around the circumference of virulence-induced cells. Proc Natl Acad Sci U S A 107: 3758–3763.
26. Cameron TA, Roper M, Zambryski PC (2012) Quantitative image analysis and modeling indicate the *Agrobacterium tumefaciens* type IV secretion system is organized in a periodic pattern of foci. PLoS One 7: e42219.
27. Lai EM, Kado CI (2002) The *Agrobacterium tumefaciens* T pilus composed of cyclic T pilin is highly resilient to extreme environments. FEMS Microbiol Lett 210: 111–114.
28. Wang YA, Yu X, Silverman PM, Harris RL, Egelman EH (2009) The structure of F-pili. J Mol Biol 385: 22–29.
29. Babic A, Lindner AB, Vulic M, Stewart EJ, Radman M (2008) Direct visualization of horizontal gene transfer. Science 319: 1533–1536.
30. Shu AC, Wu CC, Chen YY, Peng HL, Chang HY, et al. (2008) Evidence of DNA transfer through F-pilus channels during *Escherichia coli* conjugation. Langmuir 24: 6796–6802.
31. Kwok T, Zabler D, Urman S, Rohde M, Hartig R, et al. (2007) *Helicobacter* exploits integrin for type IV secretion and kinase activation. Nature 449: 862–866.
32. Banta LM, Kerr JE, Cascales E, Giuliano ME, Bailey ME, et al. (2011) An *Agrobacterium* VirB10 mutation conferring a type IV secretion system gating defect. J Bacteriol 193: 2566–2574.
33. Garza I, Christie PJ (2013) A putative transmembrane leucine zipper of *agrobacterium* VirB10 is essential for T-pilus biogenesis but not type IV secretion. J Bacteriol 195: 3022–3034.
34. Jakubowski SJ, Cascales E, Krishnamoorthy V, Christie PJ (2005) *Agrobacterium tumefaciens* VirB9, an outer-membrane-associated component of a type IV secretion system, regulates substrate selection and T-pilus biogenesis. J Bacteriol 187: 3486–3495.
35. Jakubowski SJ, Kerr JE, Garza I, Krishnamoorthy V, Bayliss R, et al. (2009) *Agrobacterium* VirB10 domain requirements for type IV secretion and T pilus biogenesis. Mol Microbiol 71: 779–794.
36. Jakubowski SJ, Krishnamoorthy V, Christie PJ (2003) *Agrobacterium tumefaciens* VirB6 protein participates in formation of VirB7 and VirB9 complexes required for type IV secretion. J Bacteriol 185: 2867–2878.
37. Berger BR, Christie PJ (1994) Genetic complementation analysis of the *Agrobacterium tumefaciens virB* operon: *virB2* through *virB11* are essential virulence genes. J Bacteriol 176: 3646–3660.
38. Bogdanov M, Xie J, Dowhan W (2009) Lipid-protein interactions drive membrane protein topogenesis in accordance with the positive inside rule. J Biol Chem 284: 9637–9641.
39. Wu HY, Liu KH, Wang YC, Wu CF, Chiu WL, et al. (2014) AGROBEST: an efficient *Agrobacterium*-mediated transient expression method for versatile gene function analyses in *Arabidopsis* seedlings. Plant Methods (In press).
40. Narasimhulu SB, Deng XB, Sarria R, Gelvin SB (1996) Early transcription of *Agrobacterium* T-DNA genes in tobacco and maize. Plant Cell 8: 873–886.
41. Li G, Brown PJ, Tang JX, Xu J, Quardokus EM, et al. (2012) Surface contact stimulates the just-in-time deployment of bacterial adhesins. Mol Microbiol 83: 41–51.
42. Aly KA, Baron C (2007) The VirB5 protein localizes to the T-pilus tips in *Agrobacterium tumefaciens*. Microbiology 153: 3766–3775.
43. Lacroix B, Citovsky V (2011) Extracellular VirB5 enhances T-DNA transfer from *Agrobacterium* to the host plant. PLoS One 6: e25578.
44. Kado CI, Heskett MG (1970) Selective media for isolation of *Agrobacterium*, *Corynebacterium*, *Erwinia*, *Pseudomonas*, and *Xanthomonas*. Trends Microbiol 60: 969–976.
45. Wu HY, Chung PC, Shih HW, Wen SR, Lai EM (2008) Secretome analysis uncovers an Hcp-family protein secreted via a type VI secretion system in *Agrobacterium tumefaciens*. J Bacteriol 190: 2841–2850.
46. Schagger H, von Jagow G (1987) Tricine-sodium dodecyl sulfate-polyacrylamide gel electrophoresis for the separation of proteins in the range from 1 to 100 kDa. Anal Biochem 166: 368–379.
47. Shirasu K, Kado CI (1993) Membrane location of the Ti plasmid VirB proteins involved in the biosynthesis of a pilin-like conjugative structure on *Agrobacterium tumefaciens*. FEMS Microbiol Lett 111: 287–294.
48. Lin JS, Ma LS, Lai EM (2013) Systematic Dissection of the *Agrobacterium* Type VI Secretion System Reveals Machinery and Secreted Components for Subcomplex Formation. PLoS One 8: e67647.
49. Julio Salinas JJS-S (2006) Arabidopsis Protocols, 2nd Edition (Methods in Molecular Biology) Humana Press.
50. Larkin MA, Blackshields G, Brown NP, Chenna R, McGettigan PA, et al. (2007) Clustal W and Clustal X version 2.0. Bioinformatics 23: 2947–2948.

7

Genomes and Transcriptomes of Partners in Plant-Fungal- Interactions between Canola (*Brassica napus*) and Two *Leptosphaeria* Species

Rohan G. T. Lowe[1], Andrew Cassin[2], Jonathan Grandaubert[3], Bethany L. Clark[1], Angela P. Van de Wouw[1], Thierry Rouxel[3], Barbara J. Howlett[1]*

1 School of Botany, The University of Melbourne, Parkville, Victoria, Australia, 2 ARC Centre of Excellence in Plant Cell Walls, School of Botany, The University of Melbourne, Parkville, Victoria, Australia, 3 INRA-Bioger, UR1290, Thiverval-Grignon, France

Abstract

Leptosphaeria maculans 'brassicae' is a damaging fungal pathogen of canola (*Brassica napus*), causing lesions on cotyledons and leaves, and cankers on the lower stem. A related species, *L. biglobosa* 'canadensis', colonises cotyledons but causes few stem cankers. We describe the complement of genes encoding carbohydrate-active enzymes (CAZys) and peptidases of these fungi, as well as of four related plant pathogens. We also report dual-organism RNA-seq transcriptomes of these two *Leptosphaeria* species and *B. napus* during disease. During the first seven days of infection *L. biglobosa* 'canadensis', a necrotroph, expressed more cell wall degrading genes than *L. maculans* 'brassicae', a hemi-biotroph. *L. maculans* 'brassicae' expressed many genes in the Carbohydrate Binding Module class of CAZy, particularly CBM50 genes, with potential roles in the evasion of basal innate immunity in the host plant. At this time, three avirulence genes were amongst the top 20 most highly upregulated *L. maculans* 'brassicae' genes *in planta*. The two fungi had a similar number of peptidase genes, and trypsin was transcribed at high levels by both fungi early in infection. *L. biglobosa* 'canadensis' infection activated the jasmonic acid and salicylic acid defence pathways in *B. napus*, consistent with defence against necrotrophs. *L. maculans* 'brassicae' triggered a high level of expression of isochorismate synthase 1, a reporter for salicylic acid signalling. *L. biglobosa* 'canadensis' infection triggered coordinated shutdown of photosynthesis genes, and a concomitant increase in transcription of cell wall remodelling genes of the host plant. Expression of particular classes of CAZy genes and the triggering of host defence and particular metabolic pathways are consistent with the necrotrophic lifestyle of *L. biglobosa* 'canadensis', and the hemibiotrophic life style of *L. maculans* 'brassicae'.

Editor: Richard A Wilson, University of Nebraska-Lincoln, United States of America

Funding: The following sources of funding supported this work: Grains Research and Development Corporation for funding RL AW BC BH, the Australian Research Council for funding AC, the Victoria Life Sciences Computation Initiative (VLSCI) for computational resources via grant RAS990 for RL, the University of Melbourne for an Early Career Researcher award to RL, and the French agency Agence Nationale de la Recherche, contract ANR-09-GENM-028 ('FungIsochores') for funding JG and TR. The funders had no role in study design, data collection and analysis, decision to publish, or preparation of the manuscript.

Competing Interests: The authors have declared that no competing interests exist.

* Email: bhowlett@unimelb.edu.au

Introduction

As more fungal genome sequences become available, it is apparent that their complement of genes and transcriptomes reflects fungal lifestyles. Lifestyles of plant pathogenic fungi are often classified into three broad categories: biotrophy, where the pathogen feeds from live host cells, necrotrophy, where the host cells are killed ahead of colonisation, and hemi-biotrophy, where the pathogen feeds from living cells before switching to a necrotrophic style of growth. These designations are imprecise and as the mechanisms of pathogenicity in a range of fungi are elucidated, lifestyle boundaries become more blurred [1].

The fungal genus *Leptosphaeria* belongs to the class Dothideomycetes, which includes a number of economically important plant pathogens that have a range of lifestyles on their hosts [2]. The *Leptosphaeria* species complex has two species *L. maculans* and *L. biglobosa* [3] and several sub-species or clades including *L. maculans* 'brassicae' and 'lepidii', and *L. biglobosa* 'canadensis', 'brassicae', 'australensis' and 'occiaustralensis' [4]. The nomenclature for these fungi is currently under review [5]. *L. maculans* 'brassicae', a hemibiotroph, causes blackleg, the most important disease of *Brassica napus* (canola) worldwide. Airborne sexual spores (ascospores) released from infested crop residues from the previous year's crop land on seedlings. Hyphae of germinated spores enter plant tissue via stomatal apertures and asymptomatically colonise the apoplastic spaces, between the plant cells. After eight to ten days, plant cells collapse and asexual sporulation begins within the necrotic leaf lesion. Hyphae then grow along the petiole and the stem, often resulting in a canker that girdles the stem causing lodging of the plant [6]. In contrast, *L. biglobosa* 'canadensis' is not well-characterised. Although it causes cotyledonary lesions, stem cankers are rarely produced [7].

Leptosphaeria maculans 'brassicae' has numerous 'gene for gene' interactions with *B. napus* whereby an avirulence allele in the fungus renders it unable to attack cultivars with the corresponding resistance gene. This 'gene for gene' resistance is expressed in seedlings; five avirulence genes have been cloned so far - *AvrLm1*, *AvrLm4-7*, *AvrLm6 AvrLm1* and *AvrLmJ1* [8–12]. Only one resistance gene, *LepR3*, has been cloned from *B. napus* [13].

During infection fungi derive nutrition from the host plant, often by enzymatic degradation of proteins and carbohydrates. These latter enzymes are classed as Carbohydrate-Active enZymes (CAZys) and they often have well characterised domains. As well as providing nutrition, CAZy activity releases cell wall products that can act as DAMPs (Damage Associated Molecular Patterns) that activate the host immune system [14]. The biotrophic plant pathogens *Blumeria graminis*, *Puccinia graminis*, *Melampsora laricus-populina*, *Hyaloperonospora arabidopsis* and *Ustilago maydis* have fewer CAZy genes than necrotrophic plant pathogens do, perhaps because biotrophs do not need to digest plant cell walls for nutrition and they must evade the host immune system [15–17].

Genomic sequences are now available for many Dothideomycetes and an extensive comparative analysis of 18 of them, including *L. maculans* 'brassicae' has been published [18]. These include the *Brassica*-infecting pathogen, *Alternaria brassicicola*, as well as three wheat-infecting pathogens, *Stagonospora nodorum*, *Pyrenophora tritici-repentis*, and *Zymoseptoria tritici*. The former three fungi, like many members of the order Pleosporales, have a necrotrophic lifestyle releasing toxins soon after invasion, whilst *Zymoseptoria tritici*, like most members of the order Capnodiales, is a hemibiotroph, with an extended period as a biotroph before causing necrosis [19–21].

The *L. maculans* 'brassicae' genome is compartmentalised into AT-rich -that are gene-poor comprising up to 35% of the genome, and gene- rich regions that have a high GC content [2], while the genome of *L. biglobosa* 'canadensis' is 30 Mb and lacks AT-rich regions (Grandaubert et al. manuscript submitted). A limited amount of oligo-array transcriptome data have been produced for *L. maculans* 'brassicae' during *in vitro* culture and infection of *B. napus* [11], but transcriptome data have not been reported for *L. biglobosa* or any closely related Brassica pathogen.

Patterns of global gene expression can be generated by RNA-seq, a technique that enables analysis of dual transcriptomes; for instance, during a plant pathogen interaction. Furthermore RNA-seq can be exploited to analyse non-model organisms for which genomic resources are not well developed [22]. Few dual transcriptomes for plants and pathogenic fungi have been reported. Two recent reports of dual RNA-seq analysis of fungal diseases are of rice blast [23] and target leaf spot of sorghum [24]. Here we describe the genomes and transcriptomes of *L. biglobosa* 'canadensis' and *L. maculans* 'brassicae' and canola (*Brassica napus*), during infection and *in vitro*. Our aim is to characterise genes that each pathogen uses to evade detection by the host, or to derive nutrition from the host, viz. the carbohydrate-active enzymes and peptidases. We also examine genes upregulated by the host during infection by each pathogen.

Methods

Fungal isolates and culture conditions

L. maculans 'brassicae' isolate IBCN18 and *L. biglobosa* 'canadensis' isolate 06J154, hereafter referred to as Lmb and Lbc, respectively, were subcultured on 10% Campbell's V8 juice agar at 22°C with a 12 h light/12 h dark light cycle. Conidia

(5×10^6) were added to 30 mL of liquid medium and incubated in still culture in a petri dish (15 cm diameter) at 22°C in the dark. Culture media were either 10% Campbell's V8 juice, or oilseed rape medium. The latter medium was prepared by homogenising leaves (200 g) of *B. napus* cv. Westar in a waring blender in a final volume of 1 L of water. The homogenate was centrifuged at 2000 g for 20 min and the resulting supernatant was filter sterilised (0.22 µm Millipore stericup filter).

Plant growth and infection conditions

Brassica napus cv. Westar was used for all infection assays; it has no known resistance genes. Seedlings were grown in a glasshouse maintained at 25°C under natural lighting. Wounded cotyledons were infected with conidia (10 uL of a 1×10^5 spores/mL suspension) or water (mock inoculum), at 10 days post sowing as described previously [25].

Extraction of RNA and gene expression analysis

For RNA-seq analysis, *B. napus* cotyledons were infected with Lmb, Lbc or water (control mock inocula), and at 7 and 14 days post inoculation (dpi) tissue around the inoculation site was harvested using a cork borer (0.5 cm diameter) and then placed into liquid nitrogen before freeze drying and subsequent grinding under liquid nitrogen. Tissue samples were prepared in biological triplicate. RNA was extracted using Trizol reagent from infected tissue and from mycelia of Lmb and Lbc from 7-day still cultures grown in oilseed rape medium. RNA was then DNAase-treated (Life Technologies) and cleaned up.

The two biological replicates of each sample with the highest RNA integrity number values (>6) were sequenced with Illumina TruSeq version 3 chemistry on an Illumina HiSeq2000 sequencer at the Australian Genome Research Facility. *In vitro* derived RNA was sequenced with 100 bp paired-end reads in order to aid gene annotation, and *in planta* derived RNA was sequenced with 100 bp single-end reads. A total of 15.5 Gbp sequence was generated from the *in vitro* libraries of the two fungi (7.75 Gbp per sample), and 72 Gbp sequence was generated from 12 *in planta* libraries (6 Gbp per sample) (Table S1 in file S2). Reads were trimmed to a minimum phred quality score of 20 using Nesoni sequence software [26], orphaned members of pairs were retained, adaptor sequences were removed, and reads shorter than 20 bp were rejected. Trimmed reads were aligned to a reference genome sequence with Tophat v1.4.1 splice-junction mapper [26]. Reference genomes were Lmb isolate v23.1.3 [2], Lbc isolate J154 (Grandaubert et al., submitted), and a *Brassica* exon array curated unigene set representing 135,201 gene models from *B. napus*, *B. rapa* and *B. oleracea* [27]. Aligned reads were quantified using Cufflinks v1.0.3 transcript assembly and quantification software and denoted as average expression levels (FPKM - fragments per kilobase of exon per million mapped reads) [28]. Cufflinks was run with bias correction to reduce variance due to sequence positional bias during Illumina sequencing. Gene expression FPKM values are listed in the relevant supplementary tables. All of the aligned RNA-seq reads have been deposited at the NCBI Sequence Read Archive (SRA), accessible under bioproject accession PRJNA230885 or sequence read archive SRP035525. SRA files may be read using the NCBI SRA toolkit (http://www.ncbi.nlm.nih.gov/sra).

Quantitative RT-PCR experiments were carried out to determine levels of expression of genes containing CBM50 (LysM) domains. Cotyledon tissue from 32 *B. napus* cv. Westar seedlings was harvested at 3, 7 and 14 days after inoculation with Lmb isolate IBCN18. RNA (2 µg) was treated with DNase 1 (Life Technologies) for 1 h at 25°C, and reverse-transcribed using oligo-

Table 1. *B. napus* defence genes analysed [50].

Gene	Full Name	Defence Signalling Pathway	*B. napus* RRES unigene ID	GenBank accession of CDS
NCED3	9-cis-epoxycarotenoid dioxygenase 3	Abscisic acid	rres040714.v1	EV137674
ACS2a	1-amino-cyclopropane-1-carboxylate synthase 2	Ethylene	rres079827.v1	HM450312
CHI	Chitinase	Ethylene/salicylic acid//jasmonic acid	rres036231.v1	X61488
HEL	Hevein-like protein	Ethylene/jasmonic acid	rres036743.v1	FG577475
ICS1	Isochorismate synthase 1	Salicylic acid	rres059693.v1	EV225528
PR-1	Pathogenesis related protein 1	Salicylic acid	rres112514.v1	BNU21849
WRKY70	WRKY transcription factor 70	Salicylic acid	rres038189.v1	EV113862
PDF1.2	Plant Defensin 1.2	Jasmonic acid	rres071321.v1	KC967203

Defence signalling pathways indicate regulatory pathway to which the gene belongs. RRES unigenes are listed from the Brassica exon array [27], to which RNA-seq reads were aligned. Corresponding GenBank accession numbers are given for each gene. The response of each gene (with the exception of PDF1.2) to its corresponding signalling pathway was confirmed by Sasek et al [50].

dT primer and Superscript III (Life Technologies) at 50°C for 1 h. Levels of gene expression were determined by qPCR using SensiMix (dT) SYBR Green PCR kit (Bioline) in a Corbett Rotor-Gene 3000 machine. Transcript levels of the gene of interest were normalized to that of Lmb actin as described previously (Gardiner et al., 2004). Primers are listed in Table S2 in file S2.

Annotation of genes and domains

Genes encoding CAZys were identified in six dothideomycetes (Lmb, Lbc, *A. brassicicola*, *S. nodorum*, *P. tritici-repentis* and *Z. tritici*) using www.cazy.org [29] and the dbCAN v3.0 HMM-based CAZy annotation server (http://csbl.bmb.uga.edu/dbCAN/) [30]. The major CAZy classes are Polysaccharide Lyases (PL), Glycosyl Transferases (GT), Glycosyl Hydrolases (GH), Carbohydrate Esterases (CE), Carbohydrate Binding Modules (CBM) and Auxiliary Activities (AA). Secretion signal peptides were predicted using the SignalP 4.0 algorithm [31]. Pfam domains (Pfam A and B) were identified using profile hidden Markov models with HMMER3.0 [32] and Pfam_scan.pl software; the e value cut off was set to 1e-5 [33]. LysM-containing genes (with CBM50 domain) were examined in more detail. Predicted gene models and aligned RNA-seq reads were viewed with the IGV browser [34], predicted intron splice sites were verified and translation start sites were checked for congruence with observed RNA-seq transcript boundaries. They were initially identified by comparison to the Pfam database [33], and then aligned to ECP6, the well

characterised LysM-containing protein from *Cladosporium fulvum* [35]. Other *Leptosphaeria* proteins with LysM motifs were identified and characterised as described in supporting information (Figure S1 in file S1).

Peptidase domains were identified by BlastP [36] comparison of predicted protein sequences to the MEROPS 'peptidase database pepunit.lib dataset' of database of peptidase and inhibitor units (http://merops.sanger.ac.uk) [37].

Analysis of expression of fungal and *B. napus* genes

The 100 most highly up regulated genes of Lmb and Lbc at 7 and 14 dpi, compared to *in vitro* growth, were identified from the RNA-seq data (Table S 3–6 in file S2). Averages of quantile-normalised log10-transformed FPKM values were calculated for each dataset for direct comparison of CAZy gene expression between Lmb and Lbc. Genes containing a CBM, GH, PL or AA domain and predicted to be secreted were analysed further. The sum of the expression values of each CAZy gene across the three treatments (*in vitro*, 7 and 14 dpi) was determined and the top 100 genes were identified. Quantile-normalisation was applied so that expression values of genes of both fungi could be directly compared and log10-transformed FPKM values were graphed. For each treatment, a heat map based on expression values was generated.

B. napus defence genes; 9-cis-epoxycarotenoid dioxygenase 3 (NCED), 1-amino-cyclopropane-1-carboxylate synthase 2(ACS2),

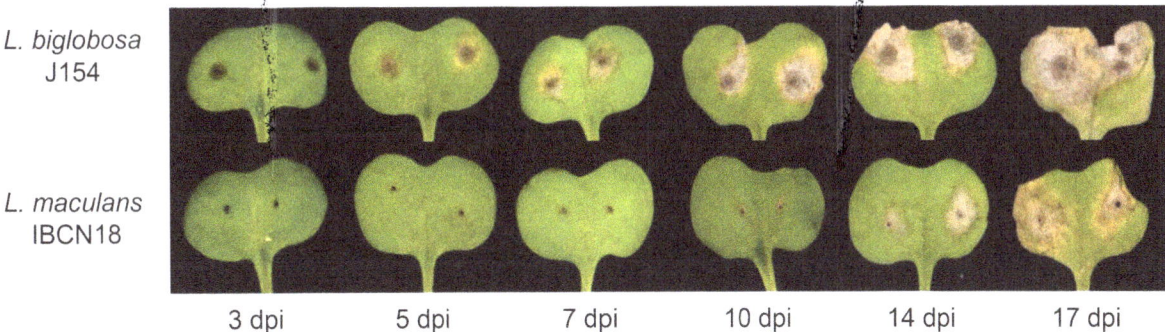

Figure 1. Symptoms on cotyledons of *B. napus* **cv. Westar infected with** *L. maculans* **'brassicae' or** *L. biglobosa* **'canadensis'.** *B. napus* cv. Westar cotyledons were wounded and inoculated with Lmb or Lbc spores and disease allowed to progress for 17 days post inoculation (dpi). Cotyledons were harvested and photographed at 3, 5, 7, 10, 14, and 17 dpi to track lesion development by the two pathogens.

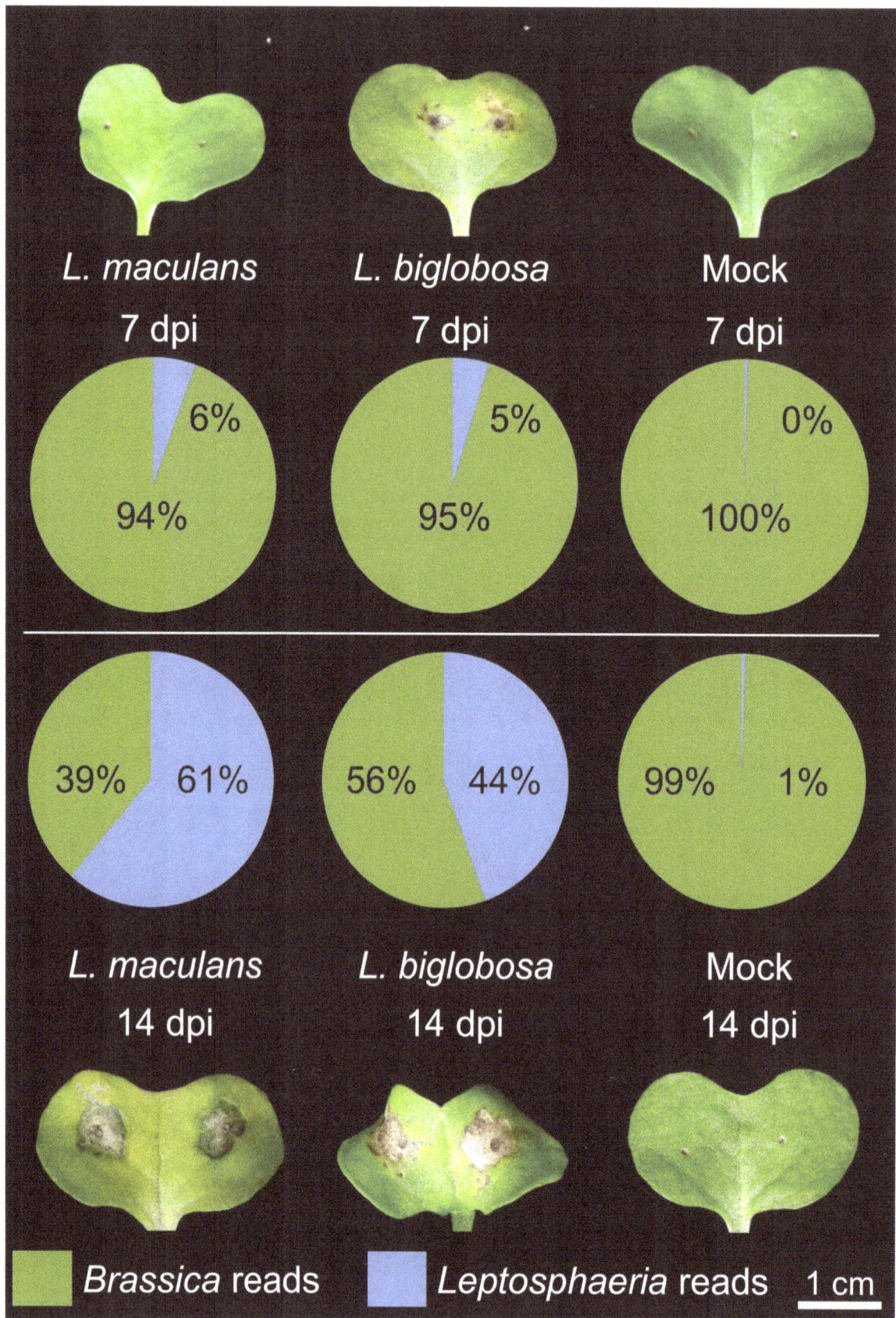

Figure 2. Percentages of plant and fungal transcripts in *B. napus* **cotyledons, uninoculated or at 7 and 14 days post-inoculation dpi with** *L. maculans* **'brassicae' (Lmb) or** *L. biglobosa* **'canadensis' (Lbc).** *B. napus* cotyledons were infected with Lmb, Lbc or water (control mock inoculum). Total RNA was extracted from lesion tissue at 7 and 14 dpi for Illumina RNA-seq sequencing. The percentage of reads aligned to the

Brassica exon array unigene set (green) or the reference genomes for Lmb or Lbc (blue) is presented, as well as a photo of each representative infection. All of the aligned sequence reads were deposited at the NCBI Sequence Read Archive (SRA), accessible under bioproject accession SRP035525.

chitinase (CHI), hevein-like protein (HEL), isochorismate synthase 1 (ICS1), Pathogenesis related protein 1(PR1), WRKY transcription factor 70, and plant defensin 1 (PDF1-2) [38] were identified (Table 1). Their RNA-seq expression was analysed at 7 and 14 dpi in inoculated and uninoculated cotyledons. Additionally, expression of *Brassica* genes involved in metabolic pathways was compared at 7 dpi by either Lmb or Lbc, and then analysed using MapMan, software that processes large gene expression datasets into metabolic pathways or other processes [39]. A Wilcoxon rank sum test with Bonferroni correction for multiple tests was used within MapMan to identify 20 functional categories of *B. napus* genes that were significantly regulated in response to infection by Lmb or Lbc.

Results and Discussion

Symptoms and lifestyles of *L. biglobosa* 'canadensis' and *L. maculans* 'brassicae' on cotyledons of *B. napus*

Leptosphaeria maculans 'brassicae' (Lmb) and *L. biglobosa* 'canadensis' (Lbc) exhibited different timing of symptom development on *B. napus* cotyledons (Figure 1). Lmb had a visually asymptomatic phase until after 7 days post inoculation (dpi), when lesions became visible. Lbc produced initial darkening of the cotyledon tissue at 3 dpi, and cell death and necrosis were apparent by 5 dpi and lesions increased in size until 17 dpi. At this time lesions caused by Lmb were of a similar size. Thus Lmb at 7 dpi appeared to be growing biotrophically, but at 14 dpi was growing necrotrophically, whilst Lbc was necrotrophic from 5 days onwards, although as previously described, its growth is usually arrested before it can colonise the stem [7]. The rapid *in planta* necrosis caused by Lbc was similar to that previously described for *L. biglobosa* 'brassicae' [40].

General features of fungal and plant transcriptomes

RNA-seq was used to define the transcriptomes of both pathogens and host during infection. A total of 87.5 Gbp of raw sequence was generated across all libraries (Table S1 in file S2). In spite of the differences in disease symptoms at 7 dpi, the percentage of Lmb and Lbc reads aligned to the Lmb and Lbc reference genomes was similar (5 and 6% of the total), while 61 and 44% of reads aligned to the reference genomes at 14 dpi (Figure 2). These data reflect that by 14 dpi the plant tissue is heavily colonised by both fungi. As expected, very few reads (less than 1%) from the mock-infected plant libraries aligned to the *Leptosphaeria* reference genomes. These libraries represent highly complex large datasets and Principal Component Analysis was carried out to compare the overall features of the transcriptomes, particularly gene identity and expression level. FPKM values were calculated for each gene in the three organisms. The duplicate sets of RNA-seq data were very similar, with the exception of the data for Lbc at 7 dpi. Gene expression of Lmb in the three conditions (7 and 14 dpi, and *in vitro* growth) was clearly distinguishable one from another (Figure 3), and expression profiles of both fungi *in vitro* were more distinctive than *in planta*. Most of the variance (around 80%) was captured in PC1 for Lmb (Figure 3A) and Lbc infections (Figure 3B), where the 7 and 14 dpi time points were distinguished from each other. The difference between *in vitro* and *in planta* samples was captured in PC2 in for both Lmb and Lbc plots. *B. napus* gene expression was markedly different at 7 dpi

after infection by Lmb compared to Lbc, but similar at 14 dpi. The response of *B. napus* to Lmb at 7 dpi was similar to that of mock inoculation, implying that at this time the plant had not mounted a strong response to infection (Figure 3C).

Since microarray expression (Nimblegen) data for another isolate of Lmb (v23.1.3) were available [2], the relative levels of expression of three avirulence genes (*AvrLm1*, *AvrLm4-7* and *AvrLm11*), present in both Lmb isolates were compared to check for broad agreement between the two technologies (microarrays and RNA-seq). These avirulence genes were ranked in the top 100 most highly expressed genes *in planta* for both microarray and RNA-seq, and both techniques reported a top 10 ranking for *AvrLm1* and *AvrLm4-7* (see below).

Only eight of the 20 most highly upregulated *in planta* genes of Lmb had Pfam domains (Table 2). These included one gene with five CBM50 (LysM) domains (see later); other genes included three cytochrome P450 monooxygenases, and a transferase present in a gene cluster containing a polyketide synthase. Several hypotheticals were amongst the top 20 genes, as well as three avirulence genes, *AvrLm1*, *AvrLm4-7* and *AvrLmJ1* [12]. In contrast, 19 of the 20 most highly upregulated *in planta* genes of Lbc had Pfam domains (Table 3), although five of these were conserved domains without functional annotation (Pfam-Bs, or Domain of Unknown Function (DUF)). Two cellulases, four other glycosyl hydrolases, as well as three peptidases were present.

At 14 dpi, 14 of the top 20 *in planta*-expressed genes of Lmb were hypothetical genes lacking any functional annotation (Table 4). Few were also in the top 20 at 7 dpi, with the exception of *AvrLm1* and LemaP114790.1, which has no Pfam domains and was the most highly expressed gene at both time points. Even though there was a high degree of necrosis on cotyledons at 14 dpi, only two hydrolytic enzymes (trypsin and a glycohydrolase 7) were included in the top 20 most highly expressed Lmb genes. Two of the top 20 genes of Lbc were CAZys; two peptidases and three dehydrogenases, genes that may have degrading roles were also present (Table 5).

Small secreted proteins (SSPs) were well represented in the most highly in planta upregulated genes of Lmb. These included hypothetical, avirulence and CAZy genes. Ohm et al (2012) compared the small secreted proteins of 18 Dothideomycete fungi including Lmb isolate 23.1.3, and found that 21.3% of all Lmb SSPs identified were singletons with no homologue in any of the other genomes [18]. A recent study found 30 of the 100 most highly expressed SSPs *in planta* at 7 dpi were unique to Lmb [52].

In summary, amongst the top 20 most highly expressed genes *in planta*, fewer encoding cell wall degrading enzymes were expressed by Lmb than by Lbc at 7 and 14 dpi. Cell wall degrading enzymes not only provide nutrition for the pathogen by hydrolysing carbohydrates, but facilitate fungal progression through the plant apoplast (intercellular spaces), during biotrophic stages of infection. The repertoire and expression of fungal CAZy genes presumably reflects the cognate carbohydrate present in the host plant and it is notable that enzymes such as those degrading cellulose are very highly expressed. The cell wall of leaves of *Arabidopsis thaliana*, which like *B. napus* is a crucifer, contain polysaccharides including rhamnogalacturonan I and II (as pectin), xyloglucan, glucuronoarabinoxylan, and cellulose (14%). Presumably the cotyledon has a similar polysaccharide profile. An additional 14% of the wall is composed of protein [41]. More than

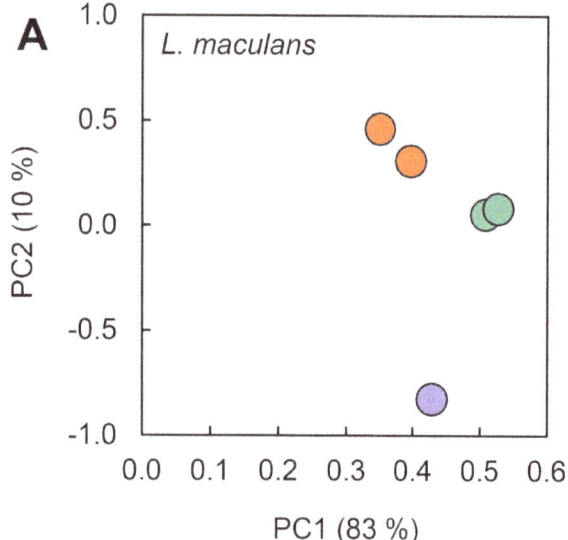

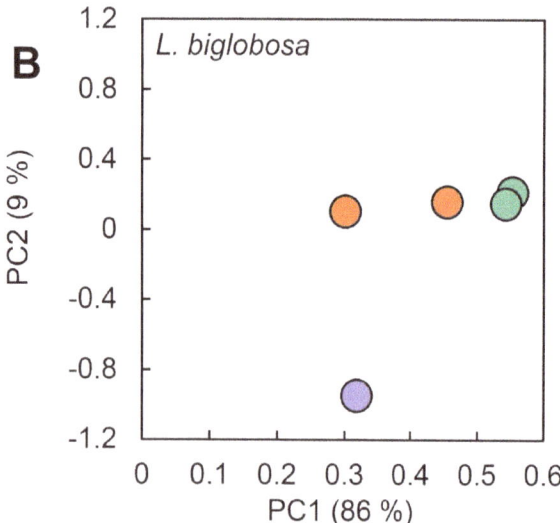

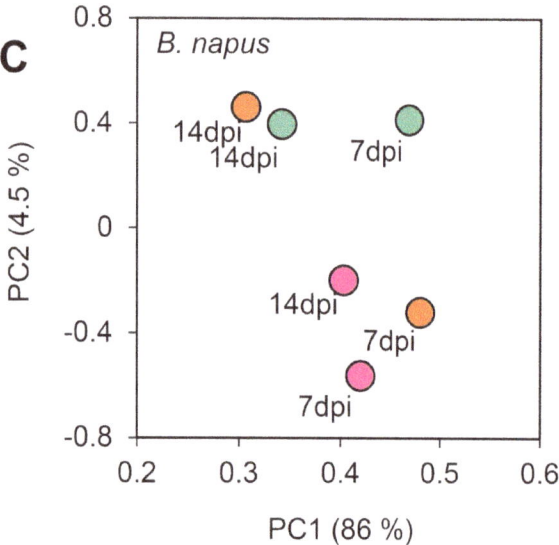

Figure 3. Principal Component Analyses of duplicate sets of RNA-seq data for *L. maculans* **'brassicae' (Lmb),** *L. biglobosa* **'canadensis' (Lbc) and** *B. napus.* Scores plots for principal component analysis of expression values (FPKM) of genes of Lmb IBCN18 (A), Lbc J154 (B), or *B. napus* (C). Gene expression values during infections of cotyledons are plotted in orange (7 dpi) and green (14 dpi). Gene expression values during growth *in vitro* are plotted in purple; those in mock-infected cotyledons are plotted in magenta. The percentage of the total variance explained is listed on each axis label.

a third of the carbohydrate is soluble in phosphate buffered saline, including half of the total pectin, suggesting it would be readily available to a pathogen growing in the apoplast [41].

Distribution of CAZy domains and their expression

In view of the dominance of CAZys in the 20 most highly expressed genes of Lbc after seven days *in planta*, CAZy domains were sought in genome sequences of Lbc, Lmb, and also in four other Dothideomycetes for comparison. The repertoire of domains derived from the dbCAN database and its HMM-based sequence similarity search was consistent with CAZy classifications carried out previously on Dothideomycetes (Table S7 in file S2) [18,42], and the www.cazy.org annotation for Lmb isolate v23.1.3. In general, the numbers of carbohydrate-binding module (CBM), glycosyl hydrolase (GH), glycosyl transferase (GT), and polysaccharide lyase (PL) domains were in agreement with previous studies (Table 6). CAZys are divided into sub classes, generally based on substrate specificity or ligand; some of these are described later. Two CAZy classes in Lmb were identified more frequently using our analyses; CE domains (109 vs 34) and AA (78 vs 27). The difference in numbers of CE domains was probably because 27 CE1 and 42 CE10 domains had been previously excluded from the CAZy expert curated dataset on the basis that they were likely to act upon non-carbohydrate substrates [43]. AA domains are a newly formed CAZy classification for ligninolytic or lytic polysaccharide monooxygenases. Our automated approach may have included closely related monooxygenases that do not degrade lignin or polysaccharides.

Most of the CAZy containing-genes had only one CAZy domain, but several had multiple CBM domains, particularly of the subclass CBM50 (see later and Figure S1 in file S1). Five of the six dothideomycetes had between 580 and 700 CAZy domains, but *Z. tritici* only had 489, which is in general agreement with previous findings [44]. Overall the two *Leptosphaeria* species had similar complements of CAZy genes. *Alternaria brassicicola* had the next most similar profile. *Pyrenophora tritici-repentis* had a profile more similar to *Stagonospora nodorum* than to *Leptosphaeria*, consistent with the host of the two former fungi being a monocotyledonous cereal, rather than a dicotyledonous oilseed plant. Dicot cell walls generally contain a higher proportion of pectin, compared to glucuronoarabinoxylan, which is predominant in a typical monocot cell wall and believed to partially substitute for low levels of pectin in monocot cell walls [15]. *Zymoseptoria tritici* had a distinct profile with many fewer CAZy genes. This is congruent with previously reported characteristics of the members of order Capnodiales, versus the Pleosporales, to which the other five fungi belong. *L maculans* 'brassicae' had fewer AA domains than Lbc, *S. nodorum* and *P. tritici-repentis* did. The number of PL domains ranged from 19 to 24 in Lmb, Lbc and *A. brassicicola*. In contrast, the three wheat pathogens, *S. nodorum*, *P. tritici repentis* and *Z. tritici* had many fewer – between 4 and 10.

Although the complement of CAZys in the two *Leptosphaeria* species was generally similar, there were some interesting

Table 2. Top 20 upregulated genes of *L. maculans* 'brassicae' isolate IBCN18 seven days after inoculation of *B. napus* cv. Westar.

Gene ID	Gene name	7dpi *in planta* (FPKM)	14dpi *in planta* (FPKM)	*in vitro* (FPKM)	Log2 (7dpi FPKM/in vitro FPKM)	Pfam description and accession	Pfam expect score
Lema_P114790.1		12645.1	1163.5	0.1	16.9		
Lema_P114800.1		4066.8	728.9	0.1	15.3		
Lema_P092260.1		1375.5	30.2	0.1	13.7		
Lema_P086290.1	AvrLm1	1831.5	38.6	0.1	13.7		
Lema_P049660.1	AvrLm4-7	2540.0	70.8	0.3	13.1		
Lema_uP037480.1		793.8	42.8	0.1	13.0		
Lema_P084480.1		2645.1	207.0	0.5	12.5	Pfam-B_14615 [PB014615]	4.2E-08
Lema_P054900.1		567.4	24.7	0.1	12.5		
Lema_uP070880.1	AvrLmJ1	1349.9	59.3	0.3	12.4		
Lema_uP082260.1		521.8	3.4	0.1	12.3		
Lema_P087720.1		498.9	3.9	0.1	12.3	Cytochrome P450 [PF00067.17]	9.0E-57
Lema_P070100.1	Lm5LysM	2672.9	105.4	0.8	11.7	LysM domain [PF01476.15]	3.9E-08
Lema_uP121480.1		295.4	19.0	0.1	11.5		
Lema_P006160.1		5011.4	1356.8	1.8	11.5	Pfam-B_2613 [PB002613]	2.1E-33
Lema_P087710.1		282.5	6.2	0.1	11.5	Transferase family [PF02458.10]	5.7E-23
Lema_P087700.1		279.5	2.1	0.1	11.4	Cytochrome P450 [PF00067.17]	6.4E-57
Lema_uP002340.1		1624.1	96.0	0.7	11.2		
Lema_P087750.1		223.7	2.1	0.1	11.1	Cytochrome P450 [PF00067.17]	4.1E-43
Lema_P037680.1		1289.1	6.9	0.6	11.0	Pfam-B_8517 [PB008517]	8.0E-140
Lema_uP123070.1		388.9	34.8	0.2	10.9		

The top 100 *in planta* upregulated genes are listed in Table S3 in file S2.

Table 3. Top 20 upregulated genes of *L. biglobosa* 'canadensis' isolate J154 seven days after inoculation of *B. napus* cv. Westar.

Gene name	Gene name	7dpi in planta (FPKM)	14dpi in planta (FPKM)	in vitro (FPKM)	Log2 (7dpi FPKM/in vitro FPKM)	Pfam description and accession	Pfam expect score
Lb_j154_P000524		886.3	18.0	0.01	17.0	Protein of unknown function (DUF3678) [PF12435.3]	1.20E-01
Lb_j154_P003557		515.7	63.8	0.01	16.2	Pfam-B_18451 [PB018451]	3.60E-69
Lb_j154_P001652		1083.8	151.2	0.1	13.5	Cellulase (glycosyl hydrolase family 5) [PF00150.13]	7.70E-20
Lb_j154_P009247		1012.1	146.1	0.1	13.2	Glycosyl hydrolases family 12 [PF01670.11]	1.20E-31
Lb_j154_P006347		341.8	76.1	0.04	13.2	Deuterolysin metalloprotease (M35) family [PF02102.10]	5.00E-86
Lb_j154_P002168		878.3	9.0	0.1	12.8	Endoribonuclease L-PSP [PF01042.16]	7.30E-11
Lb_j154_P001276		317.1	24.8	0.05	12.6	Pfam-B_19830 [PB019830]	2.60E-36
Lb_j154_P001225		605.1	114.3	0.5	10.4	Putative cyclase [PF04199.8]	2.00E-13
Lb_j154_P010735		5991.3	6460.7	4.7	10.3	Alcohol dehydrogenase GroES-like domain [PF08240.7]	1.60E-23
Lb_j154_P008673		1467.7	220.5	1.2	10.3	Pfam-B_19830 [PB019830]	1.10E-38
Lb_j154_P004138		7216.7	1370.4	6.0	10.2	Trypsin [PF00089.21]	6.60E-64
Lb_j154_P004093		3826.7	456.8	3.2	10.2	Glycosyl hydrolases family 39 [PF01229.12]	3.20E-05
Lb_j154_P006834		410.0	148.0	0.4	10.1		
Lb_j154_P005286		953.8	53.9	0.9	10.0	Pectate_lyase [PF03211.8]	1.50E-58
Lb_j154_P001719		1410.8	291.4	1.4	9.9	Trypsin [PF00089.21]	7.90E-66
Lb_j154_P000527		12641.8	8785.2	13.1	9.9	Pfam-B_14072 [PB014072]	4.30E-34
Lb_j154_P009204		694.9	99.8	0.7	9.9	Cellulase (glycosyl hydrolase family 5) [PF00150.13]	6.00E-13
Lb_j154_P006775		307.5	11.8	0.4	9.7	Putative amidotransferase (DUF4066) [PF13278.1]	1.10E-26
Lb_j154_P001246		1340.6	88.9	1.9	9.5	Glycosyl hydrolases family 43 [PF04616.9]	1.20E-59
Lb_j154_P001995		617.0	66.2	0.9	9.4	Sugar (and other) transporter [PF00083.19]	1.10E-95

The top 100 *in planta* up regulated genes are listed in Table S4 in file S2.

Table 4. Top 20 upregulated genes of *L. maculans* 'brassicae' isolate IBCN 18 fourteen days after inoculation of *B. napus* cv. Westar.

Gene ID	Gene name	7dpi in planta (FPKM)	14dpi in planta (FPKM)	*in vitro* (FPKM)	Log2 (14dpi FPKM/in vitro FPKM)	Pfam description and accession	Pfam expect score
Lema_P114790.1		12645.1	1163.5	0.1	13.51		
Lema_P114800.1		4066.8	728.9	0.1	12.83		
Lema_P082270.1		861.2	3339.4	0.9	11.86	Trypsin [PF00089.21]	3.30E-66
Lema_P043000.1		106.6	762.8	0.7	10.12	Pfam-B_11894 [PB011894]	3.90E-35
Lema_P006160.1		5011.4	1356.8	1.8	9.59	Pfam-B_2613 [PB002613]	2.10E-33
Lema_P110730.1		3.4	66.5	0.1	9.38	Flavin-containing amine oxidoreductase [PF01593.19]	6.90E-27
Lema_uP085940.1		22.6	56.5	0.1	9.14		
Lema_P085920.1		55.2	74.6	0.1	8.97		
Lema_P084480.1		2645.1	207.0	0.5	8.84	Pfam-B_14615 [PB014615]	4.20E-08
Lema_P117020.1		211.3	85.9	0.2	8.81	Pfam-B_7042 [PB007042]	6.50E-17
Lema_uP037480.1		793.8	42.8	0.1	8.74		
Lema_P077490.1		19.1	45.9	0.1	8.74	Cutinase [PF01083.17]	8.00E-49
Lema_P006340.1		14.3	249.7	0.6	8.71	Sugar (and other) transporter [PF00083.19]	2.70E-88
Lema_P060160.1		191.1	60.0	0.1	8.71		
Lema_P036670.1		113.9	213.6	0.6	8.58	Endoribonuclease L-PSP [PF01042.16]	1.10E-10
Lema_P013700.1		515.5	628.0	1.9	8.35	Glycoside hydrolase family 7 [PF00840.15]	3.30E-190
Lema_P085930.1		24.6	88.9	0.3	8.33		
Lema_P092260.1		1375.5	30.2	0.1	8.24		
Lema_P077180.1		24.1	30.2	0.1	8.24		
Lema_P086290.1	AvrLm1	1831.5	38.6	0.1	8.17		

The top 100 *in planta* up regulated genes are listed in Table S5 in file S2.

Table 5. Top 20 upregulated genes of *L. biglobosa* 'canadensis' isolate J154 fourteen days after inoculation of *B. napus* cv. Westar.

Gene ID	Gene name	7dpi in planta (FPKM)	14dpi in planta (FPKM)	*in vitro* (FPKM)	Log2 (14dpi FPKM/in vitro FPKM)	Pfam description and accession	Pfam expect score
Lb_j154_P003075		156.1	116.5	0.01	14.0		
Lb_j154_P003557		515.7	63.8	0.01	13.2	Pfam-B_18451 [PB018451]	3.6E-69
Lb_j154_P000524		886.3	18.0	0.01	11.3	Protein of unknown function (DUF3678) [PF12435.3]	1.2E-01
Lb_j154_P006347		341.8	76.1	0.04	11.0	Deuterolysin metalloprotease (M35) family [PF02102.10]	5.0E-86
Lb_j154_P001652		1083.8	151.2	0.1	10.6	Cellulase (glycosyl hydrolase family 5) [PF00150.13]	7.7E-20
Lb_j154_P009247		1012.1	146.1	0.1	10.4	Glycosyl hydrolase family 12 [PF01670.11]	1.2E-31
Lb_j154_P010735		5991.3	6460.7	4.7	10.4	Alcohol dehydrogenase GroES-like domain [PF08240.7]	1.6E-23
Lb_j154_P001707		63.1	9.2	0.01	10.4		
Lb_j154_P006327		113.6	8.5	0.01	10.2		
Lb_j154_P008556		1.4	4.9	0.01	9.4	Mediator complex subunit 27 [PF11571.3]	1.8E-01
Lb_j154_P000527		12641.8	8785.2	13.1	9.4	Pfam-B_14072 [PB014072]	4.3E-34
Lb_j154_P000350		32.9	4.3	0.01	9.3		
Lb_j154_P006232		3710.7	5194.8	8.5	9.2		
Lb_j154_P005193		1399.2	1588.4	2.8	9.2		
Lb_j154_P003830		27.8	38.5	0.1	9.1	Serine carboxypeptidase [PF00450.17]	4.2E-87
Lb_j154_P009122		5.3	3.5	0.01	9.0		
Lb_j154_P003076		158.0	118.9	0.2	9.0	Uncharacterized protein conserved in bacteria (DUF2321) [PF10083.4]	2.9E-01
Lb_j154_P001276		317.1	24.8	0.05	9.0	Pfam-B_19830 [PB019830]	2.6E-36
Lb_j154_P008991		3272.4	3034.9	6.1	9.0	Short-chain dehydrogenase [PF00106.20]	9.0E-27
Lb_j154_P003555		1776.8	1658.6	3.5	8.9	Alcohol dehydrogenase GroES-like domain [PF08240.7]	2.2E-22

The top 100 *in planta* up regulated genes are listed in Table S6 in file S2

Table 6. Distribution of CAZy domains in predicted proteins of six Dothideomycetes.

| Number of domains | | | | | | | |
| CAZy domain | AA | CBM | CE | GH | GT | PL | Total |
Fungus (lifestyle)							
Leptosphaeria biglobosa (necrotroph)	98	85	117	234	106	23	663
Leptosphaeria maculans (hemibiotroph)	78	63	109	217	100	19	586
Alternaria brassicicola (necrotroph)	92	66	120	233	92	24	627
Stagonospora nodorum (necrotroph)	122	66	143	267	96	10	704
Pyrenophora tritici-repentis (necrotroph)	114	56	124	246	105	10	655
Zymoseptoria tritici (hemibiotroph)	55	26	97	199	108	4	489

The classes of CAZy domains are Auxiliary Activities (AA) Carbohydrate Binding Modules (CBM), Carbohydrate Esterases (CE), Glycosyl Hydrolases (GH), Glycosyl Tranferases (GT) Polysaccharide lyases (PL). Some genes contain multiple CAZy domains.

differences. The major difference was in the CBM class with > 30% more domains present in Lbc than in Lmb. *L maculans* 'brassicae' had fewer CBM18-containing-genes than Lbc, but a similar number of CBM18 domains. CBM18 domains in Lmb were in a more compact gene structure (28 domains in 11 genes); indeed this domain was always present in multiple copies within a gene, five genes contained homopolymers of only CBM18. *L. biglobosa* 'canadensis' had 32 domains in 21 genes, eight of which had a single CBM18 domain. CBM50 (LysM), CE4, and GH18 CAZy domains occurred frequently in genes that had CBM18 domains. (Table S8 in file S2). In eukaryotes this domain often has a chitin-binding function and modules are often adjacent to chitinase catalytic domains, but this domain is also present in non-catalytic proteins either singly or as multiple repeats. CBM18 domains may be involved in targeting the degradative domains to particular carbohydrates, or may enhance catalysis by ensuring retention of the substrate between catalytic cycles.

About half of the genes with CAZy domains in the two *Leptosphaeria* species did not have a signal peptide and thus were not predicted as being secreted. This may be due to incorrect annotation at the 5' end of genes, genes incorrectly merged during annotation, or a false negative result from SignalP. Pectate lyases were the most frequently secreted CAZy (100% for Lbc, 89% for Lmb), while GTs were the least frequently secreted (11% for Lbc, 12% for Lmb). This is consistent with the role of GTs in synthesising carbohydrates for the fungal cell wall, and their membrane or intracellular location.

We then examined the RNA-seq data to determine expression levels of CAZys of Lbc and Lmb during infection of *B. napus* cotyledons (Table 7, Table S7 in file S2). The expression profiles of the top 100 most highly expressed (summed FPKM values at 7 dpi, 14 dpi, and *in vitro*) secreted CAZy genes are presented in Figure 4. At 7 dpi, Lmb had a high level of expression of genes with CBM domains, particularly CBM50/LysM (for instance, genes A and C in Figure 4); this level decreased by more than 50% by 14 dpi and was even lower during growth *in vitro*.

Genes with CBM domains were also among the most highly expressed CAZy genes of Lbc *in planta*, but the overall level of expression was much lower than that of Lmb genes, at the same time point. To reveal the most *in planta* specific CAZy classes, expression ratios between *in planta* and *in vitro* growth were calculated for each CAZy class (Table S9 in file S2). For Lmb, the CBM50 (chitin-binding) domain was highly upregulated *in planta* at 7 dpi, whereas at the same time point Lbc upregulated CBM6 (cellulose binding) domains more highly (Table S9 in file S2). The *in planta* upregulated RNA-seq expression profile of the CBM50-containing genes of Lmb was validated by quantitative RT-PCR (Figure S2 in file S1). These experiments showed peak expression of each LysM gene at 7 dpi *in planta* and much lower expression levels *in vitro*. At 3 dpi, LysM gene expression was higher than *in vitro* levels, but lower than levels at 7 dpi. For the CBM18-containing genes, expression was varied, with >3 orders of magnitude between the highest and lowest expressed genes. *L. biglobosa* 'canadensis' had generally higher expression compared

Table 7. Expression of CAZy genes in *L. maculans* 'brassicae' and *L. biglobosa* 'canadensis.'

| CAZy class | L. maculans | | | L. biglobosa | | |
	7dpi *in planta* (FPKM)	14dpi *in planta* (FPKM)	*in vitro* (FPKM)	7dpi *in planta* (FPKM)	14dpi *in planta* (FPKM)	*in vitro* (FPKM)
AA	86.5	109.9	117.3	104.5	89.1	159.7
CBM	508.7	203.6	158.5	163.7	174.3	341.4
CE	78.1	69.5	63.6	78.7	55.8	99.9
GH	110.2	122.3	88.1	122.7	81.3	125.5
GT	54.8	60.9	57.1	46.3	50.0	112.3
PL	56.3	23.9	6.9	121.3	37.7	29.4

RNA-seq derived gene expression values were mapped onto the identified CAZy domains for Lmb and Lbc. Average expression levels (FPKM) were then calculated for each major class of CAZy.

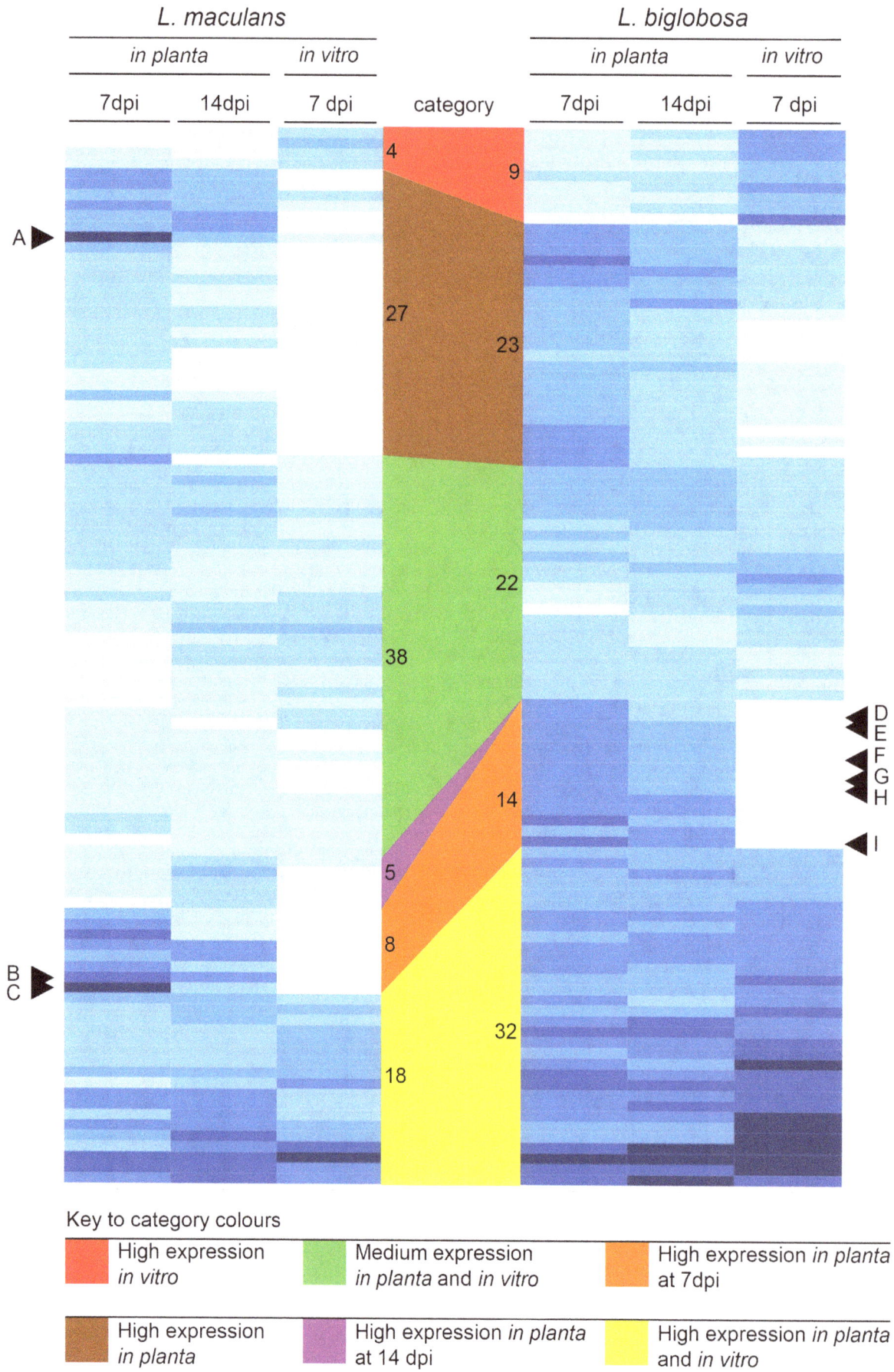

Figure 4. Expression profiles of secreted CAZys of *L. maculans* **'brassicae' (Lmb),** *L. biglobosa* **'canadensis' (Lbc)** *in planta* **and** *in vitro.*
The top 100 genes expressed across the three treatments (7, 14 dpi and *in vitro*) and predicted to be secreted and to contain a CAZy domain (CBM, GH, PL, or AA) were selected. Quantile-normalisation was applied and log10-transformed FPKM values were graphed. The intensity of blue shading is proportional to the expression level. The gene order is based on a dendrogram created from a Euclidean similarity matrix with average group distance. Letters with a triangle point to genes that were amongst the top 20 most highly expressed genes *in vitro* or *in planta* (Tables 2, 3, 4, or 5). A/ Lema_P102640.1 (Lm2LysM); B/Lema_P013700.1 (Glycoside hydrolase family 7); C/Lema_P070100.1 (Lm5LysM); D/Lb_j154_P005286 (pectate lyase); E/Lb_j154_P009204 (cellulase); F/Lb_j154_P001246 (Glycosyl hydrolase family 43); G/Lb_j154_P001652 (cellulase); H/Lb_j154_P009247 (Glycosyl hydrolases family 12); I/Lb_j154_P004093 Glycosyl hydrolases family 39). Categories of gene expression (High expression *in vitro*, High expression *in planta*, Medium expression *in planta* and *in vitro*, High expression *in planta* at 14 dpi, High expression *in planta* at 7dpi, High expression *in planta* and *in vitro*) were manually assigned and indicated by coloured polygons linking the genes in each category across Lmb and Lbc; the number of genes in each category is indicated on the vertical sides of the polygon.

to that of Lmb (Table S8 in file S2). For Lbc, the average expression value of all CBM genes was highest during *in vitro* growth, which was due to high expression of several genes that had both CBMs and hydrolytic CAZY domains (Table 7). Such genes included Lb_j154_P004089, which has three CBM18 and one CE4 domains (Table S8 in file S2).

Polysaccharide lyases (PL) were expressed highly at 7 dpi, and expression decreased by 14 dpi in both species. At 7 dpi, the expression levels of PLs in Lbc were twice those of Lmb. Expression levels then dropped by 75% at 14 dpi and were low during growth *in vitro*. *L. biglobosa* 'canadensis' had higher levels of expression of degrading CAZys such as AAs, GHs and PLs (genes D,E,F,G,H and I; Figure 4) at 7 dpi, than at 14 dpi, when lesion growth had slowed. The PL3 class (pectate lyase) was highly upregulated by Lbc at 7 dpi (Table S9 in file S2), and was also the most upregulated PL class in Lmb. This may be a reflection of the abundance of pectin in the cotyledon.

Glycosyl transferase (GT) -containing genes were expressed at similar levels in all conditions in both *Leptosphaeria* species, except for a two-fold increase in Lbc grown *in vitro* compared to *in planta* (Table 7). The class GT21, which encodes biosynthetic enzymes that glycosylate lipids, was the most *in planta* upregulated class of glycosyltransferases. Of the carbohydrate esterase enzymes, CE8 (pectin methylesterase) and CE12 (pectin acetylesterase) were the most *in planta* upregulated classes for Lbc and Lmb, respectively (Table S9 in file S2).

At 14 dpi, of the six CAZy classes, expression of only PLs was higher in Lbc than in Lmb. By 14 dpi, expression of CBM and PL-containing genes by Lmb had decreased, but expression of genes with AA and GH domains had increased. *In vitro* growth was characterised by low CBM expression and very low PL expression (Table 7). The CAZy expression profile for Lmb *in planta* may reflect early avoidance of chitin- triggered immunity via high expression of CBMs such as LysM genes, whilst the fungus feeds from soluble pectin via PL genes. At 14 dpi elevated expression of

the oxidative AA class and the hydrolytic GH class may facilitate degradation of lignin and cellulose in dead cells.

In many fungi, expression of cell wall degrading enzymes CAZys is regulated by carbon catabolite repression, a global mechanism that ensures readily assimilated carbon sources such as glucose are preferentially used. A Lmb homologue of the *Saccharomyces cerevisiae* sucrose non-fermenting protein kinase1 (*SNF1*) carbon catabolite regulator has been recently shown to regulate expression of several CAZy genes encoding pectate lyases, beta-1,3-glucanase, and a glucosidase [45]. These CAZy genes were upregulated four days after inoculation of canola cotyledons with a wild type strain of Lmb; whereas an LmSNF1 knockout strain had significantly reduced expression of these CAZys during growth on pectin *in vitro*. We recorded similar upregulation of the CAZy genes encoding two pectate lyases and the carbohydrate esterase *in planta* at 7 dpi, but not significant upregulation of the chitin deacetylase. This latter difference could be due to differing time points at which tissue was analysed and the more robust normalisation used in an RNA-seq analysis compared to a qPCR method.

In both Lmb and Lbc, the top 5% most highly expressed CAZy genes accounted for approximately 50% of the sum total CAZy expression (FPKM) at each growth stage (7 and 14 dpi and *in vitro* growth) that was analysed (data not shown). This suggests that the vast majority of CAZys are expressed at low levels or only turned on at specific times in the fungal life cycle. This phenomenon has been reported previously for *Z. tritici*, which differentially expresses CAZy genes, such as cutinases (CE5 subclass) according to biotrophic, necrotrophic or saprotrophic stages of growth [44]. The hemi-biotroph *Colletotrichum higginsianum* upregulates expression of effectors and secondary metabolite biosynthetic enzymes both before penetration and during biotrophic growth on *Arabidopsis*. *C. higginsaneum* then upregulates hydrolases and transporters at a later stage, each wave delivered according to the stage of pathogenic transition [46]. This again matches our

Table 8. Distribution of peptidase domains in six Dothideomycete genomes.

Peptidase type	A	C	G	I	M	S	T	U	Total
Species (Lifestyle)									
Leptosphaeria biglobosa (necrotroph)	18	73	1	8	144	145	23	1	413
Leptosphaeria maculans (hemibiotroph)	16	78	1	10	140	140	25	1	411
Alternaria brassicicola (necrotroph)	17	77	1	6	146	163	23	1	434
Stagonospora nodorum (necrotroph)	23	80	0	7	151	192	23	1	477
Pyrenophora tritici-repentis (necrotroph)	22	78	1	9	155	163	22	1	451
Zymoseptoria tritici (hemibiotroph)	33	67	4	8	130	191	23	1	457

Peptidase domains are classified according to the catalytic type, or inhibitor activity. Peptidase types are Aspartic (A), Cysteine (C), Glutamic (G), Inhibitor, Mettallo-(M), Serine (S), Threonine (T), Unknown (U).

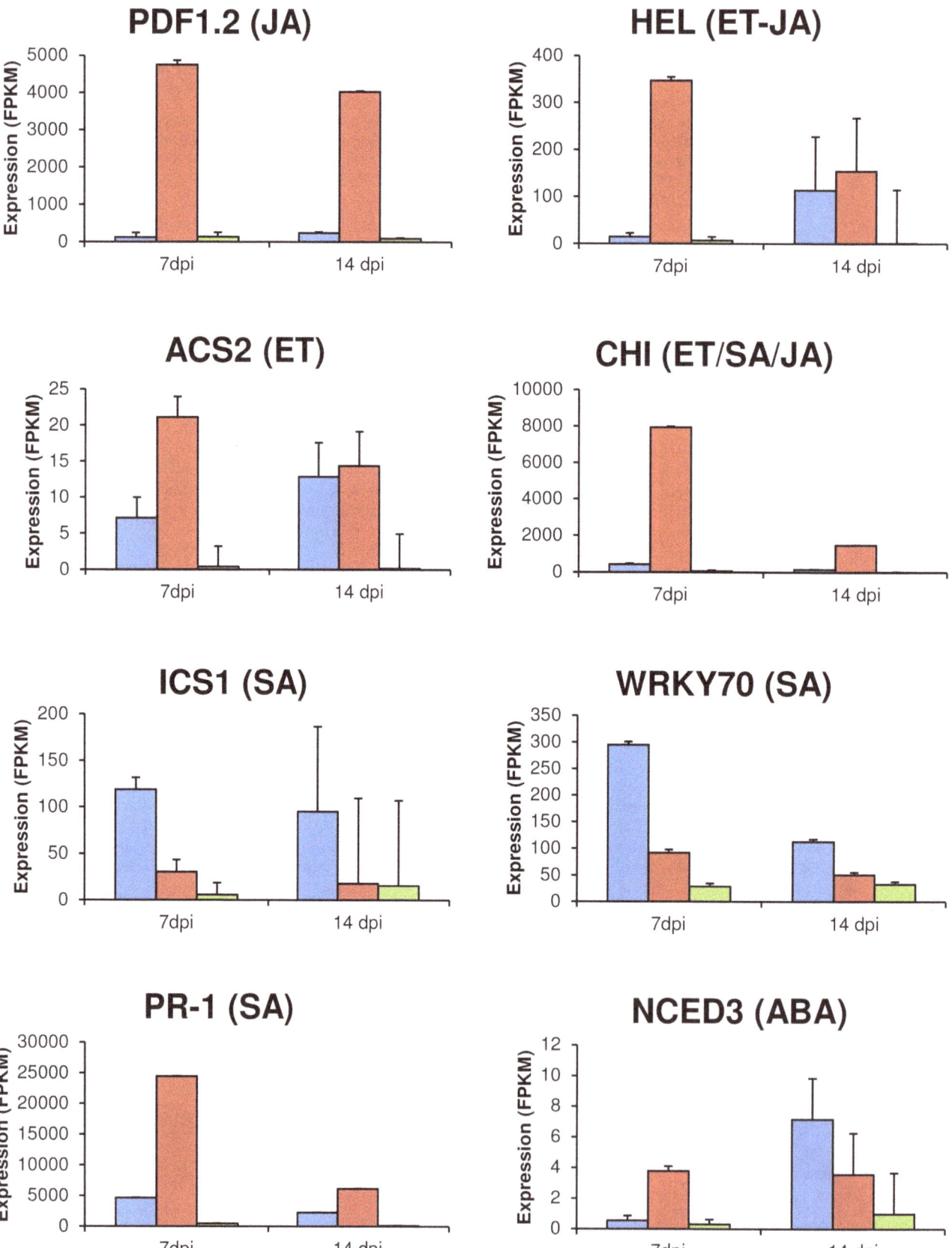

Figure 5. Expression of key defence genes of *B. napus* **at 7 or 14 dpi with either** *L. maculans* **'brassicae' (Lmb) or** *L. biglobosa* **'canadensis' (Lbc).** RNA-seq data for eight key Brassica defence genes were determined at 7 and 14 dpi. Average expression (FPKM) after infection by *L. maculans* (blue), *L. biglobosa* (red), and mock inoculum (green) is plotted. Error bars indicate the Cufflinks 95% confidence interval for each FPKM

value. The genes assayed were 9-cis-epoxycarotenoid dioxygenase 3 (NCED3), 1-amino-cyclopropane-1-carboxylate synthase 2 (ACS2), chitinase (CHI), hevein-like protein (HEL), isochorismate synthase 1 (ICS1), Pathogenesis related protein 1 (PR-1), WRKY transcription factor 70 (WRKY70), and plant defensin 1 (PDF1.2). Each gene was also classified according to hormone(s) abscisic acid (ABA), ethylene (ET), jasmonic acid (JA) and salicylic acid (SA) that induced higher expression.

observation that *L. biglobosa* lacks or only has a short biotrophic stage and expresses hydrolases earlier than Lmb does. CAZy genes of the two *Leptosphaeria* species that have low expression during development of cotyledonary lesions might be expressed more highly in another growth situation, such as earlier in penetration of the tissue, or during colonisation of petiole, stem, or during saprophytic growth on woody stubble of *B. napus*.

Peptidase distribution and expression

Since the two *Leptosphaeria* species grow between the plant cells in the apoplast, protein from the plant cell wall is a potential source of nitrogen for nutrition. A trypsin-like peptidase was highly expressed during necrotic stages of infection for both Lmb (14 dpi) and Lbc (7 dpi) (Table 2, Table 3), therefore we characterised the peptidase content of both *Leptosphaeria* genomes and compared them to four other Dothideomycete plant pathogens.

The MEROPS peptidase classifications categorises peptidases by their catalytic nucleophile [37]. For example, aspartic (A), cysteine (C), serine (S), threonine (T) peptidases are named for the corresponding catalytic residue. Exceptions are for metallo (M)

peptidases, which use a coordinated metal ion, and the classifications for unknown activities (U) and peptidase inhibitors (I). Peptidase composition and number was very similar for Lmb, Lbc and *A. brassicicola* (Table 8). Metallo- and serine-peptidase genes comprised almost 70% of the total peptidases. The next most abundant classes were the cysteine, threonine and aspartic peptidases, in that order. *L. biglobosa* 'canadensis' had seven C56 cysteine peptidases (7) whilst Lmb only had three. C56 peptidases act on peptides <20 amino acids, presumably requiring prior digestion of the substrate by another peptidase. *P. tritici-repentis* and *S. nodorum* had more peptidases in the metallo and serine classes than Lmb or Lbc. The hemi-biotroph *Z. tritici* had a larger complement of serine and aspartic peptidases than the two *Leptosphaeria* species did.

In general, the expression levels of peptidase at 7 and 14 dpi *in planta* were similar in Lmb and Lbc (Pearson correlation coefficient of 0.76 for Lmb IBCN18 and 0.85 for Lbc J154), with a few notable exceptions. A trypsin-like serine peptidase (MEROPS S01A) was the most highly *in planta* up-regulated peptidase (500-fold higher at 7 dpi *in planta* than *in vitro*) in both Lbc

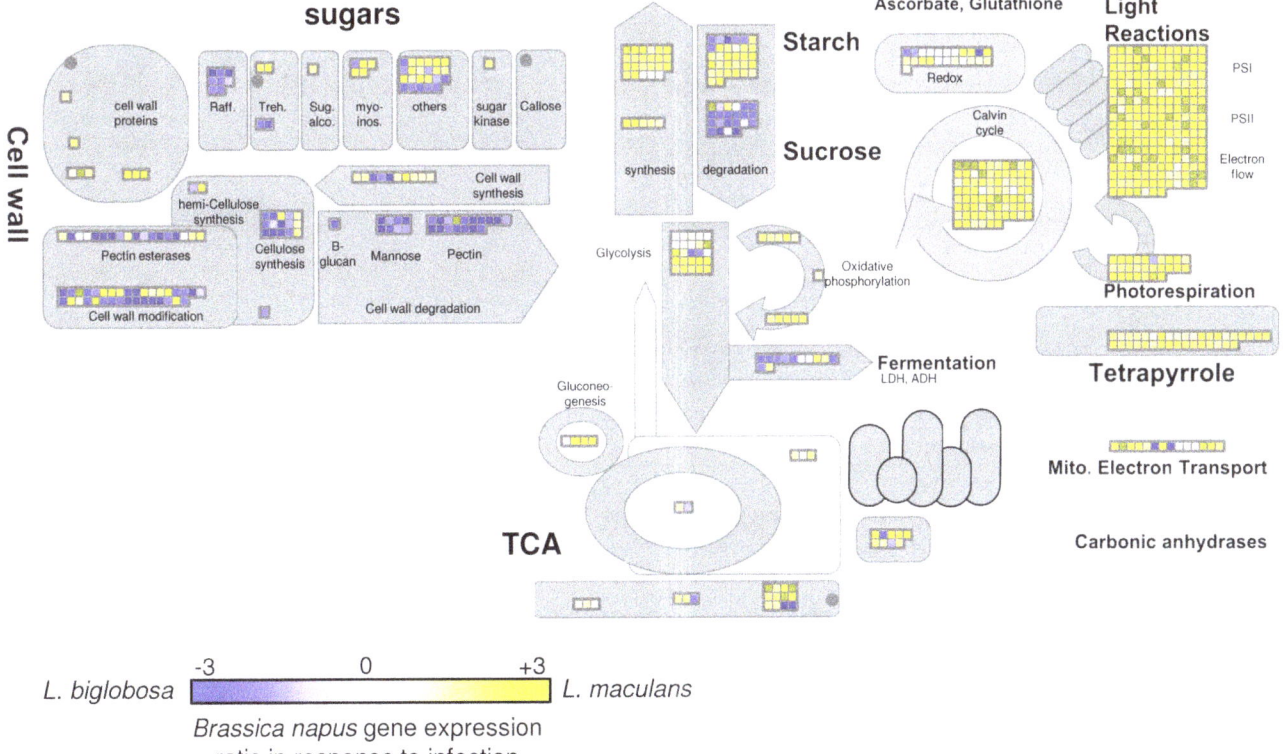

Figure 6. Transcription of *B. napus* genes involved in metabolic processes including photosynthesis seven days after inoculation with *L. maculans* 'brassicae' (Lmb) or *L. biglobosa* 'canadensis' (Lbc). RNA-seq gene expression values for a *B. napus* unigene set [27] were used to calculate a ratio of expression values (log2) for *B. napus* genes after infection by Lmb or Lbc. Ratios were plotted on major metabolic pathways with Mapman software [39]. A yellow square indicates a *B. napus* gene that is expressed more highly during Lmb infection, while a blue square indicates a *B. napus* gene with higher expression during Lbc infection. An expression ratio close to zero is shown with a white square and indicates equivalent expression during infection by either pathogen. Only genes with expression values greater than 10 FPKM were included. Abbreviations: LDH, lactate dehydrogenase; ADH, Alcohol dehydrogenases, TCA, tricarboxylic acid cycle; raff, raffinose; Treh, trehalose; PSI, photosystem one; PSII, photosystem two; ABA, abscisic acid; ET, Ethylene; SA, salicylic acid; JA, jasmonic acid.

Table 9. Top 20 functional categories that are significantly regulated in *B. napus* in response to infection by *L. maculans* 'brassicae' compared to infection by *L. biglobosa* 'canadensis.'

Mapman Function Category	Mapman Function Description	Number of genes	p-value
1	Photosynthesis	313	0.00E+00
1.1	Photosynthesis, light reactions	210	0.00E+00
1.1.1	Photosynthesis, light reactions, photosystem II	105	0.00E+00
26	Miscellaneous	365	4.42E-25
10	Cell wall	116	4.31E-12
1.1.2	Photosynthesis lightreaction photosystem I	44	5.06E-11
1.1.1.2	Photosynthesis lightreaction.photosystem II, PSII polypeptide subunits	63	5.06E-11
1.3	Photosynthesis calvin cycle	76	2.19E-10
1.1.1.1	PS lightreactions photosystem II, LHC-II	42	3.08E-10
1.1.2.2	PS lightreactions photosystem I, PSI polypeptide subunits	39	3.56E-10
20.1	Stress, biotic	123	1.98E-09
10.6	Cell wall, degradation	28	3.88E-09
26.12	Misc peroxidases	26	1.07E-08
26.8	Misc nitrilases, nitrile lyases, berberine bridge enzymes, reticuline oxidases, troponine reductases	42	3.72E-08
26.10	Misc cytochrome P450	46	1.93E-07
26.3	Misc gluco-, galacto- and mannosidases	42	1.35E-06
10.6.3	Cell wall degradation, pectate lyases and polygalacturonases	17	1.74E-05
35	Not assigned	1969	6.69E-05
35.2	Not assigned, unknown	1969	6.69E-05
17	Hormone metabolism	107	8.21E-05
2.2.1	Major CHO metabolism, degradation sucrose	22	6.47E-04
20.1.7.6	Stress biotic, PR-proteins, proteinase inhibitors	12	6.63E-04
1.2	Photosynthesis, photorespiration	27	7.32E-04
17.5	Hormone metabolism.ethylene	30	7.77E-04
34	Transport	208	8.58E-04
33	Development	107	1.62E-03
26.9	Misc glutathione-S-transferases	32	1.92E-03
10.8	Cell wall pectin esterases	18	1.92E-03
17.5.1	Hormone metabolism, ethylene synthesis-degradation	21	1.92E-03
20.1.7.6.1	Stress biotic, PR-proteins, proteinase inhibitors, trypsin inhibitor	11	1.92E-03
1.1.3	Photosynthesis, light reactions, cytochrome b6/f	8	2.25E-03
10.8.1	Cell wall,pectin esterases, PME	13	2.35E-03
31.4	Cell vesicle transport	32	2.41E-03
1.3.6	Photosynthesis, calvin cycle,aldolase	13	2.49E-03
1.1.4	Photosynthesis, lightreaction, ATP synthase	15	2.55E-03
29.5.3	Protein degradation, cysteine protease	18	3.11E-03
29.5.9	Protein degradation, AAA type	15	3.41E-03
2.2.1.3	Major CHO metabolism, degradation, sucrose invertases	13	4.18E-03
13.2.3	Amino acid metabolism, degradation, aspartate family	11	5.62E-03
1.1.5	Photosynthesis, light reaction, other electron carrier (oxidation/reduction)	18	5.77E-03

B. napus genes expressed during infection at 7 dpi were assigned to functional categories and overall pathway regulation was examined. The top 20 categories determined by MapMan analysis are listed. Abbreviations: PS (photosynthesis), LHC (light harvesting complex), CHO (carbohydrate), PR (pathogen response), PME (pectin methyl esterase), ATP (adenosine triphosphate). AAA type (ATPases Associated with diverse cellular Activities). P values indicate the likelihood of the observed pathway regulation being due to chance.

(Lb_j154_P004138) and Lmb (Lema_P082270.1) (Table 2, Table S10, Table S11 in file S2). Its role in disease is unknown. The MEROPS S01A "type" peptidase is bovine chymotrypsin but fungal homologs are usually described as "trypsin-like" because they are similar to both bovine trypsin, and trypsin from *Streptococcus bacteria* [47]. The *in vitro* growth condition resulted in a more distinct expression profile of peptidases, producing low correlation co-efficient values (0.35 and 0.38) between the *in vitro* and 7 dpi *in planta* conditions, for Lbc and Lmb, respectively. Four Lmb peptidases had moderate levels of expression at 7 dpi and were down-regulated during growth *in vitro*, in a similar manner to the trypsin homologs. Two genes, Lema_P058000.1 and Lema_P031600.1, were most similar to C56 cysteine peptidases. The other two genes, Lema_P044030.1 and Lema_P044810.1, were most similar to the S33 serine peptidases, which typically release an N-terminal proline from a peptide substrate. These peptidases may target hydroxyproline-rich glycoproteins such as extensins, which are components of the plant cell wall [48]. Average CAZy class expression for the peptidases showed Lbc expressed its peptidase inhibitors more highly than Lmb did across all treatments (Table S11 in file S2). Glutamyl peptidases were expressed at very low levels in both fungi, and threonine peptidases had the highest average expression of all peptidase classes.

Expression of Brassica genes

The infection stage in a plant-pathogen interaction is often reflected in expression of key genes implicated in host defence signalling pathways. The expression levels for eight defence reporter genes of *B. napus* were determined during invasion by Lmb and Lbc (Figure 5). In general these genes were more highly expressed at 7 dpi during infection by Lbc than by Lmb. During Lbc infection, reporter genes for ethylene signalling (ACS2, CHI, and HEL) were expressed 50- to 138-fold higher than in mock-infected controls at 7 dpi, while Lmb infection resulted in only 2–19 fold induction of these genes. *L. maculans* 'brassicae' triggered a higher expression level of a key salicylic acid reporters, ICS1 and WRKY70, but another salicylic acid reporter gene, PR-1, was not as up regulated compared to Lbc. The salicylic acid reporter gene, PR-1, but not ICS1, were similarly highly induced at 7 dpi in Lbc. The jasmonic acid reporter gene PDF1.2 was induced highly by Lbc infection, but not by Lmb. The upregulation of these genes reflected that jasmonic acid and salicylic acid defence pathways were induced by Lbc, in a pattern consistent with the timing of necrosis.

Sasek et al (2012) showed that interactions of *B. napus* and Lmb, involving the recognition of *AvrLm1* (or the mutant allele, *avrLm1*) and corresponding resistance gene, *Rlm1*, salicylic acid biosynthesis and transcription of SA-associated genes (ICS1, WRKY70 and PR-1) increased as early as 3 dpi. Expression of HEL and CHI, genes involved in ethylene signalling, increased at 7 dpi [49,50]. Although these genes were upregulated during the susceptible response compared to the uninoculated controls, they were much more highly expressed during a resistance response.

The increased level of expression of *B. napus* genes involved in ethylene, jasmonic acid and salicylic acid signalling during Lbc infections prompted further examination of biochemical pathways during infection. A ratio of expression values of *B. napus* genes at 7 days after infection with Lmb compared to Lbc was calculated and analysed by MapMan software to identify pathways co-ordinately responding to early infection of the cotyledon (Figure 6, Table 9). The major difference in host response to infection by the two fungi was in genes with photosynthesis-associated activities. *L. biglobosa* 'canadensis' infection resulted in massive down-regulation of genes involved in photosynthesis (e.g. PSI, PSII), electron transport, photorespiration and chlorophyll (tetrapyrrole) biosynthesis compared to that in Lmb. Similarly, expression of sucrose and starch biosynthesis genes was reduced, possibly as a flow-on from lack of photosynthesis. Levels of sucrose degradation genes were higher during Lbc than during Lmb infection, perhaps due to decreased amounts of photosynthate. Biosynthetic genes for raffinose, a monosaccharide osmoprotectant, were induced by

Lbc, perhaps due to water stress. In contrast, genes involved in starch metabolism (both synthesis and degradation) were transcribed at high levels during Lmb infection at 7 dpi. The cell wall remodelling genes of *B. napus* that modify β-glucans, mannans and pectin were more highly expressed during infection by Lbc than by Lmb [51]. Extensive up regulation of host cell wall remodelling genes occurred at 7 dpi as the necrotic lesion was formed by Lbc, while Lmb infection had much less impact on transcription of these genes. A cohort of genes associated with secondary metabolism in *B. napus* was also plotted using Mapman on the same dataset (Figure S3 in file S1). Isoflavone reductase genes associated with isoflavonoid biosynthesis, were more highly expressed during Lbc infection, while genes associated with carotenoid metabolism (phytoene dehydrogenase, zeta-carotene desaturase, lycopene cyclases and violaxanthin de-epoxidase) were expressed more highly during Lmb infection. These observations are in agreement with the increased necrosis observed during Lbc infection at the early stages of infection. As well as increased expression of secondary metabolism genes during Lbc infection, peroxidase, nitrilase, and cytochrome P450 genes were consistently upregulated by *B. napus* at 7 dpi (Figure S3 in file S1). Twenty four general peroxidases were upregulated suggesting a strong oxidative burst was deployed during infection by Lbc. Forty two cytochrome 450 genes, 15 oxidase genes and 34 nitrilase genes were upregulated by Lbc infection, which also suggest a strong activation of secondary metabolism.

Summary

The hemi-biotroph *L. maculans* 'brassicae' avoids triggering host defence during early infection. This is reflected by the finding that at seven days post inoculation, *L. maculans* 'brassicae' expresses a large number of genes with no known domains, many of them being small secreted proteins. One class of small-secreted protein-encoding genes that is highly expressed at this time is CBM50 (LysM) genes, which suppress chitin-triggered PAMP immunity and evade detection of the fungus by the plant. Also avirulence genes are highly upregulated; at seven days post-inoculation, two avirulence genes are amongst the top 20 most highly upregulated *L. maculans* 'brassicae' genes *in planta*. This pattern is consistent with the relatively asymptomatic growth phase of this fungus at seven days post-inoculation. In contrast, *L. biglobosa* 'canadensis' expresses a high number of cell wall degrading CAZy genes during the first seven days of infection, consistent with extensive necrosis and a high degree of activation of host defence signalling pathways.

Supporting Information

File S1 Supporting figures. Figure S1, Classifications of CBM50 (LysM) domains in *L. maculans* 'brassicae', *L. biglobosa* 'canadensis', *Cladosporium fulvum* and *Zymoseptoria tritici*. LysM domains from Lmb, Lbc and *Z. tritici* (formerly *Mycosphaerella graminicola*) with high sequence similarity to ECP6 of *C. fulvum* were aligned using ClustalW. (A) Domains are numbered by proximity to N-terminus (#1 is closest). Residues are coloured by similarity (black: 100%, dark grey: 80–100%, light grey: 60–80%, white: less than 60%). B) Phylogram based on the amino acid alignment of LysM domains from panel A. Branch numbers show % bootstrap support and scale bar shows amino acid substitutions per site. LysM domains assigned to three Positions A, B and C based on sequence similarity to the *C. fulvum* ECP6 sequence. LysM domain organisation in ECP6-like predicted proteins in Lm, Lb, *Z. tritici* and *C. fulvum* are shown in panel C. Mg LysM genes are from *Z. tritici*. **Figure S2,** Expression of three LysM-containing genes of *L. maculans* 'brassicae' grown *in vitro* and *in planta*. Quantitative RT-PCR analysis was performed on RNA from Lmb isolate IBCN18 grown in 10% Campbells V8 juice (*in vitro*), and after infection of cotyledons of *B. napus* cv. Westar at 3, 7 and 14 days post inoculation (dpi). Expression levels of Lm2LysM (Lema_P102640.1), Lm4LysM (Lema_P025400.1), Lm5LysM (Lema_P070100.1), were normalised to those of gamma-actin (Lema_P099940.1). Error bars represent one standard error of the mean (n = 2–3 biological replicates). Asterisks indicate values significantly different from *in vitro* levels (p<0.05). **Figure S3,** Response of *B. napus* secondary metabolism and large enzyme families to infection by *L. maculans* 'brassicae' or *L. biglobosa* 'canadensis.'RNA-seq gene expression values for a *B. napus* unigene set [27] were used to calculate a ratio of expression values (log2) for *B. napus* genes seven days after infection by Lmb or Lbc. Ratios were plotted on secondary metabolism gene groups using MapMan software on maps for 'Secondary metabolism', and 'Large enzyme families' maps [39]. A yellow square indicates a *B. napus* gene that is expressed more highly during Lmb infection, while a blue square indicates a *B. napus* gene with higher expression during Lbc infection. An expression ratio close to zero is shown with a white square and indicates equivalent expression during infection by either pathogen. Only genes with expression values greater than 10 FPKM were included. MVA is mevalonic acid.

(PDF)

File S2 Supporting tables. Table S1, Total number of RNA-seq reads aligned to reference genomes. **Table S2,** Oligonucleotide Primers. **Table S3,** Top 100 most highly upregulated *in planta* genes in *L. maculans* 'brassicae' at seven days post-inoculation. **Table S4,** Top 100 most highly upregulated *in planta* genes in *L. biglobosa* 'canadensis' at seven days post-inoculation. **Table S5,** Top 100 most highly upregulated *in planta* genes in *L. maculans* 'brassicae' at 14 days post-inoculation. **Table S6,** Top 100 most highly upregulated *in planta* genes in *L. biglobosa* 'canadensis' at 14 days post-inoculation. **Table S7,** Annotated CAZy domains of *L. maculans* and *L. biglobosa,* and their expression. **Table S8,** Multi-domain CBM18 genes of *L. maculans* 'brassicae' and *L. biglobosa* 'canadensis', and their expression. **Table S9,** The top 10 CAZy families of *L. maculans* 'brassicae' and *L. biglobosa* 'canadensis' based on expression ratio of *in planta* and *in vitro* growth at 7 dpi. **Table S10,** Annotated peptidases of *L. maculans* and *L. biglobosa,* and their expression. **Table S11,** Peptidase expression in *L. maculans* 'brassicae' and *L. biglobosa* 'canadensis'.

Acknowledgments

We thank Sebastian Gornik, the University of Melbourne, for bioinformatics advice.

Author Contributions

Conceived and designed the experiments: RL BH AW. Performed the experiments: RL AW BC JG. Analyzed the data: RL AC JG. Contributed reagents/materials/analysis tools: JG TR. Contributed to the writing of the manuscript: RL BH.

References

1. Oliver RP, Solomon PS (2010) New developments in pathogenicity and virulence of necrotrophs. Curr Opin Plant Biol 13: 415–419.
2. Rouxel T, Grandaubert J, Hane JK, Hoede C, van de Wouw AP, et al. (2011) Effector diversification within compartments of the *Leptosphaeria maculans* genome affected by Repeat-Induced Point mutations. Nat Comm 2: 202.
3. Shoemaker RA, Brun H (2001) The teleomorph of the weakly aggressive segregate of *Leptosphaeria maculans*. Can J Bot 79: 412–419.
4. Voigt K, Cozijnsen AJ, Kroymann J, Pöggeler S, Howlett BJ (2005) Phylogenetic relationships between members of the crucifer pathogenic *Leptosphaeria maculans* species complex as shown by mating type (MAT1-2), actin, and β-tubulin sequences. Mol Phylogenet Evol 37: 541–557.
5. de Gruyter J, Woudenberg JHC, Aveskamp MM, Verkley GJM, Groenewald JZ, et al. (2013) Redisposition of phoma-like anamorphs in Pleosporales. Stud Mycol 75: 1–36.
6. Howlett BJ (2004) Current knowledge of the interaction between *Brassica napus* and *Leptosphaeria maculans*. Can J Plant Pathol 26: 245–252.
7. Van de Wouw A, Thomas V, Cozijnsen A, Marcroft S, Salisbury P, et al. (2008) Identification of *Leptosphaeria biglobosa* 'canadensis' on *Brassica juncea* stubble from northern New South Wales, Australia. Australas Plant D Notes 3: 124–128.
8. Gout L, Fudal I, Kuhn ML, Blaise F, Eckert M, et al. (2006) Lost in the middle of nowhere: the *AvrLm1* avirulence gene of the Dothideomycete *Leptosphaeria maculans*. Mol Microbiol 60: 67–80.
9. Fudal I, Ross S, Gout L, Blaise F, Kuhn ML, et al. (2007) Heterochromatin-like regions as ecological niches for avirulence genes in the *Leptosphaeria maculans* genome: map-based cloning of *AvrLm6*. Mol Plant Microbe Interact 20: 459–470.
10. Balesdent MH, Fudal I, Ollivier B, Bally P, Grandaubert J, et al. (2013) The dispensable chromosome of *Leptosphaeria maculans* shelters an effector gene conferring avirulence towards *Brassica rapa*. New Phytol 198: 887–898.
11. Parlange F, Daverdin G, Fudal I, Kuhn ML, Balesdent MH, et al. (2009) *Leptosphaeria maculans* avirulence gene *AvrLm4-7* confers a dual recognition specificity by the *Rlm4* and *Rlm7* resistance genes of oilseed rape, and circumvents Rlm4-mediated recognition through a single amino acid change. Mol Microbiol 71: 851–863.
12. Van de Wouw AP, Lowe RGT, Elliott CE, Dubois DJ, Howlett BJ (2013) An avirulence gene, *AvrLmJ1*, from the blackleg fungus, *Leptosphaeria maculans*, confers avirulence to *Brassica juncea* cultivars. Mol Plant Pathol doi:10.1111/mpp.12105.
13. Larkan NJ, Lydiate DJ, Parkin IAP, Nelson MN, Epp DJ, et al. (2013) The *Brassica napus* blackleg resistance gene *LepR3* encodes a receptor-like protein triggered by the *Leptosphaeria maculans* effector AVRLM1. New Phytol 197: 595–605.
14. Rubartelli A, Lotze MT (2007) Inside, outside, upside down: damage-associated molecular-pattern molecules (DAMPs) and redox. Trends Immunol 28: 429–436.
15. Baxter L, Tripathy S, Ishaque N, Boot N, Cabral A, et al. (2010) Signatures of Adaptation to Obligate Biotrophy in the *Hyaloperonospora arabidopsidis* Genome. Science 330: 1549–1551.
16. Duplessis S, Cuomo CA, Lin Y, Aerts A, Tisserant E, et al. (2011) Obligate biotrophy features unraveled by the genomic analysis of rust fungi. Proc Natl Acad Sci U S A 108: 9166–9171.
17. Spanu PD, Abbott JC, Amselem J, Burgis TA, Soanes DM, et al. (2010) Genome expansion and gene loss in powdery mildew fungi reveal tradeoffs in extreme parasitism. Science 330: 1543–1546.
18. Ohm RA, Feau N, Henrissat B, Schoch CL, Horwitz BA, et al. (2012) Diverse Lifestyles and Strategies of Plant Pathogenesis Encoded in the Genomes of Eighteen *Dothideomycetes* Fungi. Plos Pathog 8: e1003037.
19. Palmer CL, Skinner W (2002) *Mycosphaerella graminicola*: latent infection, crop devastation and genomics. Mol Plant Pathol 3: 63–70.
20. Solomon PS, Lowe RG, Tan KC, Waters OD, Oliver RP (2006) *Stagonospora nodorum*: cause of *stagonospora nodorum* blotch of wheat. Mol Plant Pathol 7: 147–156.
21. Liu Z, Ellwood SR, Oliver RP, Friesen TL (2011) *Pyrenophora teres*: profile of an increasingly damaging barley pathogen. Mol Plant Pathol 12: 1–19.
22. Westermann AJ, Gorski SA, Vogel J (2012) Dual RNA-seq of pathogen and host. Nat Rev Micro 10: 618–630.
23. Kawahara Y, Oono Y, Kanamori H, Matsumoto T, Itoh T, et al. (2012) Simultaneous RNA-seq analysis of a mixed transcriptome of rice and blast fungus interaction. PLoS One 7: e49423.
24. Yazawa T, Kawahigashi H, Matsumoto T, Mizuno H (2013) Simultaneous Transcriptome Analysis of Sorghum and *Bipolaris sorghicola* by Using RNA-seq in Combination with *De Novo* Transcriptome Assembly. PLoS One 8: e62460.
25. Purwantara A, Salisbury PA, Burton WA, Howlett BJ (1998) Reaction of *Brassica juncea* (Indian mustard) lines to Australian isolates of *Leptosphaeria maculans* under glasshouse and field conditions. Eur J Plant Pathol 104: 895–902.
26. Trapnell C, Pachter L, Salzberg SL (2009) TopHat: discovering splice junctions with RNA-Seq. Bioinformatics 25: 1105–1111.
27. Love CG, Graham NS, Ó Lochlainn S, Bowen HC, May ST, et al. (2010) A *Brassica* Exon Array for Whole-Transcript Gene Expression Profiling. PLoS One 5: e12812.

28. Trapnell C, Williams BA, Pertea G, Mortazavi A, Kwan G, et al. (2010) Transcript assembly and quantification by RNA-Seq reveals unannotated transcripts and isoform switching during cell differentiation. Nat Biotech 28: 511–515.

29. Cantarel BL, Coutinho PM, Rancurel C, Bernard T, Lombard V, et al. (2009) The Carbohydrate-Active EnZymes database (CAZy): an expert resource for Glycogenomics. Nucleic Acids Res 37: D233–D238.

30. Yin Y, Mao X, Yang J, Chen X, Mao F, et al. (2012) dbCAN: a web resource for automated carbohydrate-active enzyme annotation. Nucleic Acids Res 40: W445–W451.

31. Petersen TN, Brunak S, von Heijne G, Nielsen H (2011) SignalP 4.0: discriminating signal peptides from transmembrane regions. Nat Meth 8: 785–786.

32. Eddy SR (2011) Accelerated Profile HMM Searches. Plos Comput Biol 7: e1002195.

33. Punta M, Coggill PC, Eberhardt RY, Mistry J, Tate J, et al. (2012) The Pfam protein families database. Nucleic Acids Res 40: D290–D301.

34. Robinson JT, Thorvaldsdottir H, Winckler W, Guttman M, Lander ES, et al. (2011) Integrative genomics viewer. Nat Biotech 29: 24–26.

35. de Jonge R, van Esse HP, Kombrink A, Shinya T, Desaki Y, et al. (2010) Conserved fungal LysM effector Ecp6 prevents chitin-triggered immunity in plants. Science 329: 953–955.

36. Altschul SF, Gish W, Miller W, Myers EW, Lipman DJ (1990) Basic local alignment search tool. J Mol Biol 215: 403–410.

37. Rawlings ND, Barrett AJ, Bateman A (2012) MEROPS: the database of proteolytic enzymes, their substrates and inhibitors. Nucleic Acids Res 40: D343–D350.

38. Epple P, Apel K, Bohlmann H (1997) ESTs reveal a multigene family for plant defensins in *Arabidopsis thaliana*. Febs Lett 400: 168–172.

39. Thimm O, Bläsing O, Gibon Y, Nagel A, Meyer S, et al. (2004) Mapman: a user-driven tool to display genomics data sets onto diagrams of metabolic pathways and other biological processes. Plant J 37: 914–939.

40. Eckert M, Maguire K, Urban M, Foster S, Fitt B, et al. (2005) *Agrobacterium tumefaciens*-mediated transformation of *Leptosphaeria* spp. and *Oculimacula* spp. with the reef coral gene DsRed and the jellyfish gene gfp. Fems Microbiol Lett 253: 67–74.

41. Zablackis E, Huang J, Muller B, Darvill AG, Albersheim P (1995) Characterization of the cell-wall polysaccharides of *Arabidopsis thaliana* leaves. Plant Physiol 107: 1129–1138.

42. Ipcho SV, Hane JK, Antoni EA, Ahren D, Henrissat B, et al. (2012) Transcriptome analysis of *Stagonospora nodorum*: gene models, effectors, metabolism and pantothenate dispensability. Mol Plant Pathol 13: 531–545.

43. Levasseur A, Drula E, Lombard V, Coutinho PM, Henrissat B (2013) Expansion of the enzymatic repertoire of the CAZy database to integrate auxiliary redox enzymes. Biotechnology for biofuels 6: 41.

44. Brunner PC, Torriani SFF, Croll D, Stukenbrock EH, McDonald BA (2013) Coevolution and life cycle specialization of plant cell wall degrading enzymes in a hemibiotrophic pathogen. Mol Biol Evol 30: 1337–1347.

45. Feng J, Zhang H, Strelkov SE, Hwang SF (2014) The *LmSNF1* gene is required for pathogenicity in the canola blackleg pathogen *Leptosphaeria maculans*. PLoS One 9: e92503.

46. O'Connell RJ, Thon MR, Hacquard S, Amyotte SG, Kleemann J, et al. (2012) Lifestyle transitions in plant pathogenic *Colletotrichum* fungi deciphered by genome and transcriptome analyses. Nat Genet.

47. Rypniewski WR, Hastrup S, Betzel C, Dauter M, Dauter Z, et al. (1993) The sequence and X-ray structure of the trypsin from *Fusarium oxysporum*. Protein Eng 6: 341–348.

48. Showalter AM, Keppler B, Lichtenberg J, Gu D, Welch LR (2010) A bioinformatics approach to the identification, classification, and analysis of hydroxyproline-rich glycoproteins. Plant Physiol 153: 485–513.

49. Persson M, Staal J, Oide S, Dixelius C (2009) Layers of defense responses to *Leptosphaeria maculans* below the RLM1- and camalexin-dependent resistances. New Phytol 182: 470–482.

50. Šašek V, Nováková M, Jindřichová B, Bóka K, Valentová O, et al. (2012) Recognition of Avirulence Gene *AvrLm1* from Hemibiotrophic Ascomycete *Leptosphaeria maculans* Triggers Salicylic Acid and Ethylene Signaling in *Brassica napus*. Mol Plant Microbe Interact 25: 1238–1250.

51. Vogel JP (2002) *PMR6*, a Pectate Lyase-Like Gene Required for Powdery Mildew Susceptibility in *Arabidopsis*. Plant Cell 14: 2095–2106.

52. Grandaubert J, Lowe RGT, Soyer JL, Schoch CL, Fudal I, et al. (2014) Transposable Element-assisted evolution and adaptation to host plant within the *Leptosphaeria maculans-Leptosphaeria biglobosa* species complex of fungal pathogens. Biomed Cent Genomics. In press.

8

Characterization of *Glomerella* Strains Recovered from Anthracnose Lesions on Common Bean Plants in Brazil

Quélen L. Barcelos[1], Joyce M. A. Pinto[2], Lisa J. Vaillancourt[3]*, Elaine A. Souza[1]*

1 Departamento de Biologia, Universidade Federal de Lavras, Lavras, Minas Gerais, Brazil, 2 Empresa Brasileira de Pesquisa Agropecuária (Embrapa), Sinop, Mato Grosso, Brazil, 3 Department of Plant Pathology, University of Kentucky, Lexington, Kentucky, United States of America

Abstract

Anthracnose caused by *Colletotrichum lindemuthianum* is an important disease of common bean, resulting in major economic losses worldwide. Genetic diversity of the *C. lindemuthianum* population contributes to its ability to adapt rapidly to new sources of host resistance. The origin of this diversity is unknown, but sexual recombination, via the *Glomerella* teleomorph, is one possibility. This study tested the hypothesis that *Glomerella* strains that are frequently recovered from bean anthracnose lesions represent the teleomorph of *C. lindemuthianum*. A large collection of *Glomerella* isolates could be separated into two groups based on phylogenetic analysis, morphology, and pathogenicity to beans. Both groups were unrelated to *C. lindemuthianum*. One group clustered with the *C. gloeosporioides* species complex and produced mild symptoms on bean tissues. The other group, which belonged to a clade that included the cucurbit anthracnose pathogen *C. magna*, caused no symptoms. Individual ascospores recovered from *Glomerella* perithecia gave rise to either fertile (perithecial) or infertile (conidial) colonies. Some pairings of perithecial and conidial strains resulted in induced homothallism in the conidial partner, while others led to apparent heterothallic matings. Pairings involving two perithecial, or two conidial, colonies produced neither outcome. Conidia efficiently formed conidial anastomosis tubes (CATs), but ascospores never formed CATs. The *Glomerella* strains formed appressoria and hyphae on the plant surface, but did not penetrate or form infection structures within the tissues. Their behavior was similar whether the beans were susceptible or resistant to anthracnose. These same *Glomerella* strains produced thick intracellular hyphae, and eventually acervuli, if host cell death was induced. When *Glomerella* was co-inoculated with *C. lindemuthianum*, it readily invaded anthracnose lesions. Thus, the hypothesis was not supported: *Glomerella* strains from anthracnose lesions do not represent the teleomorphic phase of *C. lindemuthianum*, and instead appear to be bean epiphytes that opportunistically invade and sporulate in the lesions.

Editor: Dongsheng Zhou, State Key Laboratory of Pathogen and Biosecurity, Beijing Institute of Microbiology and Epidemiology, China

Funding: This work was supported by grants and scholarships from the following Brazilian agencies: Fundação de Amparo à Pesquisa do Estado de Minas Gerais – FAPEMIG (CAG-APQ00635-12), Coordenação de Aperfeiçoamento de Pessoal de Nível Superior – CAPES and Conselho Nacional de Desenvolvimento Científico e Tecnológico - CNPq. The funders had no role in study design, data collection and analysis, decision to publish, or preparation of the manuscript.

Competing Interests: The authors have declared that no competing interests exist.

* E-mail: easouza@dbi.ufla.br (EAS); vaillan@uky.edu (LJV)

Introduction

Anthracnose, caused by the fungus *Colletotrichum lindemuthianum* (Sacc. & Magn.) Scribn., is one of the most important diseases on common bean (*Phaseolus vulgaris* L.) worldwide, and causes significant losses in Brazil [1,2]. Anthracnose is generally managed by the use of resistant cultivars, but the extreme genetic diversity of the pathogen population in Brazil contributes to frequent failure of resistance sources [3–8]. It is unclear how diversity arises, but genetic recombination during sexual reproduction is one possibility. The teleomorph of *C. lindemuthianum*, known as *Glomerella lindemuthiana* Shear (syn. *G. lindemuthianum*, *G. cingulata* (Stonem.) Spauld. et Schrenk f. sp. *phaseoli*), was first described in 1913 by Shear and Wood [9]. Mating in *Glomerella* has been studied in several species, including *G. lindemuthiana*, in culture. All reports agree that, although some strains of *Glomerella* can be homothallic, the majority of fertile strains are heterothallic [10–16]. All reports also agree that mating in *Glomerella* is not regulated by a single mating type locus, and that there are a large number of compatible mating types in the population, which could indicate a large

potential for outcrossing [16]. Studies of a highly fertile population of *G. cingulata* strains from morning glory in the early part of the 20th century suggested that the common occurrence of asexual strains was due to frequent mutations in fertility genes, and that multiple mating compatibilities were the result of complementation among these mutations, the so-called "unbalanced heterothallism" theory [17–28]. Previous reports suggest that sexual fertility is rare in *G. lindemuthiana* [29–31]. Sexual recombination has been demonstrated among a handful of fertile strains in the lab, but it has not been shown to occur in the field. Thus the contribution of sexual recombination to pathogen population diversity in *C. lindemuthianum* remains unclear.

We have isolated a large number of *Colletotrichum* strains from anthracnose lesions on naturally infected common beans. Some single-spored strains produced the *Glomerella* teleomorph readily when cultured alone, while others appeared to be asexual in culture [11,13,32,33]. *Glomerella* strains isolated from bean anthracnose lesions in Brazil have usually been identified as *G. cingulata* f. sp. *phaseoli* or *G. lindemuthiana*, under the assumption that they represent the teleomorphic phase of *C. lindemuthianum*

[11,13,32,34]. However, we recently reported that several of these teleomorphic strains produced no, or only mild symptoms, when reinoculated onto beans susceptible to anthracnose [5,11,13,32]. We have further reported that some sexual and asexual strains can be distinguished by morphology [35], and by molecular fingerprint [5]. Our goal in this study was to characterize the diversity and the pathogenic and sexual behavior among a larger population of these teleomorphic strains, and to determine their relationship with asexual *C. lindemuthianum* strains that cause bean anthracnose. The work described here reveals that *Glomerella* isolates from bean anthracnose lesions belong to two different genetic lineages, neither of which is closely related to *C. lindemuthianum*. Members of one lineage did not cause symptoms when inoculated on bean tissues, while those from the other lineage caused only very mild symptoms. Our conclusion is that the *Glomerella* strains are epiphytes that opportunistically colonize the anthracnose lesions produced by *C. lindemuthianum*.

Materials and Methods

Isolates and Culture Conditions

Colletotrichum isolates were collected from anthracnose lesions on pods, leaves and petioles of common bean (*P. vulgaris*) from naturally infected fields. The collections were made primarily in the cities of Lavras, Lambari, and Ribeirão Vermelho, in the southern state of Minas Gerais (Table S1). No specific permissions were required for collections in these locations and studies did not involved endangered or protected species. The study was conducted in the Department of Biology of the Universidade Federal of Lavras, Minas Gerais, Brazil and in the Plant Pathology Department of the University of Kentucky, Kentucky, United States. Small pieces of infected tissue were disinfected and deposited in Petri dishes containing M3 culture medium [36]. A total of 68 single-spored isolates, recovered from 68 individual lesions, were fertile and produced the *Glomerella* teleomorph in culture. One perithecium each from 54 of these isolates was crushed, and a strain derived from a single ascospore was recovered for each isolate. For the remaining 14 isolates (UFLAG05, UFLAG06, UFLAG07, UFLAG08, UFLAG15, UFLAG21, UFLAG30, UFLAG43, UFLAG47, UFLAG54, UFLAG73, UFLAG104, UFLAG106, and UFLAG112), between two and five monoascospore strains were recovered from a single crushed perithecium of each (Table S1). All monoascospore strains were maintained on M3 media at 22°C in the dark.

Morphological Characterization

Colony classification. The monoascospore strains were classified into four different morphological classes: 1) Conidial A strains did not produce perithecia on M3 media, but produced abundant conidia in large masses; 2) Conidial B strains did not produce perithecia, and produced conidia sparsely, scattered over the colony surface; 3) Plus strains formed perithecia in clumps; 4) Minus strains produced individual perithecia scattered across the surface of the culture (Figure 1 and Table S1). A few of the plus and minus strains also produced conidia, but production was sparse. Thus, it was not possible to collect and evaluate both ascospores and conidia from a single strain. Henceforth, the Conidial A and B strains will be referred to as "conidial" strains, and the Plus and Minus strains as "perithecial" strains. A total of 88 monoascospore strains, comprising all four classes, and including one or more representatives of 61 of the 68 original *Glomerella* single-spored isolates, were used in the experiments described below (Table S1).

Index of mycelial growth rate (IMGR) and colony diameter. The experiment was a completely randomized design (CRD) with four replicates. Each plot consisted of a single Petri dish (80 mm diameter). Mycelium plugs 6 mm in diameter were placed in the centers of individual Petri dishes containing M3 media, and cultures were incubated in the dark at 22°C. Colony diameter (millimeters) was measured twice at right angles for each colony at intervals of 24 hours over the course of 8 days, and the averages were used to estimate the IMGR according to the expression $IMGR = \sum \frac{(Dc - Dp)}{N}$, where Dc is the current average of the colony diameter, Dp is the average of the previous colony diameter, and N is the number of days after inoculation [37]. Final colony diameter (mm) was determined after 8 days of incubation.

Germination rate. The experiment was a completely randomized design (CRD), split plot in time, with two replicates. Ascospore and conidial suspensions were adjusted to 1.2×10^6 spores/mL, and 500 µl of each suspension was spread in Petri dishes containing water agar (2%). After 24 and 48 hours of incubation at 22°C in the dark, 50 spores per replicate were observed by light microscopy with an Olympus CX41 (Olympus Deutschland GmbH, Hamburg, Germany). Spores that had produced germ tubes with a length equal to or greater than the smallest diameter of the spore were considered to be germinated.

Spore measurements, septum formation and Conidial Anastomosis Tubes (CATs). For all experiments, 200 µl of a spore suspension (1.2×10^6 spores/ml) of each strain was applied to a chambered borosilicate coverglass (Nalge Nunc International, Rochester, NY) and incubated at 22°C in the dark. Samples were examined by using an inverted epifluorescence microscope (Zeiss Axio Observer Z1; Carl Zeiss Inc., Jena, Germany) after adding the dye Calcofluor White (Sigma-Aldrich, St. Louis) at a concentration of 0.12 M. The fluorescence was detected at 420/70 nm using the 40× objective. The images were captured using Zeiss Axiovision software and processed using ImageJ 1.41 software (U.S. National Institutes of Health, Bethesda, MD).

Measurements were made for 30 non-germinated spores of each strain in a CDR experiment. The width and length of the ungerminated spores (µm) were measured by using the imaging software Image Tool 3.0 (University of Texas Health Science Center, San Antonio).

The presence or absence of a septum was observed after 24 hours in two replicates of 100 germinating spores. The formation of CATs was quantified 24 hours after incubation as the percentage of spores involved in anastomosis [33]. Two replicates consisting of 200 spores each were analyzed.

Statistical analyses. Morphological data were subjected to analysis of variance (ANOVA), and means were compared by the Scott-Knott test (P = 0.05), by using the statistical program MSTAT-C 1.0 (Michigan State University, East Lansing) and SISVAR [38], respectively.

Sexual interactions. The 73 monoascospore strains were paired on M3 media in all possible combinations (2701 pairings) and incubated at 22°C in the dark. Mycelial plugs 6 mm in diameter were placed at a distance of 2 cm apart, with two repetitions per confrontation. Confrontations were evaluated after 15 days of incubation.

The formation of a line of perithecia containing ascospores at the contact zone of paired strains was an indication of a fertile interaction between the strains. To further analyze fertile interactions, dialysis membranes were used to separate the partners. Dialysis membranes do not allow physical contact between strains but permit the exchange of molecules that can induce production of fertile selfed perithecia in some strains of *Glomerella* [26]. Fertile interactions were classified as: a) induced

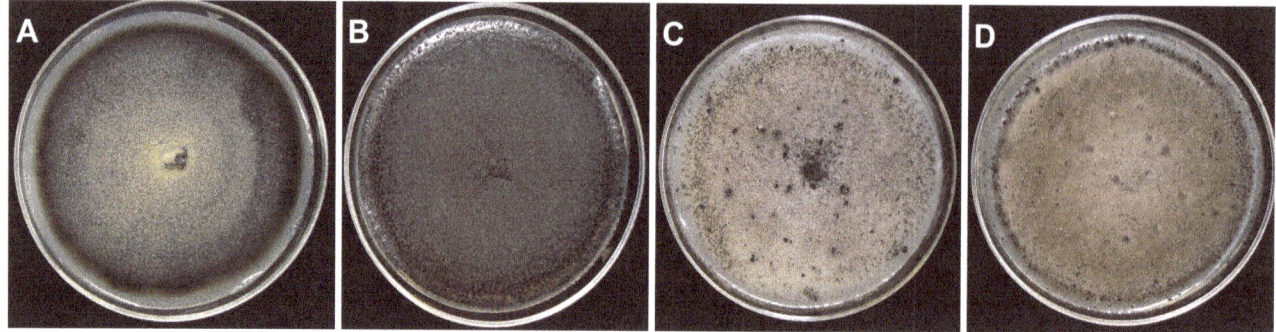

Figure 1. Morphology of colonies recovered after isolating single ascospores from *Glomerella* sp. strains. A) Black conidial A colony, producing masses of conidia; B) Black conidial B colony, with sparse production of conidia; C) White Plus colony, producing clumps of perithecia; and D) White Minus colony, producing scattered perithecia.

homothallic, when a line of perithecia was formed even when the partners were separated by a dialysis membrane; or b) likely heterothallic, when a line of perithecia was produced, but only in the absence of a dialysis membrane.

Molecular Characterization

DNA extraction, PCR reactions and sequencing. Mycelium plugs were transferred to 125 mL of M3 liquid medium in Erlenmeyer flasks. The flasks were shaken at 110 rpm/min at 22°C for 7 days in the dark. Mycelium was dried in a vacuum, and subsequently freeze-dried for 48 hours. DNA was extracted using a high-throughput DNA prep method [39], or a mini-prep method [40].

Sequences of the internal transcribed spacer (ITS variable regions) of the ribosomal DNA and the high mobility group (HMG)-encoding sequence of the MAT1-2-1 mating type gene were amplified by using polymerase chain reaction (PCR). The universal primers ITS1 and ITS4 were used for amplification of ITS variable regions [41]. PCR reactions contained 10–100 ng of genomic DNA, 2.5 mM MgCl$_2$, 1X PCR buffer, 0.2 mM each dNTP, 2.5 U of Taq DNA polymerase (Invitrogen, Carlsbad, California) and 0.5 μM of each primer. The amplification cycle consisted of denaturation at 95°C for 5 min followed by 40 cycles consisting of 30 s at 95°C, 30 s at 50°C and 1.5 min at 72°C.

To amplify HMG sequences, primers HMGCLF and HMGCLR were used. This primer pair is specific to the *C. lindemuthianum* MAT1-2-1 HMG region [10]. The amplification cycle consisted of denaturation at 95°C for 5 min followed by 40 cycles consisting of 30 s at 95°C, 30 s at 50°C, and 1.5 min at 72°C. The degenerate primers NcHMG1 and NcHMG2, and PCR conditions described previously [42], were used to amplify HMG sequences from strains that were not amplified by the HMGCLF primer pair. PCR products were cloned in pGEM-Teasy (Promega Corporation, Madison, USA) and sequenced with primers complementary to the cloning vector. A new primer pair, HMGglo1 (5′CGAGCTCCGTGTATCTCTGG3′) and HMGglo2 (5′AAAGATCACTGCGCCAA GTT3′), was developed based on these sequences. PCR reactions contained 10–100 ng of genomic DNA, 2.5 mM MgCl$_2$, 1X PCR buffer, 0.2 mM each dNTP, 2.5 U of Taq DNA polymerase (Invitrogen, Carlsbad, California) and 0.2 μM of each in 50 μL. The amplification cycle consisted of denaturation at 95°C for 5 min followed by 40 cycles consisting of 30 s at 94°C, 30 s at 55°C and 30 s at 72°C. PCR products were separated in a 1% agarose gel, stained with ethidium bromide, and viewed on a UV transilluminator.

HMG amplicons of 10 strains, and ITS amplicons of 17 strains, were recovered from the electrophoresis gels with a gel extraction system (QIAGEN Inc., Valencia, California) (Table S1). These PCR products were sequenced with the BigDye terminator cycle sequencing kit (Applied Biosystems, Foster City, California), and the sequences were analyzed on an ABI 310 automated sequencer (Applied Biosystems).

Fungal transformation. To obtain strains expressing the green fluorescent protein (GFP), an *Agrobacterium tumefaciens*-mediated transformation protocol was used [43]. The strains LV115 and UFLAG06 were transformed (Table S1). The vector used was pJF1 [43], containing the SGFP gene driven by the TOX-A promoter from *Pyrenophora tritici-repentis* [44], and with the hygromycin phosphotransferase gene as selectable marker. The transformants were selected on PDA media containing 50 μg/ml of hygromycin B, and purified by single sporing.

Phylogenetic analysis. Phylogenetic analyses were performed by using the Phylogeny.fr platform with default settings [45], and included the following steps: sequence alignment with MUSCLE (v3.7); removal of ambiguous regions with Gblocks (v0.91b); construction of a phylogenetic tree using the maximum likelihood method with PhyML (v3.0 aLRT); and evaluation of internal branch reliability using the aLRT test (SH-Like). Phylogenetic trees were drawn and edited using TreeDyn (v198.3). Trees were drawn with midpoint rooting, and all branches with values of less than 0.50 were collapsed. The ITS sequence alignment is provided in File S1.

Infection Assays

Spore solutions and inoculations. Sterile bean pods were inoculated with each strain and incubated at 22°C for 10 days in the dark. Only six of the strains sporulated on the pods (Table S1). For the other strains, six plates containing M3 culture medium were inoculated and incubated at 22°C for 10 days in the dark to produce spores. Spore solutions were made by harvesting and washing three times in sterile water with centrifugation and resuspension in water. The concentration was adjusted to 1.2×10^6 spores/mL for all experiments. Inoculations were made by spraying the spore solutions on plant seedlings to runoff. Alternatively, 5 μL drops were placed on detached hypocotyls or leaves in humidity chambers.

Pathogenicity tests. Pathogenicity tests were carried out with the 53 monoascospore strains that produced conidia on either M3 medium or bean pods (Table S1). The strains were inoculated on a differential set of 12 cultivars of common bean [46]. The susceptible cultivar Pérola was used as a control. The seeds were sown in trays filled with the growth substrate Multiplant (Terra do Paraíso Ltda., Brazil). Conidial suspensions were sprayed to runoff on ten-day-old seedlings. The inoculated bean seedlings were incubated in a fog chamber at 22°C, with a photoperiod of 12 hours, and a relative humidity of 95%, for 48 hours. The trays were then transferred to the greenhouse, and the seedlings were evaluated for severity of anthracnose ten days after inoculation using the diagrammatic scale [47]. Seedlings with ratings between 1 and 3 were considered resistant, and those with ratings greater than 3 were considered susceptible. The identification of races was made as proposed by Habgood [48].

Detached hypocotyls of the Pérola and Michelite cultivars were inoculated in order to directly observe the development of the fungi in the bean tissues. For these infection assays, five strains were used (Table S1). They included one known pathogenic, asexual *C. lindemuthianum* strain (LV115) as a control [32]. UGFLAG06-1 (Plus strain, ascospores were used for inoculation), and UFLAG06-2 (Conidial B strain, conidia were used for inoculation), were both derived from a single perithecium of the monoascospore strain UFLAG06 (Table S1). Strains UFLAG08-2 (Minus strain, ascospores were used for inoculation), and UFLAG08-3 (Conidial A strain, conidia were used for inoculation), were both derived from a single perithecium of the monoascospore strain UFLAG08 (Table S1). UFLAG06 and UFLAG08 were derived from different fertile single-spored isolates collected from two different lesions on naturally infected beans (Table S1). Detached hypocotyls from ten-day-old seedlings were placed into sterile Petri dishes lined with moistened germination paper. The Petri dishes were placed into a germination box, and the hypocotyls were inoculated with 5 µl drops of spore solutions and incubated at 22°C for seven days. The control was inoculated with drops of sterile water. Symptoms were observed and photographed 10 days after inoculation.

Light microscopy experiments. Inoculations using conidia of *C. lindemuthianum* strain LV115 and ascospores of strain UFLAG06 (Plus strain), were made on detached leaves and hypocotyls of the susceptible common bean cultivar Pérola, and the resistant cultivar G2333. Observations were made at 24, 48, 72, 96, and 120 hours after inoculation (hpi). Samples were obtained by hand-sectioning the plant tissue with a razor blade. To clarify them, the plant tissue samples were immersed in a solution of methanol: chloroform: acetic acid (60:30:10) for 30 minutes. Samples were then immersed in Trypan Blue (250 µg/ml) in lactophenol solution (lactic acid: phenol: H_2O 1: 1:1) for 20 minutes. Samples were transferred to lactophenol for 30 min, and then observed in 50% glycerol by using a Zeiss Axiosco. Images were obtained with AxioVision 4.8 software (Carl Zeiss, Oberkochen, Germany).

At 72 hpi, the percentage of spores forming appressoria was measured in a completely randomized design (CRD) experiment, split plot in time, with three replicates. Each plot consisted of a sample of 100 spores. ANOVA was conducted using MSTATC 1.0 (Michigan State University, East Lansing).

Fluorescence microscopy. The transformed fertile mono-ascospore strain tQB01, and the transformed *C. lindemuthianum* strain tQB02, both expressing GFP, were inoculated on detached hypocotyls and leaves of the susceptible cultivar Pérola and the resistant cultivar G2333. Observations were made at 72, 96 and 120 hpi.

To cause localized cell death on the plant tissues, small pieces of dry ice were placed on hypocotyls and on leaves of susceptible and resistant cultivars for three seconds. The fertile tQB01 strain was inoculated onto the killed tissues immediately after treatment. Observations were made at 24, 48 and 72 hpi. Inoculated plant tissues were sectioned and observed in 50% glycerol by using a Zeiss Axioscop equipped with epifluorescence and a GFP filter. Images were obtained by using the AxioVision 4.8 software (Carl Zeiss, Oberkochen, Germany).

Co-inoculations. Wild type and transformed GFP-expressing strains were co-inoculated on detached hypocotyls of the susceptible cultivar Pérola. The co-inoculations were made in two combinations: 1) Non-fertile *C. lindemuthianum* wild type strain LV115 and fertile transformed monoascospore strain tQB01; and 2) Fertile wild-type monoascospore strain UFLAG06 and *C. lindemuthianum* transformant strain tQB02. The wild type of each strain inoculated alone on Pérola hypocotyls, and mock inoculations with sterile water, were the controls for this experiment. Spore suspensions of each strain were prepared, combined, and then 5 µL drops of the combined suspensions were placed on the plant tissue. Evaluations were made at 120 hpi. Sections of plant tissue were observed in 50% glycerol by using a Zeiss Axioscop equipped with epifluorescence and a GFP filter. Images were obtained by using the AxioVision 4.8 software (Carl Zeiss, Oberkochen, Germany).

Results

Morphological Characterization

Colony classification. The color of the monoascospore strains in culture on M3 medium was black, brown, or white (Figure 1). The strains comprised four distinct groups based on their morphology and the production of spores: Conidial A (24 strains) and conidial B (29 strains) both produced abundant conidia in culture. In the conidial B group the conidia were scattered over the surface of the agar while in conidial A the conidia were formed in clumps. Plus (11 strains) produced perithecia in clusters while in Minus (9 strains) the perithecia were scattered across the surface of the agar. Plus and Minus strains were always white, while conidial A and B strains were either black or brown (Figure 1). More than one type could be recovered from a single perithecium (Table S1). These four groups were reminiscent of descriptions of single-ascospore strains of *G. cingulata*, published by Chilton and Wheeler [23]. Conidial A and B strains (aka "conidial strains") produced only conidia, while Plus and Minus strains (aka "perithecial strains") produced primarily ascospores, and few or no conidia. For our studies of morphological variation described below, we conducted an analysis of variance. Sources of variation were partitioned among conidial strains (conidial A and B); among perthecial strains (Plus and Minus); and among conidial vs. perithecial strains. All these sources of variation were statistically significantly different (p< 0.05).

Index of mycelial growth rate (IMGR) and colony diameter. Conidial strains could be statistically separated into two groups on the basis of IMGR, and three on the basis of colony diameter. There was no relationship between these groups and the A or B phenotypes. IMGR ranged from 8.79 mm/day to 10.39 mm/day, and colony diameter ranged from 67.75 mm to 80.0 mm. Perithecial strains could be statistically separated into two groups on the basis of IMGR and on the basis of colony diameter. There was no relationship between these groups and the Plus or Minus phenotypes. IMGR for perithecial strains ranged from 8.89 to 10.5 mm/day, and colony diameter ranged from 68.00 to 79.75 mm.

Percentage of germination. Conidial strains could be statistically separated into six groups at 24 h, and five groups at 48 h, based on the rate of conidial germination. At 24 h, the percentage of conidial germination varied from 7.0 to 89.5%, and at 48 h from 28.0 to 97.5%. Perithecial strains could be separated into three classes based on ascospore germination rate at both 24 h and 48 h. At 24 h, the ascospore germination rate ranged from 27.5 to 88.5%, and at 48 h it ranged from 51.5 to 98.5%. Ascospores from perithecial strains germinated significantly faster, on average, than conidia from conidial strains. None of the classes were related to the A, B, Plus, or Minus phenotypes.

Spores measurements, septum formation and Conidial Anatomosis Tubes (CAT's). Conidial strains formed seven groups for both spore width and length. Width varied from 2.88 to 5.62 μm, and length varied from 8.46 to 15.85 μm. Perithecial strains could be separated into five groups based on spore width and eight groups based on spore length. Width varied from 6.32 to 9.2 μm, and length varied from 20.5 to 32.78 μm. Ascospores were, on average, significantly larger than conidia.

Conidia of all conidial strains formed a single septum during germination, with the exception of UFLAG21-1. Ascospores of all perithecial strains also formed a single septum during germination (Figure 2). The percentage of spores in each strain that had formed septa after 24 hours of incubation ranged from 61.5 to 100%.

Conidia from all except two of the 53 conidial strains formed CATs. The 51 strains that formed CATs could be statistically divided into three groups, based on the rate of CATs formation, which varied from 0.75 to 78.75%. The highest percentage of CATs was formed by the conidia of the UFLAG07-2 strain (Figure 2). Ascospores from perithecial strains never formed CATs (not shown).

Sexual interactions. The 73 monoascospore strains were paired in all possible combinations (2701 pairings) on M3 medium. Pairings between conidial A and conidial B strains never resulted in the formation of perithecia (not shown). Pairings between Plus and Minus strains never resulted in the formation of a line of perithecia between the two colonies (not shown). Only 75 pairings (2.78%) resulted in the formation of a line containing perithecia at the point of contact between the strains, and these fertile interactions were only obtained when a conidial strain was paired with a perithecial strain. Twenty-six of these combinations produced only protoperithecia (not shown). The remaining 49 combinations formed fertile perithecia that contained large numbers of asci and ascospores (Table S2). These 49 pairings were repeated, but with a dialysis membrane separating the two strains. Twenty-eight of the combinations no longer formed a contact line of perithecia in the presence of the membrane (Table S2, Figure 3). The other 21 combinations still produced a line of fertile perithecia, even in the absence of physical contact between the strains. In all of these cases, the perithecia were formed on the same side of the membrane as the conidial strain (Figure 3), suggesting that the perithecial strain was inducing its conidial partner to produce fertile, selfed perithecia.

Molecular Characterization

PCR reactions. The ITS1 and ITS4 universal primers successfully amplified a product corresponding in size to the expected ITS product from the 17 perithecial and conidial

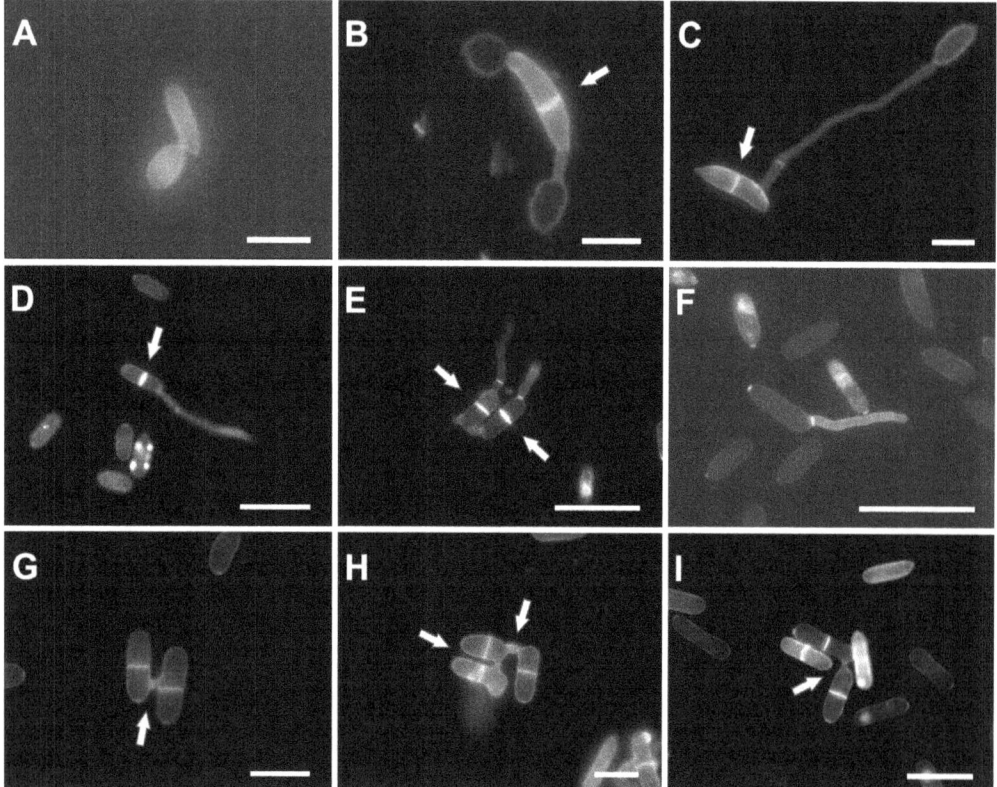

Figure 2. Formation of septa and conidial anastomosis tubes. A) *C. lindemuthianum* strain LV115 did not form a septum during germination; B) *Glomerella* UFLAG08 (Minus strain), germinated ascospore with a septum (white arrow) and two appressoria; C) *Glomerella* UFLAG06 (Plus strain), ascospore with a septum (white arrow) and an appressorium; D) *Glomerella* UFLAG68-1 (conidial A strain), conidia forming a septum (white arrow) during germination; E) *Glomerella* UFLAG36-1 (conidial B strain), two conidia forming septa (white arrows) during germination; F) *Glomerella* UFLAG21-1 (conidial A strain), conidia did not form a septum during germination. Bar: 25 μm. G) CATs formation in the conidial A strain UFLAG07-2. H) CATs formation in the conidial B strain UFLAG117-1; I) CATs formation in the conidial B strain UFLAG111-1. Bar: 20 μm.

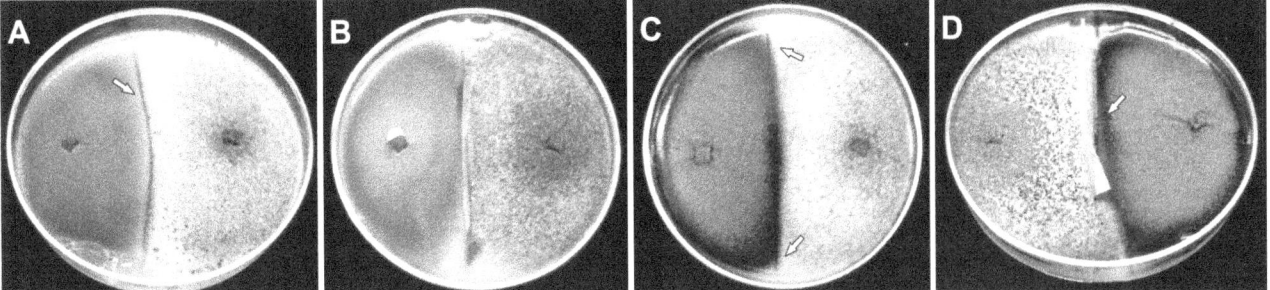

Figure 3. Pairings of monoascospore strains in culture medium with and without dialysis membrane. A) UFLAG47-1 and UFLAG47-2 strains paired without the dialysis membrane, with formation of a contact line with fertile and viable perithecia; B) UFLAG47-1 and UFLAG47-2 strains paired with the dialysis membrane preventing contact between them: the perithecial line is no longer observed; C) UFLAG21-2 and UFLAG104-2 strains paired in the absence of dialysis membrane, with formation of a contact line containing fertile perithecia and viable ascospores; D) UFLAG21-2 and UFLAG104-2 strains paired with the dialysis membrane preventing contact between them: the perithecial line is still observed.

monoascospore strains that were tested. The primers HMGCLF and HMGCLR [10] amplified a product of the expected size from three control *C. lindemuthianum* strains, but did not amplify a product from any of the 73 monoascospore strains (Figure 4A). However, the specific primers HMGglo1 and HMGglo2 did amplify a single product corresponding in size to the HMG region of the MAT2 locus from most of the monoascospore strains (Figure 4B). The exceptions were the conidial strains UFLAG85-1, UFLAG86-1, UFLAG89-1, UFLAG92-1, UFLAG93-1, and UFLAG99-1, which amplified very poorly with this primer pair (Figure 4B).

Phylogenetic analysis. A phylogenetic analysis based on ITS sequences suggested that none of the monoascospore strains are closely related to *C. lindemuthianum* (Figure 5). They also could not be positively identified as another known species of *Colletotrichum* or *Glomerella*. The largest group of isolates clustered with the cucurbit anthracnose pathogen *C. magna* (Figure 5). Other strains in this clade, identified by BLAST homology searches from the Genbank database, were named in the database entries as *C. gloeosporioides* or *G. cingulata*. However, they were distinct from the *C. gloeosporioides* clade that contained sequences of verified ex-epitype strains (Figure 5), and so they had probably been misidentified. The HMG sequences of representatives of this group of monoascospore strains matched only the HMG sequence from *C. magna* in BLAST searches of the Genbank database, and they were distinct from the HMG sequences of *C. lindemuthianum / G. lindemuthiana* (Figure 6). Three of the monoascospore strains (UFLAG85-1, UFLAG93-1, UFLAG99-1) were separate from the larger group, and were located in the ITS phylogeny within the *C. gloeosporioides* species complex (Figure 5).

Infection Assays

Pathogenicity tests. Of the 53 conidial strains tested, 47 produced no symptoms when conidia were inoculated onto leaves of a set of differential cultivars of common bean, or on the susceptible cultivar that was used as a control (not shown). Only six of the strains (UFLAG85-1, UFLAG86-1, UFLAG89-1, UFLAG92-1, UFLAG93-1 and UFLAG99-1) sporulated well on bean pods, and these same six strains produced mild symptoms on the leaves of the susceptible cultivar, but not on any of the differential cultivars (Figure 7).

Inoculation of detached hypocotyls of two susceptible bean cultivars Pérola and Michelite with spores of some of the *Glomerella* strains resulted in no development of anthracnose symptoms. Only conidia of the *C. lindemuthianum* strain LV115 produced symptoms

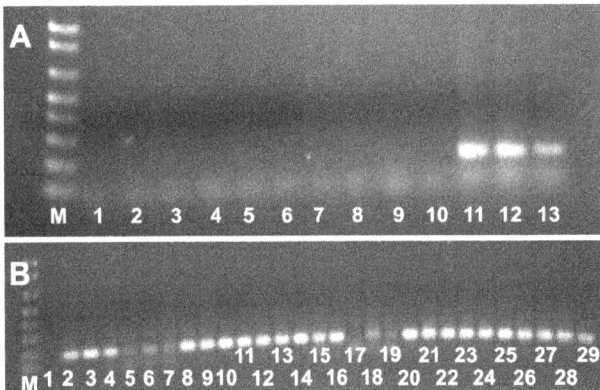

Figure 4. Amplification products of polymerase chain reaction (PCR) with representative conidial and perithecial strains. A) The HMGCLF and HMGCLR primer pair (Garcia-Serrano et al., 2008) only amplified a product from *C. lindemuthianum* strains (lanes 11–13). M) Molecular size marker-100 pb; 1) No template control; 2) Strain UFLAG21-1 (conidial A); 3) Strain UFLAGG21-2 (Plus), 4) Strain UFLAG46-1 (conidial A); 5) Strain UFLAG85-1 (conidial A); 6) Strain UFLAG93-1 (conidial A); 7) Strain UFLAG99-1 (conidial A), 8) Strain UFLAGG110-1 (conidial B); 9) Strain UFLAG112-1 (Minus); 10) Strain UFLAG113-1 (Minus); 11) *C. lindemuthianum* strain LV115; 12) *C. lindemuthianum* strain LV117; 13) *C. lindemuthianum* strain LV120. B) The primer pair HMGgloF and HMGgloR amplified products from the *Glomerella* group 1 strains but the *Glomerella* group 2 strains amplified very poorly (lanes 5–7, and 17–19). M) Molecular size marker-100 pb; 1) No template control; 2) Strain UFLAG21-1 (Conidial A); 3) Strain UFLAG21-2 (Plus); 4) Strain UFLAG23-1 (Conidial B); 5) Strain UFLAG85-1 (Conidial A); 6) Strain UFLAG93-1 (Conidial A); 7) Strain UFLAG99-1 (Conidial A); 8) Strain UFLAG110-1 (Conidial B); 9) Strain UFLAG112-1 (Minus); 10) Strain UFLAG113-1 (Minus); 11) Strain UFLAG43-1 (Minus); 12) Stain UFLAG43-2 (Plus); 13) Strain UFLAG46-1 (Conidial A); 14) Strain UFLAG49-1 (Conidial B); 15) Strain UFLAG73-1 (Plus); 16) Stain UFLAG73-2 (Minus); 17) Strain UFLAG86-1 (Conidial A); 18) UFLAG89-1 (Conidial A); 19) UFLAG92-1 (conidial B); 20) Strain UFLAG15-1 (Plus); 21) Strain UFLAG15-2 (Minus); 22) Strain UFLAG54-1 (Conidial A); 23) Strain UFLAG54-2 (Plus); 24) Strain UFLAG68-1 (Conidial A); 25) Strain UFLAG101-1 (Plus); 26) Strain UFLAG104-1 (Minus); 27) Strain UFLAG104-2 (Conidial A); 28) Strain UFLAG117-1 (Conidial B); 29) Strain UFLAG118-1 (Plus).

on both susceptible cultivars (Figure 8A and G). For some samples, pale brown discolored spots were seen at the inoculation sites (Figure 8I). Under the light microscope, these were revealed as

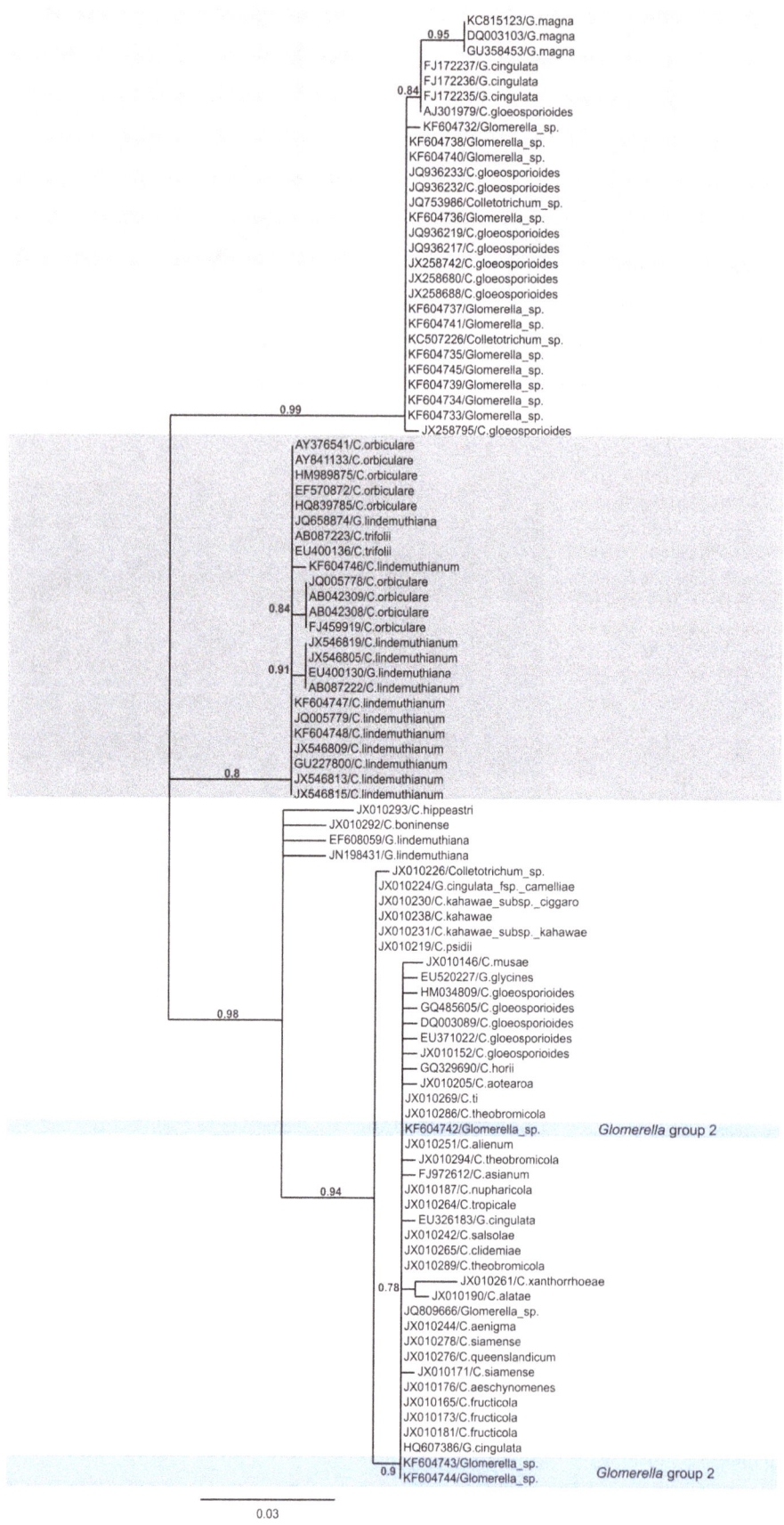

Figure 5. Phylogenetic tree based on ITS sequences of the Brazilian *Glomerella* sp. and *C. lindemuthianum* strains, and other *Colletotrichum* spp. from Genbank. The tree was constructed by using maximum likelihood analysis, and values above branches indicate the

internal branch reliability assessed using the aLRT test (SH-Like) [45]. Scale bar, 0.03 nucleotide replacements per site. Most of the sequenced *Glomerella* sp. monoascospore strains UFLAG02, UFLAG03-1, UFLAG04, UFLAG05-2, UFLAG06, UFLAG06-3, UFLAG07-3, UFLAG08, UFLAG08-4, UFLAG46-1 and UFLAG112 (KF604732, KF604733, KF604734, KF604735, KF604736, KF604737, KF604738, KF604739, KF604740, KF604741 and KF604745, respectively, *Glomerella* group I) clustered together with ITS sequences of *G.magna*, and a few strains identified as *G.cingulata* and *C.gloeosporioides* in Genbank (light gray). Both perithecial and conidial strains were represented. Three Brazilian *C.lindemuthianum* strains LV115, LV117 and LV120 (KF694746, KF604747, and KF604748, respectively) clustered in a separate clade with other sequences from *C. lindemuthianum* and other members of the *C. orbiculare* species aggregate that were available in Genbank (dark gray). ITS sequences of a few of the *Glomerella* sp. monoascospore isolates UFLAG85-1, UFLAG93-1 and UFLAG99-1 (KF604742, KF604743 and KF604744, respectively, *Glomerella* group 2, highlighted in blue) clustered together with sequences from isolates representing the *C. gloeosporioides* species aggregate, including several verified ex-epitype strains [67].

masses of superficial hyphae and melanized appressoria (Figure 9 C and D). The six strains that produced mild symptoms on leaves (above) were not tested on hypocotyls.

Cytology of infection. Typical post penetration intracellular structures, including infection vesicles and primary hyphae, were produced in infected tissues of the susceptible host by the *C. lindemuthianum* LV115 strain by 72 hours after inoculation (hpi)

(Figure 9A). Ascospores of the perithecial UFLAG06 strain germinated and produced appressoria within 24 hours (Figure 9B). The *Glomerella* UFLAG06 strain produced abundant appressoria and epiphytic mycelium on common bean tissues at 24 and 48 hpi (Figure 9C and D). By 72 hpi, ascospores of the *Glomerella* strain had produced appressoria at a much higher rate on Pérola (98.33% ±1.52) and G2333 (97% ±3.0) than conidia of

```
AY724683    --------------------CATGCCGCAGTAAAGCAAATGGACAATAGCCTCACCAAC
AY724682    ATTCTCTACCGGAGGGACCGACATGCCGCAGTAAAGCAAATGGACAATAGCCTCACCAAC
CLKY1       -----------------------------AAAGCAAATGGACAACAGCCTCACCAAC
LV120       ------------------------GCAGTAAAGCAAATGGACAACAGCCTCACCAAC
LV115       -----------------------CGCAGTAAAGCAAATGGACAACAGCCTCACCAAC
LV117       ----------------------CCGCAGTAAAGCAAATGGACAACAGCCTCACCAAC
DQ002828    --TCTCTACCGGAAAGACCACTGCGCCAAGTTGAAGAAACTCAACCCCCGCATCTCCAAC
UFLAG03-1   -----------------------------GAAGAAACTCAACCCGCGGATCTCCAAC
UFLAG08-4   -----------------------------GAAGAAACTCAACCCGCGGATCTCCAAC
UFLAG01     ----------------------------TGAAGAAACTCAACCCGCGGATCTCCAAC
UFLAG08     -----------------------------GAAGAAACTCAACCCGCGGATCTCCAAC
UFLAG06     -----------------------------GAAGAAACTCAACCCGCGGATCTCCAAC
UFLAG06-3   -----------------------------GAAGAAACTCAACCCGCGGATCTCCAAC
                                         ***  ** *   **      *   ** *****

AY724683    AATGAGATCTGTAAGT-------TGCCCCGCACATCACCGTCAATGCAGTTGCTGATCTT
AY724682    AATGAGATCTGTAAGT-------TGCCCCGCACATCACCGTCAATGCAGTTGCTGATCTT
CLKY1       AATGAGATCTGTAAGT-------TGCCCCGCACATCACCGTCAATGCAGTTGCTGATCTT
LV120       AATGAGATCTGTAAGT-------TGCCCCGCACATCACCGTCAATGCAGTTGCTGATCTT
LV115       AATGAGATCTGTAAGT-------TTCCCCGCACATCACCGTCAATGCAGTTGCTGATCTT
LV117       AATGAGATCTGTAAGT-------TGCCCCGCACATCACCGTCAATGCAGTTGCTGATCTT
DQ002828    AACGATATATGTGAGTAGATGGACTCCCCCTTTCTGAGTGTCGACTCGGATTCTAACCTG
UFLAG03-1   AACGATATATGTGAGTATACGGACTATCCAT-CCCATCTATCAACTCGGTATCTGACTTA
UFLAG08-4   AACGATATATGTGAGTATACGGACTATCCAT-CCCATCTATCAACTCGGTATCTGACTTA
UFLAG01     AACGATATATGTGAGTATACGGACTATCCAT-CCCATCTATCAACTCGGTATCTGACTTA
UFLAG08     AACGATATATGTGAGTATACGGACTATCCAT-CCCATCTATCAACTCGGTATCTGACTTA
UFLAG06     AACGATATATGTGAGTATACGGACTATCCAT-CCCATCTATCAACTCGGTATCTGACTTA
UFLAG06-3   AATGATATATGTGAGTATACGGACTATCCAT-CCCATCTATCAACTCGGTATCTGACTTA
            **  **  **  *** ***         **      ** *   * *    **  *    *

AY724683    TCAAAGCTGTCAAACTTGGCAAAGCATGGAACGCAGAGTCGCCCGCTGTCCGCGAGAGAT
AY724682    TCAAAGCTGTCAAACTTGGCAAAGCATGGAACGCAGAGTCGCCCGCTGTCCGCGAGAGAT
CLKY1       TCAAAGCTGTCAAACTTGGCAAAGCATGGAACGCAGAGTCGCCCGCTGTCCGCGAGAGAT
LV120       TCAAAGCTGTCAAACTTGGCAAAGCATGGAACGCAGAGTCGCCCGCTGTCCGCGAGAGAT
LV115       TCAAAGCTGTCAAACTTGGCAAAGCATGGAACGCAGAGTCGCCCGCTGTCCGCGAGAGAT
LV117       TCAAAGCTGTCAAACTTGGCAAAGCATGGAACGCAGAGTCGCCCGCTGTCCGCGAGAGAT
DQ002828    ATACAGCTTGCGTCTTGGGCAAGGCTTGGAACAACGAGTCGCACGAAGTCCGCGAGAGAT
UFLAG03-1   ATTCAGCTTGCGTCTTGGGCAAGGCTTGGAACAACGAGTCGM-CKAGGTCAGCC------
UFLAG08-4   ATTCAGCTTGCGTCTTGGGCAAGGCTTGGAACAACGAGTCGCACGAGGTCCG--------
UFLAG01     ATTCAGCTTGCGTCTTGGGCAAGGCTTGGAACAACGAGTCGCACGAGGTCCGA-------
UFLAG08     ATTCAGCTTGCGTCTTGGGCAAGGCTTGGAACAACGAGTCGCACGAGGTCCG--------
UFLAG06     ATTCAGCTTGCGTCTTGGGCAAGGCTTGGAACAACGAGTCGCACGAGGTCCG--------
UFLAG06-3   ATTCAGCTTGCGTCTTGGGCAAGGCTTGGAACAACGAGTCGCACGAGGTCCG--------
            ****  *     *  ***** ** ****** *   ****** *    *** *
```

Figure 6. Alignments of HMG sequences. *Glomerella* group 1 (UGLAG01, UFLAG03-1, UFLAG06, UFLAG06-3, UFLAG08 and UFLAG08-4) and *C. lindemuthianum* strains (LV115, LV117 and LV120), aligned with HMG sequences of *C. magna* (DQ002828) and *G. lindemuthiana* (AY724682 and AY724683) from Genbank.

Figure 7. Typical symptoms on Pérola susceptible cultivar (all at 10 dpi). A and B) Leaves inoculated with the conidial strain UFLAG86-1. C and D) Leaves inoculated with the conidial strain UFLAG89-1. E and F) Leaves inoculated with the *C. lindemuthianum* strain LV115.

the *C. lindemuthianum* LV115 strain (15% ±1.73 and 4% ±1.0, respectively). However, UFLAG06 inoculations were never observed to produce intracellular infection structures, even at 120 hpi. All the sources of variation in the ANOVA were significant (P<0,001).

Infection process observed with strains expressing the green fluorescent protein. The strains UFLAG06 and LV115 were transformed to produce the GFP-expressing transfomants tQB01 and tQB02, respectively. Inoculations with the GFP-expressing strains facilitated observation of post-penetration structures including infection vesicles, and primary and secondary hyphae of *C. lindemuthianum* (tQB02) (Figure 10A and B). Similar structures were never observed in inoculations of living host tissues

using strain tQB01, which was derived from the perithecial monoascospore *Glomerella* strain UFLAG06.

When strain tQB01 was inoculated on sites where localized cell death had been induced by treatment with dry ice, thick hyphae were observed inside the dead cells by 48 hpi (Figure 10C and D). Formation of acervuli was observed on the killed inoculated tissues by 72 hpi (Figure 10E and F). The formation of thick hyphae and acervuli was observed equally in the tissues of plants that were either resistant or susceptible to anthracnose (not shown).

Co-inoculations. Hyphae of the *Glomerella* strain (tQB01), expressing GFP could be seen growing within anthracnose lesions caused by the *C. lindemuthianum* wild type strain LV115 in co-inoculations (Figure 10G). In samples inoculated with the transformed *C. lindemuthianum* strain (tQB02), hyphae expressing

Figure 8. Typical symptoms at 10 dpi on hypocotyls of susceptible cultivars. A–F) Inoculations on Pérola cultivar; G–L) inoculations on Michelite cultivar; A) Inoculation with conidia of the LV115 *C. lindemuthianum* strain. B) Control mock-inoculated with sterile water; C) Inoculation with ascospores of a perithecial Plus strain (UFLAG06-1); D) Inoculation with conidia of a sibling conidial B strain (UFLAG06-2); E) Inoculation with ascospores of a perithecial Minus strain UFLAG08-2; F) Inoculation with conidia of sibling conidial A strain UFLAG08-3; G) Inoculation with conidia of the LV115 *C. lindemuthianum* strain; H) Control mock-inoculated with sterile water; I) Inoculation with ascospores of a perithecial Plus strain (UFLAG06-1); J) Inoculation with conidia of a sibling conidial B strain (UFLAG06-2); K) Inoculation with ascospores of a perithecial Minus strain UFLAG08-2; L) Inoculation with conidia of sibling conidial A strain UFLAG08-3.

GFP were observed growing in lesions, and there was an intensive formation of appressoria of the *Glomerella* wild type strain UFLAG06 around these lesions (Figure 10H and I).

Discussion

Bean anthracnose, caused by *C. lindemuthianum*, is a common and economically damaging disease in Brazil. The ability of the pathogen to rapidly overcome sources of host resistance is not well understood, but the frequent recovery of fertile *Glomerella* strains from anthracnose lesions suggested sexual recombination as one possibility. Previous studies have identified *Glomerella* strains recovered from anthracnose lesions as *Glomerella lindemuthiana* or

Glomerella cingulata f.sp *phaseoli*, assuming them to be the teleomorphs of *C. lindemuthianum* [11,13,32,34]. However, the results of the current study do not support the hypothesis that these *Glomerella* isolates are related to *C. lindemuthianum*. Instead, most of the isolates appear to belong to an unknown *Glomerella* spp. that lives on the bean as an epiphyte, and grows opportunistically in anthracnose lesions caused by *C. lindemuthianum*. A second group of isolates, belonging to the *C. gloeosporioides* species complex, appears to be comprised of weak pathogens of bean that also take advantage of the lesions caused by the more destructive *C. lindemuthianum*.

The *Glomerella* isolates could be distinguished morphologically from a previously characterized population of *C. lindemuthiaum* from common bean [33]. The *Glomerella* colonies grew faster, and their conidia germinated at a higher rate and were smaller than those of *C. lindemuthianum* [33,49,50,51,52,53]. Single ascospores recovered from *Glomerella* perithecia gave rise either to fertile strains that produced ascospores and few or no conidia, or to infertile strains that produced only conidia. Both types could be recovered from a single perithecium. The ascospores were larger than the conidia, and they germinated more quickly and at higher rates. Conidial anastomosis tubes (CATs) are a potential mechanism for horizontal transfer of genes and chromosomes, and thus for asexual recombination [32,54,55]. Ascospores never formed CATs, but the conidia of a majority of the conidial strains produced them at varying rates (depending on the strain), similar to previous observations reported for *C. lindemuthianum* strains [33]. The ability of *Glomerella* to produce both sexual and asexual progeny may confer the advantages of both: the potential for sexual recombination, and a greater ability of sexual spores to survive and germinate (due to their larger size, and presumably food reserves), combined with the ability of asexual spores to generate CATs for asexual recombination.

All but one of the *Glomerella* isolates produced a septum in the germinating spore. A lack of septum formation during germination is a trait that is considered to be diagnostic for the *C. orbiculare* species complex of *Colletotrichum*, comprised of *C. lindemuthianum*, *C. orbiculare*, *C. trifolii* and *C. malvarum* [53,56,57,58,59,60,61]. The *C. lindemuthianum* LV115 strain used as a control in the current study did not form a septum during germination. However, it should be noted that in another recent study [33], a few pathogenic, presumptive *C. lindemuthianum* strains formed a septum in the germinating conidia, so it may be that this characteristic is not universally definitive of the species.

Molecular data also supported a division between *Glomerella* and *C. lindemuthianum* strains from bean. Separation of a small subset of the strains into different groups by random amplified polymorphic DNA (RAPD) fingerprinting has been reported in earlier studies [5,62]. In the current study, primers designed to amplify the HMG region of the MAT1-2-1 gene of *C. lindemuthianum* [10] failed to amplify any of the *Glomerella* strains, although they successfully amplified the expected fragment from a large collection of asexual, pathogenic *C. lindemuthianum* strains from the same regions of Brazil [33]. A primer pair specific for the HMG sequence of the *Glomerella* strains was developed by first amplifying the region using degenerate universal primers [42]. The new specific primer pair amplified an HMG fragment from nearly all of the *Glomerella* strains. There were six strains that amplified very poorly with these primers: UFLAG85-1, UFLAG86-1, UFLAG89-1, UFLAG92-1, UFLAG93-1 and UFLAG99-1. These six strains were all conidial, and five of the six were also completely infertile in confrontations with all other *Glomerella* strains. These same six strains were the only ones that sporulated on bean pods, and they were also the only ones that produced symptoms on common bean, albeit very

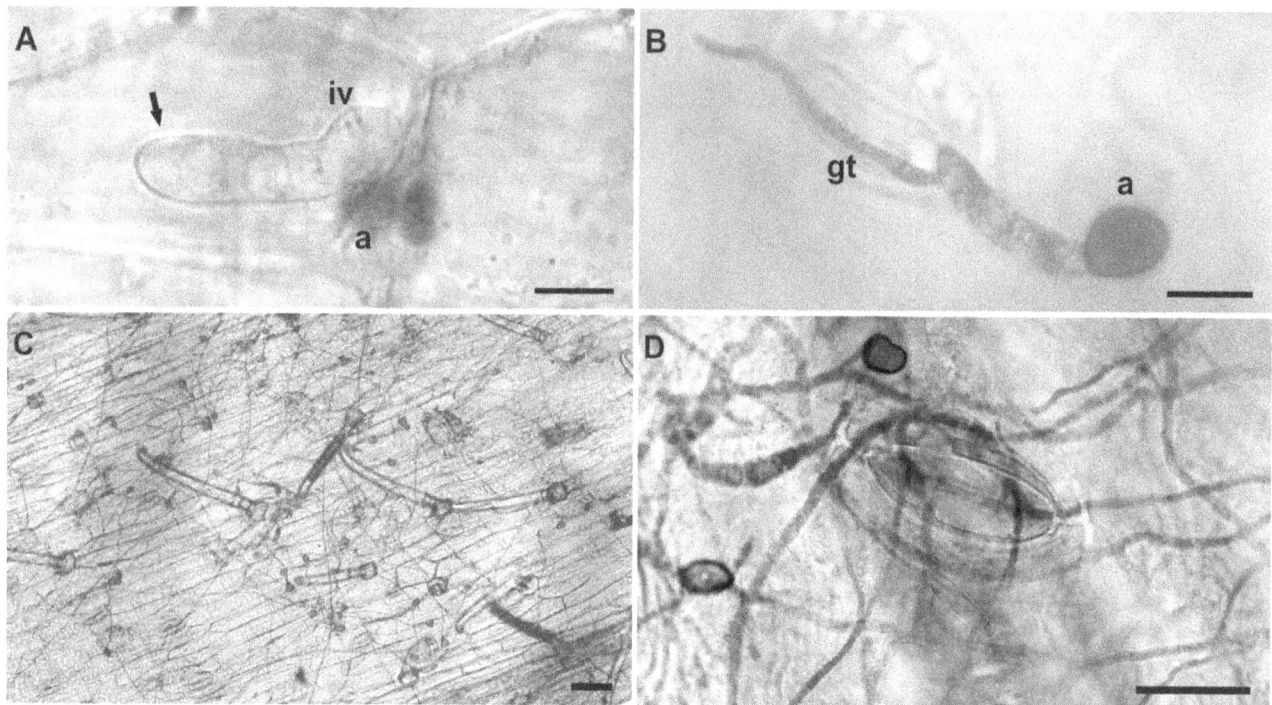

Figure 9. Infection analyses of _C. lindemuthianum_ and epiphytic growth of _Glomerella_ sp. strain on tissues of the Pérola susceptible cultivar. A) _C. lindemuthianum_ strain LV115 on a hypocotyl at 72 hpi, forming appressoria (a), infection vesicle (iv) and primary hyphae (black arrow), Bar: 10 µm. B) _Glomerella_ UFLAG06 strain forming appressorium (a) and germ tube (gt) on a leaf at 24 hpi. Bar: 10 µm. C) Epiphytic growth of _Glomerella_ sp. strain UFLAG06 on hypocotyl surface of the susceptible cultivar at 24 hpi. Bar: 50 µM. D) Epiphytic growth of _Glomerella_ sp. strain UFLAG06 on leaf surface at 48 hpi. Bar: 25 µM.

mild symptoms. These six strains (hereafter referred to as _Glomerella_ group II) thus represent a population that is distinct both from _C. lindemuthianum_, and from the remainder of the _Glomerella_ isolates in this study (hereafter referred to as _Glomerella_ group I).

Phylogenetic analysis of the HMG as well as ITS sequences confirmed that the _Glomerella_ group 1 strains are unrelated to _C. lindemuthianum_. Their closest affinity was to _C. magna_, a pathogen of cucurbits that appears to be separate from other characterized groups of _Colletotrichum_ [63]. _C. magna_ has been described as both a pathogen of cucurbits, and as an endophyte in other plant species [64,65]. The ITS sequences of _Glomerella_ group I were identical or similar, based on BLAST analyses, to those of a small number of other strains in Genbank that had been identified as either _C. gloeosporioides_ or _G. cingulata_. However, these strains were separate in the ITS phylogram from verified ex-epitype specimens of the _C. gloeosporioides_ species complex, and so it is likely that these were misidentified. Most of the isolates (JQ936233, JQ936232, JQ936217, JX258742, JX258680, JX258688, and JX258795) were identified as endophytes recovered from soybean in Brazil [66]. One of the isolates (JQ753986) was identified as an endophyte isolated from common bean leaves in Brazil (Costa, Queiroz, and Gonzaga, unpublished). Most of the rest of the isolates were also identified as plant endophytes.

The _Glomerella_ group II isolates fell within the verified _C. gloeosporioides_ species complex, based on their ITS sequences. The _C. gloeosporioides_ species complex has recently been revised [67]. A multigene phylogeny identified a large number of individual species within this complex. The ITS sequence alone is insufficient for identification of most of these species [67], but we can see that the _Glomerella_ group II isolates are distinct from a few of them, including _C. boninense_ and _C. kahawe_. Final identification of the

group I and II isolates must await a more complete phylogenetic analysis.

Sexual behavior among the _Glomerella_ strains from bean was similar in some ways to a series of descriptions of a _G. cingulata_ population isolated from morning glory, published in the first half of the 20th century [17–28]. These extensive studies resulted in a theory of sexual compatibility in _Glomerella_ known as unbalanced heterothallism, where loss of fertility occurs as a result of frequent mutations in the many required steps for full self-fertility, and compatibility is a result of genetic complementation [27]. The recovery of both fertile and infertile progeny strains, with clumped versus scattered arrangements of conidia and perithecia, from individual perithecia of the bean _Glomerella_ strains was reminiscent of the morning glory studies. A majority of the recovered progeny were infertile (53 of 73, approximately 75%) suggesting a large propensity to lose fertility. This could be due to the occurrence of a very large number of different mutations, or one or a few common mutations. About half of the infertile strains (28 of 53, 53%) could not be complemented by any of the other strains, suggesting that they shared a common mutation with all of those strains. The strains that were complemented could only be complemented by fertile strains, and not by one another, which also argues in favor of a single common mutation. Most of these strains could be complemented by diffusible substances from certain fertile strains (across membranes). Presence of diffusible hormones inducing self-fertility has been described previously in _Glomerella_ [26]. It is possible that these strains are deficient in ability to produce a sexual hormone, and only some of the fertile strains produce a version of the hormone that is recognizable to them. Some strains could be complemented only after contact with certain fertile strains, suggesting complementation by factors acting post-fusion.

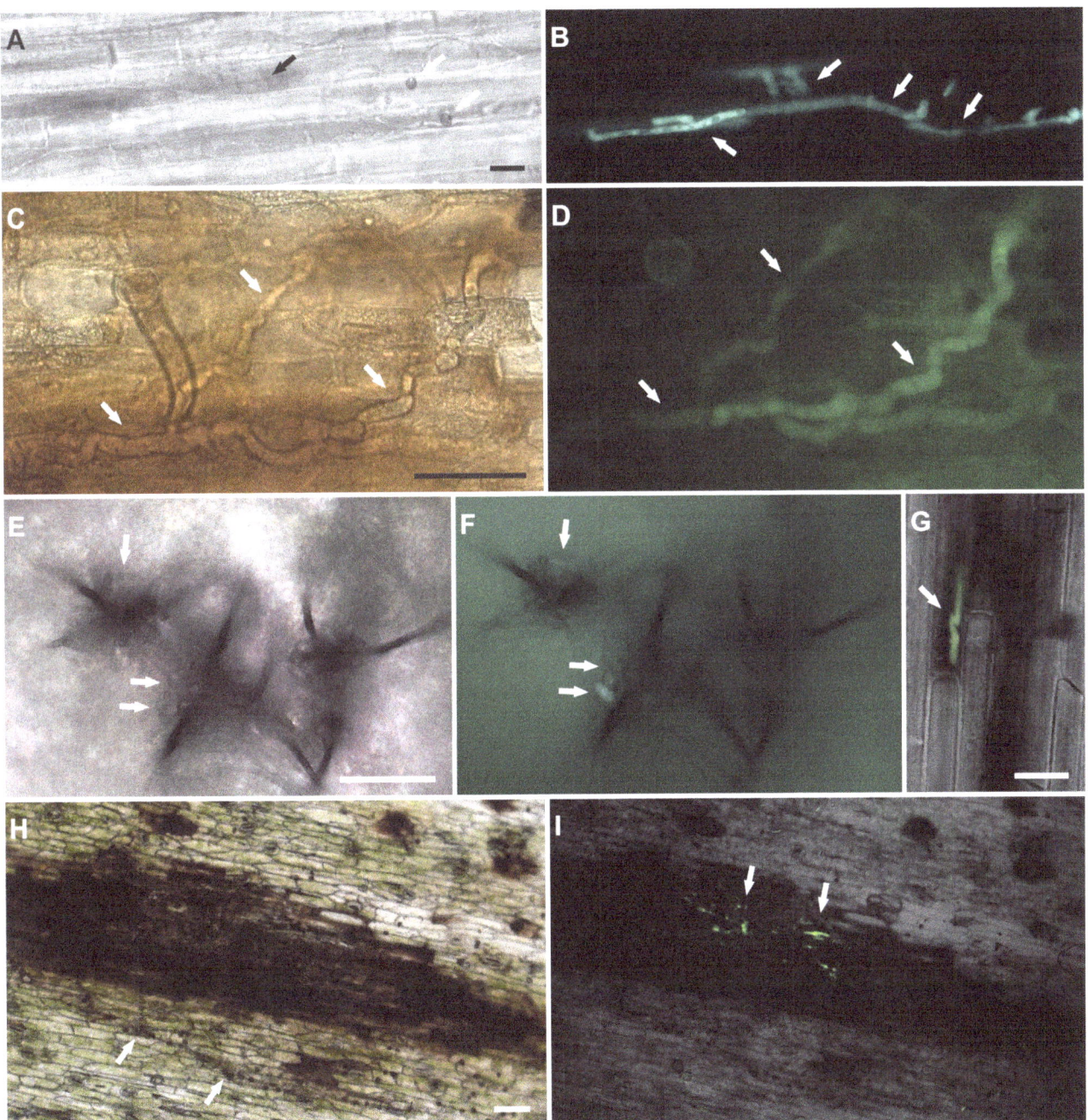

Figure 10. Comparative infection analyses of *C. lindemuthianum* **and** *Glomerella* **strain on tissues of the Pérola susceptible cultivar.** A) *C. lindemuthianum* tQB02 transformant strain on a hypocotyl at 72 hpi. White arrows indicate appressoria, and the black arrow indicates the location of primary hyphae. B) Same view as in panel A, imaged with fluorescence microscopy showing primary hyphae expressing GFP (white arrows). Bar: 20 μm. C) *Glomerella* tQB01 transformant strain inoculated on a hypocotyl at a site where cell death had been induced. Thick hyphae were observed inside the dead cells by 48 hpi (white arrows). D) Same view as in panel C, imaged with fluorescence microscopy, showing thick hyphae expressing GFP (white arrows). Bar: 20 μm. E) *Glomerella* tQB01 transformant strain forming acervuli and spores (white arrows) at 72 hpi on a leaf that had tissue been killed by treatment with dry ice. F) Same view, imaged with fluorescence microscopy showing spores (white arrows). Bar: 20 μm. G) Merged image taken with light and fluorescence microscopy showing co inoculation on hypocotyl using transformant teleomorphic strain (tQB01) and wild type anamorphic strain (LV115). tQB01 expressing GFP on hyphae (white arrow) growing on anthracnose lesion at 120 hpi. Bar: 20 μm. H) Co inoculation on hypocotyls at 120 hpi using the transformant *C. lindemuthianum* strain (tQB02) and the perithecial wild type *Glomerella* strain (UFLAG06). Appressoria of the UFLAG06 strain have formed around the lesion (white arrows). I) Same image taken by fluorescence microscopy showing hyphae of the tQB02 strain expressing GFP in the lesion (white arrows). Bar: 20 μm.

Interestingly, some strains could be complemented in both ways, by different strains. Three different infertile progeny shared a single pattern of complementation, but otherwise each pattern of complementation was unique. This suggests that each infertile strain may actually have multiple mutations, some common to all, and some unique. All of the fertile strains had different patterns of complementation, suggesting that they don't all share the same mating-related genes, even though all of the versions apparently function to condition self-fertility. If the unbalanced heterothallism theory can be applied to the *Glomerella* strains from common bean, it suggests that there are a very large number of genes, with a high level of redundancy, that condition compatibility and mating in this fungus. It will be very interesting to explore the mating behavior of this group of strains further, including establishing the existence of recombination in combinations where fertility results only after contact.

Recent studies have revealed other mechanisms that might be involved in the evolution of *mat* loci, and in transitions from heterothallism to homothallism. For example, multiple shifts in the reproductive mode have occurred during the evolutionary history of *Neurospora*, where retrotransposons within *mat* loci have facilitated unequal crossovers or translocation, resulting in relocation of genes of both mating types into the same haploid genome [68]. Transposable elements have been observed within the *mat* loci of numerous Ascomycetes and Basidiomycetes [69–72].

The *Glomerella* strains are much less aggressive to common bean than *C. lindemuthianum*. As reported in an earlier preliminary study [32], the *Glomerella* spores germinated with an extremely high efficiency on bean plants within 24 hours, producing masses of melanized appressoria and superficial hyphae. These structures resulted in visible brown flecks at the sites of inoculation, which may explain the mild symptoms that were previously reported to result from inoculation with these strains [5,11,13,32]. In the current study, post-infection processes of a representative of the Type I *Glomerella* strains including penetration and colonization of plant tissues, were compared with *C. lindemuthianum* in detail. These studies revealed that this *Glomerella* strain was completely unable to penetrate or internally colonize living common bean tissues. However, it produced abundant appressoria and epiphytic mycelium that grew along the anticlinal walls of the epidermal cells. For this reason, we do not suggest that these fungi are "endophytes" in the commonly understood sense of occupying the interior of the host tissues. There were no differences in the behavior of the perithecial versus conidial sibling strains. The anthracnose-resistant common bean cultivar significantly reduced the ability of *C. lindemuthianum* to germinate, form appressoria, and colonize the tissue. In contrast, there was no effect of host resistance to anthracnose on the *Glomerella* strains, which germinated and formed appressoria and superficial hyphae at the same rate on resistant versus susceptible cultivars. When localized injury of the common bean leaf tissue was caused by contact with dry ice, the *Glomerella* strains produced thick hyphae that rapidly colonized the dead cells, and eventually gave rise to acervuli and setae in both susceptible and resistant cultivars. The co-infection experiments demonstrated that *C. lindemuthianum* anthracnose lesions can be colonized by *Glomerella* strains. Thus, in the field, we propose that these *Glomerella* strains exist as epiphytes on the surface of the bean tissues, and opportunistically colonize and sporulate within anthracnose lesions caused by *C. lindemuthianum*, explaining the recovery of both organisms from the same lesions.

Colletotrichum (and its teleomorph *Glomerella*) is a large genus with diverse lifestyles ranging from necrotrophic pathogenic to latent and epiphytic [73]. *Colletotrichum* is a common epiphytic inhabitant of foliar tissues, where it can persist as spores, appressoria, or mycelium [74–78]. The relationship between pathogenicity, endophytism, epiphytism, and saprophytism is not clear in *Colletotrichum* although it appears that some species can exhibit more than one lifestyle in their life cycle, e.g. latent epiphytes or

endophytes can transform to pathogens or saprophytes as host status changes (tissue senescence or necrosis) [65,67,76,79,80]. Common bean in Brazil is cultivated across three seasons each year under widely varying environmental conditions, resulting in a complex biosystem and the opportunity for rapid evolution of new common bean-associated species. The specific nature of the relationship between *C. lindemuthianum* and the *Glomerella* strains on common bean remains to be determined, however the frequency of their association suggests that there may be an inter-dependence of the two species. The *Glomerella* may rely on the *C. lindemuthianum* to produce dead tissues on which it can sporulate. Formation of appressoria by *Colletotrichum* is known to trigger defenses in host species [81], so it will be important to determine whether co-inoculation with *Glomerella*, apparently an aggressive colonizer of the phylloplane, can "prime" defense mechanisms in common bean, perhaps increasing its resistance to *C. lindemuthianum*. The epiphytic *Glomerella* mycelium may compete for phylloplane resources, or they may produce inhibitory antibiotics, affecting the survival and germination of the *C. lindemuthianum* spores [78]. The *Glomerella* strains may have significant potential as biocontrol organisms. Furthermore, given that both organisms are able to form CATS, and that it has been shown previously that CATs can form between members of different species of *Colletotrichum* [82], there is at least the potential for genetic exchange between the two organisms, something that could add to the potential diversity of the *C. lindemuthianum*. This is especially true when we consider how extremely variable the population of *Glomerella* strains on common bean leaves is, possibly related to their ability to undergo sexual recombination. The recognition of these interdependent *Colletotrichum* communities associated with common bean anthracnose adds a new dimension to our understanding of the nature and potential mechanisms of phenotypic variation and adaptability of this important pathogen.

Supporting Information

Table S1 Host cultivar from which the infected tissues were collected, origin and type of the strains used in this study. * Strains used in infection assays. **transformant strains expressing green fluorescent protein and hygromycin resistance used in infection assays. Strains in bold were used for ITS sequencing; strains in italics were used for HMG sequencing. Strains in bold italics were used for both. Underlined strains were used in morphological characterization experiments. NA = not applicable; a = common bean lines from breeding programs that do not have a commercial name yet; b = single-spored field isolates; c = not applicable, transformed strains obtained in the laboratory; d = *C. lindemuthianum* strains are anamorphic, and produce only conidia.

Table S2 Fertile confrontations among conidial and perithecial strains. CA = conidial A strains; CB = conidial B strains; + = perithecial plus strains; − = perithecial minus strains; I = induced homothalism; confrontations in which lines of fertile perithecia were formed in both the presence and absence of dialysis membrane; L = likely heterothallic; confrontations in which lines of fertile perithecia were formed only in the absence of dialysis membrane.

File S1 ITS sequence alignment.

Acknowledgments

The authors thank Etta Nuckles, Doug Brown and Suellen Finamor Mota for excellent technical support, and Dr. Christopher L. Schardl for helping with the phylogenetic analyses.

Author Contributions

Conceived and designed the experiments: QLB JMAP LJV EAS. Performed the experiments: QLB JMAP LJV EAS. Analyzed the data: QLB JMAP LJV EAS. Contributed reagents/materials/analysis tools: QLB JMAP LJV EAS. Wrote the paper: QLB JMAP LJV EAS.

References

1. Peloso MJD (1992) Antracnose do feijoeiro no Estado de Minas Gerais-Brasil. In: Pastor-Corrales M, editor. La antracnosis del frijol común, *Phaseolus vulgaris*, en América Latina. CIAT:Cali. 86–108.
2. Alves MC, Pozza EA, Machado JC, Araújo DV, Talamini V, et al. (2006) Geoestatística como metodologia para estudar a dinâmica espaço-temporal de doenças associadas a *Colletotrichum* spp. transmitidos por sementes. Fitopatol Brasil 31: 557–563.
3. Balardin RS, Jarosz A, Kelly JD (1997) Virulence and molecular diversity in *Colletotrichum lindemuthianum* from South, Central and North America. Phytopathol 87: 1184–1191.
4. Menezes JR, Dianese JC (1988) Race characterization of Brazilian isolates of *Colletotrichum lidemuthianum* and deteccion of resistance to anthracnose in *Phaseolus vulgaris*. Phytopathol 78: 650–655.
5. Silva KJD, Souza EA, Ishikawa FH (2007) Characterization of *Colletothichum lindemuthianum* isolates from the state of Minas Gerais, Brazil. J Phytopathol 155: 241–247.
6. Davide LMC, Souza EA (2009) Pathogenic variability within race 65 of *Colletotrichum lindemuthianum* and its implications for common bean breeding. Crop Breed Appl Biotech 9: 23–30.
7. Ishikawa FH, Barcelos QL, Costa LC, Souza EA (2012) Investigating variability within race 81 of *Colletotrichum lindemuthianum* strains from Brazil. Ann Rep Bean Improv Coop 55: 141–142.
8. Alzate-Marin AL, Sartorato A (2004) Analysis of the pathogenic variability of *Colletotrichum lindemuthianum* in Brazil. Annu Rep Bean Improv Coop 47: 241–242.
9. Shear CL, Wood AK (1913) Studies of fungus parasites belonging to the genus *Glomerella*. Washington: USDA Bureau of Plant Industry. 110 p.
10. García-Serrano M, Laguna EA, Rodriguez-Guerra R, Simpson J (2008) Analysis of the MAT1-2-1 gene of *Colletotrichum lindemuthianum*. Mycoscience 49: 312–317.
11. Camargo Junior OA, Souza EA, Mendes-Costa MC, Santos JB, Soares MA (2007) Identification of *Glomerella cingulata* f. sp. *phaseoli* recombinants by RAPD markers. Genet. Mol. Res. 6: 607–615.
12. Rodriguez-Guerra R, Ramírez-Rueda MT, Cabral-Enciso M, García-Serrano M, Lira-Maldonado Z, et al. (2005) Heterothallic mating observed between Mexican isolates of *Glomerella lindemuthiana*. Mycologia 97: 793–803.
13. Souza EA, Camargo Junior OA, Pinto JMA (2010) Sexual recombination in *Colletotrichum lindemuthianum* occurs on a fine scale. Gen. Mol. Res. 9: 1759–1769.
14. Cisar CR, TeBeest DO (1999) Mating system of the filamentous ascomycete, *Glomerella cingulata*. Curr Genet 35: 127–133.
15. Vaillancourt IJ, Hanau RM (1994) Nitrate-nonutilizing mutants used to study heterokaryosis and vegetative compatibility in *Glomerella graminicola* (*Colletotrichum graminicola*). Exp Mycol 18: 311–319.
16. Vaillancourt LJ, Wang J, Hanau RM (2000) Genetic regulation of sexual compatibility in *Glomerella graminicola*. In: Prusky D, Reeman S, Dickman MB, editors. *Colletotrichum* : Host Specificity, Pathology, and Host-Pathogen Interaction. American Phytopathological Society. 29–44.
17. Lucas GB, Chilton SJP, Edgerton CW (1944) Genetics of *Glomerella* I: studies on the behaviour of certains strains. Am J Botany 31: 233–239.
18. Chilton SJP, Lucas GB, Edgerton CW (1945) Genetics of *Glomerella* III: crosses with a conidial strain. Am J Botany 32: 717–721.
19. Edgerton CW, Chilton SJP, Lucas GB (1945) Genetics of *Glomerella* II: fertilization between strains. Am J Botany 32: 115–118.
20. Lucas GB (1946) Genetics of *Glomerella* IV: nuclear phenomena in the ascus. Am J Botany 33: 802–806.
21. Wheeler HE, Olive LS, Ernest CT, Edgerton CW (1948) Genetics of *Glomerella* V: crozier and ascus development. Am J Botany 35: 722–728.
22. Chilton SJP, Wheeler HE (1949) Genetics of *Glomerella* VI: linkage. Am J Botany 36: 270–273.
23. Chilton SJP, Wheeler HE (1949) Genetics of *Glomerella* VII: mutation and segregation in plus cultures. Am J Botany 36: 717–721.
24. Wheeler HE (1950) Genetics of *Glomerella* VIII: a genetic basis for the occurrence of minus mutants. Am J Botany 37: 304–312.
25. McGahen JW, Wheeler HE (1951) Genetics of *Glomerella* IX: perithecial development and plasmogamy. Am J Botany 38: 610–617.
26. Driver CH, Wheeler HE (1955) A sexual hormone in *Glomerella*. Mycologia 47: 311–316.
27. Wheeler HE (1956) Linkage groups in *Glomerella*. Am J Botany 43: 1–6.
28. Wheeler HE, Driver CH; Campa C (1959) Cross-and self-fertilization in *Glomerella*. Am J Botany 46: 361–365.
29. Kimati H, Galli F (1970) *Glomerella cingulata* (Stonem.) Spauld. et v. Scherenk. f sp. *phaseoli* n.f., fase ascógena do agente causal da antracnose do feijoeiro. Anais da Escola Superior de Agricultura Luiz de Queiroz 27: 411–437.
30. Batista UG, Chaves GM (1982) Patogenicidade de culturas monoascospóricas de cruzamento entre raças de *Colletotrichum lindemuthianum* (Sacc. & Magn.) Scrib. Fitopatol Brasil 7: 285–293.
31. Bryson RJ (1990) Sexual hybridization and the genetics of pathogenic specificity in *Colletotrichum lindemuthianum*. Birmingham: University of Birmingham. 272p.
32. Ishikawa FH, Barcelos QL, Alves EA, Camargo Junior OA, Souza EA (2010) Symptoms and pre-penetration events associated with the infection of common bean by the anamorph and teleomorph of *Glomerella cingulata* f. sp. *phaseoli*. J Phytopathol 158: 270–277.
33. Pinto JMA, Pereira R, Mota SF, Ishikawa FH, Souza EA (2012) Investigating phenotypic variability in *Colletothichum lindemuthianum* populations. Phytopathol 102: 490–497.
34. Mendes-Costa MC (1996) Genetics of *Glomerella cingulata* f sp. *phaseoli* I: sexual compatibility. Rev Brasil Gen 19: 350–351.
35. Souza BO, Souza EA, Mendes-Costa MC (2007) Determinação da variabilidade em isolados de *Colletotrichum lindemuthianum* por meio de marcadores morfológicos e culturais. Cienc Agrotec 31: 1000–1006.
36. Junqueira NTV, Chaves GM, Zambolin L, Romero RS, Gasparoto L (1984) Isolamento, cultivo e esporulação de Microcylus ulei, agente etiológico do mal das folhas da seringueira. Ceres 31: 322–331.
37. Oliveira JA (1991) Efeito do tratamento fungicida em sementes no controle de tombamento de plântulas de pepino (Cucumis sativus L.) e pimentão (Capsicum annum L.). Dissertação de Mestrado, Brazil: Universidade Federal de Lavras. 111p.
38. Ferreira DF (2011) Sisvar: a computer statistical analysis system. Cienc Agrotec 35: 1039–1042.
39. Starnes JH, Thornbury DW, Novikova OS, Rehmeyer CJ, Farman ML (2012) Telomere-targeted retrotransposons in the rice blast fungus *Magnaporthe oryzae*: agents of telomere instability. Genetics 191: 389–406.
40. Thon MR, Nuckles EM, Vaillancourt LJ (2000) Restriction enzyme-mediated integration used to produce pathogenicity mutants of *Colletotrichum graminicola*. Mol Plant Microbe Interact 13: 1356–1365.
41. White TJ, Bruns T, Lee S, Taylor J (1990) Amplification and direct sequencing of fungal ribosomal RNA genes for phylogenetics. In: Innis MA, Gelfand DH, Sninsky JJ, White TJ, editor. PCR Protocols: a guide to methods and applications. Academic Press, New York, USA. 315–322.
42. Arie T, Christiansen SK, Yoder OC, Turgeon BG (1997) Efficient cloning of ascomycete mating type genes by PCR amplification of the conserved *MAT* HMG box. Fungal Genet Biol 21: 118–130.
43. Flowers JL, Vaillancourt LJ (2005) Parameters affecting the efficiency of *Agrobacterium* tumefaciens-mediated transformation of *Colletotrichum graminicola*. Curr Genet 48: 380–388.
44. Lorang JM, Tuori RP, Martinez JP, Sawyer TL, Redman RS, et al. (2001) Green fluorescent protein is lighting up fungal biology. Appl Environ Microbiol 67: 1987–1994.
45. Dereeper A, Guignon V, Blanc G, Audic S, Buffet S, et al. (2008) Phylogeny.fr: Robust phylogenetic analysis for the non-specialist. Nucleic Acids Res 36: 465–469.
46. Centro Internacional de Agricultura Tropical - CIAT (1990) Snap bean in the developing world: potential benefits of research. In: Henry G, Janssen, W, editors. Trends in CIAT commodities. CIAT: Colombia. 89–115.
47. Rava CA, Molina J, Kauffmann M, Briones I (1993) Determinación de razas fisiológicas de *C. lindemuthianum* en Nicaragua. Fitopatol Brasil 18: 388–391.
48. Habgood RM (1970) Designation of physiological races of plant pathogens. Nature 227: 1268–1269.
49. Arx JA von (1957) Die Arten der Gattung *Colletotrichum* Cda. Phytopathol Zeitschrift 29: 413–468.
50. Saccardo PA (1878) Fungi novi ex herbarium professoris Doct. P. Magnus Berolinensis. Michelia 1: 117–132.
51. Sutton BC (1992) The genus *Glomerella* and its anamorph *Colletotrichum*. In: Bailey JA, Jeger MJ, editors. *Colletotrichum*: biology, pathology and control. CAB International: Wallingford, United Kingdom. 1–26.
52. Rava AC, Sartorato A (1994) Antracnose. In: Rava AC, Sartorato, A, editors. Principais doenças do feijoeiro comum e seu controle. Embrapa: Brasilia, 17–40.
53. Liu F, Cai L, Crous PW, Damm U (2013) Circumscription of the anthracnose pathogens *Colletotrichum lindemuthianum* and *C. nigrum*. Mycologia 105: 844–860.
54. Mehrabi R, Bahkali AH, Abd-Elsalam KA, Moslem M, M'barek SB, et al. (2011) Horizontal gene and chromosome transfer in plant pathogenic fungi affecting host range. FEMS Microbiol Rev 35: 542–554.
55. Ishikawa FH, Souza EA, Shoji J, Connolly L, Freitag M, et al. (2012) Heterokaryon incompatibility is suppressed following conidial anastomosis tube fusion in a fungal plant pathogen. Plos One 7: e31175.
56. Liu B, Wasilwa LA, Morelock TE, O'Neill NR, Correll JC (2007) Comparison of *Colletotrichum orbiculare* and several allied *Colletotrichum* spp. for mtDNA RFLPs,

intron RFLP and sequence variation, vegetative compatibility and host specificity. Phytopathology 97: 1305–1314.

57. Bailey JA, Sherriff C, O'Connell RJ (1995) Identification of specific and intraspecific diversity in *Colletotrichum*. In: Leslie JF, Frederiksen RA, editors. Diverse Analysis through Genetics and Biotechnology: Interdisciplinary Bridges to Improved Sorghum and Millet Crops, Iowa State University Press. 197–211.

58. Bailey JA, Nash C, Morgan LW, O'Connel RJ, TeBeest DO (1996) Molecular taxonomy of *Colletotrichum* species causing anthracnose of Malvaceae. Phytopathology 86: 1076–1083.

59. Sreenivasaprasad S, Mills PR, Meehan BM, Brown AE (1996) Phylogeny and systematic of 18 *Colletotrichum* species based on ribosomal DNA spacer sequences. Genome 39: 499–512.

60. Sherriff C, Whelan MJ, Arnold GM, Lafay JF, Brygoo Y, et al. (1994) Ribosomal DNA sequence analysis reveals new species groupings in the genus *Colletotrichum*. Exp Mycol 18: 121–138.

61. O'Connell RJ, Nash C, Bailey JA (1992) Lectin citochesmitry: a new approach to understanding cell differentiation, pathogenesis and taxonomy in *Colletotrichum*. In: Bayley JA, Jeger MJ, editors. *Colletotrichum*: Biology, Pathology and Control. Wallingford: CAB International. 67–87.

62. Talamini V, Souza EA, Pozza EA, Silva GF, Ishikawa FH, et al. (2006) Genetic divergence among and within *Colletotrichum lindemuthianum* races assessed by RAPD. Fitopatol Brasil 31: 545–550.

63. Du M, Schardl CL, Vaillancourt LJ (2005) Using mating-type gene sequences for improved phylogenetic resolution of *Colletotrichum* species complexes. Mycologia 97: 641–658.

64. Redman RS, Freeman S, Clifton DR, Morrel J, Brown G, et al. (1999) Biochemical Analysis of Plant Protection Afforded by a Nonpathogenic Endophytic Mutant of *Colletotrichum magna*. Plant Physiol 119: 795–804.

65. Porras-Alfaro A, Bayman P (2011) Hidden fungi, emergent properties: endophytes and microbiomes. Annu Rev Phytopathol 49: 291–315.

66. Leite ST, Cnossen-Fassoni A, Pereira OL, Mizubuti ESG, Araújo EF, et al. (2013) Novel and highly diverse fungal endophytes in soybean revealed by the consortium of two different techniques. J Microbiol 51: 56–69.

67. Weir BS, Johnston PR, Damm U (2012) The *Colletotrichum gloeosporioides* species complex. Stud Mycol 73: 115–180.

68. Gioti A, Mushegian AA, Strandberg R, Stajich JE, Johannesson H (2012) Unidirectional evolutionary transitions in fungal mating systems and the role of transposable elements. Mol Biol Evol 29: 3215–26.

69. Lengeler KB, Fox DS, Fraser JA, Allen A, Forrester K, et al. (2002) Mating-type locus of *Cryptococcus neoformans*: a step in the evolution of sex chromosomes. Eukaryot Cell 1: 704–718.

70. Rydholm C, Dyer PS, Lutzoni F (2007) DNA sequence characterization and molecular evolution of MAT1 and MAT2 mating-type loci of the self-compatible ascomycete mold *Neosartorya fischeri*. Eukaryotic Cell 6: 868–874.

71. Zaffarano PL, Duo A, Gruenig CR (2010) Characterization of the mating type (MAT) locus in the *Phialocephala fortinii* s.l. - *Acephala applanata* species complex. Fungal Genet Biol 47: 761–772.

72. Poggeler S, O' Gorman CM, Hoff B, Kuck U (2011) Molecular organization of the mating-type loci in the homothallic Ascomycete *Eupenicillium crustaceum*. Fungal Biol 115: 615–624.

73. Hyde KD, Cai L, Cannon PF, Crouch JA, Crous PW, et al. (2009) *Colletotrichum* – names in current use. Fungal Divers 39: 147–182.

74. Latunde-Dada AO, O'Connell RJ, Nash C, Lucas JA (1999) Stomatal penetration of cowpea (*Vigna unguiculata*) leaves by a *Colletotrichum* species causing latent anthracnose. Plant Pathol 48: 777–785.

75. Alvindia DG, Natsuaki KT (2008) Evaluation of fungal epiphytes isolated from banana fruit surfaces for biocontrol of banana crown rot disease. Crop Prot 27: 1200–1207.

76. Freeman S, Horowitz S, Sharon A (2001) Pathogenic and nonpathogenic lifestyles in *Colletotrichum* acutatum from strawberry and other plants. Phytopathology 91: 986–992.

77. Osono T (2008) Endophytic and epiphytic phyllosphere fungi of *Camillea japonica*: seasonal and leaf-dependent variations. Mycologia 100: 387–391.

78. Santamaría J, Bayman P (2005) Fungal epiphytes and endophytes of coffee leaves (*Coffea arabica*). Microb Ecol 50: 1–8.

79. Photita W, Lumyong S, Lumyong P, Mckenzie EHC, Hyde KD (2004) Are some endophytes of *Musa acuminata* latent pathogens? Fungal Divers 16: 131–140.

80. Rojas EI, Rehner SA, Samuels GJ, Van Bael SA, Herre EA, et al. (2010) *Colletotrichum gloeosporioides* s.l. associated with *Theobroma cacao* and other plants in Panama: multilocus phylogenies distinguish host-associated pathogens from asymptomatic endophytes. Mycologia 102: 1318–38.

81. Vargas WA, Martín JMS, Rech GE, Rivera LP, Benito EP et al. (2012) Plant defense mechanisms are activated during biotrophic and necrotrophic development of *Colletotricum graminicola* in maize. Plant Physiol 158: 1342–1358.

82. Roca MG, Davide LC, Davide LMC, Mendes-Costa MC, Schwan RF (2004) Conidial anastomosis fusion between *Colletotrichum* species. Mycol Res 108: 1320–1326.

9

Direct Observation of the Photodegradation of Anthracene and Pyrene Adsorbed onto Mangrove Leaves

Ping Wang[1,2], Tun-Hua Wu[3], Yong Zhang[4]*

1 School of Environmental Science and Public Health, Wenzhou Medical University, Wenzhou, China, **2** Key Laboratory of Coastal and Wetland Ecosystems (Xiamen University), Ministry of Education, Xiamen, China, **3** School of Information and Engineering, Wenzhou Medical University, Wenzhou, China, **4** State Key Laboratory of Marine Environmental Science (Xiamen University), College of the Environment and Ecology, Xiamen University, Xiamen, China

Abstract

An established synchronous fluorimetry method was used for *in situ* investigation of the photodegradation of pyrene (PYR) and anthracene (ANT) adsorbed onto fresh leaves of the seedlings of two mangrove species, *Aegiceras corniculatum* (L.) Blanco (*Ac*) and *Kandelia obovata* (*Ko*) in multicomponent mixtures (mixture of the ANT and PYR). Experimental results indicated that photodegradation was the main transformation pathway for both ANT and PYR in multicomponent mixtures. The amount of the PAHs volatilizing from the leaf surfaces and entering the inner leaf tissues was negligible. Over a certain period of irradiation time, the photodegradation of both PYR and ANT adsorbed onto the leaves of *Ac* and *Ko* followed first-order kinetics, with faster rates being observed on *Ac* leaves. In addition, the photodegradation rate of PYR on the leaves of the mangrove species in multicomponent mixtures was much slower than that of adsorbed ANT. Compared with the PAHs adsorbed as single component, the photodegradation rate of ANT adsorbed in multicomponent mixtures was slower, while that of PYR was faster. Moreover, the photodegradation of PYR and ANT dissolved in water in multicomponent mixtures was investigated for comparison. The photodegradation rate on leaves was much slower than in water. Therefore, the physical-chemical properties of the substrate may strongly influence the photodegradation rate of adsorbed PAHs.

Editor: Jie Zheng, University of Akron, United States of America

Funding: The funding of our manuscript include National Natural Science Foundation of China (21177102, 21207103), Public Benefit Project of Zhejiang Province (2012C31025), Natural Science Foundation of Zhejiang Province (LY13H180012), Scientific Research Fund of Zhejiang Provincial Education Department (Y201222932), Specialized Research Fund for the Doctoral Program of Higher Education (20130121130005), Undergraduate Scientific and Technological Innovation project of Zhejiang Province (2013R413010) and WEL Visiting Fellowship Program. The funders had no role in study design, data collection and analysis, decision to publish, or preparation of the manuscript.

Competing Interests: The authors have declared that no competing interests exist.

* Email: fruitful@xmu.edu.cn

Introduction

Over 80% of the earth's terrestrial surface is covered by vegetation [1]. Vegetation plays a key role in the environmental fate of many polycyclic aromatic hydrocarbons (PAHs) [2]. Furthermore, leaf surfaces are covered with a complex lipid cuticle that can accumulate hydrophobic organic pollutants from the atmosphere [3]. PAHs are widely distributed persistent organic pollutants that are generated by natural combustion processes as well as by human activities [4,5]. These micropollutants in the environment pose a potential threat to aquatic organisms and humans because of their toxic, carcinogenic and mutagenic properties [6]. PAHs are ubiquitous in natural substances such as plants, soil, sediment, water and air [7–9]. Thus, it is important to study the transportation and transformation of PAHs in the environment. Many reports demonstrate that most PAHs in the environment biodegrade slowly and with difficulty because of their low water solubility [10]. Additionally, it has been shown that many PAHs exhibit photo-induced toxicity [11] and that the delocalized π bond of PAHs can absorb the visible and ultraviolet

components of sunlight [12]. Therefore, photodegradation might represent an important transformation pathway for PAHs in the environment. Investigations in this field have recently intensified and have generated some solid conclusions. At present, most studies on the photodegradation of PAHs have focused on PAHs in a liquid medium or adsorbed onto solid particles [13–16]. Only a few studies have addressed the photodegradation of PAHs adsorbed onto vegetation, although the photodegradation of PAHs adsorbed onto plant surfaces (especially leaves) plays an important role in their transfer from the atmosphere to the food chain [17–19].

Mangrove wetlands, as a buffer in estuaries, act as both a sink and a source of PAHs in coastal ecosystems [20–22]. Mangroves can accumulate lipophilic PAHs from the atmosphere because of the thick waxy or lipidic layers on the surfaces of their leaves [23]. Ke et al. argued that photodegradation is an important method for removing the PAHs adsorbed onto mangrove leaves after an oil spill accident [24]. Wang et al. and Niu et al. found that the photodegradation of some PAHs adsorbed onto spruce or pine needles might play a significant role in their environmental behavior. Therefore, it is necessary to investigate the photochemical behavior of PAHs absorbed onto mangrove leaves. However, previous research on the photodegradation of PAHs adsorbed onto plant leaves has mostly been conducted by entirely destroying the sample. Thus, these traditional methods are not compatible with the direct study of the photodegradation of PAHs adsorbed onto leaves. Moreover, in a long-term study, the quantity of PAHs volatilized from the leaf surfaces and entering the inner leaf tissues may not be negligible [17,19]. Optical fibers showing high light focalization, low weights and small sizes have been used for the direct investigation of PAHs adsorbed onto solid substrates [25–27]. In our group, a fiber-optic fluorimetry has also been established for *in situ* investigation of the photodegradation of fluoranthene adsorbed onto the leaves of three mangrove species [23,28]. Nevertheless, the mechanisms underlying the photodegradation of PAHs adsorbed onto mangrove leaves have not been fully explored. Our previous research sought to elucidate the environmental behaviors of PAHs adsorbed onto mangrove leaves using an established synchronous fluorimetry combined with a fluorescence spectrophotometer and an optical fiber to directly detect PYR and ANT adsorbed onto mangrove leaves [29]. The photodegradation of PYR and ANT adsorbed onto the leaves of two mangrove species as single component have also been studied in our laboratory [30]. However, the environmental behaviors of PAHs are very complex in the field. For example, many different types of PAHs from different sources might exist on the leaf surfaces. Because of the effects of complex factors mentioned above or others, it is difficult for *in situ* investigation the environmental behaviors of PAHs adsorbed on leaf surfaces in the field. In addition, only a few studies involve the photolytic behaviors of PAHs adsorbed on vegetation. Thus, preliminary experiments could be carried out in lab at first. And simple experimental conditions were set for study of the possible photolysis processes of some PAHs on the typical mangrove leaves. To further develop this line of research, synchronous fluorimetry was used for *in situ* studying the photodegradation of PYR and ANT adsorbed onto the leaves of two mangrove species in multicomponent mixtures. Compared with the photodegration of the PAHs adsorbed on mangrove leaves in single component, the further studies for investigation the photodegration of the PAHs adsorbed on mangrove leaves in multicomponent mixtures are more representative of real conditions for most PAHs in the environment. Seedlings of *Aegiceras corniculatum* (L.) Blanco (*Ac*) and *Kandelia obovata* (*Ko*), two of the most widespread mangrove species in China, were selected for this study, and the photodegradation of equal quantities of PYR and ANT in water as single compounds and in multicomponent mixtures was also investigated for comparison.

Materials and Methods

Preparation of PAH solution
Stock solutions of PYR and ANT (Aldrich, purity>99%, USA) were obtained based on the method described in our previous studies [22,29]. Working solutions of PYR and ANT were prepared by diluting the stock solutions with acetone just prior to being used.

Sample collection
According to our previous studies and others, the mature viviparous hypocotyls of *Ko* when off the trees could be collected for cultivation, and the non-viviparous seedlings of *Ac* who are least one-year old could be taken away from the forest and cultivated for further experiments [30,31]. Therefore, mature

mangrove hypocotyls of *Ko* and 1-year-old *Ac* seedlings were collected at the Longhai mangrove reserve in Zhangzhou, Fujian, China (longitude: 24°29′3″, latitude: 118°5′59″, altitude: 0 m above sea level). This mangrove forest is wild and does not belong to any individuals or organizations. Thus, no specific permissions were required for these activities. Only a few mangrove seedlings were collected for the experiment, which did not involve endangered or protected species or vertebrates. Hypocotyls of *Ko* and seedlings of *Ac* of approximately the same size and maturity were sampled and quickly transported to the laboratory for cultivation. The hypocotyls of *Ac* and *Ko* were cultivated on sand (collecting from the mangrove forest) in pots, partially submerged in nutrient Hoagland's solution. For a 12-hour photoperiod, propagules were illuminated by 400-W Na solar lighting. The greenhouse temperature was maintained at 25–28°C. Though, the cultivation conditions in lab were different with that in the field, our studies of past and present have attempted to simulate the living environment of the hypocotyls. The cultivation conditions in lab have their own merits. We all know that the growth of the mangrove hypocotyls in the field might be influenced by many factors. The maturity level and integrity of the hypocotyls living in the same forest might be different. Therefore, the similar maturity of hypocotyls and leaves could not be obtained in the field, which might lead to lower accuracy of the results. Thus, the possible mechanisms of environmental behaviors of the PAHs obtained in the field might be different. Taking into consideration of various factors, the cultivation conditions selected in our studies were comparatively reasonable. Ninety days later, the leaves from the two mangrove species with similar length and fresh weight were collected for the further experiment. After the collection of the leaves, experiments were carried out as soon as possible. Pretreatment of the leaves followed the steps established in our previous studies [29,30,31]. The picked fresh mangrove leaves were carefully rinsed with distilled water to remove surface silt. After air drying, six circles of 0.5 cm radius were drawn on the upper surfaces of each leaf with a pencil. The size was the same as the light circle formed by the fiber optical probe. With the use of a micropipette, a certain amount of ANT (0.25 nmol spot⁻¹) and PYR (0.25 nmol spot⁻¹) acetone solutions (5 μL) were applied as homogeneous layers to the circles of upper leaf surfaces respectively. After evaporation of the acetone from the leaf surfaces at room temperature, a series of similar-sized spots were formed, each spot indicating a sample location.

An ultrasonic cleaning device (Model KQ3200, SM, Kun Shan Ultrasonic Instruments Co., Ltd., China) was utilized to extract the lipid wax from the mangrove leaves. The wax content of the leaves was subsequently quantified using a method from described in previous work [28,31,32].

Photodegradation of PAHs absorbed onto the leaves of the two mangrove species
Mangrove leaves with adsorbed PAHs were placed under the optical fiber of a mercury lamp (CHF-XM500 W, Beijing Trusttech, Co., Ltd., China), as shown in Figure 1. Experiments began after the mercury lamp was turned on for approximately 30 minutes. The intensity of illumination, determined with a ZDS-10 illuminometer (Shanghai, China), was maintained at 3.5×10⁴ lx for the entire experiment. The light intensity was controlled by adjusting the height between the mangrove leaf surface and the optical fiber of the mercury lamp. After a certain period of irradiation, the leaves were placed under the fiber-optic probe of the spectrophotometer (Varian, Harbor City, California). In order to get a much clear comparison for the photodegration of ANT and PYR adsorbed on *Ko* and *Ac* leaves as single component and

in multicomponent mixtures. According to our previous studies and the pre-experiments, the set irradiation time interval increased gradually (5 nm, 10 nm, 15 nm and 20 nm). Thus, the optimized irradiation of the ANT and PYR adsorbed on the leaves of *Ko* and *Ac* should be set at the same sequence. That is the optimized irradiation time for the PAHs adsorbed onto the *Kc* leaves was set at 0 min, 1 min, 3 min, 7 min, 13 min, 20 min, 30 min, 45 min, 60 min, 80 min, and 100 min, 120 min and 140 min, respectively, and the optimized irradiation time for the adsorbed PAHs on the *Ac* leaves was set at 0 min, 1 min, 3 min, 7 min, 13 min, 20 min, 30 min, 45 min, 60 min, 80 min, 100 min, 120 min and 140 min, respectively [30]. Fluorescence spectra were obtained using the following instrumental setting: excitation and emission slits of 20 and 10 nm, respectively; scan speed of 120 nm min^{-1}; PMT voltage of 600 V; and the constant wavelength difference ($\Delta\lambda$) was set at 38 nm. Finally, the fluorescence intensities of the PAHs adsorbed onto the leaves were directly determined. Replicate experiments were conducted for 3 times. To avoid interference from the scattered light, the leaf had to be kept flat, and the angle between the fiber-optic probe of the spectrophotometer and the tested leaf was 45° throughout the experiment.

Photodegradation of PAHs dissolved in water

To compare and analyze the photodegradation behavior of each PAH in different media, the photodegradation of the same initial amount of PYR and ANT dissolved in water in single-component and multicomponent mixtures was also studied. A photodegradation reactor constructed in our laboratory was used to study the photodegradation of each PAH in water [33]. Using a micropipette, 0.25 nmol of PYR and 0.25 nmol of ANT dissolved in water were added to the reactor, which was then immediately placed under the optical fiber of the mercury lamp. According to our previous studies and the pre-experiments [30], the optimized irradiation time for the ANT in water as single component (0 min, 1 min, 3 min, 6 min, 11 min, 16 min, 21 min, 31 min, 36 min, 41 min, and 56 min) and in multicomponent mixtures (0 min, 1 min, 3 min, 6 min, 11 min, 16 min, 21 min, 31 min, 36 min, 41 min, and 56 min, 66 min, and 76 min) was set, respectively. In addition, the optimized irradiation time for the PYR in water as

single component (0 min, 1 min, 3 min, 6 min, 11 min, 16 min, 21 min, 31 min, 36 min, 41 min, 56 min, 66 min, 86 min, 100 min, 120 min) and in multicomponent mixtures (0 min, 1 min, 3 min, 6 min, 11 min, 16 min, 21 min, 31 min, 36 min, 41 min, 56 min, 66 min, 76 min, 86 min, and 96 min) was also set, respectively. To sum up, the set irradiation time of the PAHs in water was almost fixed (mainly 5 nm), which was different with that adsorbed on mangrove leaves (5 nm, 10 nm, 15 nm and 20 nm). Thus, after a defined period of illumination mentioned above, the reactor was placed in the fluorescence spectrophotometer. The fluorescence intensities of the PAHs dissolved in water were ultimately obtained. Because of the small volume of the self-made reactor, the working solutions of PYR and ANT did not require stirring, and we could directly detect the PAHs during photodegradation. Replicate experiments were conducted for 6 times.

Statistical analysis

In this study, the sample preparations were the same with the method of our previous studies [29–31]. And we used four mangrove leaves from each species, yielding a total of 24 locations and 72 measurements for each species. The mean values presented represent the fluorescence intensity of the PAHs adsorbed onto different locations (24 locations from 4 leaves) in the leaves of each mangrove species. Statistical analysis of the variation in fluorescence intensity was performed using the Statistical Package for the Social Science (SPSS), 13.0, for Windows. Significant differences in the results about the mean values of fluorescence intensity obtained by the adsorbed PAHs at different irradiation time were determined using the independent sample t-test. In addition, the confidence interval was utilized for estimate the reliability of the fluorescence intensity of the PAHs adsorbed onto different locations in the leaves of each mangrove species. And the confidence interval was stated at the 95% confidence level (p> 0.05 means that no significant difference existed).

When the concentration of a PAH is proportional to its fluorescence intensity, C_t (the PAH concentration at time t) and C_0 (the initial PAH concentration) can be replaced with F_t (the PAH fluorescence intensity at time t) and F_0 (the initial PAH

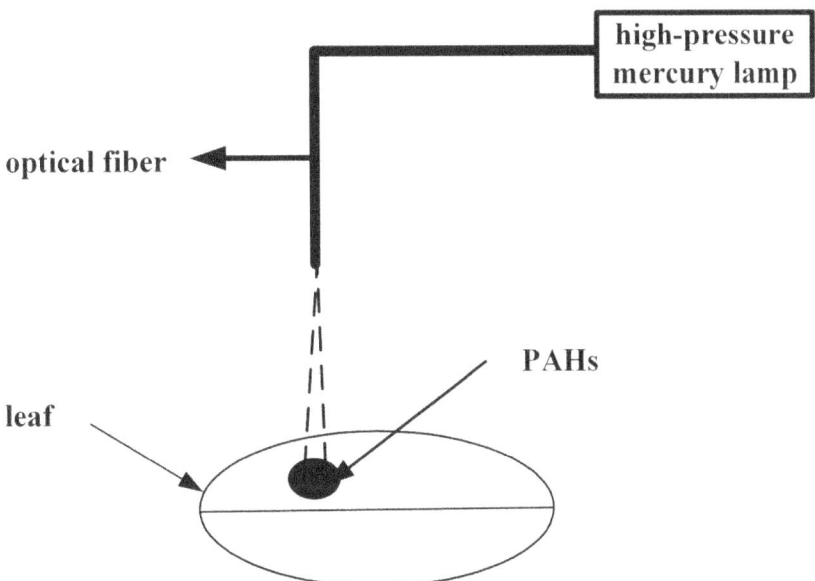

Figure 1. Schematic diagram of the photodegradation of PAHs adsorbed onto the leaves of two mangrove species.

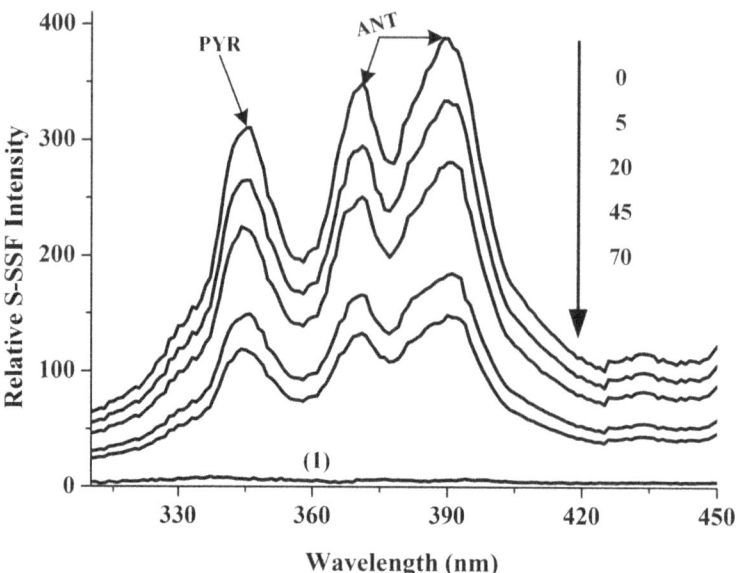

Figure 2. Synchronous fluorescence spectra of PYR and ANT adsorbed onto the leaves of *Ko* in multicomponent mixtures ($\Delta\lambda$ = 38 nm), measured at different times during irradiation. The order of the changes in fluorescence intensity over time (at 0 min, 5 min, 20 min, 45 min and 70 min) is indicated by the arrow. Wavelengths of 390 nm and 346 nm were selected for ANT and PYR, respectively. (1): Synchronous fluorescence spectra of the uncontaminated *Ko* leaves.

fluorescence intensity), respectively. The data were processed based on the same equations as our previous studies [30].

Results and Discussion

Fluorescence spectra of PYR and ANT adsorbed onto mangrove leaves

In previous studies, we established the synchronous fluorimetry method for the direct determination of PYR and ANT adsorbed onto *Ac* and *Ko* leaves both as single compounds and in multicomponent mixtures. Experimental results have indicated that when the optimized scanning wavelength difference ($\Delta\lambda$) was 38 nm, the ANT and PYR adsorbed onto the leaves of two mangrove species as single compounds and in multicomponent mixtures have their own characteristic synchronous fluorescence spectra. The fluorescence spectra of the adsorbed PAHs on mangrove leaves as single component have fluorescence maximums at 369 nm and 387 nm for ANT, and 343 nm for PYR. Thus, the wavelengths of 387 nm and 343 nm were selected for the quantification of ANT and PYR as single component, respectively. In addition, the fluorescence spectra of the PAHs adsorbed on mangrove leaves in multicomponent mixtures showed peaks at 371 nm and 390 nm for ANT, and 346 nm for PYR. Thus, the wavelengths of 390 nm and 346 nm were selected for the quantification of ANT and PYR in multicomponent mixtures, respectively [31]. More recently, the photodegradation of PYR and ANT adsorbed onto the leaves of two mangrove species as a single component has been investigated by our group [30]. However, it is reported that the photodegradation of PAHs are very complex in natural environment. For example, interaction between different types of PAHs might influence the photodegradation of the PAH in multicomponent mixtures. Thus, compared with the PAHs as single component, studies about the photodegradation of the PAHs in multicomponent mixtures are more representative of real conditions for most PAHs in natural environment. To better understand the photodegradation of these PAHs in multicomponent mixtures, evaluation of the

synchronous fluorimetry spectra of the two PAHs and their photodegradation during irradiation is useful. Figure 2 shows the fluorescence spectra of PYR and ANT adsorbed onto the leaves of *Ko* in multicomponent mixtures, measured at different times during irradiation. The relative fluorescence intensities decreased with time. As can be observed in Figure 2, the autofluorescence of *Ko* leaves without PAH was weak and was insufficient to affect the determination of adsorbed PAH in further analyses. In addition, there was no overall change in the shapes of the spectra of the PAHs or the width of the half-wide spectral band. Thus, the photolytic products did not interfere with the detection of adsorbed PYR and ANT during photodegradation (Figure 2). A similar trend was observed during the photodegradation of PYR and ANT adsorbed onto the leaves of *Ac*. Thus, it was concluded that synchronous fluorimetry could be used for the direct study of the photodegradation of PYR and ANT adsorbed onto mangrove leaves in multicomponent mixtures.

Results of the photodegradation experiments

Volatilization and absorption losses could be the most important pathways for adsorbed PAHs on plant leaves [22]. Therefore, control experiments were carried out in which the leaves were maintained in darkness. The variations in F_t/F_0 for PYR and ANT adsorbed onto the leaves of the two mangrove species in multicomponent mixtures were determined over a certain time interval during the photodegradation experiments (Figure 3). Without illumination, F_t/F_0 decreased very little as time went on (Figure 3). Hence, photodegradation was found to be the main transformation pathway for PYR and ANT adsorbed onto the mangrove leaves, and the amount of PAH volatilizing from the leaf surfaces and entering the inner leaf tissues could be ignored in the short term. In addition, the relative fluorescence intensities of PYR and ANT as single component are also shown in Figure 3 for comparison [30]. As shown in Figure 3, the curves indicated that the photodegradation of each PAH occurred via multi-decays. Over a given period of time, the photodegradation of both PYR and ANT adsorbed onto the leaves of two mangrove species in

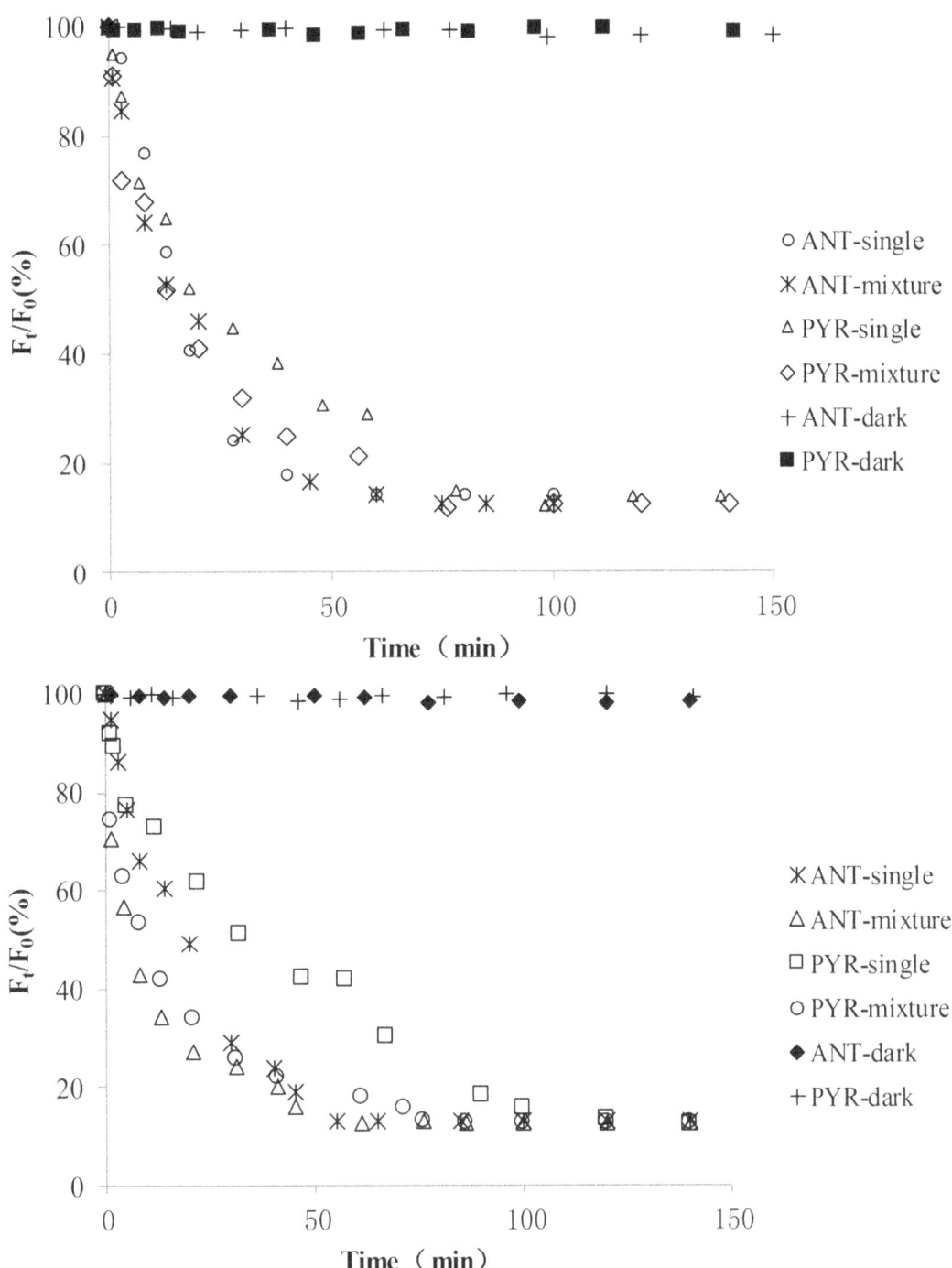

Figure 3. Change in F_t/F_0 with time during the photodegradation of ANT (0.25 nmol spot^{-1}) and PYR (0.25 nmol spot^{-1}) adsorbed onto *Ko* and *Ac* leaves as single component and in multicomponent mixtures.

multicomponent mixtures basically followed first-order reaction kinetics (Figure 4). In addition, the photodegradation of both PYR and ANT adsorbed onto the leaves of two mangrove species as single component basically followed first-order reaction kinetics in a given period of time (reference 30). Specifically, the given period of time for adsorbed ANT both on the leaves of *Ko* and *Ac* in multicomponent mixtures was from 0 min to 45 min. The given period of time for adsorbed PYR both on the leaves of *Ac* in multicomponent mixtures was from 0 min to 80 min the given period of time for adsorbed PYR on the leaves of *Ko* in

multicomponent mixtures was from 0 min to 60 min (Figure 4). The photodegradation rate of the PAHs adsorbed onto *Ac* leaves was faster than that adsorbed onto *Ko* leaves. Moreover, for a given amount of the two substances (0.25 nmol spot^{-1}), the photodegradation rate of adsorbed PYR was much lower than that of ANT. Thus, it is a reasonable assumption that the molecular structure of PAHs might be one of the important factors determining their photochemical behavior [34,35]. The experimental results also showed that for the same initial amount of each PAH, the photodegradation rate was in the order *Ac>Ko*

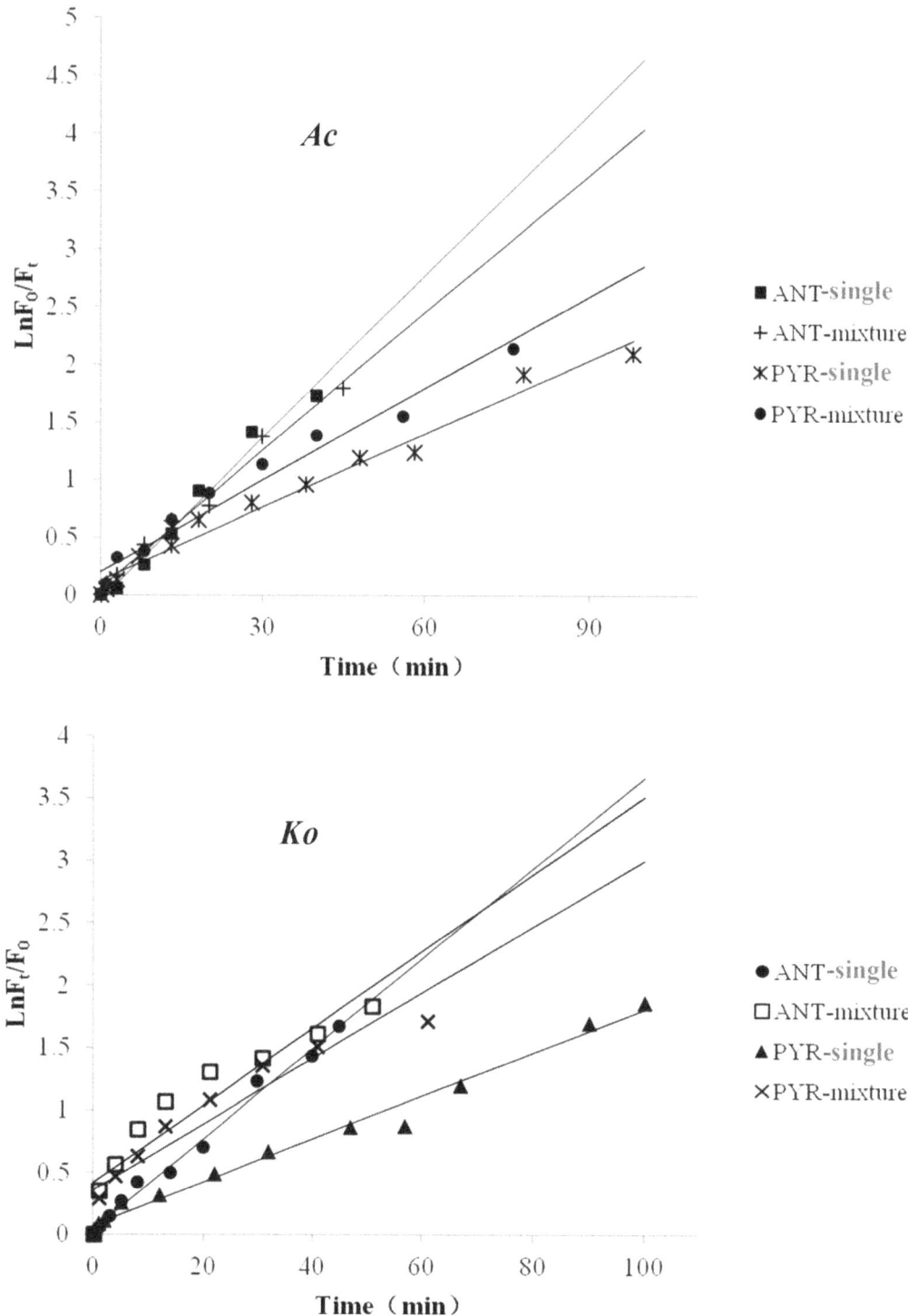

Figure 4. First-order kinetic plots of the photodegradation of ANT (0.25 nmol spot^{-1}) and PYR (0.25 nmol spot^{-1}) adsorbed onto *Ko* and *Ac* leaves as single component and in multicomponent mixtures.

(Table 1). It is assumed that the waxes and lipids on plant leaves can absorb UV photons, which might create light-filtering effects [17,19]. Thus, the number of UV photons striking each PAH adsorbed onto the leaves decreased and thereby reduced the photodegradation rate of each PAH. In this experiment, the leaf-wax content of *Ac* (4.98 mg.g^{-1}) was found to be much lower than that of *Ko* (7.63 mg.g^{-1}). Thus, the light filtering effect of *Ko*

leaves might be stronger than that of *Ac* leaves. Consequently, for the same initial amount of each substance, the number of the UV photons striking each PAH adsorbed onto *Ac* leaves would be higher. Moreover, because of the lower leaf-wax content of *Ac*, the interactions between the adsorbed PAH and *Ac* might become much weaker, making it much easier for the UV photons to strike the adsorbed PAH. These phenomena may act together to explain

Table 1. Kinetic results for the photodegradation of ANT adsorbed onto the leaves of two typical mangrove species (n = 9) and dissolved in water (n = 9) as single compounds and in multicomponent mixtures.

ANT	Substrate	Calibration carve	Correlation coefficient	K	$t^{1/2}$ (min)
single[a]	Ko	$y^b = 3.62 \times 10^{-2} x^c + 0.043$	0.9952	3.62×10^{-2}	19.2
	Ac	$y = 4.68 \times 10^{-2} x - 0.036$	0.9908	4.68×10^{-2}	14.8
	water	$y = 5.22 \times 10^{-2} x + 0.042$	0.9960	5.22×10^{-2}	13.3
mixture	Ko	$y = 3.09 \times 10^{-2} x + 0.417$	0.9356	3.09×10^{-2}	22.4
	Ac	$y = 3.98 \times 10^{-2} x + 0.063$	0.9936	3.98×10^{-2}	17.4
	water	$y = 4.81 \times 10^{-2} x - 0.017$	0.9939	4.81×10^{-2}	14.4

[a]data from our previous studies [30].
[b]the value of $\ln(F_0/F_t)$.
[c]the illumination time of ANT adsorbed onto mangrove leaves.

why the photodegradation rate of the PAHs adsorbed onto Ac leaves was faster compared with Ko leaves. (Table 1, Table 2). In addition, for both PYR and ANT, the variations in F_t/F_0 remained relatively stable when the amount of the adsorbed PAHs on the leaves decreased to approximately 12% of the initial amount (Figure 3). This finding indicates that the photodegradation of both PYR and ANT adsorbed onto the mangrove leaves could only result in partial degradation. In other words, photodegradation of the residual PAH adsorbed onto the mangrove leaves could no longer occur as the illumination time went on. Similar results were obtained regarding the photodegradation of PYR and ANT as single component in our previous work [30]. Identical results were found in studies by Xu et al. and Schuler et al., though these authors did not suggest any mechanisms for these findings [36,37]. Given previous findings by ourselves and others, a small amount of PAH might enter into the inner leaf tissues. These PAHs in the internal leaf tissues cannot absorb UV photons due to the light filtering effects of the leaves [17,19,32]. Further research will be required to better understand the possible reasons for these observations.

The experimental results also revealed that compared with PYR and ANT adsorbed onto the leaves of the same mangrove species as single component, the photodegradation rate of ANT in multicomponent mixtures was slower, while that of PYR was faster (Table 1, Table 2). Similar results were found concerning the photodegradation of PYR and ANT in water in multicomponent mixtures (Table 1, Table 2). There are several possible reasons for these results. First, for the same light intensity and irradiation time,

the total number of UV photons striking each PAH adsorbed onto the leaves as single component and multicomponent mixtures is the same. Therefore, compared with each PAH as a single compound, the number of UV photons received by each PAH in multicomponent mixtures is lower, which could slow the photodegradation rate of each PAH. Second, anthraquinone is one of the intermediate products of the photodegradation of ANT, and it has also been found to be a photosensitizer for both PYR and ANT [38]. In this experiment, the photodegradation rate of adsorbed ANT was faster compared with adsorbed PYR in multicomponent mixtures (Table 1, Table 2). Hence, in multicomponent mixtures, ANT degraded first and generated some anthraquinone. This anthraquinone might accelerate the photodegradation rate of the adsorbed PYR, and the amount of anthraquinone, as a photosensitizer of adsorbed ANT, would therefore decrease compared with that in the single-component mixture. Thus, the photodegradation rate of the adsorbed ANT would slow down in multicomponent mixtures. However, further studies are needed to confirm this hypothesis.

Studies show that the polarity of a solvent can significantly affect the photodegradation rate of PAHs in or on that solvent [14]. To further study the possible mechanisms underlying the photodegradation of PAHs adsorbed onto mangrove leaves in multicomponent mixtures, the photodegradation of PYR and ANT dissolved in water in multicomponent mixtures was also investigated for comparison (Figure 5). In this experiment, the synchronous fluorimetry method established in our previous studies was used to determine PYR and ANT levels in water as single

Table 2. Kinetic results of the photodegradation of PYR adsorbed onto the leaves of two typical mangrove species (n = 9) and dissolved in water (n = 9) as single compounds and in multicomponent mixtures.

PYR	Substrate	Calibration carve	Correlation coefficient	K	$t^{1/2}$ min
single[a]	Ko	$y^b = 1.58 \times 10^{-2} x^c + 0.114$	0.9899	1.58×10^{-2}	43.9
	Ac	$y = 2.13 \times 10^{-2} x + 0.125$	0.9865	2.13×10^{-2}	32.5
	water	$y = 2.88 \times 10^{-2} x + 0.088$	0.9937	2.88×10^{-2}	24.1
mixture	Ko	$y = 2.61 \times 10^{-2} x + 0.357$	0.9393	2.61×10^{-2}	26.6
	Ac	$y = 2.95 \times 10^{-2} x + 0.203$	0.9815	2.95×10^{-2}	23.5
	water	$y = 3.46 \times 10^{-2} x - 0.045$	0.9878	3.46×10^{-2}	20.0

[a]data from our previous studies [30].
[b]the value of $\ln(F_0/F_t)$.
[c]illumination time of the PYR adsorbed onto mangrove leaves.

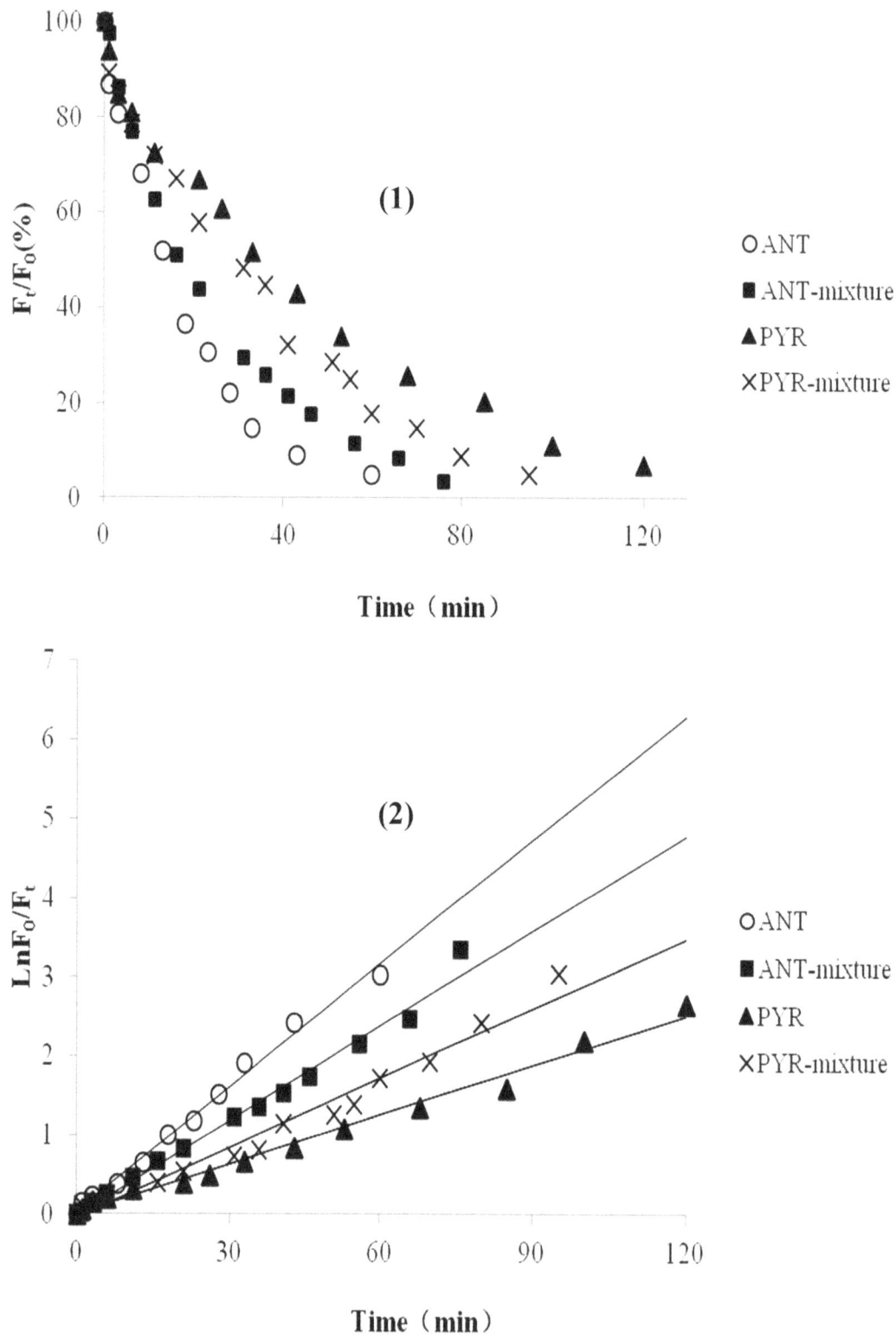

Figure 5. Photodegradations of ANT (0.25 nmol) and PYR (0.25 nmol) in water as single component and in multicomponent mixtures. (1): Change in F_t/F_0 over time during the photodegradation of the two PAHs adsorbed onto the leaves of the two mangrove species. (2): First-order kinetic plots of the photodegradation of the two PAHs adsorbed onto the leaves of the two mangrove species.

component and in multicomponent mixtures [33,39]. Experimental results showed that the photodegradation of PYR and ANT dissolved in water as single component and multicomponent mixtures followed first-order reaction kinetics (Figure 5). The photodegradation rate of each PAH in water as single component and in multicomponent mixtures was faster compared with the

PAHs adsorbed onto mangrove leaves (Table 1, Table 2). There are two possible reasons for this phenomenon. First, the wax layer on the mangrove leaves mainly consists of saturated fatty acids and unsaturated fatty acids, which could reflect or adsorb UV light. Therefore, the number of UV photons striking the adsorbed PAH might decrease and thereby reduce the photodegradation rate [17,18,37]. Second, it has been shown that the more polar the solvent, the faster photodegradation is [40,41]. In the present study, we assessed the PAHs in water, which is polar, and on the mangrove leaf surface, which is nonpolar. Thus, the photodegradation of the PAHs adsorbed onto the leaves was slower than when they were dissolved in water. In addition, it was clearly observed that the relative fluorescence intensities of PYR and ANT in water in multicomponent mixtures decreased almost to zero as the illumination time went on (Figure 5). In other words, PYR and ANT in homogeneous water could be thoroughly degraded within a certain period of time. This result differed from the photodegradation of PYR and ANT adsorbed onto the heterogeneous mangrove leaves. We can speculate that the photodegradation mechanisms for PAHs adsorbed onto mangrove leaves and dissolved in water may differ. Therefore, the photochemical behaviors of PAHs were dependent not only on their molecular structure but also on the physical-chemical properties of the substrate. Further experimental verification is needed.

Conclusions

In this work, synchronous fluorimetry was shown to be a simple, rapid and easy method for direct investigation of the photodegradation of PYR and ANT adsorbed onto the leaves of *Ac* and *Ko* in multicomponent mixtures. This method may provide a new way to study the mechanisms underlying the phytoremediation of

PAHs in mangrove wetlands or in other contaminated media. The photodegradation of both PYR and ANT on the leaves of the studied species followed first-order reaction kinetics. However, the photodegradation processes were different for the same PAHs adsorbed onto the leaves of different mangrove species. In addition, for the same amount of substance, the photodegradation rate of PYR was much slower than that of ANT. Compared with the photodegradation rates in single-component mixtures, the rates in multicomponent mixtures were different. The photodegradation mechanisms of the PAHs adsorbed onto the mangrove leaves, whether in single or multicomponent mixtures, might also be different than the mechanisms for the PAHs is dissolved in water. However, further research is needed to address this conjecture. In this experiment, only the PAHs adsorbed onto the leaf surfaces could be studied, and the PAHs entering into the internal leaf tissues could not be quantified or investigated for further photodegradation. Thus, there are still detailed and comprehensive studies to be carried out in the future. For example, in our group, two-photon laser confocal scanning microscopy has been used for direct visualization of PAHs on and within mangrove leaves. Combining these two methods with some traditional approaches may be helpful for better understanding the environmental behaviors of PAHs in real environments. Therefore, the synchronous fluorimetry method might be valuable in theoretical and practical applications.

Author Contributions

Conceived and designed the experiments: PW YZ THW. Performed the experiments: PW THW YZ. Analyzed the data: PW THW YZ. Contributed reagents/materials/analysis tools: PW THW YZ. Contributed to the writing of the manuscript: PW YZ THW.

References

1. Simonich SL, Hites A (1994) Importance of vegetation in removing polycyclic aromatic hydrocarbons from the atmosphere. Nature 370: 49–51.
2. Collins C, Fryer M, Grosso A (2006) Plant uptake of non ionic organic chemicals. Environ Sci Technol 40: 45–52.
3. Sojinu OS, Sonibare OO, Gayawan E (2013) Investigating polycyclic aromatic Hydrocarbons profiles in higher plants using statistical models. Int J Phytoremediat 15: 439–451.
4. Al-Saleh I, Alsabbahen A, Shinwari N, Billedo G, Mashhour A, et al. (2013) Polycyclic aromatic hydrocarbons (PAHs) as determinants of various anthropometric measures of birth outcome. Sci Total Environ 444: 565–578.
5. Kwon HO, Choi SD (2014) Polycyclic aromatic hydrocarbons (PAHs) in soils from a multi-industrial city, South Korea. Sci Total Environ 470: 1494–1501.
6. Pongpiachan S, Tipmanee DW, Muprasit J, Feldens P, Schwarzer K (2013) Risk assessment of the presence of polycyclic aromatic hydrocarbons (PAHs) in coastal areas of Thailand affected by the 2004 tsunami. Mar Poll Bull 76: 370–378.
7. Augusto S, Máguas C, Matos J, Pereira MJ, Branquinho C (2010) Lichens as an integrating tool for monitoring PAH atmospheric deposition: a comparison with soil, air and pine needles. Environ Pollut 158: 483–489.
8. Luiz FF, Mirian ACC, Alberto GFJ, Elisamara SS, Frederico SS, et al. (2012) Characterization and distribution of polycyclic aromatic hydrocarbons in sediments from Suruí Mangrove, Guanabara Bay, Rio de Janeiro, Brazil. J Coastal Res 28: 156–162.
9. Wang XT, Miao Y, Zhang Y, Li YC, Ming-Hong Wu MH, et al. (2013) Polycyclic aromatic hydrocarbons (PAHs) in urban soils of the megacity Shanghai: occurrence, source apportionment and potential human health risk. Sci Total Environ 447: 80–89.
10. Bamforth SM, Singleton I (2005) Bioremediation of polycyclic aromatic hydrocarbons: current knowledge and future directions. J Chem Technol Biot 80: 723–736.
11. Wang Y, Chen JW, Li F, Qin H, Qiao XL, et al. (2009) Modeling photoinduced toxicity of PAHs based on DFT-calculated descriptors. Chemosphere 76: 999–1005.
12. Dittmar T, Koch BP (2006) Thermogenic organic matter dissolved in the abyssal ocean. Mar Chem 102: 208–217.
13. Debestani R, Ellis KJ, Sigman ME (1995) Photodecomposition of anthracene on dry surface: products and mechanism. J Photoch Photobio A 86: 231–239.
14. Kahan TF, Donaldson DJ (2007) Photolysis of polycyclic aromatic hydrocarbons on water and ice surfaces. J Phys Chem A 111: 1277–1285.
15. Lehto KM, Vuorimaa E, Lemmetyinen H (2000) Photolysis of polycyclic aromatic hydrocarbons (PAHs) in dilute aqueous solutions detected by fluorescence. J Photoch Photobio A 136: 53–60.
16. Reyes CA, Medina M, Crespo-Hernandez C, Cedeno MZ, Arce R, et al. (2000) Photochemistry of pyrene on unactivated and activated silica surfaces. Environ Sci Technol 34: 415–421.
17. Niu JF, Chen JW, Martens D, Quan X, Yang FL, et al. (2003) Photolysis of polycyclic aromatic hydrocarbons adsorbed on spruce [Picea abies (L.) Karst.] needles under sunlight irradiation. Environ Pollut 123: 39–45.
18. Huang XD, Dixon DG, Greenberg BM (1993) Impact of UV radiation and photo-modification on the toxicity of PAHs to the higher plant Lemna gibba (Duckweed). Environ Toxicol Chem 12: 1067–1077.
19. Wang DG, Chen JW, Zhen X, Qiao XL, Huang LP (2005) Disappearance of polycyclic aromatic hydrocarbons sorbed on surfaces of pine [Pinua thunbergii] needles under irradiation of sunlight: volatilization and photolysis. Atmos Environ 39: 4583–4591.
20. Henry KM, Twilley RR (2013) Soil development in a coastal Louisiana wetland during a climate-induced vegetation shift from salt marsh to mangrove. J Coastal Res 29: 1273–1283.
21. Tansel B, Lee M, Tansel DZ (2013) Comparison of fate profiles of PAHs in soil, sediments and mangrove leaves after oil spills by QSAR and QSPR. Mar Poll Bull 73: 258–262.
22. Wang P, Zhang Y, Wu TH (2010) Novel method for in situ visualization of polycyclic aromatic hydrocarbons in mangrove plants. Toxicol Environ Chem 92: 1825–1829.
23. Chen L, Zhang Y, Liu BB (2010) In situ simultaneous determination the photolysis of multi-component PAHs adsorbed on the leaf surfaces of living Kandelia candel seedlings. Talanta 83: 324–331.
24. Ke L, Wong TWY, Wong YS, Tam NF (2002) Fate of polycyclic aromatic hydrocarbon (PAH) contamination in a mangrove swamp in Hong Kong following an oil spill. Mar Pollut Bull 45: 339–347.
25. Niessner R, Panne U, Schroeder H (1991) Fiber-optic sensor for the determination of polynuclear aromatic hydrocarbons with time-resolved laser-induced fluorescence. Anal Chim Acta 255: 231–243.

26. Panne U, Niessner R (1993) A fiber-optical sensor for polynuclear aromatic hydrocarbons based on multidimensional fluorescence. Sensors Actuators B - Chem 13: 288–292.

27. Rogers KR, Poziomek EJ (1996) Fiber optic sensors for environmental monitoring. Chemosphere 33:1151–1174.

28. Chen L, Wang P, Liu JB, Liu BB, Zhang Y, et al. (2011) In situ monitoring the photolysis of fluoranthene adsorbed on mangrove leaves using fiber-optic fluorimetry. J Fluoresc 21: 765–773.

29. Wang P, Wu TH, Zhang Y (2012) Monitoring and visualizing of PAHs into mangrove plant by two-photon laser confocal scanning microscopy. Mar Poll Bull 64: 1654–1658.

30. Wang P, Wu TH, Zhang Y (2014) In situ investigation the photodegradation of the PAHs adsorbed on mangrove leaf surfaces by synchronous solid surface fluorimetry. Plos One 9: e84296.

31. Wang P, Wu TH, Wang XD, Zhang Y (2012) Novel method for in situ investigation of PAH adsorption onto mangrove leaves. J Coast Res 28: 499–504.

32. Wang P, Du KZ, Zhu YX, Zhang Y (2008) A novel analytical approach for investigation of anthracene adsorption onto mangrove leaves. Talanta 76:1177–1182.

33. Xiao X (2009) Photodegradation of Dissolved Anthracene and Pyrene in a Mixture and the Effect of Surfactant on the Photodegradation of Dissolved Anthracene or Pyrene [D]. Xiamen: Xiamen University, Ph.D. thesis, 18p.

34. Feilberg A, Nielsen T (2001) Photodegradation of nitro-PAHs in viscous organic media used as models of organic aerosols. Environ Sci Technol 35: 108–113.

35. Mallakin A, Dixon DG, Bruce MG (2000) Pathway of anthracene modification under simulated solar radiation. Chemosphere 40: 1435–1441.

36. Xu Z (2004) Investigation on the Chemical and Photochemical Transformation of the Some Organic Compounds in the Atmosphere. Dalian, China: Dalian University of Technology, Ph.D. thesis, 53p.

37. Schuler F, Schmid P, Schlatter CH (1997) Photodegradation of polychlorinated dibenzo-P- dioxins and dibenzofurans in cuticular waxes of laurel cherry (Prunus laurocerasus). Chemosphere 36: 21–34.

38. Ehrhardt M, Weber RR (1991) Formation of low molecular weight carbonyl compounds by sensitized photochemical decomposition of aliphatic hydrocarbons in seawater. Fresenius J Anal Chem 339: 772–776.

39. Cai ZQ, Lu ZB, Zhu YX, Hong LY, Zhang Y (2006) Simultaneous determination of dissolved anthracene and pyrene in an aqueous solution by synchronous fluorimetry. J Clin Lab Anal 25: 61–64.

40. Low GKC, Batley GB, Brockbank CI (1987) Solvent-induced photodegradation as a source of error in the analysis of polycyclic aromatic hydrocarbons. J Chromatogr A 392: 199–210.

41. Sigman ME, Chevis EA, Brown A, Barbas JT, Dabestani R, et al. (1996) Enhanced photoreactivity of acenaphthylene in water: a product and mechanism study. J Photoch Photobia A 93: 149–155.

10

Systems and Evolutionary Characterization of MicroRNAs and Their Underlying Regulatory Networks in Soybean Cotyledons

Wolfgang Goettel[1,9], Zongrang Liu[2], Jing Xia[3], Weixiong Zhang[3], Patrick X. Zhao[4], Yong-Qiang (Charles) An[1,9]*

1 United States Department of Agriculture, Agricultural Research Service, Plant Genetics Research Unit, Donald Danforth Plant Science Center, Saint Louis, Missouri, United States of America, 2 United States Department of Agriculture, Agricultural Research Service, Appalachian Fruit Research Station, Kearneysville, West Virginia, United States of America, 3 Department of Computer Science and Engineering, Washington University, Saint Louis, Missouri, United States of America, 4 Plant Biology Division, The Samuel Roberts Noble Foundation, Ardmore, Oklahoma, United States of America

Abstract

MicroRNAs (miRNAs) are an emerging class of small RNAs regulating a wide range of biological processes. Soybean cotyledons evolved as sink tissues to synthesize and store seed reserves which directly affect soybean seed yield and quality. However, little is known about miRNAs and their regulatory networks in soybean cotyledons. We sequenced 292 million small RNA reads expressed in soybean cotyledons, and discovered 130 novel miRNA genes and 72 novel miRNA families. The cotyledon miRNAs arose at various stages of land plant evolution. Evolutionary analysis of the miRNA genes in duplicated genome segments from the recent *Glycine* whole genome duplication revealed that the majority of novel soybean cotyledon miRNAs were young, and likely arose after the duplication event 13 million years ago. We revealed the evolutionary pathway of a soybean cotyledon miRNA family (soy-miR15/49) that evolved from a neutral invertase gene through an inverted duplication and a series of DNA amplification and deletion events. A total of 304 miRNA genes were expressed in soybean cotyledons. The miRNAs were predicted to target 1910 genes, and form complex miRNA networks regulating a wide range of biological pathways in cotyledons. The comprehensive characterization of the miRNAs and their underlying regulatory networks at gene, pathway and system levels provides a foundation for further studies of miRNAs in cotyledons.

Editor: Leonardo Mariño-Ramírez, National Institutes of Health, United States of America

Funding: This research is supported by funds from USDA-ARS and the United Soybean Board to Yong-qiang Charles An. The funders had no role in study design, data collection and analysis, decision to publish, or preparation of the manuscript.

Competing Interests: The authors have declared that no competing interests exist.

* E-mail: yong-qiang.an@ars.usda.gov

9 These authors contributed equally to this work.

Introduction

MicroRNAs (miRNAs) are a class of ~21 nt long non-coding RNAs that repress expression of their targeted genes in eukaryotes, predominantly at the post-transcriptional level. Canonical miRNAs are produced from their own miRNA genes (*MIR*), which are mostly transcribed by RNA polymerase II (Pol II) [1,2]. The resulting single-stranded primary miRNAs (pri-miRNAs) then fold into stem-loop structures that are recognized by the RNase III type enzymes Dicer-like (DCL). DCL proteins (mostly DCL1) typically cleave first at the base of the stems generating the miRNA precursor hairpins (pre-miRNAs) and then at the loop regions of the precursors to liberate the miRNA/miRNA* duplexes from the hairpins [3–5]. The 3′ overhangs of the duplex that result from the staggered dicing activity are methylated by Hua Enhancer 1 (HEN1) to protect the duplex from degradation. The mature miRNAs are incorporated into Argonaute (AGO) family proteins, mostly AGO1 and the RISC effector complex, that target mRNAs for slicing through perfect or partially complementary base pairing [6]. miRNA* are generally considered as by-product or non-functional [3–5], but a few miRNA* are functional and can interact with AGO proteins to exert their function [7]. In plants, most miRNAs perfectly or nearly perfectly match coding regions or 5′ and 3′ UTRs of their mRNA targets resulting in transcript cleavage [8–11]. miRNAs may also imperfectly bind to 3′ UTRs of target genes to cause translational inhibition [11,12]. miRNAs, especially 22nt long miRNA species, are also able to trigger phased RNA (PhasiRNA)/trans-acting siRNA (tasiRNA) production in both non-coding (e.g. TAS genes) and coding gene loci [13,14].

Plant miRNAs are known to regulate diverse biological processes, including development, organ identity, metabolism, and stress response [15–18]. Although plant miRNAs regulate genes with diverse biological functions, they preferentially target transcription factor genes, suggesting that miRNAs are at key positions in gene regulatory networks underlying those biological processes. Some miRNAs and their targets are evolutionarily conserved across many orders of divergent plant species while others are young and species-specific. Conserved miRNAs tend to have large families with well-defined targets and accumulate at high levels. In contrast, non-conserved and young miRNAs tend to have small families with less-defined targets and accumulate at low levels [19–22]. Young miRNAs are often excised less precisely, and their length deviates more from the typical 21nt size compared to ancient and conserved miRNAs [20]. Accumulation of miRNAs is often differentially regulated with respect to specific tissues, developmental stages and environmental signals, and may differ by several orders of magnitude [23–26]. It often requires high-depth sequencing of small RNAs to identify young miRNAs in any given tissue.

Soybean is an important dual-purpose crop, which provides both seed protein and oil for animal feed and human consumption. Soybean cotyledons evolved as specialized sink tissue for the synthesis and storage of protein and oil. Soybean is a paleopolyploid with a genome of approximately 1.1 gigabases [27]. Its genome went through two rounds of whole genome duplications occurring at approximately 59 and 13 million years ago, which resulted in a highly duplicated genome with nearly 75% of its genes present in multiple copies [27]. The highly duplicated genome and the dynamic genome rearrangements provide opportunities for the evolution of novel miRNAs.

Computational prediction and sequencing of small RNAs have been used to discover miRNAs in a number of soybean tissues [28–34]. To gain insight into both evolution of miRNAs and their underlying regulatory networks in soybean cotyledons, we carried out an extensive and comprehensive characterization of miRNAs expressed in soybean cotyledons. Eighteen small RNA libraries from cotyledons at various stages were sequenced to generate a total of 292 million small RNAs, which represents one of the largest collections of small RNA reads in a single tissue type in plants. The high-depth sequencing project led us to discover 130 cotyledon miRNAs and 72 cotyledon miRNA families, which have not been reported previously. The majority of soybean cotyledon miRNAs arose after the *Glycine* whole genome duplication event 13 million years ago. We showed that a novel cotyledon miRNA family recently arose through an inverted duplication of a protein-coding gene fragment. Further, we depicted its evolutionary pathway and provided strong evidence for the hypothesis that miRNAs can originate from inverted duplications of their target genes. Cotyledon miRNAs regulate a large number of genes and biological pathways with diverse biological functions, suggesting that they play central roles in gene regulatory networks in cotyledons. The regulatory networks underlying interactions of cotyledon miRNAs with their targeted genes and biological pathways were inferred to reveal their regulatory functions in soybean cotyledons.

Results and Discussion

Complex Composition of small RNAs in Soybean Cotyledons

A total of 18 small RNA libraries constructed with 17–30nt RNAs from cotyledons at various stages were sequenced. A total of 292 million raw sequencing reads were generated. Reads without 3′ sequencing adaptors, with ambiguous sequences or shorter than 17nt, which accounted for 13% of the raw reads, were removed from further analysis. A total of 253 million reads accounting for 87% of the total raw sequence reads were considered as qualified reads, and used for miRNA prediction and characterization. To our knowledge, these small RNA reads represent one of the largest collections of small RNA reads generated from a single tissue type in plants to date. The high-depth sequencing of small RNAs in cotyledons greatly increases the chances to identify rare and low abundant miRNAs. Eighty-five percent of the qualified small RNA reads could be mapped to the soybean genome and transcript sequences with no mismatches while the remainders were excluded from this analysis. In total, 32, 23, and 26% of the

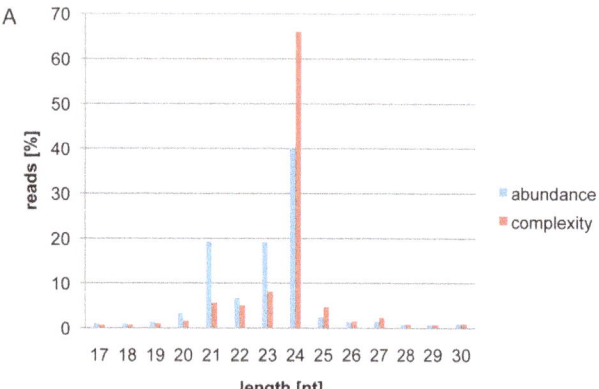

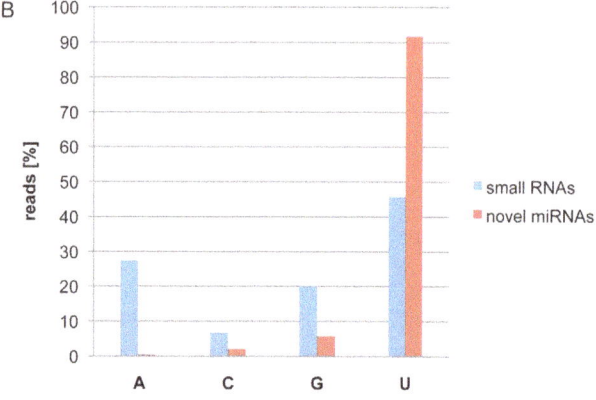

Figure 1. Characterization of Small RNAs in Soybean Cotyledons. (A) Abundance and complexity distribution of small RNAs at each size in soybean cotyledons: The percentages of 253 million qualified small RNA reads at each given size (abundance, in blue) and the percentages of 41 million unique small RNA species at each given size (complexity, in red) are indicated on the Y-axis. Sizes of the small RNA reads are displayed on the X-axis. (B) Highly biased presence of U at 5′ ends of newly discovered cotyledon miRNAs: The percentages of 253 million qualified reads for each nucleotide at their 5′ end (blue column) and the percentages of newly discovered miRNA reads for each nucleotide at their 5′ end in cotyledons (red column) are shown on the Y-axis. The X-axis denotes nucleotides at the 5′ end of small RNAs or newly discovered miRNAs.

Table 1. Summary of Cotyledon miRNA Genes and Families.

	Conserved	Non-conserved	Total
No. of cotyledon miRNA genes	178	126	304
No. of cotyledon miRNA families	35	113	148
Median miRNA family size	4	1	
Median miRNA expression	2518	246	
Median precision rate (%)	87.39	63.49	
No. of known miRNA genes	140	34	174
No. of novel miRNA genes	38	92	130
No. of known miRNA families	35	41	76
No. of novel miRNA families	0	72	72

small RNAs could be perfectly mapped to protein-coding transcript, rRNA and repeated sequences, respectively. All miRNAs identified previously and in this work accounted for 24% of the small RNA population. Thus, miRNAs only contributed a small portion to small RNAs accumulated in soybean cotyledons.

The sequencing data revealed that 21, 22, 23 and 24nt RNAs were the predominant small RNAs in cotyledons, accounting for 85% of small RNA reads. The 24nt RNAs were the most abundant small RNAs, and represented 40% of all small RNA reads. A total of 41 million unique small RNA species were identified in the 253 million qualified reads, indicative of an extreme complexity in the small RNA population of cotyledons. Twenty-four nucleotide long RNAs had the highest complexity, which made up 66% of the small RNA complexity (Figure 1A), but had the lowest accumulation levels in cotyledons with 2 counts per small RNA on average (Figure S1). They most likely represent heterochromatic siRNAs [35]. In contrast, 21nt long RNAs had the highest accumulation level per small RNA species at 13 counts per small RNA on average (Figure S1), but only accounted for 6%

of small RNA complexity. The extreme complexity of the small RNA population with a large percentage of low-abundant small RNAs in cotyledons suggests that high-depth sequencing and systems characterization in a plant tissue are required to gain comprehensive insight into complex small RNA networks. Small RNA compositions in diverse plant tissues and species vary greatly. Like in soybean cotyledons, it was observed that 24nt RNAs are the predominant size of small RNAs in soybean roots, nodules, flowers and developing seeds [30,31]. However, Subramanian et al. showed that 20nt long RNAs are the most abundant in soybean nodules [29]. Lelandais-Brière et al. showed that 24nt RNAs are the most abundant in Medicago truncatula nodules while 21nt RNAs are the most abundant in root apex. The set of 21nt RNAs are around threefold more abundant in root tips than in nodules. Interestingly, the expression levels of MtDCL1 and MtDCL3 homologs in these tissues, which are expected to produce essentially 21 and 24nt RNAs, respectively, correlate with their accumulation levels, suggesting that the composition of small RNAs are mediated through differential expression of DCL1 and DCL3 activities [36]. It remains to be determined if higher

Table 2. Ratio of miRNA and miRNA* for *MIRs* with Both miRNA-3P and 5P.

miRNA in cotyledon	miRNA read count	miRNA* in cotyledon	miRNA* read count	Ratio of miRNA/miRNA*
gma-miR160a-5p	30656	gma-miR160a-3p	1688	18.16
gma-miR166a-3p	234260	gma-miR166a-5p	157794	1.48
gma-miR172b-3p	499	gma-miR172b-5p	42	11.88
gma-miR396a-5p	1287	gma-miR396a-3p	93	13.84
gma-miR396b-5p	7621	gma-miR396b-3p	795	9.59
gma-miR390a-5p	4190	gma-miR390a-3p	2330	1.80
gma-miR390b-5p	3476	gma-miR390b-3p	0	
gma-miR1510b-3p	980289	gma-miR1510b-5p	28181	34.79
gma-miR482b-3p	470351	gma-miR482b-5p	18650	25.22
gma-miR4397-5p	126	gma-miR4397-3p	0	
gma-miR394b-5p	1088	gma-miR394b-3p	2	544.00
gma-miR4412-5p	37449	gma-miR4412-3p	23	1,628.22
gma-miR394a-5p	1088	gma-miR394a-3p	1	1,088.00
gma-miR4415a-3p	1625	gma-miR4415a-5p	62	26.21
gma-miR166c-3p	234260	gma-miR166c-5p	157794	1.48
gma-miR171i-3p	457	gma-miR171i-5p	0	
gma-miR5371-5p	24	gma-miR5371-3p	5	4.80
gma-miR171j-3p	2941	gma-miR171j-5p	2026	1.45
gma-miR397b-5p	2618	gma-miR397b-3p	307	8.53
gma-miR408a-3p	1123	gma-miR408a-5p	73	15.38
gma-miR408b-3p	1123	gma-miR408b-5p	33	34.03
gma-miR408c-3p	1123	gma-miR408c-5p	73	15.38
gma-miR159e-3p	16954526	gma-miR159e-5p	10467	1,619.81
gma-miR166i-3p	234260	gma-miR166i-5p	84	2,788.81
gma-miR169j-5p	163	gma-miR169j-3p	91	1.79
gma-miR169l-5p	53	gma-miR169l-3p	1	53.00
gma-miR171k-5p	346	gma-miR171k-3p	93	3.72
gma-miR172h-3p	499	gma-miR172h-5p	35	14.26
gma-miR172i-5p	35	gma-miR172i-3p	17	2.06
gma-miR396i-3p	2369	gma-miR396i-5p	1287	1.84
gma-miR482d-3p	470351	gma-miR482d-5p	18650	25.22

accumulation of 24nt small RNAs were attributed by a higher DCL3 activity in cotyledon tissue.

A Small Portion of Known miRNA Genes Were Expressed in Cotyledons

To identify the known miRNAs that accumulate in cotyledons, we compared qualified small RNA reads with the mature soybean miRNA sequences in miRBase release 20. We revealed that 174 known soybean miRNA genes representing 60 soybean miRNA families were expressed in cotyledons (Table 1 and Table S1 in File S1). At least 10 reads of a known miRNA had to be detected from our collection of small RNAs to increase the confidence of the miRNA assignment. A total of 530 individual miRNA genes and 219 miRNA families have been identified in soybean (based on miRBase release 20) [34]. Thus, only 33% of known soybean miRNAs and 27% of known soybean miRNA families were expressed in soybean cotyledons. It may reflect that cotyledons evolved into a distinct biological network system highly specialized to synthesize and deposit seed storage reserves. One hundred forty of the 174 known miRNA genes were conserved in other species and belonged to 31 miRNA families (Table 1 and Table S1 in File S1). The remaining 34 non-conserved soybean-specific miRNA genes belonged to 30 distinct families.

There is a significant number of miRNA genes that produce equivalent accumulation levels of miRNA and miRNA* and are named as miRNA-3P and -5P in miRBase release 20 [34]. Thirty-one of those miRNA genes were expressed in cotyledon tissue. Table 2 lists read counts for each pair of miRNA sequences in cotyledons. In our data set, only 6 of the 31 miRNA genes produced miRNA-3P and -5P reads with less than 2 fold differences in their accumulation levels. Twenty-two miRNA genes produced miRNA-3P and -5P reads with 2.1 to 2788 fold differences at their accumulation. The remaining 3 miRNA genes, *GMA-MIR1711*, *390B* and *4397* had no detectable read counts for one of the two forms in our large collection of small RNA reads. Thus, most of the miRNA-3P and -5P accumulated in cotyledon tissue at highly biased levels. We assigned more abundant small RNA sequence reads as mature miRNAs. For example, there were 1088 small RNA read counts for gma-miR394a-5P, while we only detected 1 read for gma-miR394a-3P from our collection of small RNA reads. Consequently, the miRNA-5P likely represents the mature miRNA while the miRNA-3P is a minor miRNA or a miRNA* in soybean cotyledons. Mature miRNAs are normally more abundant than miRNAs* and are incorporated into the RISC effector complex that targets mRNAs for slicing through perfect or partially complementary base pairing [3–6]. However, there are also cases where miRNA* strands reside in AGO1 complexes and possess gene-regulatory activity [37]. Therefore, miRNA* strands have fates other than default degradation. It has also been demonstrated that miRNA* can be preferentially sorted into AGO2 complexes in Drosophila [7].

A Large Number of Novel miRNA Genes Were Expressed in Cotyledons

Stringent criteria were applied to identify novel miRNA genes expressed in cotyledons, principally following the standards proposed by Meyers *et al.* [38]. We determined the miRNA processing precision rates for each predicted hairpin structure and removed miRNAs with a low precision rate. Each newly identified miRNA had to be detected in multiple small RNA libraries with a minimum of 10 reads in total, which was the identical cut-off used in identifying known miRNAs. We identified a total of 130 novel miRNA genes (Table 1 and Table S2 in File S1). The newly

identified miRNAs have an obvious bias for uracil (U) at their 5′ ends. Ninety-two percent of the novel cotyledon miRNAs had a 5′ terminal uracil, while only 46% of all qualified small RNA reads had a uracil at their 5′ ends (Figure 1B). It has been shown in Arabidopsis that AGO1 proteins have a higher affinity to miRNAs containing a uracil in their first position, which could result in the enrichment of mature miRNA sequences with uracil at their 5′ termini [39,40]. A highly biased presence of 5′ terminal uracils in miRNAs was also reported in other plant species such as maize and Medicago [36,41], supporting the high effectiveness in our identification of bona fide soybean miRNAs.

Thirty-eight novel soybean miRNA genes belong to 20 previously identified conserved miRNA families. In addition, 13 novel miRNAs belonged to non-conserved miRNA families only identified previously in soybean. Seven miRNAs, soy-*MIR15A-F* genes and soy-*MIR49* genes, were grouped as a new family, soy-miRNA15/49 (see below). Two miRNAs, soy-miR245a and b were grouped in a new family soy-miR245. The remaining 70 *MIR* loci had mature miRNA sequences distinct from one another and from known mature miRNA sequences. Each of these miRNAs should represent a novel miRNA family with a recent origin. Thus, 130 novel soybean miRNA genes and 72 novel miRNA families were discovered and added to the current collection of soybean miRNA genes (Table 1). Together with previously identified miRNA genes, a total of 304 miRNA genes were expressed in cotyledons, of which 178 were conserved and 126 non-conserved. They belong to 35 conserved and 113 non-conserved miRNA families. The conserved cotyledon miRNAs had larger family sizes, higher accumulation levels and higher precision rates than non-conserved cotyledon miRNAs. Interestingly, 140 out of the 174 previously identified cotyledon miRNAs were conserved while only 38 out of 130 newly identified cotyledon miRNAs were conserved miRNAs. Thus, the majority of the newly identified miRNAs arose after the soybean speciation. Some of the non-conserved miRNAs are likely to have evolved soybean-specific functions. It is intriguing to understand their functions in soybean.

Conservation of Soybean Cotyledon miRNAs across Land Plants

A total of 35 conserved miRNA families were expressed in soybean cotyledons. To estimate their conservation across the plant kingdom, we determined their presence in 13 orders of plant species that diverged from the soybean lineage between less than 1 million years ago and more than 1 billion years ago (Figure 2). None of the miRNA families were found in the unicellular green alga *Chlamydomonas reinhardtii*, which was consistent with previous reports (Molnar et al., 2007; Zhao et al., 2007; Cuperus et al 2011). miR156, 160, 166, 171, 319, 390, and 408 represented the most ancient miRNA families. They were present in almost all land plant orders (embryophyta) including funariales, which experienced more than 400 million years of divergent evolution (Figure 2). Seven miRNA families (miR159, miR162, miR164, miR169, miR395, miR396, and miR397) most likely evolved in the common ancestor of spermatophytes, as they were present in both gymnosperm and angiosperm lineages. Seven miRNA families (miR167, miR168, 172, 393, 394, 398 and 399) were angiosperm-specific and present in both eudicots and monocots. miR828, miR403, miR2111, and miR482 were identified in core eudicots, and miR482 is likely restricted to the rosids. Ten miRNA families (miR530, 1507, 1508, 1509, 1510, 1514, 2118, 2119, 3522, and 4414) were only observed in species belonging to the fabales order such as wild soybean, common bean and Medicago, suggesting that these are the youngest conserved miRNA families

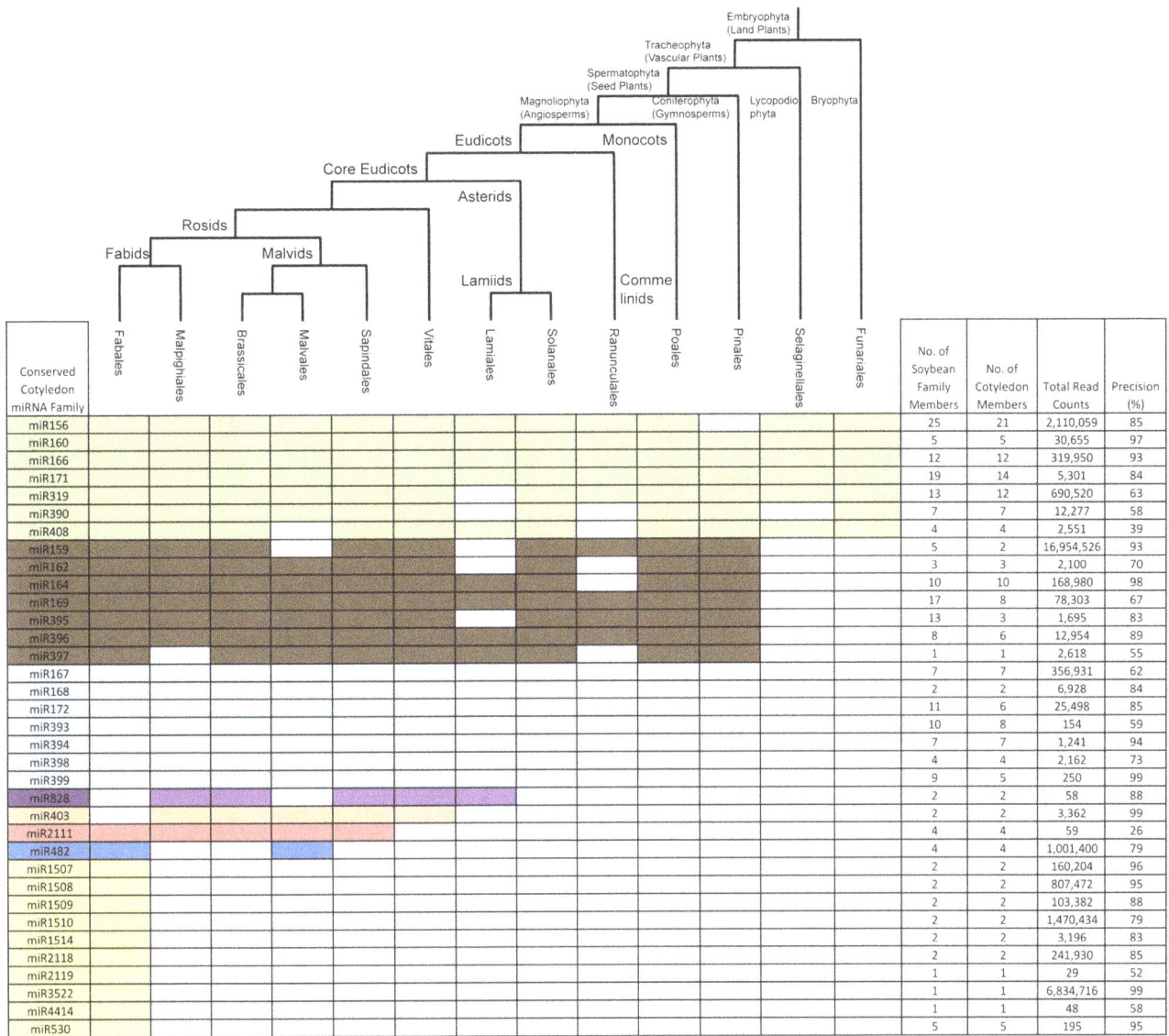

Figure 2. Conserved Soybean Cotyledon miRNAs in Divergent Orders of Land Plants. The conservation of soybean cotyledon miRNA families across plant species belonging to 12 orders is presented. The presence of a miRNA family that is conserved with a soybean cotyledon miRNA in at least one plant species in a given order is shown by differently colored cells. Green cells indicate conservation in land plants, brown cells in seed plants, light blue cells in angiosperms, purple cells in core eudicots, tan cells in vitales, red cells in sapindales, blue cells in rosids and yellow cells in fabales. The evolutionary relationship among the 13 orders is illustrated by the phylogenetic tree placed on top of the miRNA table. The number of family members in the soybean genome and the expressed family members in cotyledons, the count of miRNA reads in cotyledons, and the processing precision are shown for each conserved miRNA family.

in soybean. miR403, and miR828 had not been identified in any other species in fabales, but were present in plant species belonging to other eudicot orders. Those miRNAs likely had an ancient origin, but were lost or had not been identified yet in other species in fabales. Considering that miRNAs function as one of key regulatory components in gene regulatory networks, the angiosperm and fabales-specific miRNAs could play important roles in evolving their distinct biological characteristics in cotyledons.

The size of miRNA families in soybean tended to correlate with their evolutionary age (Figure 2). The most ancient miRNA families that evolved in the common ancestors of land plants had a median of 12 miRNA genes per family while the fabales-specific miRNA families only contained a median of 2 miRNA members per family, suggesting that the size of miRNA families expanded

over their long evolutionary history. Such an association had been previously observed in Arabidopsis [42,43]. Although most miRNA genes in each conserved miRNA family were expressed in soybean cotyledons, some of the miRNA genes did not have miRNA reads accumulating in cotyledon tissue. For example, we detected reads for only 2 out of 5 miR159 loci in soybean cotyledons (Table S1 in File S1 and Table S2 in File S1). Although conserved miRNAs had higher accumulation levels than non-conserved miRNAs in soybean cotyledons as shown earlier (Table 1), there was no strong correlation between evolutionary age and accumulation of miRNAs for the conserved miRNA families in cotyledons. Fabales-specific miRNA families had a median accumulation of 131,793 reads per miRNA family while the miRNA families conserved in all orders of embryophyta

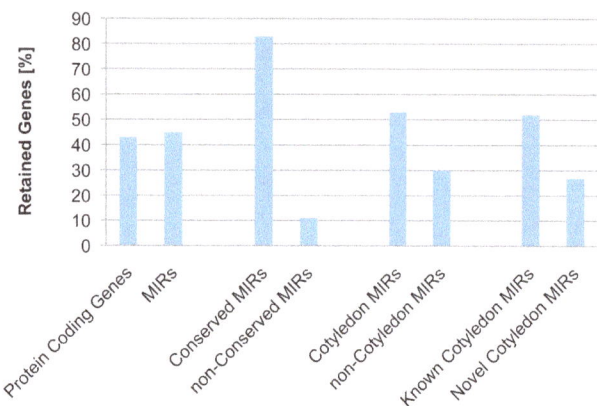

Figure 3. Retention of miRNAs on Recently Duplicated Soybean Genome Segments. The percentage of miRNA genes located on duplicated genome segments that are retained in their paired homologous genome segments from the *Glycine* whole genome duplication is shown on the Y-axis. Each category of miRNAs is indicated on the X-axis.

had a median accumulation of 30,656 reads per miRNA family (Figure 2). miRNA processing has been shown to be more precise in ancient miRNAs than recently arisen miRNA families [43,44]. We did not find such an obvious correlation between evolutionary age and miRNA processing precision in conserved soybean cotyledon miRNA families. However, as previously reported, more miRNAs from ancient families were 21nt long compared to young, fabales-specific miRNA families [33,43,45].

Characterization of miRNA Loci in the Recently Duplicated Homoeologous Genome Segments

Soybean is an allotetraploid species. The most recent whole genome duplication occurred about 13 million years ago. One hundred fifty pairs of homoeologous genome segments have been retained from the whole genome duplication event. A total of 419 miRNA loci were located on 172 homoeologous segments belonging to 104 pairs (Table S3 in File S1). Out of the these 419 miRNA loci, 190 miRNA loci (45%) were retained on both duplicated segments while 229 miRNA loci (55%) lost a corresponding miRNA on a homoeologous segment or were newly evolved after the whole genome duplication (Table S4 in File S1). It was observed that 43.4% of protein-coding genes were

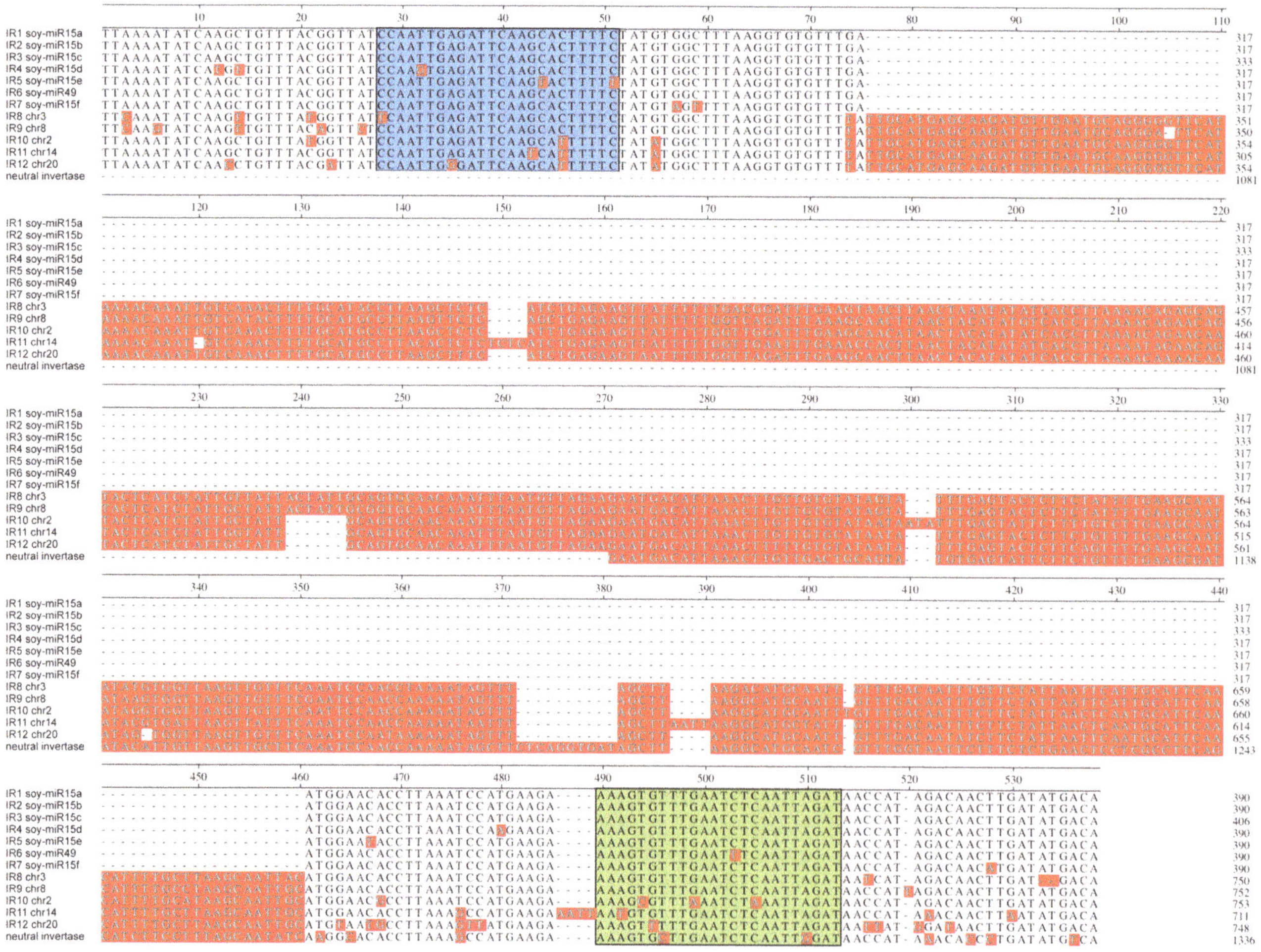

Figure 4. Sequence Alignment of Inverted Repeats and a Neutral Invertase Gene. The central regions of the inverted repeat sequences 1 to 8, their homologous IR sequences from chromosomes 8, 2, 14 and 20 and a fragment of a neutral invertase gene (Glyma12g02690) were aligned. The miRNA and miRNA star sequences are highlighted in green and blue, respectively. Nucleotides that differ from the consensus sequence are shaded in red.

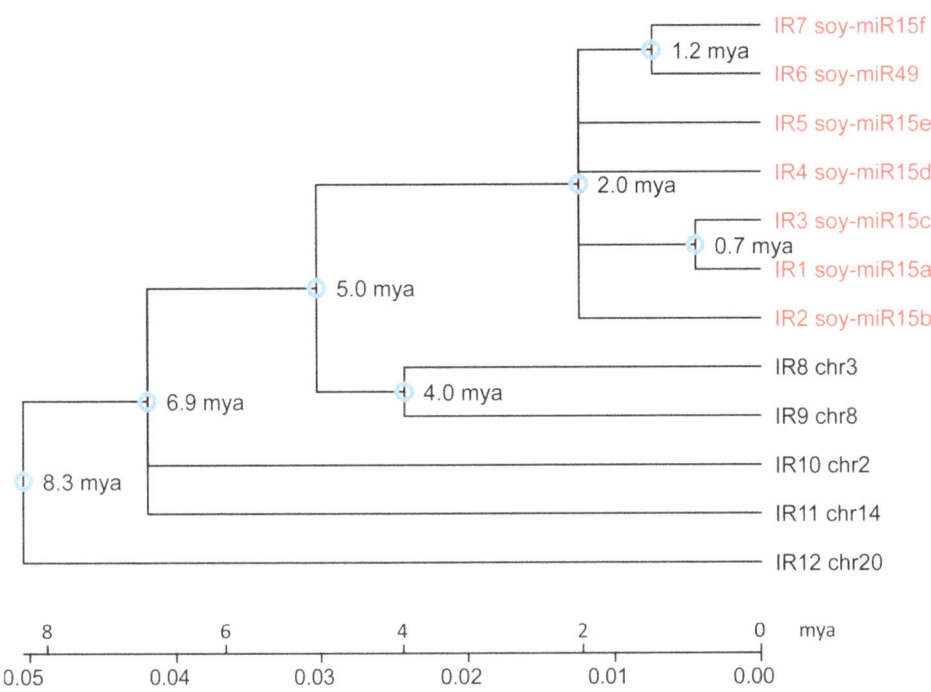

Figure 5. Phylogenetic Relationship of the Inverted Repeat Sequences. The total length of the inverted repeat (IR) sequences 1 to 8 and their homologous IR sequences located on chromosome 2, 8, 14 and 20 were used to construct the phylogenetic tree. IR sequences 1–7 potentially give rise to miRNAs in cotyledons. The time of divergence among the inverted repeats is shown at the tree nodes.

retained in duplicated segments [27]. Thus, miRNA genes had a rate of loss/gain similar to that of protein-coding genes (Figure 3).

We further determined if conserved miRNAs and non-conserved miRNAs were biasedly retained in duplicated genome segments. One hundred ninety-eight out of the 417 miRNA loci located on duplicated segments were conserved miRNAs. Approximately 83% of the conserved miRNA loci were retained in both paired segments. In contrast, a total of 221 non-conserved miRNA loci were mapped to the duplicated segments, but only 11% of them were retained in both paired segments (Figure 3). Thus, a larger number of non-conserved miRNA loci were either gained or lost after the whole genome duplications. This is consistent with the observation that miRNA loci undergo a frequent birth and death [46]. In addition, 53% of cotyledon miRNAs and 30% of non-cotyledon miRNAs from duplicated segments were retained, which suggests there is a bias in gaining/losing cotyledon miRNA loci in the homoeologous segments. Interestingly, only 27% of novel cotyledon miRNA loci and 52% of known cotyledon miRNA loci from the duplicated segments were retained in both paired segments (Figure 3). Therefore, the newly identified novel cotyledon miRNA loci were preferentially lost or gained after the whole genome duplication. Considering that the non-conserved miRNAs had no orthologs in its closest species such as *Phaseolus vulgaris*, which shared the most recent ancestor with soybean 19 million years ago [47], and majority of those novel miRNAs were non-conserved, it is likely that most of the non-conserved novel miRNAs were gained through a recent birth after the whole genome duplication.

Recent Birth of the Soy-miR15/49 Family from an Inverted Duplication of a Neutral Invertase Gene

We discovered a novel soybean cotyledon miRNA family, soy-miR15/49. soy-miR15 and soy-miR49 mature sequences differed in one nucleotide. All six soy-miR15 genes and the soy-miR49

gene were located on a 70 kb long region on chromosome 3 (Figure S2). Each of these miRNA genes was situated within a long inverted repeat (IR), and shared strong sequence similarities with each other (Figure 4). The seven inverted repeats (IR1-7) were approximately 647 nucleotides long. In addition, we identified a longer inverted repeat (IR8) and six truncated sequences homologous to IR1-7 that were interspersed with IR1-7 on a 110 kb region of chromosome 3 (Figure S2). The highly repetitive nature of this 110 kb region was further supported by the presence of a 12 kb direct repeat containing a solo LTR and a gene encoding a reticulon protein (Figure S2). Thus, the 110 kb sequence region represents a hot spot of DNA duplication and rearrangement. A total of 657 soy-miR15 reads and 2324 soy-miR49 reads were detected in cotyledons (Table S5 in File S1). All soy-miR15/49 genes had high miRNA processing precision rates, which range from 77% to 92%, further supporting the authenticity of soy-miR15 and 49 (Table S5 in File S1 and Table S6 in File S1).

We identified four additional longer inverted repeats (IR9-12) on chromosome 2, 8, 14 and 20 that were nearly identical to IR8 on chromosome 3. All of the longer inverted repeats had strong sequence similarities with IR1-7, but contain a 363 bp insertion sequence at their center (Figure 4). The insertion also formed an inverted repeat sequence, which increased the stem length of their hairpin structures. Our phylogenetic analysis revealed that these IRs arose at least 8.3 million years ago (Figure 5). IR1-7, which lacked the insertion/deletion (indel), clustered in one clade. They probably amplified on chromosome 3 through multiple recombination events over the past two million years. All other IRs, which contained the indel and diverged earlier, likely represent the ancestral IR structure. However, the indel-containing IR8 and IR9 were in the same clade and closely related to the IR1-7. Both clades originated about 5 million years ago. Thus, the indel was likely deleted from the ancestral IR1-7 sequence between 2 to

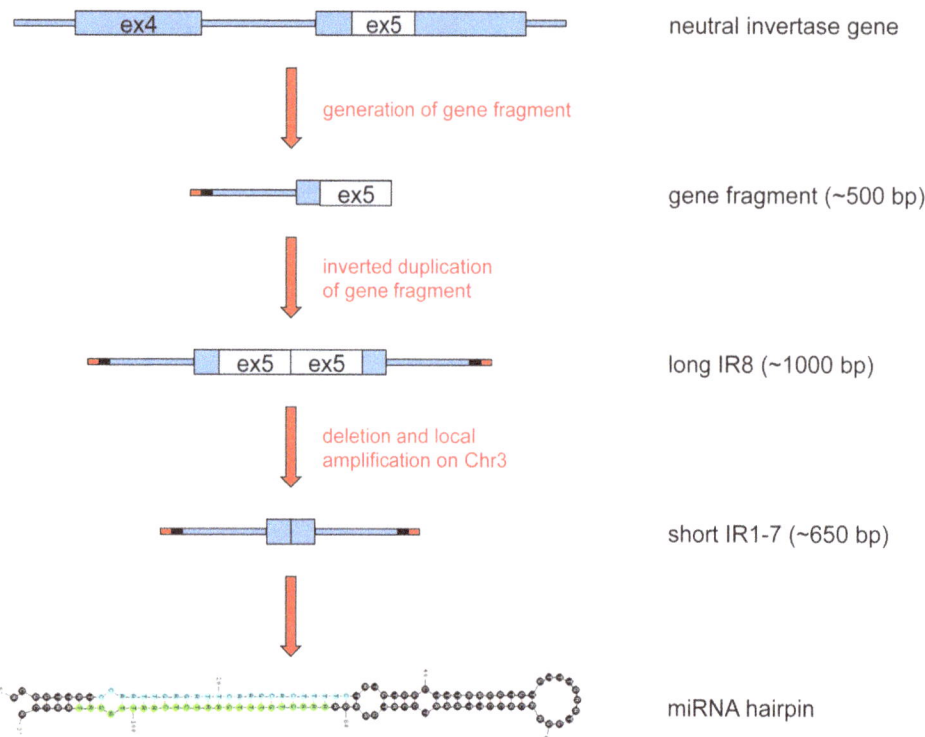

Figure 6. Model Depicting the Evolution of the Soy-miR15/49 Family from an Inverted Duplication of a Neutral Invertase Gene Fragment. A neutral invertase gene fragment containing parts of its intron 4 (blue line) and exon 5 (blue and light blue rectangle) was linked to a short LTR terminal fragment (red line) and a short sequence of unknown origin (black line). This complex DNA arrangement underwent an inverted duplication event, which resulted in a long inverted repeat fragment (IR). Subsequently, this long IR was amplified and transposed to different chromosomes. One copy on chromosome 3 experienced a deletion of a short inverted sequence (light blue rectangles) at its center region, thereby generating a shortened IR structure. This shortened IR went through a series of local amplification events to produce IR1-7, which further evolved to the soy-miR15/49 family. The hairpin structure of *MIR15a* and its miRNA and miRNA* are shown at the bottom.

5 million years ago. Although the inverted repeat (IR1-8) cluster on chromosome 3 was located on one of the duplicated genome segments (ID: 22835160) retained from the *Glycine* whole genome duplication, which occurred 13 million years ago, no homologous IR was observed in the corresponding duplicated genome segment. A gain of the IRs after the whole genome duplication event on just one of the duplicated segments was consistent with the phylogenetic analysis that soy-miR15/49 family arose after the soybean speciation, which occurred 5–10 million years ago [27,48].

Interestingly, the inverted repeat units containing the insertion were highly homologous to part of exon 5 (~240 bp) and intron 4 (~210 bp) of a gene encoding a neutral invertase (Glyma12g02690). Their terminal sequences contained a 38nt sequence of unknown origin and a 25nt sequence homologous to the same LTR family that was found in the 12 kb direct repeats (Figure 6 and Figure S1). This patchwork of sequences was reminiscent of filler DNA that results from the repair of chromosome breaks through non-homologous end-joining. Thus, we propose that the soy-miR15/49 family most likely evolved recently from the neutral invertase gene (Glyma12g02690) through complex DNA rearrangements and amplifications. The DNA fragment was duplicated and joined in a tail-to-tail fashion to form an inverted repeat structure. This inverted repeat was amplified and spread to chromosomes 20, 14, 2 and 8 and 3 over the past 8.3 million years. Subsequently, the ancestral inverted repeat of IR1-7 lost the 360 bp indel sequence between 2 to 5 million years ago and was later amplified through a series of

local DNA duplications until the IR cluster reached its current shape. We did not detect miRNA reads aligning to IR10-12. It is likely that the miR15/49 family arose after the split of the IR1-7 clade from the IR8-9 clade. Deletion of the 360 bp long indel significantly decreased the distance between the miR15/miR15* duplex and the loop of the hairpin structure, and might promote the birth of the soy-miR15/49 family. soy-miR15/49 represented a young and recently evolved miRNA family, and still maintained a strong sequence similarity to the neutral invertase gene. The length of the duplicated soy-miR15 gene sequences extends beyond their miRNA fold-back structure, which has been reported for many Arabidopsis miRNA genes as well [19]. It has also been proposed that *MIR* genes can evolve from inverted duplicates of protein-coding genes or gene segments [19,21,46,49–52]. Initially, these perfect inverted repeats would produce hairpin transcripts that are processed by DCL4 and DCL3 to yield 24nt siRNAs. Over time, drift mutations would disrupt the perfect hairpin structures that would cause the transition from DCL4 to DCL1 cleavage resulting in the production of 21nt miRNAs. Further accumulation of mutations over time in regions surrounding the miRNA/miRNA* sequence could generate miRNA genes that have lost similarities with the parental donor sequence and therefore may appear unrelated to their locus of origin. Our in-depth analysis of the soy-miR15/49 family not only provided strong evidence for the hypothesis that miRNAs can evolve from inverted duplications of protein-coding gene fragments, but also illustrated the evolutionary pathway that gave birth to the miR15/49 family.

Complex Networks Underlying Interaction of Cotyledon MicroRNAs and Their Target Transcripts

Plant miRNAs mainly interact with protein-coding transcripts containing perfect or near-perfect complement sequences, and further cause the degradation of their target transcripts in plant cells. Such miRNA-transcript interactions form a core regulatory network in gene regulatory programs. A TargetFinder algorithm was used to query all known soybean protein-coding transcripts with 148 cotyledon miRNA families to predict genes that are regulated by cotyledon miRNAs [46]. Out of 148 cotyledon miRNA families, 130 miRNA families were predicted to target at least one gene. Approximately 16% of the miRNA-transcript interactions occurred at 5′ or 3′ UTRs (Table S7 in File S1). The 130 cotyledon miRNA families were predicted to target 1910 protein-coding transcripts, which form a complex network system containing 2021 interactions (Figure S3). Cotyledon miRNA families had an average of 16 targets with a range of 1 to 124 targets per miRNA family (Table S8 in File S1). The newly identified and non-conserved soy-miR38 family had the highest number of predicted targets (Table S9 in File S1).

The majority of targets specifically interacted with a single miRNA family. However, 81 genes were targeted by multiple miRNA families, which formed 192 distinct gene-miRNA family interactions. Thirty-seven genes were targeted by 42 cotyledon miRNA families at more than two distinct sites (Table S10 in File S1). Shared target genes connected miRNA modules together to form compound modules (Figure 7 and Figure S3), which increased the complexity of the miRNA-target network system. For example, miR319 shared two targets with miR395 and seven targets with miR159 (Figure 7). All three miRNA families are ancient (Figure 2). miR319 and miR159 shared significant sequence similarity and had been shown to evolve from a common ancestral miRNA gene [53]. Their shared targets were a beta-galactosidase gene, five MYB33 and one MYB65 genes (Table S11 in File S1). It is likely that these interactions predated the divergence of miR159 and miR319 and were conserved through their long evolutionary history. However, thirty-eight genes with a variety of functions were specifically targeted by miR319 while 18 genes were only targeted by miR159. miR159-specific or miR319-specific interactions contributed to their distinct biological functions in soybean cotyledons and are possible outcomes from neo-functionalization or sub-functionalization after the gene duplication event. It has been shown in Arabidopsis that the interaction of miR159 and its target MYB genes is involved in the regulation of vegetative growth, flowering time, seed shape and germination [54–56]. In contrast to miR159, miR319 regulates embryonic patterning, jasmonate synthesis, leaf morphogenesis and senescence in Arabidopsis and tomatoes [57–59]. Expression patterns of miR159 and miR319 are significantly different in Arabidopsis. Accumulation of miR159 is abundant and widespread over the whole plants, while accumulation of miR319 is mainly confined to specific tissues and developmental stages [53,57]. Their diverse temporal and spatial-specific expression also contributes to their functional specialization in plant growth and development.

Networks Underlying Interaction of Cotyledon MicroRNAs and Biological Pathways

We predicted biological pathways and functional groups that are regulated by cotyledon miRNAs. In the analysis, soybean genes targeted by cotyledon miRNAs were assigned into hierarchical functional groups based on their molecular function in metabolic and signaling pathways [60]. A Fisher's Exact Test was used to determine the significance of over- and under-represented miRNA target genes in those functional groups. Cotyledon miRNAs preferentially targeted genes involved in RNA metabolism and the regulation of transcription (Table 3). A large number of transcription factor gene families such as CCAAT box binding factor, TCP, and ARF transcription factor families were preferentially targeted by cotyledon miRNAs. In addition, we also observed that cotyledon miRNAs preferentially targeted a number of other functional groups such as lipid metabolism and major CHO metabolism.

Representation analysis of genes targeted by individual cotyledon miRNA families in functional groups was conducted to predict biological pathways a given family preferentially regulated. A total of 101 miRNA families preferentially targeted at least one biological pathway and combine to form complex miRNA-pathway networks (Table S12 in File S1 and Figure S4). We observed nine cotyledon miRNA families, each preferentially regulating one specific biological pathway. The rest of the cotyledon miRNA families preferentially targeted two or more pathways. Additionally, several biological pathways were preferentially targeted by multiple miRNA families (Table S12 in File S1). The one-to-one, one-to-many, and many-to-one miRNA pathway interactions led to a complex regulatory network system composed of interaction modules with variable size in soybean cotyledons (Figure S4).

Conclusions

Soybean cotyledons serve as specialized sink tissues for the synthesis and storage of seed reserves. They provide nutrients and energy through reserve mobilization for seed germination and seedling growth. We generated and analyzed one of the largest collections of small RNAs with a total of 292 million raw sequencing reads. Soybean cotyledons accumulated at least 41 million unique small RNA species, and therefore had an extremely complex small RNA population. A total of 130 novel miRNAs and 72 novel miRNA families were identified, which significantly increased the collection of currently known soybean miRNAs. We showed that 304 miRNA genes were expressed in soybean cotyledons. Examining the presence of conserved cotyledon miRNAs in 13 divergent orders of land plant species showed that the cotyledon miRNAs arose at various stages over the course of land plant evolution. The size of conserved miRNA families tend to correlate with their evolutionary age in soybean. Ancient miRNAs typically had a larger family size than young miRNAs. Although conserved cotyledon miRNAs had the tendency to be accumulated at higher read levels than non-conserved cotyledon miRNAs, there was no obvious correlation between accumulation level and age of conserved miRNAs.

It has been proposed that miRNA genes could evolve from inverted duplications of their target genes. However, evolutionary mechanisms for such a miRNA origin largely remain unknown. We conducted a detailed structural and phylogenetic analysis of the soy-miR15/49 family and illustrated that the soy-miR15/49 family probably evolved from an inverted duplication of a protein-coding gene fragment through transposition of the inverted repeats to different chromosomes, local tandem amplifications, fragment deletion and nucleotide mutations. Bioinformatic and statistical analysis revealed that cotyledon miRNAs target genes encoding proteins with diverse biological functions. The genes coding for proteins involved in RNA metabolism, regulation of transcription and signaling pathways are preferentially targeted by miRNAs. Thus, miRNAs are likely to regulate a variety of biological pathways in cotyledons and play central roles in their underlying

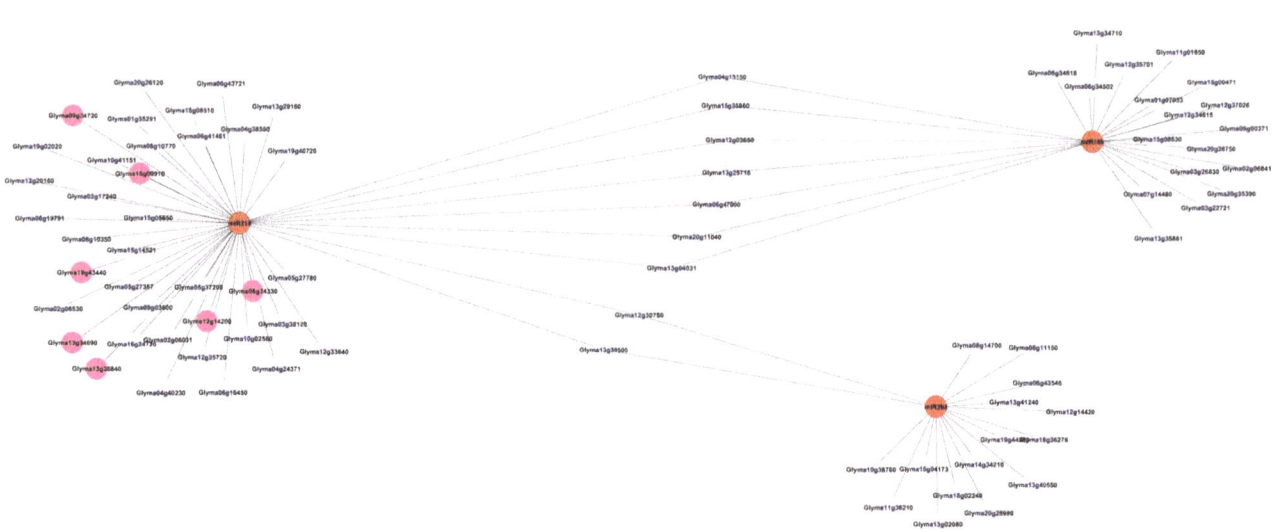

Figure 7. Interactions of miR395, 319 and 159 with Their Target Genes. Interactions of miR395, 319 and 159 (red circles) with their target genes (light blue circles and pink circles) are shown. Two targets are commonly regulated by miR319 and miR395 while seven targets are commonly regulated by miR319 and miR159.

gene regulatory networks. One-to-one, one-to-many, many-to-one and many-to-many interactions of miRNA-target gene/pathways were identified, which could further form a complex regulatory network system composed of interaction modules of variable size in soybean cotyledons. The topology of the miRNA-gene and miRNA-pathway networks provides a comprehensive view into

Table 3. Over-representation of Cotyledon miRNA Targeted Genes in Biological Pathways.

Functional Groups	p-value	odds ratio
RNA	2.04E-02	1.28
RNA.regulation_of_transcription	2.14E-04	1.57
RNA.regulation_of_transcription.General_Transcription	3.00E-04	7.91
RNA.regulation_of_transcription.ARF_Auxin_Response_Factor_family	3.02E-08	10.31
RNA.regulation_of_transcription.CCAAT_box_binding_factor_family_HAP2	2.57E-05	21.94
RNA.regulation_of_transcription.HBHomeobox_transcription_factor_family	1.02E-03	3.58
RNA.regulation_of_transcription.TCP_transcription_factor_family	8.55E-03	8.75
hormone_metabolism.auxin.signal_transduction	6.64E-03	6.18
signalling.receptor_kinases.misc	2.32E-02	2.63
protein.degradation.ubiquitin.E2	2.11E-02	2.98
amino_acid_metabolism.degradation.glutamate_family.proline	1.26E-02	17.46
fermentation.ADH	1.34E-02	7.15
lipid_metabolism	2.38E-02	1.68
lipid_metabolism.FA_synthesis_and_FA_elongation	3.65E-02	2.22
major_CHO_metabolism.degradation.sucrose.invertases	2.31E-02	5.62
major_CHO_metabolism.degradation.sucrose.invertases.neutral	3.55E-03	13.11
minor_CHO_metabolism.galactose.galactokinases	2.52E-02	10.48
misc.oxidases___copper_flavone_etc	1.08E-02	2.64
misc.plastocyanin_like	2.34E-02	4.04
protein.synthesis.misc_ribososomal_protein.BRIX	3.28E-02	8.73
protein.targeting.secretory_pathway.golgi	2.31E-02	5.62
S_assimilation.APS	1.26E-02	17.46
secondary_metabolism.simple_phenols	1.36E-11	20.42
transport.sulphate	2.69E-02	5.25

their complex and intricate interactions at a systems level and lays a foundation to design a variety of wet-bench experiments for functional validation.

Materials and Methods

Plant Materials and Library Construction

Soybean (*Glycine max* (L.) Merrill, cv. Jack) plants were grown in growth rooms with supplemental lighting. Cotyledons were isolated from seeds at six developmental stages (S2, 3, 4, 6, and 8) with three replications. Total RNA was isolated as described by Chen and An [61] with minor modification. 17–30nt long RNA was purified by polyacrylamide gel electrophoresis to construct cDNA libraries. A total of 18 cDNA libraries were generated and sequenced by Expression Analysis, Inc. (Durham, NC) using the SBS (sequencing by synthesis) technology.

Initial processing of sequencing libraries

Raw sequence reads with no 3′ sequencing adaptor, of low quality, or shorter than 17nt were removed. The adaptor trimming was done by an in-house method that recursively searches for the longest substring of the adaptor appearing within a sequence read. If a raw sequence read did not have a substring of the adaptor longer than 6nt, it was considered as having no adaptor. The adaptor-trimmed sequences with no ambiguous reads, referred to as qualified reads, were then mapped, allowing for 0 mismatches, to the soybean genome using Bowtie [62].

Identification of novel miRNAs

Novel miRNAs were identified by following the previously described method with minor revisions [63] [64]. Briefly, qualified reads in all libraries that mapped to known soybean miRNAs in miRBase release 20, rRNA, and tRNA were excluded from the analysis. We then mapped the remaining reads to the soybean genome using Bowtie [27,62,65] and merged neighboring loci if they shared overlapping reads. Subsequently, we examined the folding structures of the (merged) loci. Since the average length of a miRNA precursor in plants is often ~200nt, we took 250nt as the length of putative pre-miRNAs in our analysis. At each genomic locus to be analyzed, a series of DNA sequence segments covering the sequence reads were extracted for secondary structure analysis. The starting sequence segment extended 220nt upstream of the sequence reads, and subsequent segments were extracted by a sliding window of 250nt, with an increment of 10nt, until the window reached 220nt downstream of the sequence reads. Each of these individual segments was folded by the RNA-fold program [66]. A segment lacking a stem of at least 18nt or a segment lacking sequencing reads that map to any of its stems was excluded. A representative segment was chosen from those that have the same or a similar folding structure. The top five folding structures with a free energy no greater than -18 kCal/mol were further visually inspected. We retained those segments that formed a typical hairpin fold-back structure with up to 5 mismatches on one stem region. Finally, a program, which encoded the four miRNA-annotation rules described below, was adopted to scan each segment. Candidate miRNAs were chosen based on the criteria: 1) occurrence of miRNA reads on the arms of predicted hairpin structures; 2) presence of no less than 10 miRNA reads of the highest frequency on predicted hairpins; 3) presence of possible miRNA* sequencing reads unless specifically stated (see below) 4) presence of 2–3nt 3′ overhangs on miRNA/miRNA* duplexes [38]. After miRNA were derived computationally, they were subjected to a manual review. All small RNA reads were aligned to their corresponding hairpin sequence and visually inspected. The

read with the highest read count, cumulatively from all libraries, was preferentially selected as the mature miRNA sequence. The count of a miRNA or a miRNA* was calculated solely based on the reads aligning to its mature miRNA sequence or miRNA* sequence. The iso-miRNA reads were not counted. The secondary structure for each hairpin was also visualized to verify the star sequence. Any hairpins with more than 5 mismatches in their miRNA/miRNA* duplexes were removed from the set. Neither the mature nor the star sequence could be located in the loop. The hairpin loci were compared to one another to see which were overlapping. The hairpin with the highest expression and/or greatest precision was selected in the cases where overlaps were noted and the others were removed from the set. A low stringency cut off of 10% processing precision was applied to all miRNA. miRNAs with no star sequences were removed if they showed less than 45% precision. Processing precision was calculated as ((mature read count) + (star read count)/total count of reads aligned to hairpin) * 100.

Target Gene Prediction and Validation

The TargetFinder program release 1.6 (http://jcclab.science. oregonstate.edu/node/view/56334) was used to query all known and novel soybean miRNA sequences against all annotated soybean transcript sequences from Gmax_109_transcriptome.fa in Phytozome v9.1 at the cutoff alignment score of 4. Soybean genes were binned in MapMan functional groups by converting the respective gene ID to Affymetrix probeset IDs. The total number of binned target genes and soybean genes in each of the functional groups were input to a Fisher's exact test using an R script to determine the p-values for over or under-representation of target genes. Custom Perl scripts were used to parse the odds ratios and p-values.

Mapping and displaying miRNA loci and targeted genes

pre-miRNA coordinates for all known and novel soybean miRNA loci were determined by BLAST searches against the genomic soybean reference sequence in Phytozome v9.1. The pre-miRNA genomic sequence coordinates were written in a gff file format and then loaded to and viewed in the SoyBase Genome Browser (http://www.soybase.org/gb2/gbrowse/gmax1.01/). A custom Perl script was used to identify pre-miRNA sequences on regions of recent segmental duplication defined in ftp://ftp.jgi-psf. org/pub/JGI_data/phytozome/v6.0/Gmax/related_data/ recentDuplSegments.gff3.

Phylogenetic analysis

Sequences were aligned using ClustalW in MEGA5 [67], and the resulting alignment was manually adjusted. The evolutionary history was inferred using the Neighbor-Joining method [68]. The optimal tree with the sum of branch length = 0.30150857 is shown. The evolutionary distances were computed using the Kimura 2-parameter method [69] and are in the units of the number of base substitutions per site. Evolutionary analyses were conducted in MEGA5 [67]). The divergence time points of the inverted repeats were calculated in MEGA5 [67] using the substitution rate R of 6.1×10^{-9} mutations per site per year [70].

Supporting Information

Figure S1 Average Accumulation Distribution of Small RNAs at Each Size in Soybean Cotyledons. The average count of reads per unique small RNA at each given size in cotyledon tissues is indicated on the Y-axis. The X-axis indicates the size of small RNAs.

Figure S2 Soy-miR15/49 Gene Cluster Region. A 110 kb region on chromosome 3 contains multiple miRNA genes of the soy-miR15/49 family. Each of their miRNA fold-back structures are embedded in inverted repeat sequences (IR1-7) drawn as blue diamonds to represent their inverted structures. The inverted repeat sequence containing an indel in its middle region is illustrated as an orange diamond (IR8). The phylogenetic tree presented on the top indicates the evolutionarily relationship among those IRs. Truncated IRs homologous to the full-length IRs are shown as green triangles. Four genes (gray and purple rectangles) are located in this repetitive region. The reticulon protein genes are embedded in the 12 kb duplicated regions (light blue rectangle). The direct repeat region also contains a solo LTR (black rectangle).

Figure S3 Global Topology of miRNA-target Networks. The conserved and non-conserved miRNAs are represented by red and blue circles, respectively. Target genes that were categorized into functional bins are shown. Targets encoding proteins related to RNA metabolism (pink circles), Ubiquitin based protein degradation (green circles), receptor kinase families (orange circles), secondary metabolism (turquoise circles), lipid metabolism (yellow circles) are distinguished from all other targets (light blue circles). Thick black edges connecting nodes indicate network overrepresented by cotyledon miRNA targets in preferentially regulated biological pathways.

Figure S4 Global Topology of miRNA-Biological Pathway Networks. The conserved and non-conserved miRNAs were indicated by red and blue circles, respectively. Their targeted biological pathways are shown by pink circles representing biological pathways related to RNA metabolism, green circles standing for Ubiquitin based protein degradation pathways, orange circles denoting receptor kinase families and light blue circles representing all remaining pathways.

File S1 This file contains: Table S1. Known miRNAs Expressed in Soybean Cotyledon. Table S2. Novel Cotyledon miRNAs. Table S3. Summary of miRNAs in Duplicated Genome Segments. Table S4. Summary of miRNA loci in Recently Duplicated Genome Segments. Table S5. Read Count and Precision Rate for Soy-mir15/49. Table S6. Reads Aligned to the Hairpin Sequences of Each Soy-miR15/49. Table S7. Cotyledon Soybean miRNA Families and Their Putative Target Genes. Table S8. Summary of Cotyledon miRNA Family Targeted Genes. Table S9. Summary of Soy-Mir38 Targeted Genes. Table S10. Genes Targeted by Multiple Cotyledon miRNA at Distinct Locations. Table S11. Genes Targeted by Mirna319, 395 and 159 Families. Table S12. Over- or Under-Representation of Genes Targeted by Cotyledon miRNA Family in Each Functional Bin.

Acknowledgments

The authors would like to thank Rick Meyer for his wonderful technical support in computational data processing and analysis and Dr. Xinbin Dai for his input in initial data analysis.

Disclaimer Note

Names are necessary to report factually on available data; however, the USDA neither guarantees nor warrants the standard of the product, and the use of the name by USDA implies no approval of the product to the exclusion of others that may also be suitable. USDA is an equal opportunity provider and employer.

Author Contributions

Conceived and designed the experiments: YQA. Performed the experiments: YQA. Analyzed the data: YQA WG JX WZ PZ ZL. Wrote the paper: YQA WG.

References

1. Lee Y, Kim M, Han J, Yeom KH, Lee S, et al. (2004) MicroRNA genes are transcribed by RNA polymerase II. EMBO J 23: 4051–4060.
2. Zhou X, Ruan J, Wang G, Zhang W (2007) Characterization and identification of microRNA core promoters in four model species. PLoS Comput Biol 3: e37.
3. Carthew RW, Sontheimer EJ (2009) Origins and Mechanisms of miRNAs and siRNAs. Cell 136: 642–655.
4. Bartel DP (2004) MicroRNAs: genomics, biogenesis, mechanism, and function. Cell 116: 281–297.
5. Kim VN (2005) MicroRNA biogenesis: coordinated cropping and dicing. Nat Rev Mol Cell Biol 6: 376–385.
6. Vaucheret H (2008) Plant ARGONAUTES. Trends in Plant Science 13: 350–358.
7. Okamura K, Liu N, Eric C, Lai EC (2009) Distinct Mechanisms for MicroRNA Strand Selection by Drosophila Argonautes. Mol Cell 36: 431–444.
8. Rhoades MW, Reinhart BJ, Lim LP, Burge CB, Bartel B, et al. (2002) Prediction of plant microRNA targets. Cell 110: 513–520.
9. Carrington JC, Ambros V (2003) Role of microRNAs in plant and animal development. Science 301: 336–338.
10. Bartel DP (2004) MicroRNAs: genomics, biogenesis, mechanism, and function. Cell 116: 281–297.
11. Dugas DV, Bartel B (2008) Sucrose induction of Arabidopsis miR398 represses two Cu/Zn superoxide dismutases. Plant Molecular Biology 67: 403–417.
12. Brodersen P, Sakvarelidze-Achard L, Bruun-Rasmussen M, Dunoyer P, Yamamoto YY, et al. (2008) Widespread translational inhibition by plant miRNAs and siRNAs. Science 320: 1185–1190.
13. Zhai J, Jeong DH, De Paoli E, Park S, Rosen BD, et al. (2011) MicroRNAs as master regulators of the plant NB-LRR defense gene family via the production of phased, trans-acting siRNAs. Genes & Development 25: 2540–2553.
14. Xia R, Zhu H, An YQ, Beers EP, Liu Z (2012) Apple miRNAs and tasiRNAs with novel regulatory networks. Genome biology 13: R47.
15. Palatnik JF, Allen E, Wu X, Schommer C, Schwab R, et al. (2003) Control of leaf morphogenesis by microRNAs. Nature 425: 257–263.
16. Aukerman MJ, Sakai H (2003) Regulation of flowering time and floral organ identity by a MicroRNA and its APETALA2-like target genes. Plant Cell 15: 2730–2741.
17. Park W, Li J, Song R, Messing J, Chen X (2002) CARPEL FACTORY, a Dicer homolog, and HEN1, a novel protein, act in microRNA metabolism in Arabidopsis thaliana. Curr Biol 12: 1484–1495.
18. Sunkar R, Zhu JK (2004) Novel and stress-regulated microRNAs and other small RNAs from Arabidopsis. Plant Cell 16: 2001–2019.
19. Fahlgren N, Jogdeo S, Kasschau KD, Sullivan CM, Chapman EJ, et al. (2010) MicroRNA gene evolution in Arabidopsis lyrata and Arabidopsis thaliana. Plant Cell 22: 1074–1089.
20. Ma Z, Coruh C, Axtell MJ (2010) Arabidopsis lyrata small RNAs: transient MIRNA and small interfering RNA loci within the Arabidopsis genus. Plant Cell 22: 1090–1103.
21. Rajagopalan R, Vaucheret H, Trejo J, Bartel DP (2006) A diverse and evolutionarily fluid set of microRNAs in Arabidopsis thaliana. Genes Dev 20: 3407–3425.
22. Xie Z, Allen E, Fahlgren N, Calamar A, Givan SA, et al. (2005) Expression of Arabidopsis MIRNA genes. Plant Physiol 138: 2145–2154.
23. Chen X (2004) A microRNA as a translational repressor of APETALA2 in Arabidopsis flower development. Science 303: 2022–2025.
24. Li H, Deng Y, Wu T, Subramanian S, Yu O (2010) Misexpression of miR482, miR1512, and miR1515 Increases Soybean Nodulation. Plant Physiol 153: 1759–1770.
25. Jones-Rhoades MW, Bartel DP, Bartel B (2006) MicroRNAs and Their Regulatory Roles in Plants Annual Review of Plant Biology 57: 19–53.
26. Shamimuzzaman M, Vodkin L (2012) Identification of soybean seed developmental stage-specific and tissue-specific miRNA targets by degradome sequencing. BMC Genomics 13: 310.
27. Schmutz J, Cannon SB, Schlueter J, Ma J, Mitros T, et al. (2010) Genome sequence of the palaeopolyploid soybean. Nature 463: 178–183.
28. Zhang B, Pan X, Stellwag E (2008) Identification of soybean microRNAs and their targets. Planta 229: 161–182.

29. Subramanian S, Fu Y, Sunkar R, Barbazuk WB, Zhu JK, et al. (2008) Novel and nodulation-regulated microRNAs in soybean roots. BMC Genomics 9: 160.

30. Joshi T, Yan Z, Libault M, Jeong D-H, Park S, et al. (2010) Prediction of novel miRNAs and associated target genes in Glycine max. BMC Bioinformatics 11: S14.

31. Song Q-X, Liu Y-F, Hu X-Y, Zhang W-K, Ma B, et al. (2011) Identification of miRNAs and their target genes in developing soybean seeds by deep sequencing. BMC plant biology 11: 5.

32. Kulcheski F, de Oliveira L, Molina L, Almerao M, Rodrigues F, et al. (2011) Identification of novel soybean microRNAs involved in abiotic and biotic stresses. BMC Genomics 12: 307.

33. Turner M, Yu O, Subramanian S (2012) Genome organization and characteristics of soybean microRNAs. BMC Genomics 13: 169.

34. Kozomara A, Griffiths-Jones S (2011) miRBase: integrating microRNA annotation and deep-sequencing data. Nucleic Acids Research 39: D152–D157.

35. Axtell MJ (2013) Classification and Comparison of Small RNAs from Plants. Annual Review of Plant Biology 64: 137–159.

36. Lelandais-Briere C, Naya L, Sallet E, Calenge F, Frugier F, et al. (2009) Genome-Wide Medicago truncatula Small RNA Analysis Revealed Novel MicroRNAs and Isoforms Differentially Regulated in Roots and Nodules. Plant Cell 21: 2780–2796.

37. Okamura K, Phillips MD, Tyler DM, Duan H, Chou YT, et al. (2008) The regulatory activity of microRNA* species has substantial influence on microRNA and 3′ UTR evolution. Nature structural & molecular biology 15: 354–363.

38. Meyers BC, Axtell MJ, Bartel B, Bartel DP, Baulcombe D, et al. (2008) Criteria for Annotation of Plant MicroRNAs. Plant Cell 20: 3186–3190.

39. Mi S, Cai T, Hu Y, Chen Y, Hodges E, et al. (2008) Sorting of small RNAs into Arabidopsis argonaute complexes is directed by the 5′ terminal nucleotide. Cell 133: 116–127.

40. Li B, Qin Y, Duan H, Yin W, Xia X (2011) Genome-wide characterization of new and drought stress responsive microRNAs in Populus euphratica. Journal of experimental botany 62: 3765–3779.

41. Zhang L, Chia J-M, Kumari S, Stein JC, Liu Z, et al. (2009) A Genome-Wide Characterization of MicroRNA Genes in Maize. PLoS Genet 5: e1000716.

42. Nozawa M, Miura S, Nei M (2012) Origins and evolution of microRNA genes in plant species. Genome biology and evolution 4: 230–239.

43. Cuperus JT, Fahlgren N, Carrington JC (2011) Evolution and Functional Diversification of MIRNA Genes. The Plant Cell 23: 431–442.

44. Cui H, Levesque MP, Vernoux T, Jung JW, Paquette AJ, et al. (2007) An evolutionarily conserved mechanism delimiting SHR movement defines a single layer of endodermis in plants. Science 316: 421–425.

45. Ma Z, Coruh C, Axtell MJ (2010) Arabidopsis lyrata small RNAs: transient MIRNA and small interfering RNA loci within the Arabidopsis genus. The Plant cell 22: 1090–1103.

46. Fahlgren N, Howell MD, Kasschau KD, Chapman EJ, Sullivan CM, et al. (2007) High-throughput sequencing of Arabidopsis microRNAs: evidence for frequent birth and death of MIRNA genes. PLoS ONE 2: e219.

47. McClean PE, Mamidi S, McConnell M, Chikara S, Lee R (2010) Synteny mapping between common bean and soybean reveals extensive blocks of shared loci. BMC Genomics 11: 184.

48. Lin JY, Stupar RM, Hans C, Hyten DL, Jackson SA (2010) Structural and functional divergence of a 1-Mb duplicated region in the soybean (Glycine max) genome and comparison to an orthologous region from Phaseolus vulgaris. Plant Cell 22: 2545–2561.

49. Allen E, Xie Z, Gustafson AM, Sung GH, Spatafora JW, et al. (2004) Evolution of microRNA genes by inverted duplication of target gene sequences in Arabidopsis thaliana. Nat Genet 36: 1282–1290.

50. Nozawa M, Miura S, Nei M (2012) Origins and Evolution of MicroRNA Genes in Plant Species. Genome Biol Evol 4: 230–239.

51. Axtell MJ, Snyder JA, Bartel DP (2007) Common functions for diverse small RNAs of land plants. Plant Cell 19: 1750–1769.

52. Chen K, Rajewsky N (2007) The evolution of gene regulation by transcription factors and microRNAs. Nat Rev Genet 8: 93–103.

53. Li Y, Li C, Ding G, Jin Y (2011) Evolution of MIR159/319 microRNA genes and their post-transcriptional regulatory link to siRNA pathways. BMC evolutionary biology 11: 122.

54. Millar AA, Gubler F (2005) The Arabidopsis GAMYB-like genes, MYB33 and MYB65, are microRNA-regulated genes that redundantly facilitate anther development. The Plant cell 17: 705–721.

55. Allen RS, Li J, Stahle MI, Dubroue A, Gubler F, et al. (2007) Genetic analysis reveals functional redundancy and the major target genes of the Arabidopsis miR159 family. Proceedings of the National Academy of Sciences of the United States of America 104: 16371–16376.

56. Reyes JL, Chua NH (2007) ABA induction of miR159 controls transcript levels of two MYB factors during Arabidopsis seed germination. The Plant journal: for cell and molecular biology 49: 592–606.

57. Palatnik JF, Wollmann H, Schommer C, Schwab R, Boisbouvier J, et al. (2007) Sequence and expression differences underlie functional specialization of Arabidopsis microRNAs miR159 and miR319. Dev Cell 13: 115–125.

58. Ori N, Cohen AR, Etzioni A, Brand A, Yanai O, et al. (2007) Regulation of LANCEOLATE by miR319 is required for compound-leaf development in tomato. Nature genetics 39: 787–791.

59. Schommer C, Palatnik JF, Aggarwal P, Chetelat A, Cubas P, et al. (2008) Control of jasmonate biosynthesis and senescence by miR319 targets. PLoS Biology 6: e230.

60. Thimm O, Bläsing O, Gibon Y, Nagel A, Meyer S, et al. (2004) Mapman: a user-driven tool to display genomics data sets onto diagrams of metabolic pathways and other biological processes. Plant Journal 37: 914–939.

61. Chen K, An Y-Q (2006) Transcriptional Responses to Gibberellin and Abscisic Acid in Barley Aleurone. Journal of Integrative Plant Biology 48: 591–612.

62. Langmead B, Trapnell C, Pop M, Salzberg SL (2009) Ultrafast and memory-efficient alignment of short DNA sequences to the human genome. Genome Biol 10: R25.

63. Zhang W, Zhou X, Xia J (2012) Identification of microRNAs and natural antisense transcript-originated endogenous siRNAs from small-RNA deep sequencing data. Methods in molecular biology 883: 221–227.

64. Reese TA, Xia J, Johnson LS, Zhou X, Zhang W, et al. (2010) Identification of novel microRNA-like molecules generated from herpesvirus and host tRNA transcripts. J Virol 84: 10344–10353.

65. Griffiths-Jones S, Saini HK, van Dongen S, Enright AJ (2008) miRBase: tools for microRNA genomics. Nucl Acids Res 36: D154–158.

66. Hofacker IL, Fontana W, Stadler PF, Bonhoeffer LS, Tacker M, et al. (1994) Fast folding and comparison of RNA secondary structures. Monatshefte für Chemie/Chemical Monthly 125: 167–188.

67. Tamura K, Peterson D, Peterson N, Stecher G, Nei M, et al. (2011) MEGA5: molecular evolutionary genetics analysis using maximum likelihood, evolutionary distance, and maximum parsimony methods. Molecular biology and evolution 28: 2731–2739.

68. Saitou N, Nei M (1987) The neighbor-joining method: a new method for reconstructing phylogenetic trees. Molecular biology and evolution 4: 406–425.

69. Kimura M (1980) A simple method for estimating evolutionary rates of base substitutions through comparative studies of nucleotide sequences. Journal of Molecular Evolution 16: 111–120.

70. Lin JY, Stupar RM, Hans C, Hyten DL, Jackson SA (2010) Structural and functional divergence of a 1-Mb duplicated region in the soybean (Glycine max) genome and comparison to an orthologous region from Phaseolus vulgaris. The Plant cell 22: 2545–2561.

11

Self-Sterility in *Camellia oleifera* May Be Due to the Prezygotic Late-Acting Self-Incompatibility

Ting Liao, De-Yi Yuan*, Feng Zou, Chao Gao, Ya Yang, Lin Zhang, Xiao-Feng Tan

Key Laboratory of Cultivation and Protection for Non-Wood Forest Trees, Ministry of Education, The Key Lab of Non-Wood Forest Products of Forestry Ministry, Central South University of Forestry and Technology, Changsha, Hunan, China

Abstract

In this report, self-sterility in *Camellia oleifera* was explored by comparing structural and statistical characteristics following self-pollination (SP) and cross-pollination (CP). Although slightly delayed pollen germination and pollen tube growth in selfed ovaries compared to crossed ovaries was observed, there was no significant difference in the percentages of pollen that germinated and pollen tubes that grew to the base of the style. There was also no difference in morphological structure after the two pollination treatments. However, the proportions of ovule penetration and double fertilization in selfed ovules were significantly lower than in crossed ovules, indicating that a prezygotic late-acting self-incompatible mechanism may exist in *C. oleifera*. Callose deposition was observed in selfed abortive ovules, but not in normal. Ovules did not show differences in anatomic structure during embryonic development, whereas significant differences were observed in the final fruit and seed set. In addition, aborted ovules in selfed ovaries occurred within 35 days after SP and prior to zygote division. However, this process did not occur continuously throughout the life cycle, and no zygotes were observed in the selfed abortive ovules. These results indicated that the self-sterility in *C. oleifera* may be caused by prezygotic late-acting self-incompatibility (LSI).

Editor: Randall P. Niedz, United States Department of Agriculture, United States of America

Funding: This study was supported by the National Natural Science Foundation of China (No. 31170639). The funders had no role in study design, data collection and analysis, decision to publish, or preparation of the manuscript.

Competing Interests: The authors have declared that no competing interests exist.

* E-mail: yuan-deyi@163.com

Introduction

Camellia oleifera (Theaceae), an evergreen shrub species, is widely cultivated in southern China. Oil from its seeds is commonly used as an edible oil with known health benefits. At this time, there are approximately three million hectares of cultivated area of *C. oleifera* in China, with 35.2 kg of oil per acre [1]. The large number of flowers with relatively low fruit/seed sets is a serious problem and has restricted development of the oil tea industry [1]. To address this issue, previous studies focusing on flowering biology [2], pollination biology [2], pollen viability [3–6], flower bud differentiation [7], megaspore and microspore development [8], male and female gamete development [8], and embryological development [9] have been performed in *C. oleifera*. However, these studies did not explore flower development or the large number of ovules that fail to mature into seeds during fruit development in nature [1,9].

Self-sterility is a common reproductive phenomenon in Theaceae, which is mainly the result of self-incompatibility (SI) and early-acting inbreeding depression (EID) [10–16]. SI, an initiative abortion process, can prevent reproduction after selfing and enhance heterozygosity in plants [15–17]. SI is involved in several processes, including pollen inhibition on the stigma or in the style in some species of Brassicaceae, Polemoniaceae, Rosaceae, and Solanaceae [17–21]. Another type of SI that occurs in the ovary is called late-acting SI (LSI) or ovarian SI [17,22]. Abortion in LSI may occur before or after zygote formation, and postzygotic mechanisms are more complicated

than prezygotic mechanisms. It is known that prezygotic SI occurs in *Theobroma cacao* [23], *Acacia retinodes* [24], and *Camellia sinensis* [25], while postzygotic SI occurs in *Caesalpinia calycina* [26], *Pseudowintera axillaris* [27], *Tabebuia* [28], *Citrus grandis* Osbeck [29], and *Xanthoceras sorbifolium* [30]. In contrast, EID is caused by the expression of recessive alleles during seed development [17,22]. This may occur throughout the plant life cycle, and aborting ovules/seeds may be observed in ovaries/fruits. It is currently difficult to distinguish between postzygotic SI and EID mechanisms in the study species.

In this report, we compared pollen germination, pollen tube growth, ovule penetration, double fertilization, embryonic development, and fruit/seed sets with a focus on morphology and anatomical structures following SP and CP. In addition, we characterized a possible mechanism of self-sterility in *C. oleifera*.

Materials and Methods

Ethics Statement

No specific permits were required for the described field studies, and the field observations did not collect any animal, endangered or protected specimen.

Plant material

Populations of two *C. oleifera* cultivars ('Xianglin XLC15' and 'Xianglin XLJ14') were used in this study. The plants were mixed planted in Taoshui (27°05′06″N, 113°13′11″E), You County,

Zhuzhou, Hunan Province, China. 'Xianglin XLC15' bloomed from late October to mid-December, with about 70 mm in diameter of the flower, while C. oleifera 'Xianglin XLJ14' bloomed from late October to late November, and the diameter of the flower was about 55 mm. Both of them were hermaphroditic and homogamous. Fruits of both cultivars matured in mid-to-late October of the next year [31]. The plants were 2.2-m high and 12-years-old, they were situated at 150 m above sea level on red soil and subjected to a typical subtropical moist climate. The mean annual temperature was 17.8°C, and the mean annual rainfall was about 1410 mm. Rainfall occurred primarily from April to June, accounting for 45% of the yearly total.

Pollination treatments

Controlled pollination included an SP combination ('Xianglin XLC15' × 'Xianglin XLC15) with pollen from a mix of several individuals from the same clone and a CP combination ('Xianglin XLC15' × 'Xianglin XLJ14') that was performed in November of 2011. Moreover, non-pollination (NP) was included. In addition to the three pollination treatments (SP, CP and NP), a control open-pollinated (OP) sample was also included to evaluate fruit and seed sets 1 year after pollination (AP). In SP and CP, the flowers were emasculated and pollinated during the bud stage so that the stigmas were not stained by any pollen until they became receptive. In NP, the stamens of the flower buds were picked before anthers shed pollens. Afterwards, all the pistils were bagged in sulfate paper bags for 7 days after pollination (DAP) and marked with signs. In OP, no action except signing the bud randomly was being taken. Self- and cross-pollinated pistils were harvested at 2, 4, 8, 12, 24, 36, 48, 60, 72, 84, 96, 108, and 120 h, and further every day from 6–10 DAP, every 2 days from 10–30 DAP, every 5 days from 30–50 DAP, and every 20 days from 50–210 DAP, with 15 pistils collected at each time point. Unpollinated and bagged samples were harvested every 2 days from the beginning of the study until the 30th day, then every 20 days until they dropped off, with 10 pistils harvested each time. All collected samples were fixed in formalin: acetic acid: ethyl alcohol (5: 5: 90, V/V) and stored at 4°C prior to sectioning [32]. In each pollination combination, 1000 flowers were pollinated and used for fluorescence analysis and sectioning in the SP and CP experiments. 300 non-pollinated flowers were used for sectioning and statistics. In addition to the previously explained pollination, other 300 pollinated flowers with three replicates (each 100 flowers) were used to investigate fruit and seed sets after SP, CP, and NP.

Pollen germination, pollen tube growth, and ovule penetration following CP and SP

Pollen germination was observed using the method of Hiratsuka et al. [33] with minor modifications. Briefly, the style 2–120 h after SP and CP was macerated in 8 M NaOH for 3–4 h to soften the materials, after which it was thoroughly washed with deionized water several times with no residual chemical substances. The samples were stained for at least 4 h in 0.05% aniline blue dissolved in 0.15 M K$_2$HPO$_4$. The stigma was placed on a glass slide and then slowly flattened with a cover slip, as described by Chen et al. [25]. The length of the pollen tubes was measured using Motic Images Plus 2.0 software. Pollen germination and pollen tube growth were observed and photographs were taken using a fluorescence microscope (Olympus BX-51, Tokyo, Japan).

Ovule penetration into the embryo sac in both the CP and SP experiments was examined from 60 h to 25 DAP through sectioning and aniline blue staining. 3–5 ovaries which contained 16–25 ovules each ovary were observed per sample. Aniline blue staining was performed as described above. Each slice was cut to 15 µm. Several serial sections in one ovule were stained with aniline blue for 1 h after eliminating the paraffin adhered to the glass slide to detect callose [34]. Ovule penetration into the embryo sac was demonstrated based on the presence of pollen tubes penetrating the micropyle [17]. An Olympus BX-51 fluorescence microscope was also used to observe and take photos of penetrating ovules.

Double fertilization, embryo sac, and ovule development following CP and SP

Ovaries between 60 h and 210 days after SP and CP, and ovaries at all stages following NP were dehydrated in ethyl alcohol, embedded in paraffin with a 58–60°C melting point, sectioned every 10 µm, and stained with hematoxylin-eosin Y to detect double fertilization, embryo sacs, ovules, and seed development [35]. Double fertilization was identified based on the presence of a degenerated and intensively colored synergid or the formation of endosperm nuclei, as well as the presence of zygotes or embryos [36]. To assess the area, as well as the thickness of the inner and outer integuments, we examined stained sections in the center of the ovary using Motic Images Plus 2.0 software, by which quantitative differences between various stages following CP, SP, and NP could be determined. Microscopic observations and photography were performed as described above.

Fruit set, seed set, and fruit characteristics

SP, CP, OP, and NP were performed to evaluate the fruit set, seed set, and fruit characteristics. The number of subsistent fruits and seeds were counted. Seeds were considered fertile if they contained a complete seed composition and were full in appearance compared to incomplete, flat, and sterile seeds. The seed set was then calculated as the number of fertile seeds divided by the number of total ovules in mature fruits. Pollinated flowers that failed to form mature fruits or fell off before maturity were not included in the seed set (but were included in the fruit set) [37]. The fruit characteristics examined included mean single fruit transverse diameter, vertical diameter, and weight. The index of self-incompatibility (ISI) was used to assess the degree of SI in C. oleifera, which was calculated by dividing the seed set after SP by the set after CP [38–40].

Statistical analysis

The SPSS statistical package, version 19.0, was used for most of the statistical analysis. Comparative variables between the CP and SP treatments, including pollen tube length, percent of pollen germination on the stigma, no. of pollen tubes in the style, percent of pollen tubes at the style base, percent of penetrated ovules, and ovule size, were examined. Comparisons were performed using a one-way ANOVA with 95% confidence interval to determine whether there were any significant differences in pollen tube length, percentages of pollen and pollen tubes at various stages, double fertilization and ovule development after SP and CP. Comparisons of the mean area of embryo sacs, fruit set, seed set and fruit characteristics after four pollination treatments were evaluated based on Duncan's multiple range test at the 5% level. Statistical data with 0 were read as N/A and they were treated as missing values when statistical analysis. All linear and scatter plots were generated using Origin 8.5 software. The Design Expert software, version 8.0, was used to evaluate the effects of three fertilization treatments, the time after pollination, and their interaction in integument thickness with the general factorial method. The factor of DAP was treated as a quantitative factor, and the fertilization treatment as a qualitative factor, thus, a 1-factor response surface analysis was generated.

Results

Characterization of pollen germination, pollen tube growth, and ovule penetration following SP and CP

Significant differences in pollen tube length at every stage over 2–48 h following SP compared to CP were observed in *C. oleifera*, although the length increased continuously during the life cycle (Fig. 1; 2 h AP, $SS_B = 0.032$, df = 1, F = 21.275, P<0.05; 4 h AP, $SS_B = 1.696$, df = 1, F = 104.585, p<0.05; 8 h AP, $SS_B = 9.45$, df = 1, F = 432.501, p<0.05; 12 h AP, $SS_B = 12.586$, df = 1, F = 793.236; p<0.05; 24 h AP, $SS_B = 17.785$, df = 1, F = 257.13, p<0.05; 36 h AP, $SS_B = 19.984$, df = 1, F = 211.022, p<0.05; 48 h AP, $SS_B = 22.157$, df = 1, F = 14.31, p<0.05). However, no morphological or structural differences were observed in pollen germination and pollen tube growth in the style following SP and CP. At 2 h AP, a small number of selfed pollen grains had germinated compared with the large number of pollen grains that germinated 2 h after CP (Fig. 2A, D). At 36 h after SP, the pollen tube length reached about ½ of the style length, while 24 h was required to reach this length following CP (Fig. 2B, E). Afterwards, it took approximately 60 h for selfed tubes to grow at the base of the style, and only 48 h for crossed tubes (Fig. 2C, F). The first selfed pollen tube penetrated the embryo sac at 84 and 60 h after SP and CP, respectively (Fig. 2I, J). The pollen grains could germinate normally and the pollen tubes could grow normally at the base of the style, and even penetrate the ovule, regardless of whether they were produced by SP or CP.

Differences were observed based on a one-way ANOVA regarding the percentages of germinated pollen (Fig. 3A), pollen tubes in the middle of the style (Fig. 3B), pollen tubes at the base of the style (Fig. 3C), and penetrated ovules (Fig. 3D) at various times following SP compared to CP. With the increasing time AP, the percentages of germinated pollen, pollen tubes in the middle of the style, and pollen tubes at the base of the style increased initially, then they maintained a steady level. Nonetheless, the percentage of pollen tubes penetrating the ovule decreased significantly 8 DAP following both SP and CP (Fig. 3D). Pollen germination peaked at 72.3% 72 h after SP vs. 74% 48 h after CP. Significant differences in the percent of germinated pollen at 2 h

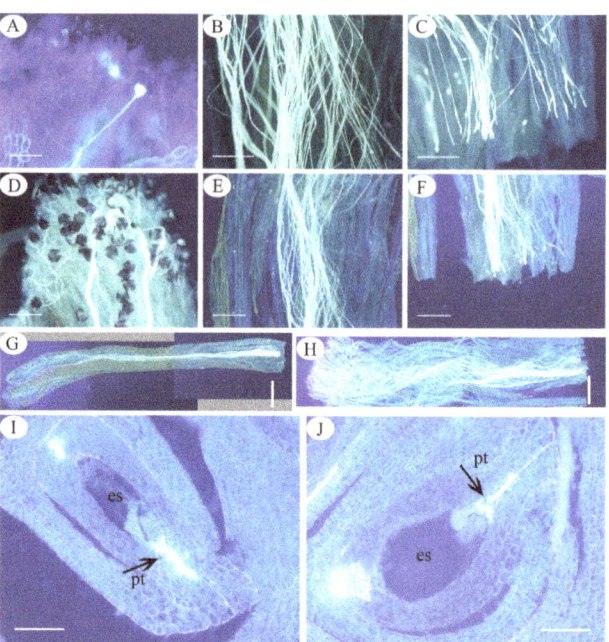

Figure 2. Pollen germination, pollen tube growth, and ovule penetration after SP and CP in *C. oleifera*. (A). Pollen germination 2 h after SP. (B). Pollen tubes in the middle of the style 36 h after SP. (C). Pollen tubes at the base of style 60 h after SP. (D). Pollen germination 2 h after CP. (E). Pollen tubes in the middle of the style 24 h after CP. (F). Pollen tubes at the base of style 48 h after CP. (G). Overall growth 60 h after SP. (H). Overall growth 48 h after CP. (I). A pollen tube penetrating an embryo sac 84 h after SP. (J). Ovule penetration 60 h after CP. Abbreviations: es, embryo sac; pt, pollen tube. Bars: A, D, I, J = 200 μm; B, C = 500 μm; E, F = 1000 μm; G, H = 2000 μm.

($SS_B = 170.667$, df = 1, F = 20.48, p<0.05) and 12 h ($SS_B = 96$, df = 1, F = 0.041, p<0.05) were observed following SP compared to CP, while no differences were observed at 24, 36, 48, 60, and 72 h (p>0.05 in these treatments). At 72 h AP ($SS_B = 0.167$, df = 1, F = 0.018, p>0.05), there were no significant differences in the percentage of pollen tubes in the middle of the style, similar to the percentage of pollen tubes at the base of the style ($SS_B = 42.667$, df = 1, F = 5.953, p>0.05, 72 h AP).

Although the growth of pollen tubes in the style showed no structural differences between SP and CP, abnormal pollen tubes were observed in the ovary after SP (Fig. 4). In our study, distorted pollen tubes containing reversal tubes (Fig. 4B), swelling tube tips with callose deposits (Fig. 4B), irregular tubes (Fig. 4A), furcal tubes (Fig. 4C), and abortive ovules were observed in the SP samples. However, furcal tubes were the most common.

Nevertheless, the percentage of penetrated ovules was significantly greater following CP than SP. The pollen tubes entered embryo sacs through a degenerated synergid, and the other synergid disappeared after a period of time in normal ovules. Based on our observations, at 8 DAP the maximum ratio of penetrated ovules after SP was approximately 4.33±1.15% compared with 21.33±3.06% after CP ($SS_B = 433.5$, df = 1, F = 81.281, p<0.05).

Double fertilization, embryo sac, and ovule development following SP and CP Double fertilization after SP and CP

Double fertilization was demonstrated based on the fusion between sperm and egg cells or between a polar nucleus and sperm cells, as indicated by the presence of zygotes and a primary

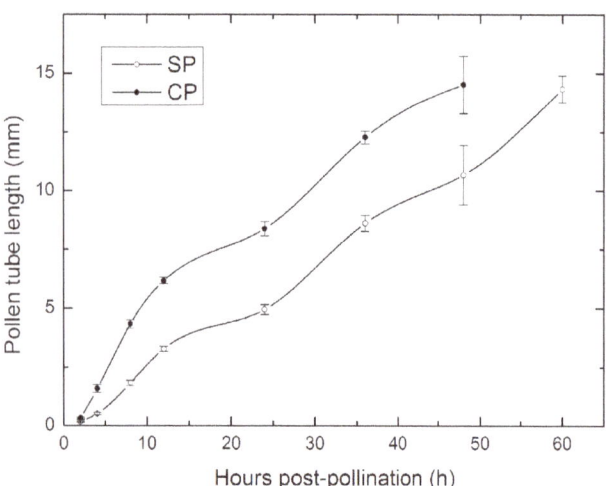

Figure 1. Mean length of pollen tubes at various times following SP and CP. Vertical bars represent the standard deviation.

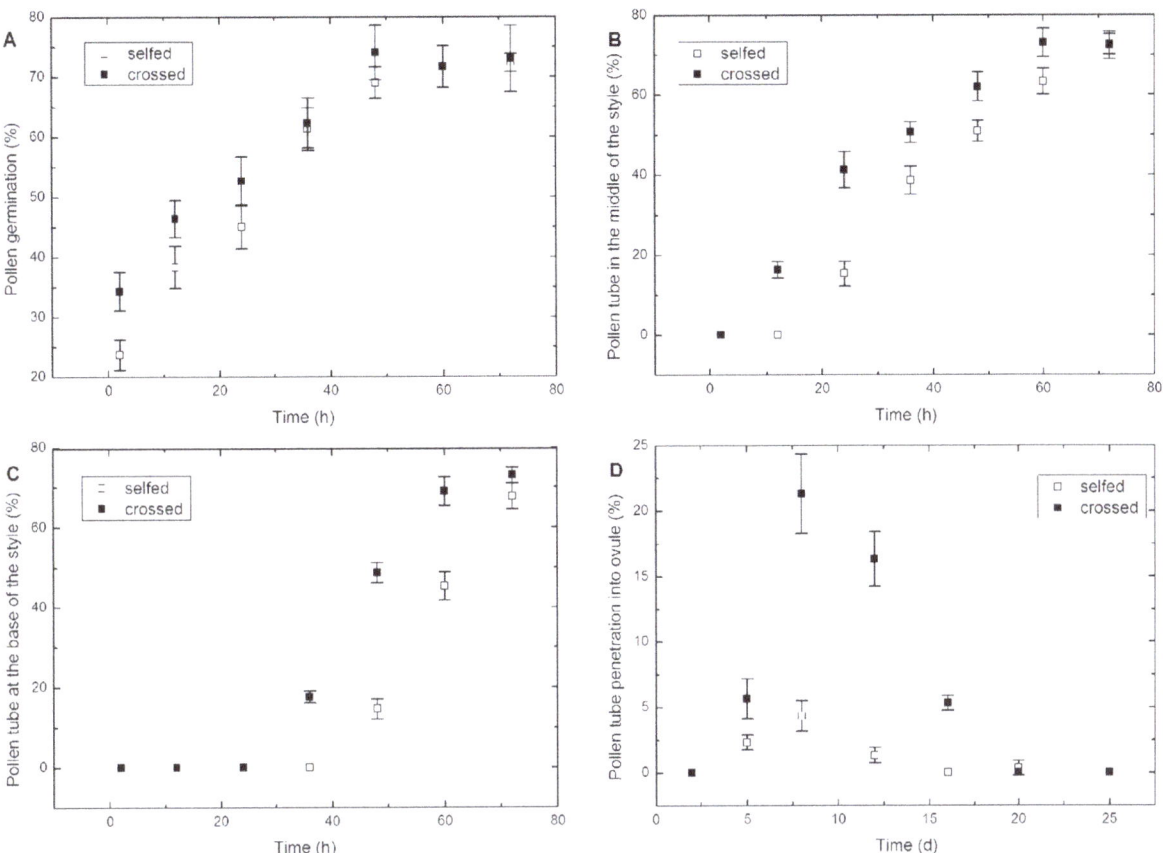

Figure 3. Percentages of pollen and pollen tubes at various stages after SP and CP. (A). Germinated pollen on the stigma. (B). Pollen tubes in the middle of the style. (C). Pollen tubes at the base of the style. (D). Pollen tubes penetrating into an ovule. Vertical bars represent the standard deviation.

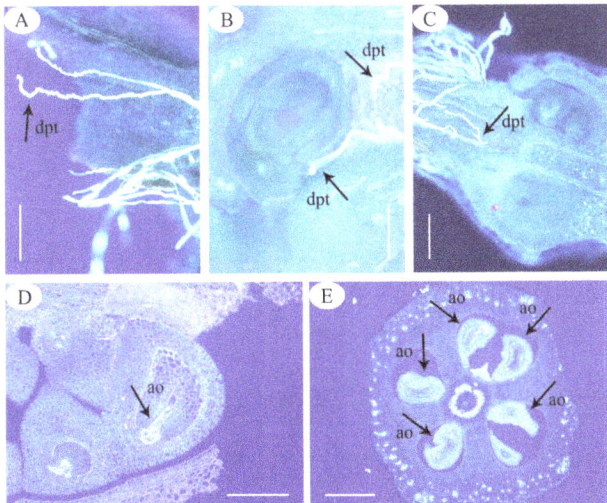

Figure 4. Distorted pollen tube grown in the ovary after SP in *C. oleifera.* (A). Distorted pollen tube with irregular ovaries 60 h after SP. (B). Distorted pollen tube with a reserving tube (the arrow above) and swelling tip (the arrow below) outside the micropyle 78 h after SP. (C). Distorted pollen tube with fortification of the ovaries 96 h after SP. (D). An abortive ovule with bright spots in the embryo sac 10 days after SP. (E). An abortive ovule with bright spots throughout the ovary 25 days after SP. Abbreviations: ao, abortive ovule; dpt, distorted pollen tube. Bars: A, B, D = 500 μm; C, E = 1000 μm.

endosperm nucleus in embryo sacs after both SP and CP (Fig. 5). There were no structural differences between normal selfed and crossed embryos based on anatomical observations, but differences were observed in the frequency of double fertilization and ovule sterility (Fig. 5; Table 1). The highest ratio of double fertilization observed after SP (5.08%) was significantly lower than after CP (24.39%) at the same time AP, which corresponded to ovule penetration in SP compared to CP ovaries. After both SP and CP, delayed ovule penetration and double fertilization were observed in *C. oleifera* (although this was not statistically significant). SP resulted in a smaller proportion of double fertilization at all times compared with CP, suggesting that certain fertilization barriers existed. The level of sterility increased in 4 weeks AP. The process of double fertilization was rarely observed in selfed or crossed ovules, although a relatively higher percentage of double fertilization was observed after CP than SP.

Embryo sac area and integument thickness

To examine the irregular anatropous ovule in *C. oleifera*, the mean area and integument thickness (rather than ovule length and width) were measured. The mean embryo sac area (Table 2) and variance analysis of integument thickness (Table 3, Fig. S1) were examined at various stages following different pollination treatments. There were no significant differences in mean area between the selfed ovules and crossed ovules by 90 days, but significant differences were observed in the non-fertilized ovules, which indicates that ovule growth was determined based on whether

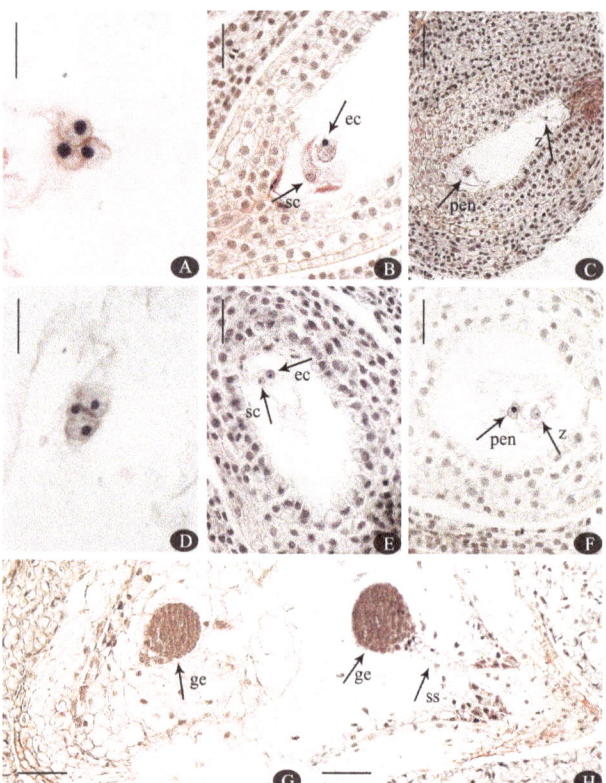

Figure 5. Normal double fertilization and early embryonic development after SP vs. CP in *C. oleifera*. (A). Polar nucleus fusion with a sperm nucleus 120 h after SP. (B). Sperm cell approaching an egg cell 96 h after SP. (C). A zygote and primary endosperm nucleus in an embryo sac 14 days after SP. (D). Polar nucleus fertilization 720 h after CP. (E). Sperm cell fused with an egg cell 6 days after CP. (F). A zygote and primary endosperm nucleus 10 days after CP. (G). A globular embryo 210 days after SP. (H). A globular embryo and suspensor 200 days after CP. Abbreviations: ec, egg cell; ge, globular embryo; pen, primary endosperm nucleus; sc, sperm cell; ss, suspensor; z, zygote. Bars: A, D = 1000 μm; B, E, F = 100 μm; C, G, H = 200 μm.

fertilization occurred, rather than whether pollination occurred before zygote development. When the zygote began to divide, significant differences were gradually observed among self-, cross-, and non-fertilized ovules. Finally, the area of selfed embryo sacs was larger than the area of crossed embryo sacs. However, under the same pollination conditions, we observed an increase in both SP and CP selfed seeds and crossed seeds. Within the first 50 days, an increased embryo sac area in non-fertilized ovules was observed, while the area decreased 50 days after harvesting (with a final area of 0), indicating abortive ovules.

Both DAP and FT had significant differences in inner integument thickness ($p<0.0001$) (Table 3). Additionally, significant differences were observed in the interaction of DAP and FT in the inner integument thickness ($p<0.0001$) (Fig. S2). The effect of DAP was significant in outer integument thickness ($p<0.0001$), similar to the inner integument. However, no significant differences in the outer integument thickness were observed in FT ($p>0.05$) and the interaction of DAP and FT ($p>0.05$) (Fig. S3). It meant that the presence or absence of fertilization, or the type of fertilization, did not affect the outer integument growth. Moreover, the F value of DAP was greater than FT both in the inner and outer integument thickness, which revealed that the DAP was a more important factor than FT in integument thickness.

Table 1. Anatomical and statistical details of ovule and seed development following self-pollination (SP) and cross-pollination (CP) at 1 week ovules, 2 weeks, 3 weeks, 4 weeks, and 1 year seeds after pollination.

Flower harvested	Pollination treatment	No. of samples observed	Normal ovules/seeds	Sterile ovules/seeds	Double fertilization	Zygote	Endosperm
			%				
1 week	SP	103	79.61	20.39	2.91	0.97	1.94
	CP	135	71.11	28.89	4.44	1.48	2.22
2 weeks	SP	142	65.49	34.51	3.52	0.70	1.41
	CP	104	66.35	33.65	24.04	11.54	7.69
3 weeks	SP	118	27.97	72.03	5.08	1.69	1.69
	CP	123	53.66	46.34	24.39	4.07	15.45
4 weeks	SP	181	29.83	70.17	4.42	1.10	2.21
	CP	113	46.90	53.10	18.58	5.31	10.62
1 year	SP	171	10.53	89.47	2.92	1.75	1.75
	CP	123	54.47	45.53	23.58	17.07	23.58

Table 2. The mean (±SD) area (μm^2) of embryo sacs in self-fertilized (SF), cross-fertilized (CF), and non-fertilized (NF) ovules at various times.

DAP	SF	CF	NF
1	20789.91 (591.58) Ba	20608.85 (497.57) Ba	1944.27 (112.04) Aa
14	22849.99 (742.68) Bab	22204.87 (718.09) Bab	6111.42 (179.29) Ad
30	24529.11 (612.24) Bab	24069.40 (525.73) Bbc	14928.28 (301.08) Af
50	25829.85 (1352.91) Bab	26134.31 (1167.62) Bcd	17313.16 (211.37) Ag
70	25877.17 (518.59) Bab	26048.65 (292.39) Bcd	12807.90 (240.35) Ae
90	28603.46 (784.15) Bb	28295.16 (738.38) Bd	5250.70 (238.61) Ac
130	36884.32 (330.61) Cc	35226.40 (916.50) Be	3090.72 (235.74) Ab
170	111299.25 (8333.01) Bd	97441.08 (3406.03) Af	N/A

Different capital letters within the same line represent significant differences at the 5% level as determined by Duncan's multiple range test, while the same letter represents no significant differences. In addition, completely different small letters after capital letters within the same column represent significant differences at the 5% level based on Duncan's multiple range test, while the same small letter represents no significant difference at the 5% level. Abbreviation: DAP, days after pollination; SD, Standard Deviation.

Ovule/seed abortion

A small number of SP vs. CP ovules developed to mature seeds normally through double fertilization. Ovules were mostly observed aborting during seed formation, especially prior to the zygote development phase (Fig. 6). Abortion was judged based on structural characteristics, with a strongly stained egg apparatus and abnormal embryo sacs. Ovules with shriveling embryo sacs but normal integument development (Fig. 6B, C) at early stages, embryo sacs with screwy shapes and a degenerated egg apparatus (Fig. 6D–H), shrinking ovules with dead whole tissues (Fig. 6I, J, L), and even completely dead ovaries (Fig. 6K) were observed at different stages following SP. Two sets of egg apparatus were observed in one of the selfed ovules 72 h after SP (Fig. 6A), but there was no evidence of polyembryony in mature seeds in this study, while Cao observed a small number of polyembryonic seeds in *C. oleifera* [9].

Based on the paraffin section statistics of 214 ovaries and 4242 ovules after SP, and 180 ovaries and 3736 ovules after CP, the ratio of aborted ovules remained significant by 16 days after SP vs. CP, whereas it became distinct 16 days post-pollination (Fig. 7). The percentage of aborted ovules reached 85.5% 35 days after SP and 46.3% 22 days after CP, and was then relatively stable until the seeds reached maturity, which demonstrates that the greatest incidence of self-sterility occurred prior to zygote division. On the other hand, ultimately there was a higher proportion of seed abortion following SP than CP.

Fruit set, seed set, and fruit characteristics following CP, SP, OP, and NP

Comparisons between fruit sets, seed sets, and fruit characteristics after CP, SP, OP, and NP over time revealed quantitative differences (Table 4). Significant differences in mean fruit sets ($SS_B = 2882.889$, df = 2, $MS_B = 1441.444$, F = 123.552, $p<0.05$), seed sets ($SS_B = 9448.32$, df = 2, $MS_B = 4724.16$, F = 167.315, $p<0.05$), transverse diameter ($SS_B = 2333.301$, df = 2, $MS_B = 1166.65$, F = 43.366, $p<0.05$), vertical diameter ($SS_B = 924.121$, df = 2, $MS_B = 462.06$, F = 33.96, $p<0.05$), and weight ($SS_B = 9660.556$, df = 2, $MS_B = 4830.278$, F = 52.841, $p<0.05$) per fruit were observed following SP, CP and OP. Fruit size after CP was larger and heavier than after the other treatments, but seed size after SP was visibly larger than after CP (Fig. 8). Fruit size and weight for OP differed between SP and CP (Table 4). NP with no fruit revealed no apomixes in *C. oleifera*. Similarly, there were significant differences in the fruit sets among SP and CP. On the contrary, no significant differences were observed between SP and OP in fruit sets, indicating a high incidence of SP in nature.

Table 3. The ANOVA (IIT/OIT) for response surface of integument thickness (μm) in three fertilization treatments (FT) at various times.

Source	Sum of Squares	df	Mean Square	F Value	*p*-value Prob >F
Model	75077/25709	9/9	8342/2857	274/234	<0.0001/<0.0001
A-DAP	53514/24164	1/1	53514/24164	1760/1979	<0.0001/<0.0001
B-FT	2977/19	2/2	1488/9.47	49/0.78	<0.0001/0.465
AB	7535/50	2/2	3768/25	124/2.06	<0.0001/0.136
A^2	3394/620	1/1	3394/620	112/51	<0.0001/<0.0001
A^{2}B	3003/15	2/2	1502/7.25	49/0.59	<0.0001/0.555
A^3	4654/841	1/1	4654/841	153/69	<0.0001/<0.0001
Residual	1885/757	62/62	30/12	N/A	N/A

p-value <0.05 represents significant differences at the 95% confidence interval based on response surface analysis. Abbreviations: DAP, days after pollination; IIT, inner integument thickness; OIT, outer integument thickness; FT, fertilization treatments.

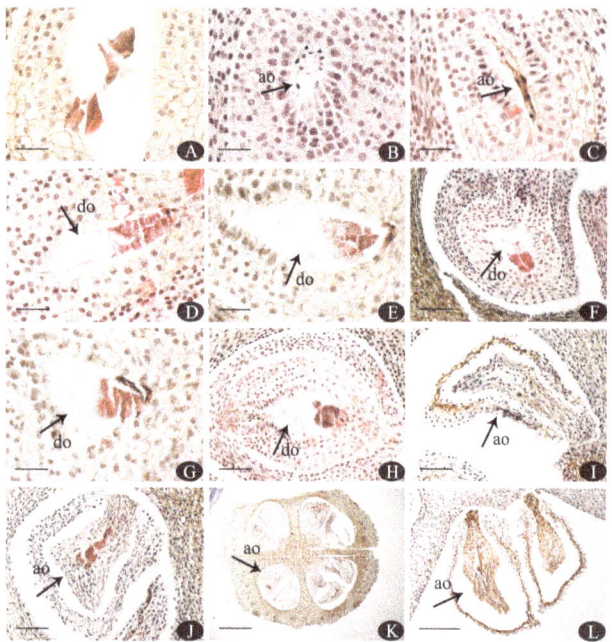

Figure 6. Degenerate and abortive ovules in self-pollinated ovaries at various stages. (A). Two sets of egg apparatus in embryo sacs 72 h after SP. (B–C). Abortive ovules with shriveling embryo sacs but normal development of the integument 8 and 10 days after SP, respectively. (D–H). Embryo sacs with an abnormal shape and degenerated egg apparatus 16, 20, 25, 30, and 35 days after SP, respectively. (I–L). Abortive ovules with dead whole tissues 40, 45, 50, and 170 days after SP, respectively. Abbreviations: ao, abortive ovule; do, degenerate ovule. Bars: A–E, G = 100 μm; F, H–J = 200 μm; K = 1000 μm; L = 500 μm.

The ISI of *C. oleifera* was 0.19 (10.12/53.41). This corresponds to the most self-incompatible category according to Zapata and Arroyo's classification system [38], and may explain the high percentage of aborted ovules (corresponding to the high level of self-sterility) observed in the present study.

Discussion

In this report, we demonstrated that *C. oleifera* 'XianglinXLC15' is a self-sterile cultivar since the fruit and seed sets after SP (pollen from the same genotype) were significantly reduced compared to after CP with a different cultivar. However, pollen tube growth to the ovary after selfing and the incidence of ovule penetration was indicative of LSI in selfed pistils. In addition, the significantly lower percentages of ovule penetration and double fertilization in selfed ovules vs. crossed ovules support a prezygotic LSI mechanism in *C. oleifera* 'XianglinXLC15'. We then discussed the possible causes of self-sterility in detail in *C. oleifera* based on a comparison to other species with similar characteristics.

Pollen tube inhibition and callose deposition in *C. oleifera*

Based on histological fluorescence microscopic observations of the pistil structure and a detailed analysis of the percentage of pollen tubes at different locations in pistils after SP compared to CP, no significant differences in morphological structure and the percentage of germinated pollen grains, pollen tubes in the middle of the style, or pollen tubes at the base of the style were observed. This was also reported for *C. sinensis*, with pollen tubes successfully reaching the ovary 48 h AP following both SP and CP [25]. In the present study, pollen tubes reached the base of the style 60 h after SP and 48 h after CP, and the growth speed of crossed tubes was slightly greater than of selfed tubes, similar to our previous findings in *C. oleifera* 'Huashuo' [6].

Furthermore, the majority of the pollen tubes did not penetrate into embryo sacs after SP. The present study shows that the percentage of penetrated ovules following SP was significantly lower than following CP. Also, a mass of bright spots considered to be callose was detected in the ovules after SP; this is probably related to ovule abortion, ovule non-fertilization, and ovule death [41]. The inhibition of pollen tubes has been reported in other species of the Theaceae. For example, Yang et al. reported inhibition at the base of the style [42]. Wang et al. also observed this phenomenon and suggested that parts of selfed tubes could grow to the ovary [43]. However, this was also documented in *Dolichandra cynanchoides* and *Tabebuia nodosa*, with pollen tubes growing successfully to the ovary with ovule penetration [44]. Based on our results, callose deposition was observed in abortive ovules 10 and 25 days etc. after SP. This was observed in a

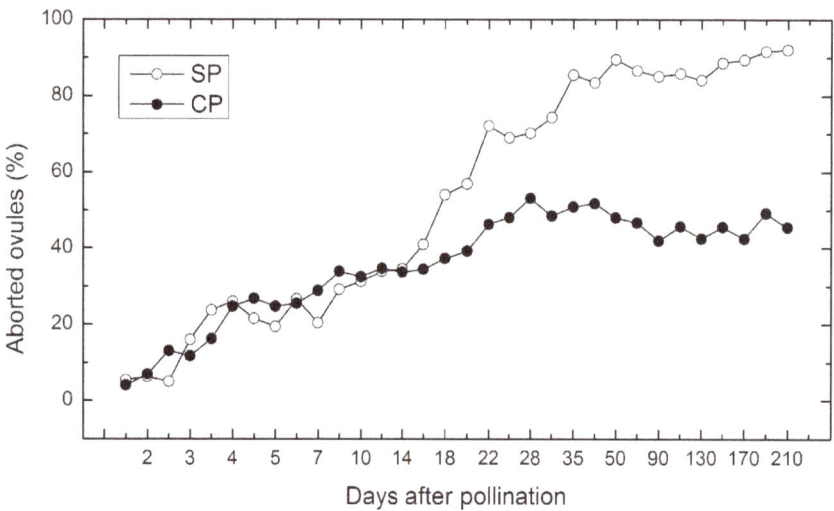

Figure 7. Percentage of aborted ovules 210 days after SP and CP at various stages.

Figure 8. Mature fruits, mature seeds, and abortive seeds after SP and CP in C. oleifera. Abbreviations: cpf, CP fruit; cps, CP seed; spf, SP fruit; sps, SP seed; sss, self-sterile seed.

significant number of self-sterile species, with different degrees of fluorescence in abortive ovules but not in fertile ovules [45]. Li et al. reported that some *C. oleifera* cultivars were self-compatible since no callose deposition was observed [5]. In *C. sinensis*, callose deposition was not observed in ovules, but was observed outside the micropyle, indicating that the pollen tubes were prevented to penetrate micropyle [25].

Mechanism of prezygotic SI in *C. oleifera*

Self-sterility in angiosperms results in a reduced fruit/seed set following SP compared to CP, and is mainly achieved through SI and EID [14,15]. The abortion period was used to distinguish between SI and EID. For SI, uniform abortion occurred at a single stage during seed development, while continuously abortive development was indicative of EID [22,27,46]. Thus, if the self-sterility was caused by postzygotic SI, fertilized ovules would be consistently rejected, and uniform rejection and failure would be observed [22,47,48]. Self-sterility resulting from EID is expected to act at any stage during embryonic development [22,37,49].

In the present study, double fertilization in a small number of selfed ovules was demonstrated. The selfed ovules also showed a smaller proportion of double fertilization 3 weeks AP, revealing that some fertilized ovules had degenerated, similar to *Eucalyptus globules* [50]. For comparison, zygotes were not observed in whole abortive ovules. The presence of two intensively stained synergids in embryo sacs suggests that the postzygotic phenomenon LSI did not occur. Various sizes (the sizes of the unfertilized ovules differed significantly) of selfed unfertilized ovaries were observed during seed development. Thus, seed abortion occurred at various stages, but the maximal area of unfertilized ovules was much smaller than in normal ovules, which is indicative of a pseudomorphic incidence in unfertilized ovules. In addition, the level of sterility increased 4 weeks AP, which indirectly indicates that self-sterility was not associated with subsequent embryonic development, which started after approximately 3 months of dormancy in *C. oleifera* [9]. And the percentage of abortive ovules 35 days after SP was similar to the final seed abortion rate (the selfed seed set was 10.12%). Additionally, the percentage of abortive ovules remained stable during subsequent development, demonstrating that self-

Table 4. Mean (±SD) comparisons of fruit set, seed set, and fruit characteristics of four pollination treatments.

Pollination treatment	Fruit set (%)	Seed set (%)	Transverse diameter (mm)	Vertical diameter (mm)	Weight (g)
SP	31.67 (2.52) a	10.12 (3.76) a	32.68 (7.70) a	30.72 (4.49) a	20.30 (12.54) a
CP	71.33 (3.51) b	53.41 (4.62) c	45.06 (2.33) c	38.55 (2.82) c	44.97 (7.02) c
OP	35.33 (4.04) a	35.22 (7.02) b	37.60 (3.99) b	34.16 (3.56) b	27.47 (8.22) b
NP	N/A	N/A	N/A	N/A	N/A

Different letters within the same column represent significant differences at the 5% level based on Duncan's multiple range test, while the same letter represents no significant differences. Abbreviations: SP, self-pollination; CP, cross-pollination; OP, open-pollination; NP, non-pollination; SD, Standard Deviation.

sterility did not occur continuously throughout the life cycle. These results support an LSI-based mechanism. LSI after ovule penetration has been reported in numerous plants [44,50]. LSI has been documented in *C. sinensis (L.) O. Kuntze*, with selfed tubes penetrating the ovule, but with fewer seed sets after SP compared to CP [51].

Although double fertilization was observed in both SP and CP ovules, with no structural or histological differences, the frequency after SP was much lower than that after CP, which could be attributed to reduced ovule penetration in selfed ovules. According to Rodríguez-Riaño, prezygotic SI is represented by low rates of ovule penetration and double fertilization in selfed ovules compared with crossed ovules [52]. In the present study, the significantly reduced rate of ovule penetration and double fertilization after SP than CP in *C. oleifera* 'XianglinXLC15' is suggestive of a prezygotic SI-based mechanism. Similar observations were reported in *Cytisus. multiflorus*, *Ipomopsis aggregate*, and *Tectona grandis* [16,17,39]. After double fertilization, the structural characteristics (excluding the area of the embryo sac 130 DAP) were similar between the selfed and crossed ovules. However, consistent suppression of self-unfertilized ovule enlargement AP suggests that self-sterility occurred at the prezygotic stage. This was also observed in *C. multiflorus* and *Cytisus. striatus* [17]. Overall, ovule development is significantly most influenced by the factor of DAP, but not the pollen type.

Sage et al. reported a reduction in fertile ovules in the absence of a stimulus in selfed ovules, which resulted in prezygotic SI [53]. However, stimulation produced by carpel tissue development after CP has been documented in many species [27,54–57]. According to Sage et al., stimulation after CP that resulted in the development of crossed ovules (but not selfed ovules) was detected, indicating that prezygotic abortion in *Narcissus triandrus* was related to post-pollination signaling [53]. A similar phenomenon has been documented in *I. aggregate* [58]. The long-distance signaling that results from self-recognition and self-rejection is lower than in the stigma-ovule in selfed ovaries of *I. aggregate*, and changes in stimulatory function are thought to be caused by this long-distance signaling [16,59]. Previous studies of long-distance signaling affecting ovule development have been conducted in other plants [54,60–63]. However, in our study, prezygotic recognition and the absence of or change in the stimulus in selfed ovules of *C. oleifera* was not observed. Physiological and histochemical studies of SI and failed ovule penetration in *C. oleifera* based on long-distance signaling in stigma-ovules are currently underway.

Fruit/seed sets are indicative of reproduction

Our data show that SP significantly reduced fruit/seed sets compared to CP in *C. oleifera*. This was also reported in *I. aggregate* and *Ziziphus* (Chinese jujubes) [58,64]. Sage et al. hypothesized that this reduced seed set resulted from physiological asynchrony in ovules. Additionally, many hermaphroditic plants were reported to contain a reduced fruit/seed set [26,39,65,66]. In our study, the selfed pollen grains germination and pollen tubes reaching the base of style were observed, similar to CP. The arrival time with them was delayed in SP than CP. After that, lower percentage of ovule penetration after SP than CP was observed and the aborted ovules after SP were greater than those after CP. All these results provided a chain of evidence that the low fruit/seed set after SP treatment was probably due to the delayed time. And this probably occurred between pollen tubes penetrating into the ovaries and the ovule viability. A similar situation was reported in

C. sinensis (L.) O. Kuntze, another species in Theaceae [51]. However, the fruit sets after OP and SP showed no significant differences. This may be due to self-pollen mixing with cross-pollen in OP, according to other cultivated species [16,23,51,58,67]. Thus, other cultivars of *C. oleifera* could be planted with *C. oleifera* 'XianglinXLC15' to generate an increased fruit/seed set.

Although the fruit/seed set after SP was lower than after CP, the size of the selfed seeds tended to be larger than of the crossed seeds. This was also reported in *C. striatus*, possibly due to greater availability of spatial resources for selfed seeds [17]. It was reported that both pollen resource availability limited fruit set and the number of seeds produced, even the size of seeds in plant [68,69]. In this study, One or several normal seeds are present in each selfed fruit, compared with the abundant developing seeds found in crossed fruit. The lower level of competition for nutrition and space makes larger seeds possible in selfed fruit of *C. oleifera*. Pollinators are an important factor for seed setting in entomophilous flowers. According to the pollination biology of *C. oleifera*, bees are effective pollinators, but they are seldom active during the flowering phase of *C. oleifera* since the temperature in late autumn is low in southern China [2]. Hence, pollen restriction in *C. oleifera* may have resulted in the low seed set. This has also been reported for *Camellia azalea* [70] and *Cyrtanthus breviflorus* [71].

Conclusions

In this research, we observed the self-sterility and low fruit/seed set after SP in *Camellia oleifera*, and explained the phenomenon in anatomy and statistics. The results indicated the existence of LSI, which supported by selfed pollen grains could germinated on the stigmas, selfed pollen tube growth in style and ovary, then pollen tubes penetrated into ovules, ovule development and fruit/seed set following SP. Further, significantly lower percentages of ovule penetration and double fertilization following SP than CP, in addition, selfed aborted ovules were prior to zygotes division, preliminary confirmed part of prezygotic LSI existing in *Camellia oleifera*.

Acknowledgments

The authors thank Professor Zhaokun Wan for the analysis of reproductive biology, Luhong Wang for assistance with field work, and Zhiqiang Han, Hui Xiang, Jing Tang for experimental assistance during the course of work.

Supporting Information

Figure S1   Thickness of integument under three fertilization treatments at various days after pollination.

Figure S2   Interaction between DAP and FT on IIT.

Figure S3   Interaction between DAP and FT on OIT.

Author Contributions

Conceived and designed the experiments: TL DYY. Performed the experiments: TL FZ CG YY. Analyzed the data: TL. Contributed reagents/materials/analysis tools: DYY FZ. Wrote the paper: TL. Revise the article critically for important intellectual content: LZ XFT TL.

References

1. Zhuang RL (2008) *Camellia oleifera* in China. Beijing: China Forestry Press.
2. Deng YY (2009) Study on pollination biology of *Camellia oleifera*. Master's thesis, Central South University of Forestry & Technology, Changsha, Hunan, China.
3. Yuan DY, Wang R, Yuan J, Liao T, Cui X, et al. (2010) The influence of nutrient elements on pollen germination percentage in *Camellia oleifera*. Journal of Fujian Agriculture and Forestry University (Natural Science Edition) 39: 471–474.
4. Tan XF, Yuan DY, Yuan J, Liao T (2010) Pollen germination in *Camellia oleifera* with ascorbic acid and plant growth regulators. Journal of Zhejiang Forestry College 27: 941–944.
5. Li CL, Yao XH, Yang SP, Ren HD, Cao YQ (2011) Research on pollen morphology and germination of *Camellia oleifera* Able. Chinese journal of oil crop sciences 33: 242–246.
6. Liao T, Yuan DY, Peng SF, Zou F (2012) A fluorescence microscope observation on self and cross-pollination of pollen tubes in *Camellia oleifera*. Journal of Central South University of Forestry & Technology 32: 34–37.
7. He CY (2009) The anatomical studies on the progress of sexual reproduction of *Camellia oleifera*. Master's thesis, Central South University of Forestry & Technology, Changsha, Hunan, China.
8. Yuan DY, Zou F, Tan XF, He CY, Yuan J, et al. (2011) Flower bud differentiation and development of male and female gametophytes in *Camellia oleifera*. Journal of Central South University of Forestry & Technology 31: 15–20.
9. Cao HJ (1965) Embryological observation on *Camellia oleifera*. Acta Bot Sin 13: 44–53.
10. Jiang CJ, Wang ZH (1987) Anatomical studies on the development of the embryo of tea plant. J Tea Sci 7: 23–28.
11. Liu ZS, Liang YR (1987) Study on fertilization of tea plant. J Tea Sci 7: 19–22.
12. Hwang YJ, Okubo H, Fujieda K (1992) Pollen Tube Growth, Fertilization and Embryo Development of *Camellia japonica* L. ×C. *chrysantha* (Hu) Tuyama. J Jpn Soc Hortic Sci 60: 955–961.
13. Sethi SB (1965) Structure and development of seed in *Camellia Sinensis* (L.) O. Kuntze. Proc Natn Inst Sci India 31: 24–42.
14. de Nettancourt D (1977) Incompatibility in angiosperms. Berlin: Springer-Verlag.
15. de Nettancourt D (1997) Incompatibility in angiosperms. Sex Plant Reprod 10: 185–199.
16. Sage TL, Price MV, Waser NM (2006) Self-steritily in *Ipomopsis aggregate* (Polemoniaceae) is due to prezygotic ovule degeneration. Am J Bot 93: 254–262.
17. Valtueña FJ, Rodríguez-Riaño T, Espinosa F, Ortega-Olivencia A (2010) Self-sterility in two *Cytisus* species (Leguminosae, Papilionoideae) due to early-acting inbreeding depression. Am J Bot 97: 123–135.
18. de Nettancourt D (2001) Incompatibility and incongruity in wild and cultivated plants. New York: Springer-Verlag.
19. Silva NF, Goring DR (2001) Mechanisms of self-incompatibility in flowering plants. Cell Mol Life Sci 58: 1988–2007.
20. Franklin-Tong VE, Franklin FCH (2003) The different mechanisms of gametophytic self-incompatibility. Phil Trans R Soc Lond B, Biological Sciences 358: 1025–1032.
21. Hiscock SJ, McInnis SM (2003) Pollen recognition and rejection during the sporophytic self-incompatibility response: *Brassica* and beyond. Trends Plant Sci 8: 606–612.
22. Seavey SR, Bawa KS (1986) Late-acting self-incompatibility in angiosperms. Bot Rev 52: 195–219.
23. Cope FW (1962) The mechanism of pollen incompatibility in *Theobroma cacao* L. Heredity 17: 157–182.
24. Kenrick J, Kaul V, Williams EG (1986) Self-incompatibility in *Acacia retinodes*: site of pollen-tube arrest is the nucellus. Planta 169: 245–250.
25. Chen X, Hao S, Wang L, Fang W, Wang Y, et al. (2012) Late-acting self-incompatibility in tea plant (*Camellia sinensis*). Biologia 67: 347–351.
26. Lewis G, Gibbs P (1999) Reproductive biology of *Caesalpinia calycina* and C. *pluviosa* (Leguminosae) of the cattanga of northeastern Brazil. Plant Syst and Evol 217: 43–53.
27. Sage TL, Sampson FB (2003) Evidence for ovarian self-incompatibility as a cause of self-sterility in the relictual woody angiosperm, *Pseudowintera axillaris* (Winteraceae). Ann Bot 91: 807–816.
28. Bittencourt NSJr, Semir J (2005) Late-acting self-incompatibility and other breeding systems in Tabebuia (Bignoniaceae). Int J Plant Sci 166: 493–506.
29. Chai LJ, Ge XX, Biswas MK, Xu Q, Deng XX (2011) Self-sterility in the mutant 'Zigui shatian' pummel (*Citrus grandis* Osbeck) is due to abnormal post-zygotic embryo development and not self-incompatibility. Plant Cell Tiss Org 104: 1–11.
30. Zhou QY, Liu GS (2011) The embryology of *Xanthoceras* and its phylogenetic implications. Plant Syst Evol 298: 457–468.
31. Wang XN, Chen YZ, Peng SF, Yang XH (2008) Five elite varieties of *Cameillia oleifera*. Scientia Silvae Sinicae 44: 173–174.
32. Hsu HW, Kuo SR, Chuang NJ, Liang YC (2002) Phenology of growth and development of strobili of *Taiwania cryptomerioides* Hay. Taiwan J For Sci 17: 241–255.
33. Hiratsuka S, Takahashi E, Hirata N (1987) Pollen tube growth in hibitors involved in the ovary of self-incompatible Japanese pear. Plant Cell Physiol 28: 293–299.
34. Martin FW (1959) Staining and observing pollen tubes in the style by means of fluorescence. Biotech Histochem 34: 125–128.
35. Chung JD, Kuo SR (2005) Reproductive Cycles of *Calocedrus formosana*. Taiwan J For Sci 20: 315–329.
36. Bittencourt NSJr, Gibbs PE, Semir J (2003) Histological study of post-pollination events in *Spathodea campanulata* Beauv. (Bignoniaceae), a species with late-acting self-incompatibility. Ann Bot 91: 827–834.
37. Mahy G, Jacquemart AL (1999) Early inbreeding depression and pollen competition in *Calluna vulgaris* (L.) Hull. Ann Bot 83: 697–704.
38. Zapata TR, Arroyo MTK (1978) Plant reproductive ecology of a secondary deciduous tropical forest in Venezuela. Biotropica 10: 221–230.
39. Tangmitcharoen S, Owens JN (1997) Pollen viability and pollen-tube growth following controlled pollination and their relation to low fruit production in Teak (*Tectona grandis* Linn. f.). Ann Bot 80: 401–410.
40. Stephenson AG, Good SV, Vogler DW (2000) Interrelationships among inbreeding depression, plasticity in the self-incompatibility system, and the breeding system of *Campanula rapunculoides* L. (Campanulaceae). Ann Bot 85: 211–219.
41. Dumas C, Knox RB (1983) Callose and determination of pistil viability and incompatibility. Theor Appl Genet 67: 1–10.
42. Yang MZ, Chen YR (2000) Observation of self-incompatibility in tea. Chinese Horticulture Abstracts 46: 83–92.
43. Wang Y, Jiang CJ, Zhang HY (2008) Observation on the self-incompatibility of pollen tubes in self-pollination of tea plant in style in vivo. J Tea Sci 28: 429–435.
44. Gibbs PE, Bianchi MB (1999) Does late-acting self-incompatibility (LSI) show family clustering? Two more species of Bignoniaceae with LSI: *Dolichandra cynanchoides* and *Tabebuia nodosa*. Ann Bot 84: 449–457.
45. Vishnyakova MA (1991) Callose as an indicator of sterile ovules. Phytomorphology 41: 245–252.
46. Pound LM, Wallwork MAB, Potts BM, Sedgley M (2003) Pollen tube growth and early ovule development following self- and cross-pollination in *Eucalyptus nitens*. Sex Plant Reprod 16: 59–69.
47. Seavey SR, Carter SK (1994) Self-sterility in *Epilobium obcordatum* (Onagraceae). Am J Bot 81: 331–338.
48. Gibbs PE, Bianchi MB (1993) Post-pollination events in species of *Chorisia* (Bombaceae) and *Tabebuia* (Bignoniaceae) with late-acting self-incompatibility. Botanica Acta 106: 64–71.
49. Husband BC, Schemske DW (1995) Magnitude and timing of inbreeding depression in a diploid population of *Epilobium angustifolium* (Onagraceae). Heredity 75: 206–215.
50. Pound LM, Wallwork MAB, Potts BM, Sedgley M (2002) Early ovule development following self- and cross-pollinations in *Eucalyptus globules* Labill. ssp. *globules*. Ann Bot 89: 613–620.
51. Wachira FN, Kamunya SK (2005) Pseudo-self-incompatibility in some tea clones (*Camellia sinensis* (L.) O. Kuntze). J Hortic Sci Biotech 80: 716–720.
52. Rodriguez-Riaño T, Ortega-Olivencia A, Devesa JA (1999) Reproductive biology in two *Genisteae* (Papilionoideae) endemic of the western Mediterranean region: *Cytisus striatus* and *Retama sphaerocarpa*. Can J Bot 77: 809–820.
53. Sage TL, Strumas F, Cole WW, Barrett SCH (1999) Differential ovule development following self- and cross-pollination: the basis of self-sterility in Narcissus triandrus (Amaryllidaceae). Am J Bot 86: 855–870.
54. Pimienta E, Polito VS (1983) Embryo sac development in almond (*Prunus dulcis* (Mill.) D. A. Webb) as affected by cross-, self- and non-pollination. Ann Bot 71: 469–479.
55. Sekhar KNC, Heij EG (1995) Changes in proteins and peroxidases induced by compatible pollination in the ovary of *Nicotiana tabacum* L. Sex Plant Reprod 8: 369–374.
56. O'Neil SD (1997) Pollination regulation of flower development. Annual Review of Plant Physiology and Plant Molecular Biology 48: 547–574.
57. Pontieri V, Sage TL (1999) Evidence for stigmatic self-incompatibility, pollination-induced ovule enlargement and transmitting tissue exudates in the palaeoherb, *Saururus cernuus* L. (Saururaceae). Ann Bot 84: 507–519.
58. Waser NM, Price MV (1991) Reproductive costs of self pollination in *Ipomopsis aggregate* (Polemoniaceae): are ovules usurped? Am J Bot 78: 1036–1043.
59. Waser NM, Fugate ML (1986) Pollen precedence and stigma closure: a mechanism of competition for pollination between *Delphinium nelsonii* and *Ipomopsis aggregata*. Oecologia 70: 573–577.
60. Pimienta E, Polito VS (1982) Ovule abortion in 'Nonpareil' almond (*Prunus dulcis* (Mill.) D.A. Webb). Am J Bot 69: 913–920.
61. Herrero M, Araeloa A (1989) Influence of the pistil on pollen tube kinetics in peach (*Prunus persica*). Am J Bot 76: 1441–1447.
62. Herrero M, Gascon M (1987) Prolongation of embryo sac viability in pear (*Pyrus communis*) following pollination or treatment with gibberellic acid. Ann Bot 60: 287–293.
63. Koehl V (2002) Functional reproductive biology of *Illicium floridanum* (Illiciaceae). Master's thesis, University of Toronto, Toronto, Ontario, Canada.
64. Ackerman WL (1961) Flowering, pollination, self-sterility and seed development of Chinese jujube. J Am Soc Hortic Sci 77: 265–269.
65. Bryndum K, Hedegart T (1969) Pollination of teak (*Tectona grandis* L. f.). Silvae Genet 18: 77–80.

66. Stephenson AG (1981) Flower and fruit abortion: proximate causes and ultimate functions. Annu Rev Ecol Syst 12: 253–280.

67. Broyles SB, Wyatt R (1993) The consequences of self-pollination in *Asclepias exaltata*, a self-incompatible milkweed. Am J Bot 80: 41–44.

68. Asikainen E, Mutikainen P (2005) Pollen and resource limitation in a gynodioecious species. Am J Bot 92: 487–494.

69. Obeso JR (2002) The costs of reproduction in plants. New Phytol 155: 321–348.

70. Luo XY, Tang GD, Mo IJ, Zhuang XY (2011) Pollination biology of *Camellia changii*. Chinese Journal of Ecology 30: 552–557.

71. Vaughton G, Ramsey M, Johnson SD (2010) Pollination and late-acting self-incompatibility in *Cyrtanthus breviflorus* (Amaryllidaceae): implications for seed production. Ann Bot. 106: 547–555.

12

Ras GTPase Activating Protein Colra1 Is Involved in Infection-Related Morphogenesis by Regulating cAMP and MAPK Signaling Pathways through CoRas2 in *Colletotrichum orbiculare*

Ken Harata, Yasuyuki Kubo*

Laboratory of Plant Pathology, Graduate School of Life and Environmental Sciences, Kyoto Prefectural University, Kyoto, Japan

Abstract

Colletotrichum orbiculare is the causative agent of anthracnose disease on cucurbitaceous plants. Several signaling pathways, including cAMP–PKA and mitogen-activating protein kinase (MAPK) pathways are involved in the infection-related morphogenesis and pathogenicity of *C. orbiculare*. However, upstream regulators of these pathways for this species remain unidentified. In this study, *ColRA1*, encoding RAS GTPase activating protein, was identified by screening the *Agrobacterium tumefaciens*-mediated transformation (AtMT) mutant, which was defective in the pathogenesis of *C. orbiculare*. The *coira1* disrupted mutant showed an abnormal infection-related morphogenesis and attenuated pathogenesis. In *Saccharomyces cerevisiae*, Ira1/2 inactivates Ras1/2, which activates adenylate cyclase, leading to the synthesis of cAMP. Increase in the intracellular cAMP levels in *coira1* mutants and dominant active forms of *CoRAS2* introduced transformants indicated that Colra1 regulates intracellular cAMP levels through CoRas2. Moreover, the phenotypic analysis of transformants that express dominant active form *CoRAS2* in the *comekk1* mutant or a dominant active form *CoMEKK1* in the *coras2* mutant indicated that CoRas2 regulates the MAPK CoMekk1–Cmk1 signaling pathway. The CoRas2 localization pattern in vegetative hyphae of the *coira1* mutant was similar to that of the wild-type, expressing a dominant active form of *RFP–CoRAS2*. Moreover, we demonstrated that bimolecular fluorescence complementation (BiFC) signals between Colra1 and CoRas2 were detected in the plasma membrane of vegetative hyphae. Therefore, it is likely that Colra1 negatively regulates CoRas2 in vegetative hyphae. Furthermore, cytological analysis of the localization of Colral and CoRas2 revealed the dynamic cellular localization of the proteins that leads to proper assembly of F-actin at appressorial pore required for successful penetration peg formation through the pore. Thus, our results indicated that Colra1 is involved in infection-related morphogenesis and pathogenicity by proper regulation of cAMP and MAPK signaling pathways through CoRas2.

Editor: Zhengyi Wang, Zhejiang University, China

Funding: Grants-in-Aid for scientific Research from the Ministry of Education, Culture, Sports, Science and Technology (Grant number 24248009). The funders had no role in study design, data collection and analysis, decision to publish, or preparation of the manuscript.

Competing Interests: The authors have declared that no competing interests exist.

* Email: y_kubo@kpu.ac.jp

Introduction

Colletotrichum orbiculare is the causative agent of cucumber anthracnose disease. The infection process is initiated by the recognition of an appropriate surface. A series of changes in the morphology upon recognizing the appropriate signals, including the formation of a specialized infection structure called appressorium, is important for the successful infection of the host plants.

Several signal-transduction related genes associated with these morphological changes have been characterized in *C. orbiculare* [1]. It also has been shown that the MAPK and cyclic AMP (cAMP) signaling pathways are linked to infection-related morphological changes in this fungus. A yeast MAPK kinase (MAPKK) kinase *STE11* homolog, *CoMEKK1*, is involved in the appressorium development in *C. orbiculare* [2]. *C. orbiculare* mutant with the adenylate cyclase encoding gene *cac1* shows

defectiveness in conidial germination [3], however, no upstream regulators of cAMP and MAPK signaling pathways have been identified in *C. orbiculare*.

Ras is the prototypical member of the small GTP binding protein superfamily that plays a pivotal role in proliferation and differentiation in eukaryotic cells. In *Magnaporthe oryzae*, MoRas1 and MoRas2 interact with Mst50, which interacts directly with the mitogen-activated proteins Mst11 and Mst7 [4]. In *Fusarium graminearum*, FgRas2 is involved in hyphal growth and pathogenicity [5]. In *C. trifolii*, CtRas is involved in conidial germination and hyphal growth [6]. Ras acts as a molecular switch that exists both in the active (GTP-bound) and the inactive (GTPase-bound) states. Cycling between these two states is aided by the interaction of a GTPase activating protein and guanine-nucleotide exchange factors. In *Saccharomyces cerevisiae*, the Ras activity is controlled positively by the guanine nucleotide exchange factor (GEF) Cdc25 and negatively by the GTPase activating proteins Ira1 and Ira2 (GAPs) [7]. In *S. cerevisiae*, the RAS GTPase activating proteins Ira1 and Ira2 inactivate Ras1 and Ras2, which in turn activate adenylate cyclase, promoting the synthesis of cAMP from ATP. Ira1 and Ira2 are important factors for controlling cAMP production. The cAMP signaling pathway plays a pivotal role in transducing environmental cues during cell development in *Magnaporthe oryzae* [8]. Moreover, the cAMP signaling pathway of *C. orbiculare* plays a critical role in regulating conidial germination and pathogenicity [9]. Adenylate cyclase gene *MAC1* encodes a Ras-association domain [10], [11]. In terms of protein structure, the Ras protein putatively interacts with Mac1 in *M. oryzae*. It was shown that cyclase-associated protein Cap1 of *M. oryzae* directly interacts with Mac1 and plays a role in the activation of Mac1, which may function downstream of Ras [12]. In *Ustilago maydis*, *RAS2* promotes filamentous growth through a MAP kinase cascade, however, the function of *RAS2* in the cAMP signal transduction remains unknown [13]. Therefore, direct evidences for the involvement of the Ras protein in the cAMP signal transduction are not sufficient in phytopathogenic fungi.

Because Ras proteins transduce signals from the cell surface to various intracellular effectors, they are located and function only at the plasma membrane [14], [15]. Conversely, data provided by the yeast GFP fusion localization database [16] and other published data [17], [18] indicate that Ras2, Cyr1, Cdc25, and Ira proteins mainly localize at the internal membranes of the endoplasmic reticulum and mitochondrial membranes, but only marginally at the plasma membrane. Recently, it was reported that active Ras accumulates mainly in the plasma membrane and nucleus when the cells are grown on medium containing glucose, whereas it accumulates mainly in the mitochondria in glucose-starved cells [19]. Moreover, PKA activity causes Ira2 to move away from the mitochondria [19]. Therefore, it is considered that the behavior of Ras proteins and Ras GTPase activating proteins and their localization depends on various growth conditions. In plant pathogenic fungi, subcellular dynamics of Ras proteins and Ras GTPase activating proteins during infection and the location of this interaction remain unclear.

In this study, we identified a novel *S. cerevisiae* homolog gene *CoIRA1*, encoding the RAS GTPase-activating protein (RAS-GAP) of *C. orbiculare* and characterized the *CoIRA1* function in relation to the activation of cAMP-PKA and MAPK signaling pathways. By phenotypic analysis of the *coira1* mutant, cytological analysis of CoRas2 localization and BiFC assay, we showed that CoIra1 regulates these signaling pathways through CoRas2 as a negative regulator. Conclusively, we presented CoIra1 is involved in infection-related morphogenesis by regulating cAMP and MAPK signaling pathways through CoRas2 in *C. orbiculare*.

Materials and Methods

Fungal and bacterial strains

The 104-T (MAFF240422) *C. orbiculare* (Berk. & Mont.) Arx [syn. *C. lagenarium* (pass.); Ellis & Halst.] strain was used as the wild-type. All the *C. orbiculare* strains used in this study are listed in Table 1 and were cultured on PDA media (3.9% [w/v] PDA; Difco laboratories, Detroit) at 24°C. One shop TOP10 chemically competent *E. coli* cultured in Luria-Bartani media [20] at 37°C was used as a host for gene manipulation. When required, the supplement kanamycin was added to the medium at 50 μg/ml. *A. tumefaciens* C58C1 cultured in Luria-Bartani media at 28°C was used to transform *C. orbiculare* by AtMT.

Fungal transformation

For the fungal transformation, we used an AtMT protocol that was previously described [21] with slight modifications. The hygromycin-resistant transformants were selected on the PDA medium containing 100 μg/ml of hygromycin B (Wako Chemicals, Osaka, Japan), 50 μg/ml of cefotaxim (Wako Chemicals, Osaka, Japan), and 50 μg/ml of spectinomycin (Wako Chemicals, Osaka, Japan). The bialaphos-resistant transformants were selected on an SD medium containing 10 μg/mL of bialaphos (Meiji Seika Kaisha, Ltd., Tokyo, Japan), 100 μg/ml of cefotaxim, and 100 μg/ml of spectinomycin. The sulfonylurea-resistant transformants were selected on an SD medium containing 4 μg/ml of chlorimuron-ethyl (Chem Service West Chester, PA, USA.), 100 μg/ml of cefotaxim, and 100 μg/ml of spectinomycin.

Genomic DNA blot analysis

The total DNA from the mycelia of *C. orbiculare* was isolated, and a DNA blot analysis was performed using a previously described method [22]. DNA digestion, gel electrophoresis, labeling of probes, and hybridization were performed according to the manufacturer's instructions following standard methods [20]. DNA probes were labeled with DIG-dUTP using a BcaBESTTM DIG labeling kit (Takara Bio, Ohtsu, Japan). Hybridized DNA was detected with anti-Digoxygenin antibody Fab fragments conjugated to alkaline phosphatase (Roche Diagnostics, Tokyo, Japan). Light emission from the enzymatic dephosphorylation of the CDP-Star Detection Reagent (GE Healthcare, Tokyo, Japan) was detected using the Fujifilm LAS-1000 Plus Gel Documentation System (Fujifilm, Tokyo).

Construction of the *CoIRA1* gene replacement vector and *CoIRA1* complementation vector

To replace the *CoIRA1* gene with the hygromycin-resistance gene, we constructed a *CoIRA1* gene-replacement vector pBIG4MRBrev-coira1. We first amplified the upstream region of the *CoIRA1* gene, the hygromycin-resistance gene, and the downstream region of the *CoIRA1* gene by PCR using the primer pairs CoIRA1F1A–CoIRA1R2D, CoIRA1hphF1C–CoIRA1hphR1D, and CoIRA1F2C–CoIRA1R1B, respectively. The primers used are listed in Table S1. Next, the pBIG4MRBrev–coira1 vector was constructed using the GeneART seamless cloning and assembly kit (Life Technologies, Carlsbad, California USA) with the amplified product and the *A. tumefaciens* binary vector pBIG4MRBrev. This vector, which contains a bialaphos resistance gene, was used as the gene replacement plasmid.

To perform a complementation assay of the *coira1* mutant, we constructed the CoIRA1 complementation vector pBIG4MRBrev–CoIRA1. We first amplified the upstream region of the *CoIRA1* gene, the middle region of the *CoIRA1* gene, and the downstream region of the *CoIRA1* gene by PCR using the primer pairs CoIRA1F3A–CoIRA1R4D, CoIRA1F5D–CoIRA1R5C, and CoIRA1F4C–CoIRA1R4D, respectively. Next, the pBIG4MRBrev–CoIRA1 vector was constructed with the amplified product, and the *A. tumefaciens* binary vector pBIG4MRBrev was constructed using the GeneART seamless cloning and assembly kit. This vector, which contains a bialaphos resistance gene, was used as the gene replacement plasmid.

Construction of dominant active and negative *CoRAS1* vectors

To construct a dominant active form of the *CoRAS1* vector, we amplified a 1.4-kb fragment containing the upstream region of the *CoRAS1* gene, a 2.0-kb fragment containing the downstream

Table 1. Fungal strains used in this study.

Strain	Description	References
WT	Wild-type strain of *Colletorichum orbiculare*	This study
Dl1–1	*coira1* disruptant of WT	This study
Dl1–2	*coira1* disruptant of WT	This study
El1–1	*CoIRA1* ectopic transformant WT	This study
Cl1–1	Dl1–1 complemented with *CoIRA1*	This study
Cl1–2	Dl1–2 complemented with *CoIRA1*	This study
DC1	*cac1* disruptant of WT	Yamauchi et al. 2004
DARS1	WT transformed with a dominant active form *CoRAS1*	This study
DARS2	WT transformed with a dominant active form *CoRAS2*	This study
iDNRA1	Dl1–1 transformed with a dominant negative form *CoRAS1*	This study
iDNRA2	Dl1–1 transformed with a dominant negative form *CoRAS2*	This study
DMK1	*comekk1* disruptant of WT	Sakaguchi et al. 2010
DRS2–1	*coras2* disruptant of WT	This study
DRS2–2	*coras2* disruptant of WT	This study
CRS2–1	DRS2–1 complemented with *CoRAS2*	This study
CRS2–2	DRS2–2 complemented with *CoRAS2*	This study
DRS2/DAMK1	DRS2 transformed with a dominant active form *CoMEKK1*	This study
DMK/DARS2	DMK1 transformed with a dominant negative form *CoRAS2*	This study
DRS2/RFP-RS2	DRS2 transformed with *RFP-CoRAS2*	This study
WT/RFP-RS2	WT transformed with *RFP-CoRAS2*	This study
WT/RFP-DARS2	WT transformed with RFP-*DACoRAS2*	This study
iRFP-RS2	Dl1–1 transformed with *RFP-CoRAS2*	This study
Vc-RS2	WT transformed with *VENUS*(1–158aa)*-CoRAS2*	This study
IRA1-Vn	WT transformed with *CoIRA1-VENUS*(159aa-238aa)	This study
Vc-RS2/IRA1-Vn	IRA-Vn transformed with *VENUS*(1–158aa)*-CoRAS2*	This study
IRA1-VENUS	WT transformed with *CoIRA1-VENUS*	This study
RFP-RS2/IRA1-VENUS	WT/RFP-RS2 transformed with *CoIRA1-VENUS*	This study
RFP-DARS2/IRA1-VENUS	WT/RFP-DARS2 transformed with *CoIRA1-VENUS*	This study
LA/IRA1-VENUS	IRA1-VENUS transformed with *LifeAct-RFP*	This study
WT/LA	WT with *LifeAct-RFP*	This study
iLA	Dl1–1 transformed with *LifeAct-RFP*	This study
DPS1	*pks1* disruptant of WT	Takano et al. 1995
DSD1	*ssd1* disruptant of WT	Tanaka et al. 2007

region of the *CoRAS1* gene, and the linearized pBIG4MRBrev vector by PCR using the primer pairs CoRAS1F1A–CoRAS1^{G17V}R1B, CoRAS1^{G17V}F1A–CoRAS1R1B, and CoRAS1p-BIF1A–R1B. Next, the pBIG4MRBrev–CoRAS1^{G17V} vector was constructed with the amplified product using the GeneART seamless cloning and assembly kit. To construct a dominant negative form of the *CoRAS1* vector, we amplified a 1.4-kb fragment containing the upstream region of the *CoRAS1* gene, a 2.0-kb fragment containing the downstream region of the *CoRAS1* gene, and the linearized pBIG4MRBrev vector by PCR using the primer pairs RAS1^{S22N}F1A–RAS1R1B, RAS1F1A–RAS1^{S22N}R1B, and CoRAS1pBIF1A–R1B, respectively. Next, the pBIG4MRBrev–CoRAS1^{S22N} vector was constructed with the

amplified product using the GeneART seamless cloning and assembly kit.

Construction of the *CoRAS2* gene replacement, complementation, dominant active or negative vectors

To replace the *CoRAS2* gene for a hygromycin-resistance gene, we constructed the *CoRAS2* gene-replacement vector pBIG4MRBrev–coras2. We first amplified the upstream region of the *CoRAS2* gene, the hygromycin-resistance gene, the downstream region of the *CoRAS2* gene, and the linearized pBIG4MRBrev vector by PCR using the primer pairs CoRAS2F1A–CoRAS2R2D, CoRAS2hphF1C–CoRAS2hphR1D, CoRAS2F2C–CoRAS2R1B, and CoRAS2pBIF1–R1, respectively. Next, the pBIG4MRBrev–coras2 vector was constructed with the amplified product using the GeneART seamless cloning and assembly kit. To perform a complementation assay for the *coras2* mutant, we constructed the *CoRAS2* complementation vector pBIG4MR-Brev–CoRAS2. We amplified a 3.3-kb fragment containing the *CoRAS2* gene and the linearized pBIG4MRBrev vector by PCR using the primer pairs CoRAS2F1A–R1B and CoRAS2pBIF1A–R1B, respectively. Next, the pBIG4MRBrev–CoIRA1 vector was constructed with the amplified product using the GeneART seamless cloning and assembly kit.

To construct a dominant active form of the *CoRAS2* vector, we amplified a 1.4-kb fragment containing the upstream region of the *CoRAS2* gene, a 1.9-kb fragment containing the downstream region of the *CoRAS2* gene, and the linearized pBIG4MRBrev vector by PCR using the primer pairs CoRAS2F1A–CoRAS2^{Q65L}R1B, CoRAS2^{Q65L}F1A–CoRAS2R1B, and CoRAS2pBIF1A–R1B, respectively. Next, the pBIG4MRBrev–CoRAS2^{Q65L} vector was constructed with the amplified product using the GeneART seamless cloning and assembly kit. To construct a dominant negative form of the CoRas2 vector, we amplified a 1.2-kb fragment containing the upstream region of the *CoRAS2* gene, a 2.2-kb fragment containing the downstream region of the *CoRAS2* gene, and the linearized pBIG4MRBrev vector by PCR using the primer pairs CoRAS2F1A–CoRAS2^{G19A}R1B, CoRAS2^{G19A}F1A–CoRAS2R1B, and CoRAS2pBIF1A–R1B, respectively. Next, the pBIG4MRBrev–CoRAS2^{G19A} vector was constructed with the amplified product using the GeneART seamless cloning and assembly kit.

Construction of the *CoIRA1*–VENUS vector

To construct the pBITEF–VENUS vector, we amplified a 0.9-kb fragment containing the TEF promoter and the upstream region of the *GFP* gene, a 0.6-kb fragment containing the downstream region of the *GFP* gene, and the linearized pBIG4MRBrev vector by PCR using the primer pairs TEFF1–VENUSR1, VENUSF1–glyGFPR1, and pBIG4VENUSF1–R1, respectively. Next, the pBIG4MRBrev–TEF–VENUS vector was constructed with the amplified product using the GeneART seamless cloning and assembly kit. To construct the CoIRA1–VENUS vector, we amplified the *VENUS* gene, the hygromycin-resistance gene, and the linearized pBIG4MRBrev–CoIRA1 vector by PCR using the primer pairs glyGFPF1–GFPR1, VENUShphF1–VENUSR1, and pBICoIRA1–VENUSF1–CoIRApBIcomR1–GFP, respectively. Next, the pBIG4MRBrev–CoIRA1–VENUS–hph vector was constructed with the amplified product using the GeneART seamless cloning and assembly kit.

Construction of the *RFP* fused *CoRAS2* and the *RFP* fused *CoRAS2*^{Q65L} vectors

To construct the *RFP* fused *CoRAS2* vector, we amplified the *RFP* gene and linearized pBIG4MRBrev–CoRAS2 vector by PCR using the primer pairs RFPF1–glyRFPR1 and CoRAS2pBIF1–R1, respectively. Next, the pBIG4MRBrev–RFP–CoRAS2 vector was constructed with the amplified product using the GeneART seamless cloning and assembly kit. To construct the *RFP*–fused *CoRAS2*^{Q65L} vector, we amplified the *RFP* gene and linearized pBIG4MRBrev–CoRAS2^{Q65L} vector using the primer pairs RFPF1–glyRFPR1 and CoRAS2pBIF1–R1, respectively. Next, the pBIG4MRBrev–RFP–CoRAS2^{Q65L} vector was constructed with the amplified product using the GeneART seamless cloning and assembly kit.

Construction of the VENUS–N (1–158 aa) fused CoIRA1 and VENUS–C (159–238 aa) fused CoRAS2 vectors for BiFC assays

To construct the *VENUS*-N fused *CoIRA1* vector, we amplified the *VENUS*–N fragment, the hygromycin-resistance gene, and the linearized pBIG4MRBrev–CoIRA1 vector by PCR using the primer pairs glyGFPF1–αGFPR1, VENUShphF1–VENUSR1, and pBICoIRA1–VENUSF1–CoIRApBIcomR1–GFP respectively. Next, the pBIG4MRBrev–CoIRA1–nVENUS–hph vector was constructed with the amplified product using the GeneART seamless cloning and assembly kit. To construct the *VENUS*–C fused *CoRAS2* vector, we amplified the *VENUS*–C fragment and linearized pBIG4MRBrev–CoRAS2 vector by PCR using the primer pair BGFPF2–glyGFPR1 and CoRAS2PBIcompF3–c–CoRAS2PBIcompR3, respectively. Next, the pBIG4MRBrev–cVENUS–CoRAS2 vector was constructed with the amplified product using the GeneART seamless cloning and assembly kit.

Intracellular cAMP measurements

Mycelia were collected from three-day old PSY liquid cultures and frozen in liquid nitrogen. All samples were lyophilized for 24 h and weighed. For every 10 mg of mycelia, 200 μl of ice-cold 6% trichloroacetic acid was added. The precipitate was removed by centrifugation at 2000×g for 15 min at 4°C, the supernatant was transferred to a new tube, and the TCA was extracted four times with five volumes of water-saturated ether. The concentration of cAMP was determined using the cAMP Biotrak Enzyme immunoassay system (GE Health Life Science, UK) according to the manufacturer's instructions.

Western blot

The total protein was isolated from vegetative hyphae using a previously described method [23]. The protein separated on SDS–PAGE gels was transferred onto a PVDF membrane using an Xcell SureLock Mini-Cell. The phosphorylation activation of Maf1 and Cmk1 MAPK kinase was detected using a PhosphoPlus p44/42 MAP kinase antibody kit (Cell Signaling Technology). Alkaline Phosphatase-conjugated secondary antibody and light emission from the enzymatic dephosphorylation of the CDP-Star Detection Reagent (GE Healthcare, Tokyo, Japan) was detected using the Fujifilm LAS-1000 Plus Gel Documentation System (Fujifilm, Tokyo). Anti-actin antibodies (Wako, Japan) were used at a 1:1000 dilution for Western blot analysis.

Pathogenicity tests

An inoculation assay on cucumber cotyledons (*Cucumis sativus* L. "Suyo") was performed as described by Tsuji et al. (1997) [24]. The conidia of *C. orbiculare* were obtained from seven-day old cultures, and drops of 10-μl conidial suspension (1×10^5 conidia per ml) were added on the surface of cucumber cotyledons at different locations. To assess invasive growth ability, 10-μl drops of spore suspension were added on wounded sites that were created by scratching the leaf surface with a sterile toothpick. After inoculation, the cotyledons were incubated at 24°C for seven days.

Light microscopy

For appressorium formation and penetration assays *in vitro*, conidia were harvested from seven-day old PDA cultures and suspended in distilled water. The conidial suspension, adjusted to 1×10^5 conidia per ml, was placed on a multiwell glass slide (eight-well multi-test slide; ICN Biomedicals, Aurora, OH, U.S.A.) and incubated in humid boxes at 24°C. Germlings were observed by a Nikon ECLIPSE E600 microscope with differential interference contrast optics (Nikon, Tokyo, Japan). To observe the formation of infectious hyphae in cucumber leaves, the conidial suspension was inoculated on the abaxial surface of cucumber cotyledons and incubated at 24°C for three days. Then, the inoculation site was cut off and stained with 0.1% lactophenol-aniline blue. VENUS and RFP fluorescence was observed by a Carl ZEISS Axio Imager M2 microscope (Zeiss, Gottingen, Germany) with 470 and 595 nm of excitation wavelength, respectively.

Results

Identification of an *IRA1/2* homolog, *CoIRA1*, in *C. orbiculare*

CoIRA1 was first identified as a mutant gene in the *Agrobacterium tumefaciens*-mediated transformation (AtMT) T-DNA mutant AA4510 of *C. orbiculare*, which shows an attenuated pathogenicity on cucumber leaves. DNA flanking regions, adjacent to the inserted plasmid, were isolated from the mutant AA4510 by thermal asymmetrical interlaced-polymerase chain reaction (TAIL-PCR) and the amplified products were subsequently sequenced. The TAIL-PCR result showed that the T-DNA fragment was inserted probable 10-bp downstream from the translational origin of the gene ENH81573, based on *C. orbiculare* genomic information [25]. This gene putatively encodes a 2255-amino acid protein with a predicted RAS GTPase-activating protein (RASGAP) domain (Figure S1A). We named this gene *CoIRA1*.

A blast search of the *CoIRA1* fungal genome homologs in non-redundant protein database of the National Center for Biotechnology Information indicated that the derived amino acid sequence from this gene is significantly similar to that of the *IRA1/2* homolog in *S. cerevisiae* and filamentous fungi, including *C. graminicola*, *Neurospora crassa,* and *M. oryzae* (Figure S1B).

CoIRA1 is required for infection-related morphogenesis

To analyze the function of *CoIRA1*, the disruption vector pBIG4MRBrevcoira1 was designed to replace the wild-type *CoIRA1* gene with the hygromycin phosphotransferase (*hph*) gene, by double crossover homologous recombination. Successful replacement of the targeted gene in the transformants was confirmed by genomic DNA blot analysis (Figure S2). A single 3.4-kb band was detected in the wild type, whereas a single 5.6-kb band was detected in *coira1* mutants, as expected from such a targeted gene replacement event (Figure S2). In ectopic transformants, the 3.4-kb band and several additional bands were detected, indicating ectopic insertion events.

The hyphal growth of the *coira1* mutant was similar to that of the wild-type, but the *coira1* mutant formed a rather dark colony compared with the wild-type (Figure S3). The *coira1* mutant also showed reduced conidiation compared with the wild-type. To investigate infection-related morphogenesis in the *coira1* mutant, we observed conidial germination and appressorium development on a glass slide, and infection hyphae development on cellulose membranes. In the wild-type, *coira1* mutant, ectopic transformants, and *CoIRA1* reintroduced transformants, approximately 80% of the conidia germinated and formed darkly melanized appressoria after 24 h, but the frequency of abnormal appressorium formation in the *coira1* mutant was higher compared with the wild-type, as observed on glass slides (Figures 1A and B). The appressorium turgor pressure is required to penetrate the plant surface mechanically during infection [26], therefore, we evaluated the appressorium turgor in the *coira1* mutant using a cytorrhysis assay [27]. The proportion of the collapsed appressoria at each glycerol concentration in the *coira1* mutant was similar to that of

the wild-type (Figure S4), indicating that Colra1 is not involved in the generation of the appressorium turgor pressure. The wild-type, ectopic transformants, and *CoIRA1* reintroduced transformants formed normal infection-hyphae that penetrated the cellulose membranes with high frequency. The *coira1* mutant effectively formed infection hyphae as did the wild-type. Interestingly, the *coira1* mutant hyphae had a bulbous shape, which was quite different from those of the wild-type (Figures 2A and B). These results indicated that *CoIRA1* is involved in the progression of normal morphogenesis during infection.

CoIRA1 is involved in the pathogenicity of the host plant

To investigate whether the pathogenicity of *coira1* mutants is attenuated, conidial suspensions were inoculated onto cucumber cotyledons. The pathogenicity of the *coira1* mutant was found to be attenuated during the infection of the cucumber plants compared with that of the wild-type (Figure 3A). Moreover, when conidial suspensions were applied directly on wounded sites, the *coira1* mutant caused the formation of smaller lesions compared with the wild-type (Figure S5). Microscopic observations of the infection process of the *coira1* mutant showed that it formed normal darkly melanized appressoria and infection hyphae, however, its frequency of infection hyphae formation was lower than that of the wild-type (Figures 3B and C). These data indicated that during the infection of the cucumber, the *coira1* mutants had a defective infection-related morphogenesis. In *C. orbiculare*, we showed that the *ssd1* mutant that are defective in proper fungal cell walls constitution failed to penetrate into host epidermal cells due to the increased defence reaction of the host plant with rapidly induced callose deposition at at the attempted penetration site from appressoria [28]. Therefore, to investigate whether the observed low frequency of the infection hyphae formation in the *coira1* mutant was induced by a host defense mechanism, we monitored the callose deposition under the appressorium of the *coira1* mutant on the cucumber cotyledons and found that the extent and frequency of the deposition was similar to that of the wild-type (Figure S6A). Moreover, the *coira1* mutant caused the formation of smaller lesions than the wild-type on heat-shocked cucumber cotyledons (Figure S6B), which impaired the host defense responses, indicating that the induction of plant defense responses is not involved in the attenuated pathogenicity observed for *coira1* mutants.

CoIRA1 is involved in cAMP signaling

In *S. cerevisiae*, Ira1/2 inactivates Ras1/2, which in turn activates adenylate cyclase, leading to the synthesis of cAMP from ATP [29]. We therefore examined whether the appressorium morphogenesis of the *coira1* mutants is affected by exogenous cAMP signals. In the presence of 10 mM cAMP, the frequency of abnormal appressorium formation observed in the *coira1* mutant was higher compared with that of the wild-type (Figures 4A and B). This data suggested that Colra1 is involved in cAMP signaling pathway during the process of the appressorium formation. We identified two Ras genes and named *CoRAS1* (ENH84705) and *CoRAS2* (ENH80898). The comparative amino acid sequence analysis of Ras proteins showed that *CoRAS1* was 90% identical to *M. oryzae* MoRAS1 and *CoRAS2* was 80% identical to *M. oryzae* MoRAS2. To better understand relationship between Colra1 and Ras proteins, we set up a hypothesis that the wild-type expressing a dominant active form *CoRAS1* and *CoRAS2* increases abnormal appressorium formation compared with the wild-type in the presence of excess cAMP, while the *coira1* mutant expressing a dominant active form *CoRAS1* and *CoRAS2* decrease abnormal appressorium compared with the *coira1* mutant in the presence of

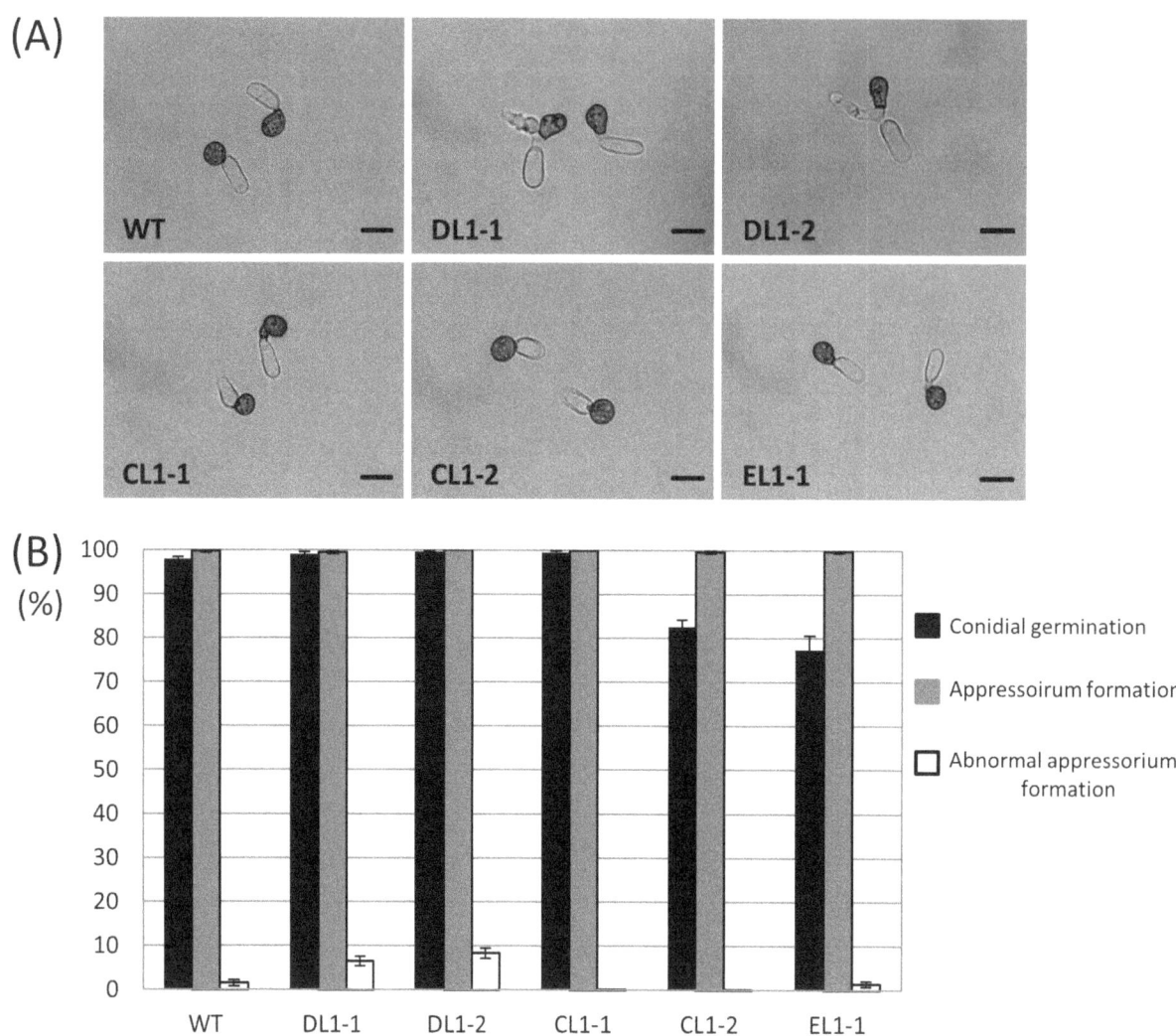

Figure 1. Appressorium formation in *coira1* mutants of *C. orbiculare* on glass slides. (A) Conidial suspensions of each strain in distilled water were incubated on multiwell glass slides at 24°C for 24 h. WT, the wild-type 104-T; DL1-1 and DL1-2, the *coira1* mutants; CL1-1, the *CoIRA1*-complemented transformant of DL1-1; CL1-2, the *CoIRA1*-complemented transformant of DL1-2; EL1-1, ectopic strain. Scale bar, 10 μm. (B) Percentages of conidial germination, appressorium formation, and abnormal appressorium formation in the *C. orbiculare* WT and *coira1* mutants on multiwell glass slides. Approximately 100 conidia of each strain were observed per well on multiwell slide glass. Three replicates were examined. Three independent experiments were conducted, and standard errors are shown. Black bar, conidial germination; gray bar, appressorium formation; white bar, abnormal appressorium formation.

excess cAMP. Therefore, we generated these transformants. The Ras protein shows a high degree of amino acid conservation. On the basis of the human Ras gene mutation information, we constructed the dominant active forms CoRas1 (*RAS1*^G17V) and CoRas2 (*RAS2*^Q65L) alleles by replacing Gln-65 with Leu and Gly-17 with Val, respectively. Moreover, based on *C. trifolii* and *Candida albicans* Ras gene mutation information, we constructed the dominant negative forms CoRas1 (*RAS1*^S22N) and CoRas2 (*RAS2*^G19A) alleles by replacing Ser-22 with Asn and Gly-19 with Ala, respectively [6], [30]–[32]. Next, we generated the dominant active and negative forms of CoRAS1 and CoRAS2 of *C. orbiculare*. DARS1 (Dominant Active form *CoRAS1*^G17V) is the wild-type transformant, expressing the dominant active form *CoRAS1*^G17V and DARS2 (Dominant Active form *CoRAS2*^Q65L) is the wild-type transformant, expressing the dominant active form *CoRAS2*^Q65L. On the other hand, iDNRS1 (the *coira1*/Dominant Negative form *CoRAS1*^S22N) is the *coira1* transformant, expressing the dominant negative form *CoRAS1*^S22N and iDNRS2 (the

coira1/Dominant Negative form *CoRAS2*^G19A) is the *coira1* transformant, expressing the dominant negative form *CoRAS2*^G19A.

DARS1 and DARS2 showed normal appressorium formation on the glass slides in distilled water (Figure 4) and normal penetration hyphae similar to the wild-type on the cellulose membranes (Figure S7), however, DARS2 caused the formation of smaller lesions compared with the wild-type and DARS1 on the cucumber cotyledons (Figure S8). In the presence of 10 mM cAMP, DARS2 formed abnormal appressoria, similar to those observed for the *coira1* mutants. In contrast, iDNRS2 suppressed the abnormal appressorium formation compared with that of the *coira1* mutants (Figure 4). These data indicated that CoIra1 could control the cAMP signaling pathway through CoRas2 during the process of appressorium formation. We also checked the intracellular cAMP accumulation in the *coira1* mutant and the dominant active and negative *CoRAS1* and *CoRAS2* introduced transformants in vegetative hyphae whether the cAMP signaling

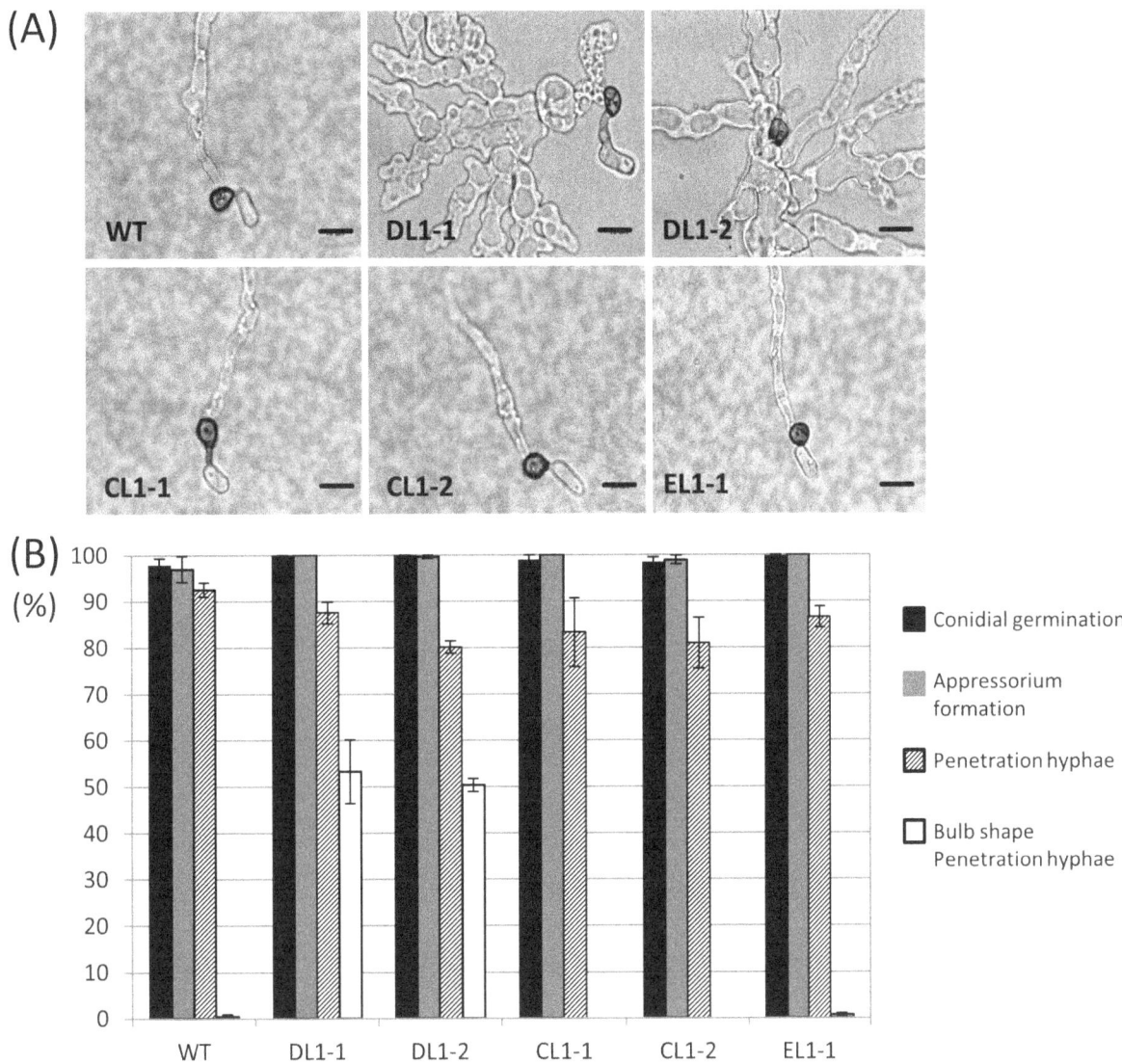

Figure 2. Penetration hyphae formation of the *coira1* mutants of *C. orbiculare* on cellulose membranes. Conidial suspensions of each strain in distilled water were incubated on cellulose membranes at 24°C for 48 h. WT, the wild-type 104-T; DL1-1 and DL1-2, the *coira1* mutant; CL1-1, the *CoIRA1*-complemented transformant of DL1-1; CL1-2, the *CoIRA1*-complemented transformant of DL1-2; EL1-1, the ectopic strain. Scale bar, 10 μm. (B) Percentages of conidial germination, appressorium formation, penetration hyphae formation, and bulb-shaped penetration-hyphae formation of *C. orbiculare* WT and *coira1* mutants on cellulose membranes. Approximately 200 conidia of each strain were observed on cellulose membranes. Three replicates were examined. Three independent experiments were conducted, and standard errors are shown. Black bar, conidial germination; gray bar, appressorium formation; slash bar, penetration hyphae; white bar, bulb-shape penetration formation.

pathway was regulated by those genes. The *coira1* mutants accumulated higher levels of cAMP compared with the wild-type (Figure 5). Moreover, intracellular cAMP levels in the DARS1 and DARS2 mutants were similar to those of the *coira1* mutant. These data indicated that intracellular cAMP levels in the vegetative hyphae were controlled by Colra1, CoRas1, and CoRas2.

CoRAS2 is involved in conidial germination and pathogenicity

Colra1 regulates intracellular cAMP levels through Ras proteins. Therefore, to analyze the functional roles of *CoRAS1* and *CoRAS2*, we aimed to generate *coras1* and *coras2* mutants by AtMT. Whereas we successfully developed *coras2* mutants, the generation of *coras1* mutants was challenging, indicating that *CoRAS1* could be an essential gene in *C. orbiculare*. To investigate

whether *CoRAS2* is involved in infection-related morphogenesis, we observed conidial germination and appressorium formation on glass slides. Strikingly, conidial germination was not observed on glass slides for *coras2* mutants, indicating that *CoRAS2* is involved in conidial germination (Figure S9). To investigate whether the pathogenicity of the *coras2* mutant was attenuated during the infection of the cucumber cotyledons, conidial suspensions were inoculated onto the cucumber cotyledons. The *coras2* mutant was defective in pathogenesis to the cucumber cotyledons, forming small speck lesion (Figure S10). Microscopic observation revealed that most conidia in the *coras2* mutant were defective in conidial germination as *in vitro* condition, indicating that the attenuated pathogenicity of this mutant was due to a defect in conidial germination and that small speck lesion was apparently caused by small numbers of appressorium forming conidia.

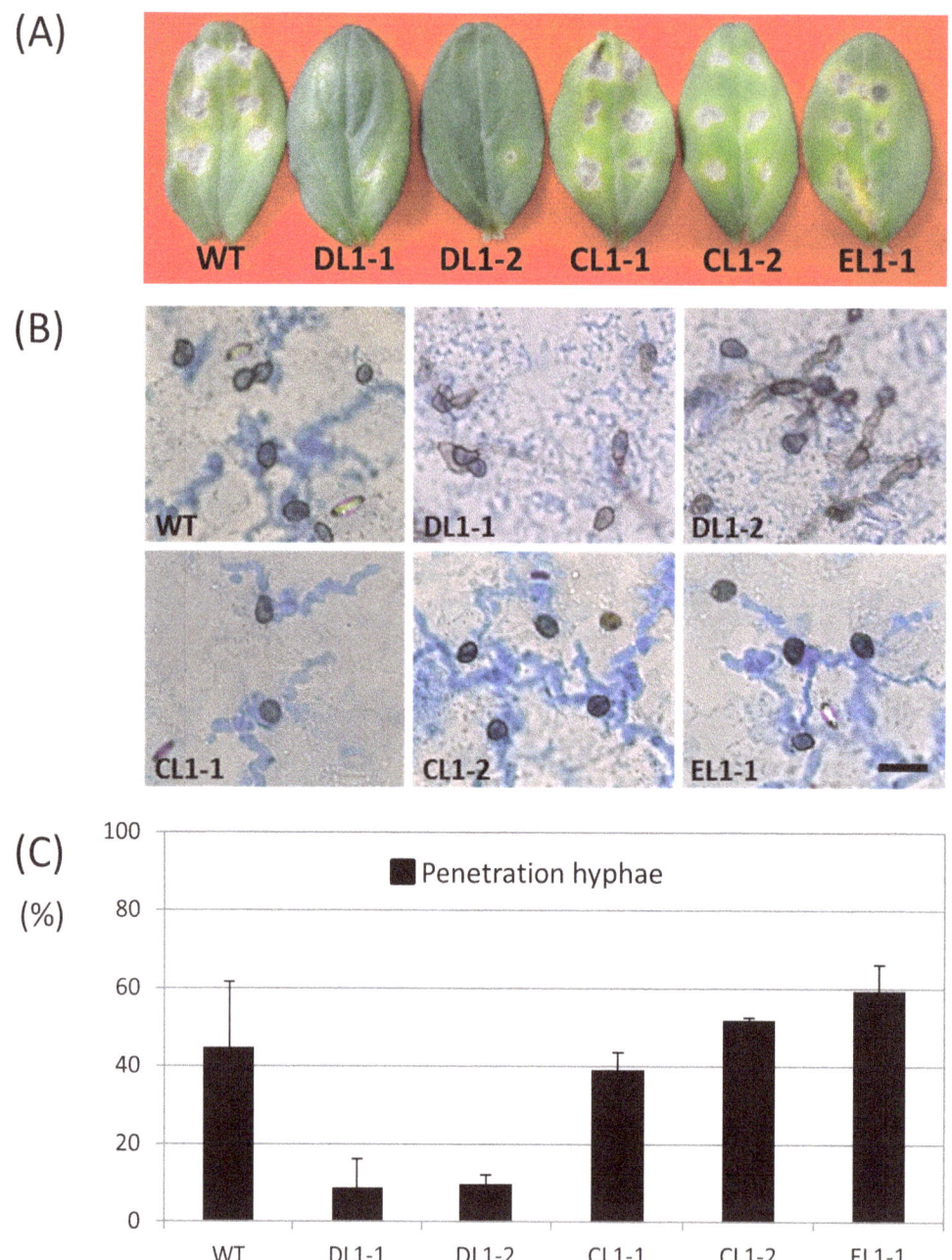

Figure 3. Pathogenicity assay and penetration ability of *coira1* mutants of *C. orbiculare* on the cucumber cotyledons. (A) Conidial suspensions of each strain (10 μl) placed on detached cucumber cotyledons and leaves incubated at 24°C for three days. The figure shows the leaves after incubation with the following strains: WT, the wild-type 104-T; DL1-1 and DL1-2, the *coira1* mutant; CL1-1, the *CoIRA1*-complemented transformant of DL1-1; CL1-2, the *CoIRA1*-complemented transformant of DL1-2; EL1-1, the ectopic strain. (B) Penetration hyphae development of each strain on the cucumber cotyledons. Conidial suspensions (10 μl) were applied to the abaxial surface of the cucumber cotyledons and incubated at 24°C for 72 h. Scale bar, 20 μm. (C) Percentage of penetration hyphae of the *coira1* mutants on the abaxial surface of cucumber cotyledons. Approximately 100 appressorium were observed per incubated site. Three replicates were examined. Three independent experiments were conducted, and standard errors are shown. Black bar, penetration hyphae. Scale bar, 20 μm.

CoRas2 is an upstream regulator of the MAPK signaling pathway

From previous reports, the MAPK CoMekk1–Cmk1 signaling pathway has been shown to play pivotal roles in conidial germination and appressorium formation in several *Colletotrichum* speciese [2], [33]. In *C. orbiculare*, the Ste11 homolog *CoMEKK1* encodes a Ras-association domain. Thus, to analyze whether CoRas2 is an upstream regulator of CoMekk1, we generated a

comekk1 transformant DMK1/DARS2 expressing a dominant active form of the *CoRAS2* allele. In addition, we generated the *coras2* mutant DRS2/DAMK1 expressing a dominant active form of the *CoMEKK1* allele. A normal conidial germination and appressorium formation was observed on the glass slide for the DRS2/DAMK1 mutant (Figure 6). On the other hand, the defective conidial germination of the *coras2* mutant remained in the DMK1/DARS2 mutant, showing a similar frequency of

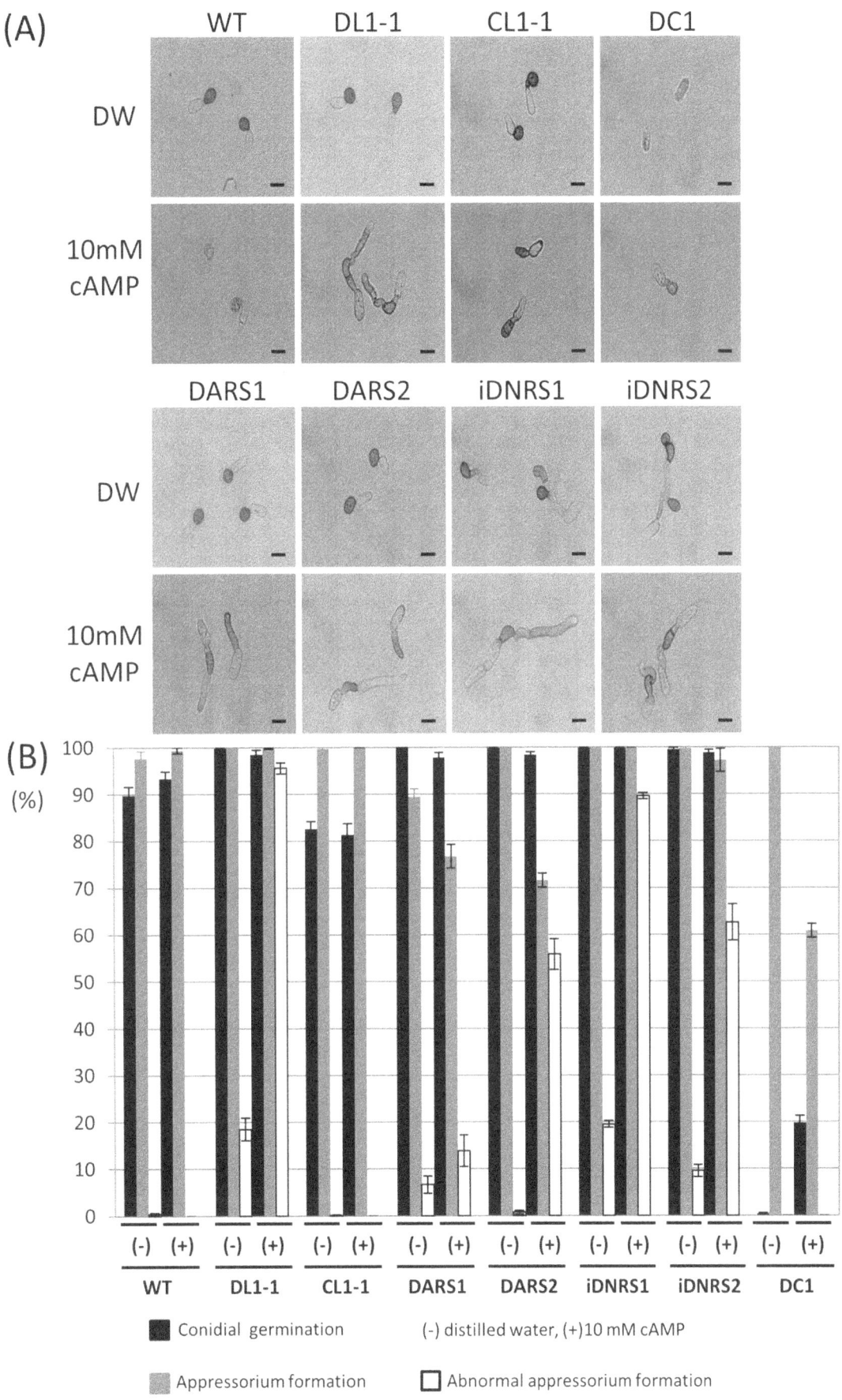

Figure 4. Appressorium formation by *coira1* mutants on glass slide in the presence of 10-mM cAMP. (A) Conidial suspensions of each strain in distilled water or 10-mM cAMP were incubated on the multiwell glass slide at 24°C for 24 h. WT, the wild-type 104-T; DL1-1, the *coira1* mutant; CL1-1, the *CoIRA1*-complemented transformant of DL1-1; DC1, the *cac1* mutant; DARS1, WT transformed with a dominant active form *CoRAS1*; DARS2, WT transformed with a dominant active form *CoRAS2*; iDNRS1, DL1-1 transformed with a dominant negative form *CoRAS1*; iDNRS2, DL1-1 transformed with a dominant negative form *CoRAS2*. Scale bar, 10 μm. (B) Percentages of conidial germination, appressorium formation, and abnormal appressorium formation of *C. orbiculare* on multiwell glass slides in the presence of 10-mM cAMP. Approximately 100 conidia of each strain were observed on multiwell glass slides. Three replicate experiments were examined. Three independent experiments were conducted, and standard errors are shown. Black bar, conidial germination; gray bar, appressorium formation that includes normal appressorium and abnormal appressorium; white bar, abnormal appressorium formation. (–) distilled water, (+) 10-mM cAMP.

conidial germination as that in the *coras2* mutant. These data suggested that *CoRAS2* is an upstream regulator of the MAPK CoMekk1–Cmk1 signaling pathway.

The MAP kinase Cmk1 is activated through threonine/tyrosine phosphorylation catalyzed by MAPKK, which at the same time is activated through serine phosphorylation catalyzed by CoMekk1. To determine whether the phosphorylation of Cmk1 in the *coira1* and *coras2* mutants was affected in the vegetative hyphae, we investigated the Thr–Gln–Tyr residue phosphorylation of Cmk1 using an anti-TpEY specific antibody. The phosphorylation levels of Cmk1 in the *coira1* mutant and DARS2 were higher compared with those in the wild-type, whereas phosphorylation levels of Cmk1 in iDNRS2 were lower compared with those in the *coira1* mutant, which was similar to those of the wild-type (Figure 7). These data indicated that CoIra1 negatively regulates the phosphorylation of Cmk1 through CoRas2. Interestingly, the phosphorylation level of Cmk1 in DARS1 was higher compared with that in the wild-type, indicating that CoRas1 positively regulates the phosphorylation of MAPK Cmk1.

CoRas2 localization in the *coira1* mutant was similar to the active CoRas2 localization pattern in vegetative hyphae

To analyze the cellular CoRas2 localization, a red fluorescent protein (RFP) gene was fused to the C-terminus of the *CoRAS2* gene. In *S. cerevisiae* and *C. albicans*, the Ras protein conserves a CAAX motif (C, cysteine; A, aliphatic amino-acids; X; methionine or serine), which is important for post-translational modification, including farnesylation and palmitoylation, to ensure specific

membrane localization and function [29], [34]. We assumed that the *RFP gene* fused C-terminus of *CoRAS2* gene may block the localization and function of CoRas2. Thus, the *RFP* gene fused to the N-terminus of *CoRAS2* gene was constructed, and the *RFP–CoRAS2* gene controlled under the *CoRAS2* native promoter was transformed into the *coras2* mutant (DRS2/RFP–RS2) and the wild-type (WT/RFP–RS2). Both DRS2/RFP–RS2 and WT/RFP–RS2 retained normal infection-related morphogenesis and pathogenicity. The signals of RFP–CoRas2 were detected within vesicle-like structures in conidia (Figure 8). During the process of premature appressoria formation, signals of the RFP–CoRas2 were localized mainly in subcellular compartments resembling vacuole-like structures in conidia, but these signals were not detected in matured appressoria. In vegetative hyphae, RFP–CoRas2 proteins showed no specific localization and their signals were uniformly distributed throughout the entire cell (Figure 8). To analyze the cellular localization of the active form of CoRas2, the *RFP* gene was fused to the N-terminus of the dominant active form of the *CoRAS2* gene. The *RFP–DACoRAS2* gene controlled under the native *CoRAS2* promoter was transformed into the wild-type (WT/RFP–DARS2). During the process of appressoria formation, the signal pattern of the RFP–DACoRas2 was similar to that of the RFP-CoRas2 (Figure S11), however, the signals of the RFP–DACoRas2 in vegetative hyphae were detected predominantly in the plasma membrane, unlike those of native CoRas2. To analyze whether CoIra1 negatively regulates CoRas2, we generated an iRFP–RS2 transformant expressing *RFP–CoRAS2* in the *coira1* mutant and observed the cellular localization of CoRas2 in this mutant. During appressoria formation, the signals of RFP-CoRas2 in the *coira1* mutant were detected as vesicle like structures as well as RFP-CoRas2 and RFP-DACoRas2. In vegetative hyphae, the signals of RFP-CoRas2 in the *coira1* mutant were detected at the plasma membrane as well as RFP-DACoRas2 localization (Figure 9). Engagement of CoIra1 in the regulation of CoRas2 during appressorium formation was not directly suggested by these data from the viewpoint of the cellular localization of Ras protein, whereas it was suggested that CoIra1 functions as a negative regulator for CoRas2 in vegetative hyphae. Furthermore, to analyze whether CoIra1 regulates the CoRas2 in the plasma membrane in vegetative hyphae, we generated the Vc–RS2/IRA–Vn transformant, expressing the C-terminal domain (159–238) *VENUS* that was fused with *CoRAS2* and the N-terminal domain (1–158) of *VENUS* fused with *CoIRA1* in the wild-type for bimolecular fluorescence complementation (BiFC) assays [35]. BiFC fluorescence was detected at the plasma membrane in vegetative hyphae. This result indicated that CoIra1 regulates CoRas2 at the plasma membrane in vegetative hyphae (Figure 10).

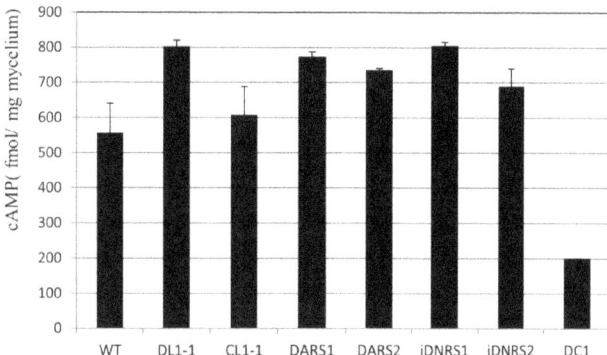

Figure 5. Intracellular cAMP levels in the *coira1* mutant. Intracellular cAMP levels were measured in tissue collection from three-day old liquid cultures of each strain. WT, the wild-type 104-T; DL1-1, the *coira1* mutant; CL1-1, the *CoIRA1*-complemented transformant of DL1-1; DARS1, WT transformed with a dominant active form *CoRAS1*; DARS2, WT transformed with a dominant active form *CoRAS2*; iDNRS1, DI1-1 transformed with a dominant negative form *CoRAS1*; iDNRS2, DI1-1 transformed with a dominant negative form *CoRAS2*; DC1, the *cac1* mutant. Three independent experiments were conducted, and standard errors are shown.

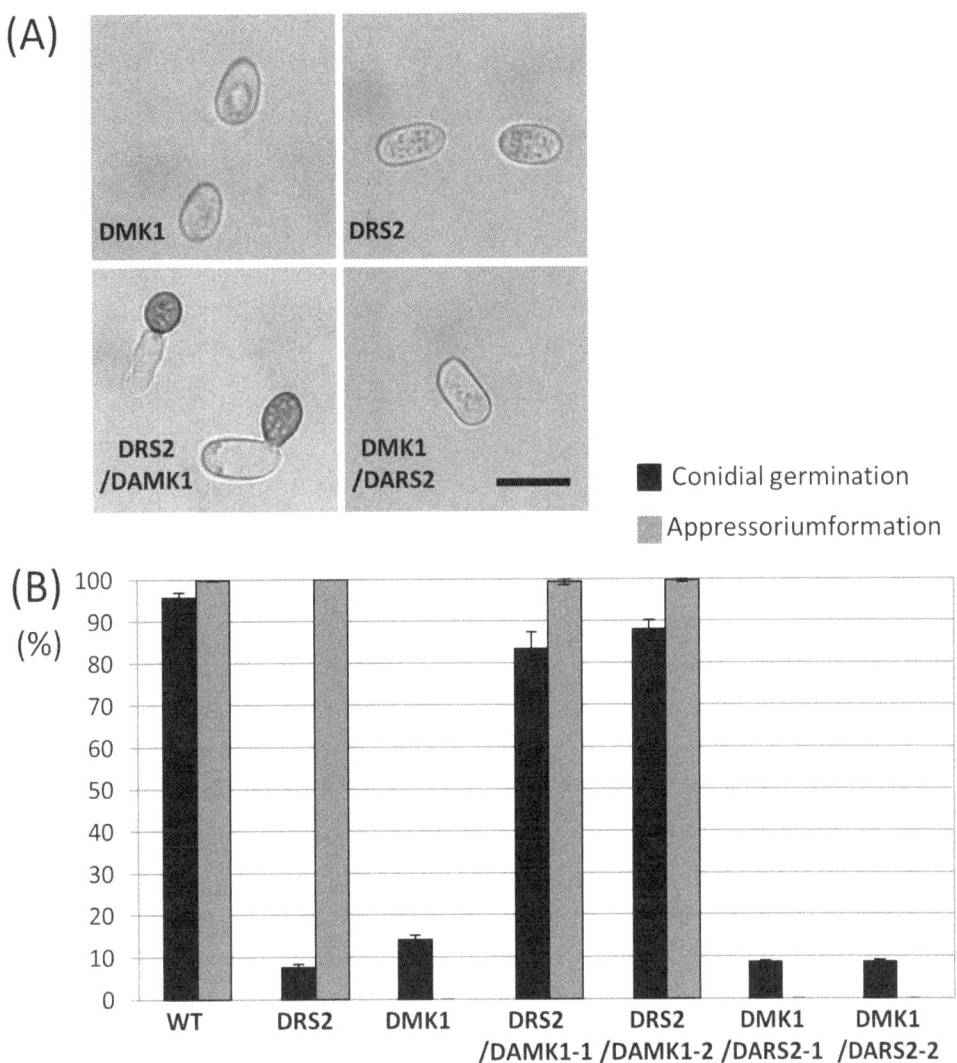

Figure 6. Appressorium formation assay of DRS2/DAMK1 and DMK1/DARS2 on the glass slides. (A) Conidial suspensions of each strain in distilled water incubated on multiwell glass slides at 24°C for 24 h. DRS2, the *coras2* mutant; DMK1, the *comekk1* mutant; DRS2/DAMK1, DRS2 transformed with a dominant active form *CoMEKK1*; DMK1/DARS2, DMK1 transformed with a dominant active form *CoRAS2*. Scale bar, 10 μm. (B) Percentages of conidial germination and appressorium formation in *C. orbiculare* on multiwell glass slides. Approximately 100 conidia of each strain were observed per well on the multiwell slide glass. Three replicates were examined. Three independent experiments were conducted, and standard errors are shown. Black bar, conidial germination; gray bar, appressorium formation.

Colra1 colocalizes with CoRas2 in pregerminated conidia and appressoria that initiate the differentiation of infection hyphae

To elucidate the cellular localization of Colra1, the *VENUS* fluorescence gene was fused to the C-terminal of the *CoIRA1* gene [36]. *VENUS* fused with *CoIRA1*, expressing under a native *CoIRA1* promoter was transformed into the wild-type (IRA1–VENUS). Colra1–VENUS was detected in a vesicle-like structure of conidia (Figure 11). In the germinated conidia, Colra1–VENUS was detected at the tip of the germ tubes and distributed uniformly in the conidial cells. In developing appressoria, Colra1–VENUS did not show any specific localization, and uniform signals were distributed throughout the appressoria except the presumable lipid bodies, similar to those on cucumber leaves. In developing infection hyphae, Colra1–VENUS showed a uniform distribution either in the initial or late infection hyphae in cucumber cotyledons (Figure S12). Then, to elucidate the

intracellular colocalization of Colra1 and CoRas2, we generated the RFP–RS2/IRA1–VENUS and RFP–DARS2/IRA1–VENUS transformants by introducing the *RFP* fused with *CoRAS2* and *VENUS* fused with *CoIRA1* genes into the wild-type, and the *RFP* fused with a dominant active form *CoRAS2* and *VENUS* fused with *CoIRA1* genes into the wild-type, respectively. RFP–CoRas2 and Colra1–VENUS colocalized at vesicle-like structures in conidia and germinated conidia (Figure 11), however, the RFP–CoRas2 and Colra1–VENUS did not show specific colocalization in developing appressoria, similar to RFP–DARS2/IRA1–VE-NUS (Figure S11). Surprisingly, after 48 h of conidial incubation on glass slides, RFP–CoRas2 and Colra1–VENUS were colocalized in a vesicle-like structures in appressoria (Figure 12). Conclusively, Colra1 colocalizes with CoRas2 in pregerminated conidia and in appressoria that initiate differentiation of infection hyphae.

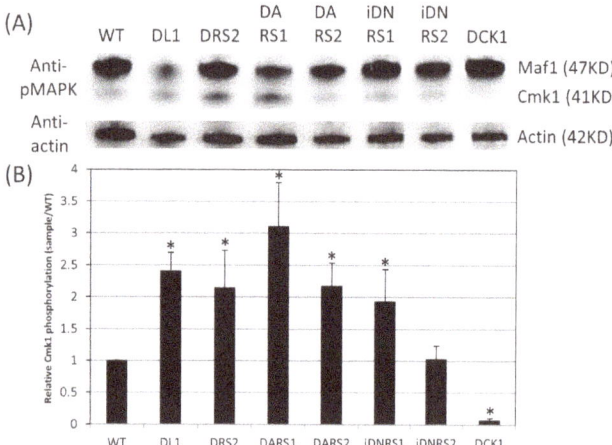

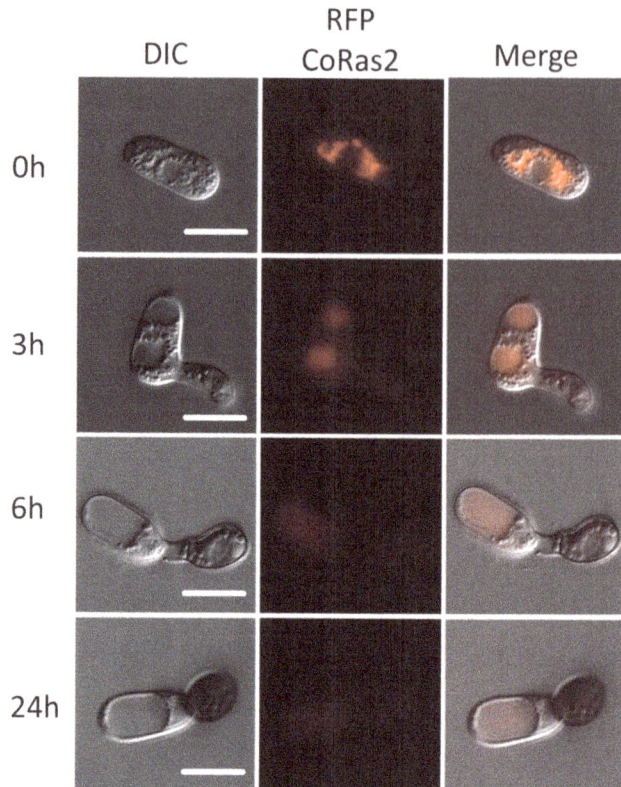

Figure 7. The phosphorylation of MAPK Cmk1 in the *coira1* mutant. (A) The total protein isolated from mycelia of each strain. WT; the wild-type, DL1; the *coira1* mutant, DRS2; the *coras2* mutant, DARS1, WT transformed with a dominant active form *CoRAS1*, DARS2; WT transformed with a dominant active form *CoRAS2*, iDNRS1; the *coira1* mutant transformed with a dominant negative form *CoRAS1*, iDNRS2; the *coira1* mutant transformed with a dominant negative form *CoRAS2*, DCK1; the *cmk1* mutant The anti-phospho p44/42 MAPK antibody detected a 41-KD Cmk1 and 47-KD Maf1. The anti-actin antibody detected a 42-KD actin. (B) Relative activity of MAPK Cmk1 phosphorylation of each mutant was calculated by comparison of signal intensity with that of the wild-type, normalized by actin signal. The quantitative analysis of phosphorylated Cmk1 was performed by four replicated experiments. Asterisk represents significant differences between the wild type and each mutant. (Student's t test: *indicate $P < 0.05$).

CoIra1 regulates the assembly of actin at the appressorium pore

In *M. oryzae*, it has been reported that proper assembly of the F-actin network in the appressorium pore is required for host infection and MAPK *Mst12*, the *C. orbiculare Cst1* homolog, is known to be involved in the proper assembly of the F-actin network in the appressorium pore [37], [38]. Moreover, it has been reported that cAMP signaling is involved in the remodeling of the actin structure in *S. cerevisiae* [39]. Our data showed that CoIra1 regulated the cAMP and MAPK signaling pathways (Figure 5 and 7). To elucidate whether CoIra1 colocalize with F-actin, we generated the LA/IRA1–VENUS transformant by introducing Lifeact–*RFP* into it. In *C. orbiculare*, Lifeact, which binds to F-actin, forms vesicle-like structures in pregerminated conidia and then Lifeact–*RFP* mainly localizes in the appressorium pore during appressoria development. We examined the localization of CoIra1–VENUS and Lifeact–*RFP* in pregerminated conidia and appressorium pores. CoIra1–VENUS was colocalized with Lifeact–*RFP* in pregerminated conidia, however, CoIra1–VENUS was not clearly colocalized with Lifeact–RFP in the appressorium pore (Figure 13). To elucidate whether CoIra1 is involved in the assembly of the F-actin in the appressorium pore, we generated the iLA transformant by introducing Lifeact–*RFP* into the *coira1* mutant. In *C. orbiculare*, the wild-type showed a specific assembly of F-actin in the appressorium pore on the cucumber cotyledons, however, the frequency of this assembly in the *coira1* mutant was lower than that in the wild-type (Figure 14). These data indicated that CoIra1 is involved in the assembly of the F-actin in the appressorium pore.

Figure 8. Localization of a functional RFP–CoRas2 fusion protein in *C. orbiculare* during conidial germination and appressorium formation. Conidial suspensions of the DRS2/RFP–RS2strain in distilled water were incubated in glass slides at 24°C for 0 h, 3 h, 6 h, and 24 h. After incubation, RFP fluorescence was observed by fluorescence microscopy. DRS2/RFP–RS2, the *coras2* mutant expressing *RFP–CoRAS2*. Scale bar, 10 µm.

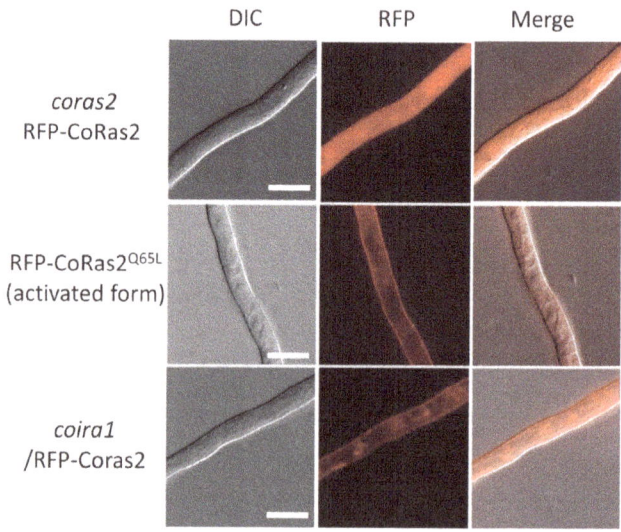

Figure 9. CoRas2 localization was regulated by CoIra1 in vegetative hyphae. Conidia harvested from each strain were observed on glass slides by fluorescent microscopy. *coras2*/RFP–CoRAS2, the *coras2* mutant expressing the *RFP–CoRAS2*; RFP–CoRas2^{Q65L}, the wild-type strain expressing *RFP–CoRAS2*^{Q65L}; and *coira1*/RFP–CoRas2, the *coira1* mutant expressing *RFP–CoRAS2*. Scale bar, 10 µm.

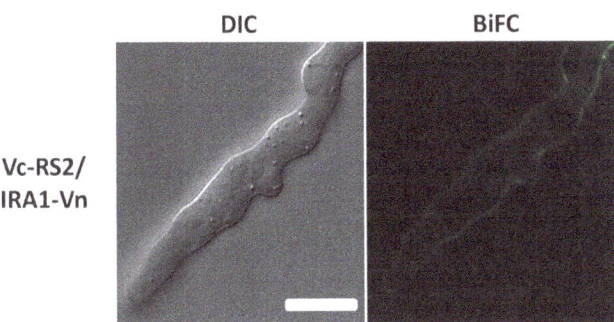

Figure 10. BiFC assays for Colra1 and CoRas2 interactions in vegetative hyphae. Conidial suspensions of Vc-CoRas2/Colra1-n transformant in liquid PSY medium were incubated at 28°C for 24 h and BiFC fluorescence was observed by fluorescent microscopy in vegetative hyphae. Vc–CoRas2/Colra1-Vn; the wild-type strain expressing Vc–*CoRAS2* and Vn–*ColRA1*. Scale bar, 10 μm.

Discussion

Colra1 is involved in the crosstalk between cAMP and MAPK signaling pathways through CoRas2 in *C. orbiculare*

In *S. cerevisiae*, the activation of the cAMP-dependent pathway causes cells to undergo unipolar growth, which is coupled with an elongated growth that is controlled by the filamentous MAPK pathway [40]. In *C. orbiculare*, cAMP–PKA signal transduction is involved in conidial germination, and the MAPK cascade is involved in appressorium development [1]. In *S. cerevisiae*, an increase in the intracellular levels of RAS–GTP against RAS–GDP is observed in *ira1* and *ira2* mutants, activating various target effectors [41]. Our data indicated that intracellular cAMP levels and the phosphorylation of MAPK Cmk1 in the *coira1* mutant, DARS1, and DARS2 was higher than that in the wild-type. Interestingly, the phosphorylation of Cmk1 in the *coras2* mutant was higher compared with the wild-type, although CoRas2 is an upstream regulator of CoMekk1–Cmk1. In *M. oryzae*, the intracellular cAMP levels of the dominant active MEK $Mst7^{S212D\ T216E}$ strain are lower than those of the wild-type [42]. Recently, the functional analysis of the adenylate cyclase-associated protein encoding *CAP1*, the ortholog of yeast *Srv2*, Cap1 may play a role in the feedback inhibition of MoRas2 signaling when Pmk1 MAP kinase is activated [12]. In yeast and phytopathogenic fungi, cAMP signaling is intimately associated with a MAPK homologous with PMK1 for regulating various developmental and plant-infection processes [43]–[45]. We assumed that the constitutively active forms CoRas1 and CoRas2 induce excessive activation of the cAMP–PKA and MAPK CoMekk1–Cmk1 signaling pathways, resulting in the interruption of the cAMP–PKA and MAPK signaling pathways (Figure 15). Interestingly, the level of cAMP and the phosphorylation of Cmk1 in DARS1 was higher than that in the DARS2; however, the pathogenicity of DARS1 was not attenuated during the infection of cucumber leaves (Figure S11). Therefore, CoRas1 may be involved in the cAMP–PKA and MAPK CoMekk1–Cmk1 signaling pathways in vegetative hyphae. To understand Ras-mediated signaling pathways clearly, further functional analysis of the relationship between CoRas1 and CoRas2 during infection is required.

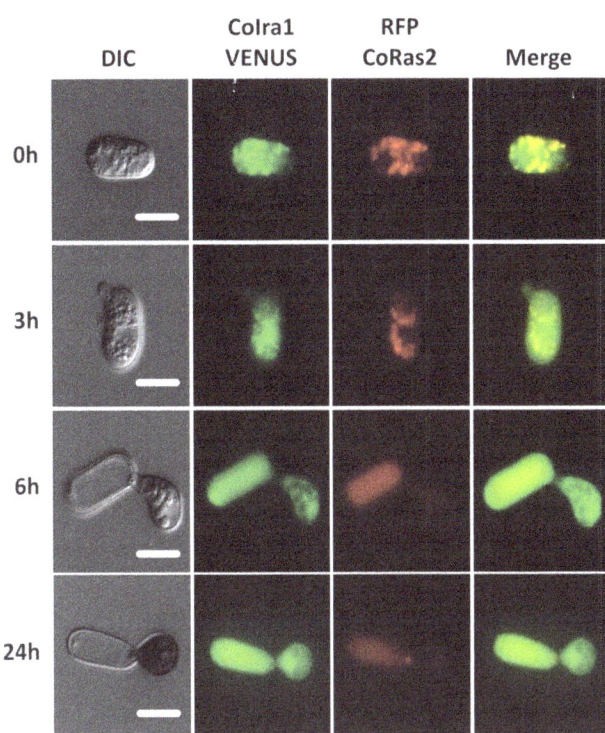

Figure 11. Assay for colocalization of Colra1 and CoRas2. Conidial suspensions of the RFP–RS2/IRA1–VENUS strain were incubated on glass slides at 24°C for 0 h, 3 h, 6 h, and 24 h and observed by fluorescent microscopy. RFP–RS2/IRA1–VENUS, the wild-type strain expressing *ColRA1*–VENUS and *RFP–CoRAS2*. Scale bar, 10 μm.

Colra1 interacts with CoRas2 and negatively regulates CoRas2

Ras serves as a molecular switch, coupling activated membrane receptors to the downstream signaling molecule, by alternating between the GTP-bound (active) and the GDP bound (inactive) conformations. In *S. cerevisiae*, Ira1/2 negatively regulates Ras1/2; therefore, the ira1/2 mutants accumulate high amounts of Ras–GTP compared with the wild-type [7]. In the present study, the localization pattern of CoRas2 in the *coira1* mutant was similar to that in the active form CoRas2 in vegetative hyphae. Moreover, BiFC assays supported that Colra1 regulates CoRas2 at the plasma membrane of vegetative hyphae. It is likely that Colra1 negatively regulates CoRas2 in vegetative hyphae. In *Aspergillus fumigatus*, RasA localizes in plasma membranes in hyphae where it associates with and stimulates targeted effectors, resulting in the

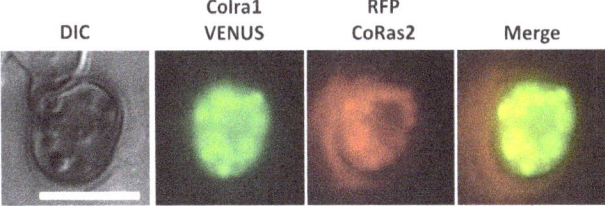

Figure 12. Assay for the colocalization of Colra1 and CoRas2 in appressoria at 48 h after inoculation on cucumber leaves. Conidial suspensions of RFP–RS2/IRA1-strain were incubated in cucumber leaves at 24°C for 48 h and observed under fluorescent microscopy. RFP–RS2/IRA1–VENUS, the wild-type strain expressing *ColRA1*–VENUS and *RFP–CoRAS2*. Scale bar, 10 μm.

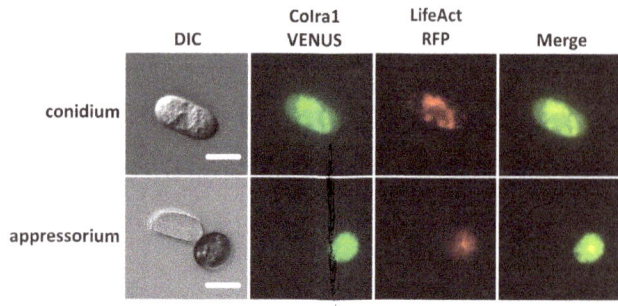

Figure 13. Assay for colocalization of Colra1 and F-actin. Conidial suspensions of the LA/IRA1-VENUS strain were incubated on glass slides at 24°C for 0 h and 24 h and observed under fluorescent microscopy. LA/IRA1-VENUS, the wild-type strain expressing *CoIRA1–VENUS* and Lifeact–*RFP*. Scale bar, 10 μm.

regulation of polarized morphogenesis [46]. We considered that CoIra1 negatively regulates CoRas2, which is required for the proper morphogenesis of vegetative hyphae. However, this morphogenesis in the *coira1* mutant and the dominant active form CoRas2 expressed strain was similar to that in the wild-type. In *C. orbiculare*, the conidia in the MAPK *cmk1* disruption mutant fails to germinate, however, by adding 0.1-% yeast extract, conidial germination efficiency can be restored [33]. Thus, there is a strong possibility that a nutrient-specific signaling pathway regulates conidial germination in *C. orbiculare*. Therefore, we speculated the presence of a bypass pathway that regulates vegetative hyphae morphogenesis independent of the CoIra1–CoRas2 signaling pathway in the presence of nutrients.

CoIRA1 regulates proper infection-related morphogenesis

Our data indicated that the pathogenicity of the *coira1* mutant was attenuated during the infection of the cucumber cotyledons.

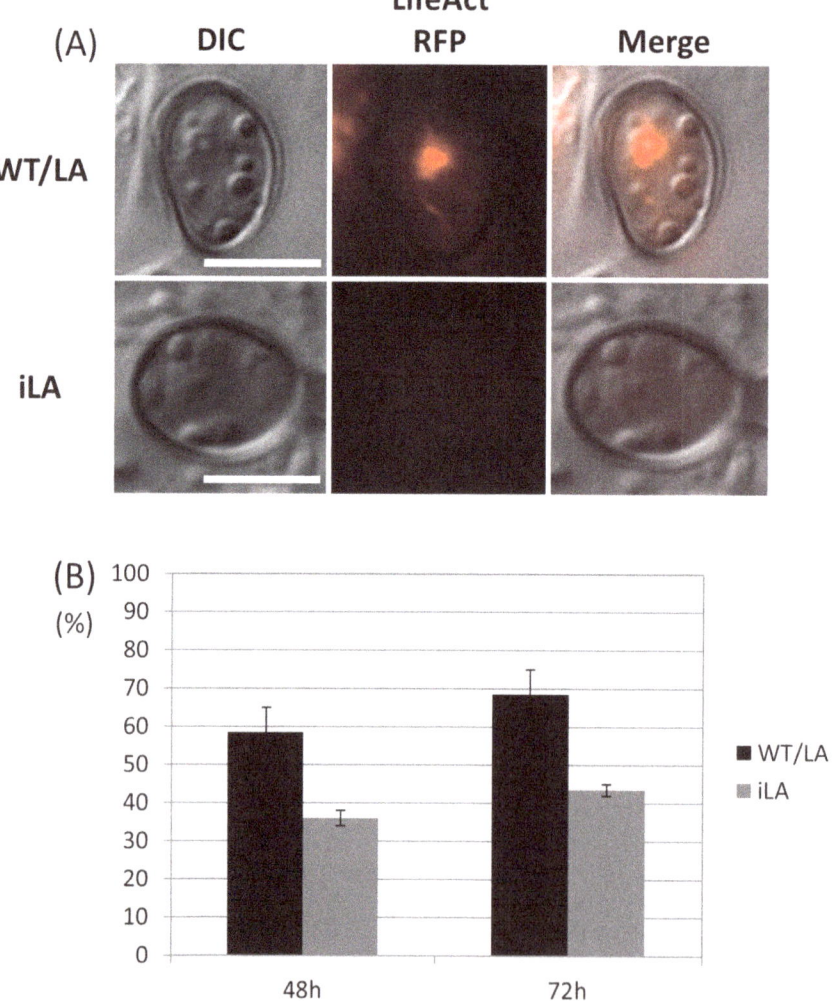

Figure 14. Assembly of F-actin in the appressorium pores of the *coira1* mutant. (A) Micrograph of F-actin organization in the appressorium pore visualized by the expression of Lifeact–RFP in the wild-type and in the *coira1* mutant. Conidial suspensions (10 μl) of each strain were applied to the abaxial surface of the cucumber cotyledons and incubated at 24°C for 48 h. WT/RA, the wild type expressing Lifeact–RFP; iRA, the *coira1* mutant expressing Lifeact–RFP (B) Percentage of the assembly of F-actin in the appressorium pore of the *coira1* mutants on the abaxial surface of cucumber cotyledons. Approximately 100 appressoria of each strain were observed per incubated site for 48 h, 72 h post-inoculation. Two replicates were examined. Three independent experiments were conducted, and standard errors are shown. Black bar, WT/RA; gray bar, Ira; WT/RA, the wild type expressing LifeAct-RFP; iRA, the *coira1* mutant expressing LifeAct-RFP.

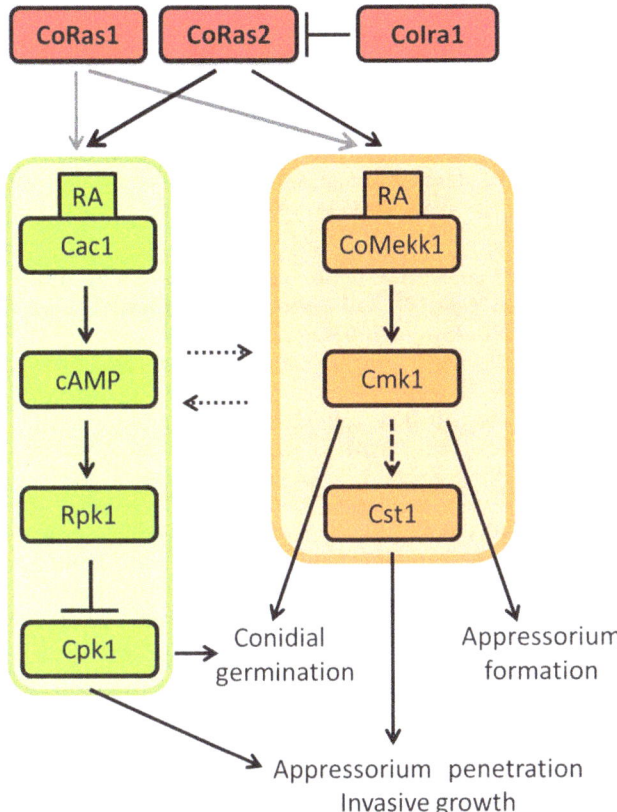

Figure 15. The hypothetical model for Colra1 and CoRas1/2 mediated signaling transduction pathway in *C. orbiculare*. The cAMP-PKA signaling pathway is involved in conidial germination, appressorium penetration and invasive hyphae formation. The MAPK CoMekk1–Cmk1 signaling pathway is involved in conidial germination and appressorium formation. The CoRas2 localization pattern in the *coira1* mutant was similar to that in DARS2. Moreover, BiFC assays supported that Colra1 interacted with CoRas2 in the plasma membrane. Therefore, Colra1 negatively regulates CoRas2. The *coira1* mutant and DARS2 significantly induced abnormal appressorium formation and the frequency of abnormal appressorium formation in iDNRS2 was lower compared with that in the *coira1* mutant. Moreover, intracellular cAMP levels in the *coira1* mutant and DARS2 was high compared with those in the wild type. Therefore, Colra1 regulates cAMP-PKA signaling pathway through CoRas2. The conidia of the *coras2* mutant failed to germinate, whereas DRS2/DAMK1 restored the phenotype of the *coras2* mutant. Therefore, CoRas2 is an upstream regulator of the MAPK CoMekk1–Cmk1 signaling pathway. Interestingly, the phosphorylation of MAPK Cmk1 in the *coras2* mutant was higher compared with that in the wild-type, although CoRas2 is an upstream regulator of CoMekk1–Cmk1. Therefore, that Colra1 may be a key factor for regulating the crosstalk between the cAMP–PKA and CoMekk1–Cmk1 MAPK signaling pathway through CoRas2. Intracellular cAMP levels in DARS1 were higher compared with those in the wild-type in vegetative hyphae. However, DARS1 showed lower sensitivity to exogenous cAMP in appressorium development compared with DARS2. Moreover, the intensity of pathogenesis in DARS1 was similar to that in the wild-type. Therefore, CoRas1 could be involved in the cAMP–PKA signaling pathway in vegetative hyphae but not during infection related morphogenesis.

Plants have evolved with a variety of defense mechanisms against attacking phytopathogenic fungi. These include the deposition of cell wall reinforcement components (papillae), hypersensitive cell death, and the synthesis of an antimicrobial secondary metabolite. However, callose deposition was not observed following attempted penetration from the appressoria of the *coira1* mutant. Further-

more, the pathogenesis of the *coira1* mutant was not restored during the infection of heat-shocked cucumber cotyledons. Therefore, there is a small possibility that the observed reduction of pathogenicity in the *coira1* mutant was caused by the plant defense response. In *C. orbiculare*, the generation of the appressorium turgor pressure is required for mechanical penetration of the plant surface [47], however, appressorial turgor of the *coira1* mutant was not affected, as shown by the cytorrhysis assay [27]. The pathogenicity of the *coira1* mutant was attenuated compared with that of the wild-type during infection of wounded leaves, thereby resulting in the reduction in the development of hyphae causing infection in the host plants. In *C. orbiculare*, the catalytic subunit of PKA Cpk1 and adenylate cyclase Cac1 is involved in infectious growth in plants [3]. We postulate that excessive cAMP levels in the *coira1* mutant would affect the process of the infection hyphae development in the host plant. The vegetative hyphae of the *coira1* mutant and the wild-type showed normal morphology; however, the *coira1* mutant formed bulb-shaped hyphae on the cellulose membranes. In response to nitrogen starvation, diploid *S. cerevisiae* cells undergo pseudohyphal growth, which is enhanced by the expression of the dominant active allele *RAS2* [48]. Further investigation has revealed that the pseudohyphal growth is controlled by both Kss1 MAPK and cAMP–PKA signaling [49]. Similarly, *C. albicans* strains carrying the activating $RAS1^{V13}$ allele formed more abundant hyphae in a shorter time than that in the wild-type strain. Thus, activated Ras is a key factor for regulating hyphal morphogenesis in fungi. In this study, the molecular mechanism of bulb-shape penetration hyphae caused by the *coira1* mutation remains unclear. Therefore, further functional analysis of Colra1 will be helpful for understanding the regulation of penetration hyphae morphogenesis for the requirement of proper infection of the host plant.

Colra1 and CoRas2 is involved in the appressoria-mediated differentiation of infection hyphae development

CoRas2 and Colra1 did not show specific colocalization in developing appressoria. Consistently, Colra1 and the active form CoRas2 did not colocalize in developing appressoria. Therefore, during appressorium formation, CoRas2 may be regulated by other factors different from Colra1. In *C. orbiculare*, 48 to 72 h post-inoculation is a crucial phase for appressorium-mediated penetration. Our data indicated that Colra1 and CoRas2 colocalized in a vesicle-like structure in the appressoria 48 h post-inoculation. Moreover, in *M. oryzae*, the correct organization of F-actin in the appressorium pore is important for penetration peg development [37]. Our data also indicated that Colra1 is involved in the assembly of the F-actin in the appressorium pore. Therefore, we assume that Colra1 could be involved in the F-actin organization in appressoria, which is required for penetration peg emergence. However, it is likely that Colra1 does not directly regulate F-actin, because it does not colocalize with the F-actin in the appressoria pore. In *M. oryzae*, NADPH oxidases regulate the septin-mediated assembly of F-actin in the appressorium pore [50]. Moreover, in *C. neoformans*, the GTP-bound form of Ras1 interacts with Rho–GEF Cdc24 that mediates the activation of Cdc42 and Rac proteins, and Cdc42 is involved in cytokinesis and bud morphogenesis through the organization of septin proteins [51]. In *Epichloë festucae*, NoxA is activated by a small GTPase RacA [52]. Therefore, in *C. orbiculare*, Ras–Cdc42–septin proteins and the Ras–Rac–Nox proteins signaling pathway may regulate the assembly of F-actin in the appressorium pore. Our future work will identify interaction factors between Colra1 and

CoRas2, elucidating those that are involved in the assembly of F-actin in the appressorial pore.

Supporting Information

Figure S1 Organization of the *CoIRA1* gene in *C. orbiculare*. (A) Schematic representation of *CoIRA1*. Exons are indicated by gray boxes. The predicted RASGAP domains are indicated by slashed boxes. Eleven exons of *CoIRA1* are indicated by a gray square. Ten introns of *CoIRA1* are indicated by a black bar among 11 exons. (B) RASGAP domain in CoIra1. Amino acid sequence alignment of the predicted *CoIRA1* gene product with homologs from *Saccharomyces cerevisiae* IRA1, IRA2, and *Magnaporthe oryzae* (M.o.). Identical amino acids are indicated by a black background, similar residues are indicated by a gray background, and gaps introduced for alignments are indicated by a hyphen. The predicted RASGAP domain is indicated by a black line.

Figure S2 Gene disruption of the *CoIRA1* of *C. orbiculare*. (A) *CoIRA1* gene disruption by homologous recombination with the *CoIRA1* disruption vector in which a *hph* fragment was inserted into the *CoIRA1* gene. Through double crossover, an *Eco*RV fragment of approximately 3.4 kb containing wild-type *CoIRA1* was predicted to be replaced by a fragment of approximately 5.6 kb containing the *hph* fragment. (B) *CoIRA1* gene disruption was confirmed by Southern blot analysis. Genomic DNAs from the wild-type 104-T and transformants were digested with *Eco*RV and probe with an upstream 1.0-kb fragment of the *CoIRA1* gene. WT, wild-type; dis1-5, *coira1* mutants; ec1-2, ectopic strains.

Figure S3 Hyphal growth and conidia number of the *coira1* mutant on PDA. (A) Each strain was grown on the PDA medium at 24°C for five days and the number of conidia harvested from a 9-cm PDA plate at 5 days after incubation at 24°C.

Figure S4 Appressorium cytorrhysis assay for the *coira1* mutant. Appressoria formed on the multiwell slides were exposed to glycerol solutions ranging from 0 M to 4 M and the percentage of collapsed spherical appressorium was counted. Approximately 200 conidia were observed per well. Three replicates were examined. Three independent experiments were conducted, and standard errors are shown. Solid line, the wild-type; dotted line, the *coira1* mutant.

Figure S5 Pathogenicity assay of the *coira1* mutant in wounded leaves. Conidial suspensions of each strain were inoculated on wounded sites on the cotyledon of the cucumber prepared by scratching the leaves with a sterile toothpick. The leaves were incubated at 24°C for seven days. Strains: WT, wild-type 104-T; DL1, the *coira1* mutant; CL1, the *CoIRA1*-complemented transformant of DL1; DPS1, the *pks1* mutant.

Figure S6 Host defense response was not induced by the penetration of the *coira1* mutant. (A) Quantification of papilla formation at sites of attempted penetration by appressorium in *C. orbiculare*. At three days, leaf epidermal strips inoculated with each strain was stained with Aniline blue to reveal the papilla and observed with epi-fluorescence microscopy. Strains: WT, wild-type; DL1, the *coira1* mutant; CL1, the *CoIRA1*-complemented transformant of DL1; DSD1, the *ssd1*

mutant. At least 200 appressoria were counted for each strain and standard deviations were calculated from three replicated experiments. (B) Pathogenicity assay of the *coira1* mutant on heat-shock cotyledons after the heat treatment at 50°C for 30 s, cucumber cotyledons were inoculated with conidial suspensions. Strains: WT, wild-type; DL1, *coira1* mutant; CL1, the *CoIRA1*-complemented transformant of DL1; DSD1, the *ssd1* mutant. Controls were not exposed to heat shock.

Figure S7 Penetration hyphae formation of a dominant active form *CoRAS1* and *CoRAS2* introduced transformants on cellulose membranes. Conidial suspensions of each strain in distilled water were incubated on cellulose membranes at 24°C for 48 h. WT, the wild-type 104-T; DARS1, WT transformed with a dominant active form *CoRAS1*; DARS2, WT transformed with a dominant active form *CoRAS2*. Scale bar, 10 μm. (B) Percentages of penetration hyphae formation, and bulb-shaped penetration-hyphae formation of *C. orbiculare* WT, DARS1 and DARS2 on cellulose membranes. Approximately 200 conidia of each strain were observed on cellulose membranes. Three replicates were examined. Three independent experiments were conducted, and standard errors are shown. black bar, penetration hyphae; gray bar, bulb-shape penetration hyphae formation.

Figure S8 Pathogenicity assay of a dominant active form *CoRAS1* and *CoRAS2* introduced transformants on cucumber cotyledons. Conidial suspensions of each strain were inoculated on the detached cucumber cotyledons, and the leaves were incubated at 24°C for seven days. WT, the wild-type; DARS1, WT transformed with a dominant active form *CoRAS1*; DARS2, WT transformed with a dominant active form *CoRAS2*.

Figure S9 Appressorium formation in *coira1* mutants of *C. orbiculare* on glass slides. (A) Conidial suspensions of each strain in distilled water were incubated on multiwell glass slides at 24°C for 24 h. WT, wild-type; DRS2-1 and DRS2-2, the *coras2* mutant; CRS2-1, the *CoRAS2*-complemented transformant of DRS2-1; CR2-2, the *CoRAS2*-complemented transformant of DRS2-2. Scale bar, 10 μm. (B) Percentages of conidial germination, appressorium formation in *C. orbiculare* WT and *coras2* mutants on multiwell glass slides. Approximately 100 conidia of each strain were observed per well on multiwell glass slides. Three replicates were examined. Three independent experiments were conducted, and standard errors are shown. Black bar, conidial germination; gray bar, appressorium formation.

Figure S10 Pathogenicity assay of *coras2* mutants of *C. orbiculare* on the cucumber cotyledons. Conidial suspensions of each strain were placed on detached cotyledons of the cucumber, and the leaves were incubated at 24°C for seven days. Shown are the leaves after incubation with the following strains: WT, wild-type 104-T; DRS2-1 and DSR2-2, the *coras2* mutant; CRS2-1, the *CoRAS2*-complemented transformant of DRS2-1; the CRS2-2, *CoRAS2*-complemented transformant of DRS2-2.

Figure S11 Assay for colocalizations of CoIra1 and active form CoRas2. Conidial suspensions of RFP–DARS2/IRA1–VENUS strain were incubated on glass slides at 24°C for 0 h, 3 h, 6 h and 24 h and observed by fluorescent microscopy. RFP–DARS2/IRA1–VENUS, the wild-type strain expressing

RFP fused with a dominant active form *CoRAS2* and *CoIRA1*–VENUS. Scale bar, 10 μm.

Figure S12   Localization of a functional CoIra1–VENUS fusion protein in *C. orbiculare* in initial and late infection hyphae in the cucumber leaves. The wild-type strain expressing *CoIRA1*–VENUS was inoculated on cucumber leaves and incubated at 48 h, 72 h and CoIra1–VENUS was observed using fluorescent microscopy.

Table S1   PCR primers used in this study.

References

1. Kubo Y, Takano Y (2013) Dynamics of infection-related morphogenesis and pathogenesis in *Colletotrichum orbiculare*. J Gen Plant Pathol 79: 233–242. doi:10.1007/s10327-013-0451-9.

2. Sakaguchi A, Tsuji G, Kubo Y (2010) A yeast *STE11* homologue *CoMEKK1* is essential for pathogenesis-related morphogenesis in *Colletotrichum orbiculare*. Mol Plant Microbe Interact 23: 1563–1572. doi:1 0.1094/MPMI-03-10-0051.

3. Yamauchi J, Takayanagi N, Komeda K, Takano Y, Okuno T (2004) cAMP-PKA signaling regulates multiple steps of fungal infection cooperatively with Cmk1 MAP kinase in *Colletotrichum lagenarium*. Mol Plant Microbe Interact 17: 1355–1365. doi:10.1094/MPMI.2004.17.12.1355.

4. Park G, Xue C, Zhao S, Kim Y, Orbach M, et al. (2006) Multiple upstream signals converge on the adaptor protein Mst50 in *Magnaporthe grisea*. Plant Cell 18: 2822–2835. doi:10.1105/tpc.105.038422.

5. Bluhm BH, Zhao X, Flaherty JE, Xu JR, Dunkle LD (2007) *RAS2* regulates growth and pathogenesis in *Fusarium graminnearum*. Mol Plant Microbe Interact 20: 627–636. doi:10.1094/MPMI-20-6-0627.

6. Ha YS, Memmott SD, Dickman MB (2003) Function analysis of Ras in *Colletotrichum trifolii*. FEMS Microbiol Lett 26: 315–321.

7. Harashima T, Anderson S, Yates JR 3rd, Heitman J (2006) The kelch proteins Gbp1 and Gpb2 inhibit Ras activity via association with the yeast RasGAP neurofibromin homologs Ira1 and Ira2. Mol Cell 22: 819–830. doi:10.1016/j.molcel.2006.05.011.

8. Lee YH, Dean RA (1993) cAMP regulates infection structure formation in the plant pathogenic fungus *Magnaporthe grisea*. Plant Cell 5: 693–700. doi:10.1105/tpc.5.6.693.

9. Takano Y, Komeda K, Kojima K, Okuno T (2001) Proper regulation of cyclic AMP-dependent protein kinase is required for growth, conidiation, and appressorium formation in the anthracnose fungus *Colletotrichum lagenarium*. Mol Plant Microbe Interact 14: 1149–1157. doi:10.1094/MPMI.2001.14.10.1149.

10. Adachi K, Hamer JE (1998) Divergent cAMP signaling pathways regulate growth and pathogenesis in the rice blast fungus *Magnaporthe grisea*. Plant Cell 10: 1361–1374. doi:10.1105/tpc.10.8.1361.

11. Choi W, Dean RA (1997) The adenylate cyclase gene *MAC1* of *Magnaporthe grisea* controls appressorium formation and other aspects of growth and development. Plant Cell 9: 1973–1983. doi:10.1105/tpc.9.11.1973.

12. Zhou X, Zhang H, Li G, Shaw B, Xu JR (2012) The Cyclase-associated protein Cap1 is important for proper regulation of infection-related morphogenesis in *Magnaporthe oryzae*. PLOS Pathogen 8: e1002911. doi:10.1371/journal.ppat.1002911.

13. Lee N, Kronstad JW (2002) ras2 controls morphogenesis, pheromone response, and pathogenicity in the fungal pathogen *Ustilago maydis*. Eukaryot Cell 1: 954–966. doi:10.1128/EC.1.6.954-966.2002.

14. Gibbs JB (1991) Ras C-terminal processing enzymes-new drug targets? Cell 65: 1–4.

15. Magee T, Marshall C (1999) New insights into the interaction of Ras with the plasma membrane. Cell 98: 9–12. doi:10.1016/S0092-8674(00)80601-7.

16. Ghaemmaghami S, Huh WK, Bower K, Howson RW, Belle A, et al. (2003) Global analysis of protein expression in yeast. Nature 16: 737–741. doi:10.1038/nature02046.

17. Belotti F, Tisi R, Paiardi C, Groppi S, Martegani E (2011) PKA-dependent regulation of Cdc25 RasGEF localization in budding yeast. FEBS Lett 15: 3914–3920. doi:10.1016/j.febslet.2011.10.032.

18. Belotti F, Tisi R, Paiardi C, Rigamonti M, Groppi S, et al. (2012) Localization of Ras signaling complex in budding yeast. Biochim Biophys Acta 1823: 1208–1216. doi:10.1016/j.bbamcr.2012.04.016.

19. Broggi S, Martegani E, Colombo S (2013) Live-cell imaging of endogenous Ras-GTP shows predominant Ras activation at the plasma membrane and in the nucleus in the *Saccharomyces cerevisiae*. Int J Biochem Cell Biol 45: 384–394. doi:10.1016/j.biocel.2012.10.013.

20. Sambrook J, Fritsch EF, Maniatis T (1989) Molecular Cloning: A laboratory Manual. Cold Spring Harbor Laboratory Press, Cold Spring Harbor, NY.

21. Tsuji G, Fujii S, Fujihara N, Hirose C, Tsuge S, et al. (2003) Agrobacterium tumefaciens-mediated transformation for random insertional mutagenesis in Colletotrichum lagenarium. J Gen Plant Pathol 69: 230–239. doi:10.1007/s10327-003-0040-4.

22. Takano Y, Kubo Y, Kawamura C, Tsuge T, Furusawa I (1997) The *Alternaria alternata* melanin biosynthesis gene restores appressorial melanization and penetration of cellulose membranes in the melanin-deficient albino mutant of *Colletotrichum lagenarium*. Fungal Genet Biol 21: 131–140. doi:10.1006/fgbi.1997.0963.

23. Bruno KS, Tenjo F, Li L, Hamer JE, Xu JR (2004) Cellular localization and role of kinase activity of PMK1 in *Magnaporthe grisea*. Eukaryot Cell 3: 1525–1532. doi:10.1128/EC.3.6.1525-1532.2004.

24. Tsuji G, Takeda T, Furusawa I, Horino O, Kubo Y (1997) Carpropamid, an anti-rice blastfungicide, inhibits scytalone dehydratase activity and appressorial penetration in *Colletotrichum lagenarium*. Pestic Biochem Physiol 57: 211–219.

25. Gan P, Ikeda K, Irieda H, Narusaka M, O'Connel RJ, et al. (2013) Comparative genomic and transcriptomic analyses reveal the hemibiotrophic stage shift of *Colletotrichum* fungi. New Phytol 197: 1236–1249. doi:10.1111/nph.12085.

26. Bechinger C, Giebel KF, Schnell M, Leiderer P, Deising HB, et al. (1999) Optical measurements of invasive forces exerted by appressoria of a plant pathogenic fungus. Science 17: 1896–1899. doi:10.1126/science.285.5435.1896.

27. Howard RJ, Ferrari MA, Roach DH, Money NP (1991) Penetration of hard substrates by a fungus employing enormous turgor pressures. Proc Natl Acad Sci U S A 15: 11281–11284.

28. Tanaka S, Ishihama N, Yoshioka H, Huser A, O'Connell R, et al. (2009) The *Colletotrichum orbiculare SSD1* mutant enhances *Nicotiana benthamiana* basal resistance by activating a mitogen-activated protein kinase pathway. Plant Cell 8: 2517–2526 doi:10.1105/tpc.109.068023.

29. Tamanoi F (2011) Ras signaling in yeast. Genes Cancer 2: 210–215. doi:10.1177/1947601911407322.

30. Seeburg PH, Colby WW, Capon DJ, Goeddel DV, Levinson AD (1984) Biological properties of human c-Ha-ras1 genes mutated at codon 12. Nature 312: 71–75. doi:10.1038/312071a0.

31. Der CJ, Finkel T, Cooper GM (1986) Biological and biochemical properties of human rasH genes mutated at codon 61. Cell 17: 167–176.

32. Feng Q, Summers E, Guo B, Fink G (1999) Ras signaling is required for serum-induced hyphal differentiation in *Candida albicans*. J Bacteriol 181: 6339–6346.

33. Takano Y, Kikuchi T, Kubo Y, Hamer JE, Mise K, et al. (2000) The *Colletotrichum lagenarium* MAP kinase gene CMK1 regulates diverse aspects of fungal pathogenesis. Mol Plant Microbe Interact 13: 374–383. doi:10.1094/MPMI.2000.13.4.374.

34. Piispanen AE, Bonnefoi O, Carden S, Deveau A, Bassilana M Hogan DA (2011) Roles of Ras1 membrane localization during *Candida albicans* hyphal growth and farnesol response. Eukaryot Cell 10: 1473–1484. doi:10.1128/EC.05153-11.

35. Herrera F, Tenreiro S, Miller-Fleming L, Outeiro TF (2011) Visualization of cell-to-cell transmission of mutant huntingtin oligomers. PLoS Curr 3: RRN1210. doi:10.1371/currents.RRN1210.

36. Nagai T, Ibata K, Park ES, Kubota M, Mikoshiba K, et al. (2002) A variant of yellow fluorescent protein with fast and efficient maturation for cell-biological applications. Nat Biotechnol 20: 87–90. doi:10.1038/nbt0102-87.

37. Dagdas YF, Yoshino K, Dagdas G, Ryder LS, Bielska E, et al. (2012) Septin-mediated plant cell invasion by the rice blast fungus, *Magnaporthe oryzae*. Science 336: 1590–1595. doi:10.1126/science.1222934.

38. Tsuji G, Fujii S, Tsuge S, Shiraishi T, Kubo Y (2003) The *Colletotrichum lagenarium* Ste12-like gene *CST1* is essential for appressorium penetration. Mol Plant Microbe Interact 16: 315–325. doi.org/10.1094/MPMI.2003.16.4.315.

39. Gourlay CW, Ayscough KR (2006) Actin-induced hyperactivation of the Ras signaling pathway leads to apoptosis in *Saccharomyces cerevisiae*. Mol Cell Biol 17: 6487–6501. doi:10.1128/MCB.00117-06.

40. Vinod PK, Sengupta N, Bhat PJ, Venkatesh KV (2008) Integration of global signaling pathways, cAMP-PKA, MAPK and TOR in the regulation of FLO11. PLoS One 3: e1663. doi:10.1371/journal.pone.0001663.

41. Phan VT, Ding VW, Li F, Chalkley RJ, Burlingame A, et al. (2010) The RasGAP proteins Ira2 and neurofibromin are negatively regulated by Gpb1 in

Acknowledgments

We are grateful to Dr. Y. Takano (Kyoto University) for providing the *cac1* and *maf1* mutants of *C. orbiculare*. We thank Dr. G. Tsuji (Kyoto Prefectural University) for providing the Lifeact–RFP vector.

Author Contributions

Conceived and designed the experiments: KH YK. Performed the experiments: KH. Analyzed the data: KH YK. Contributed reagents/materials/analysis tools: KH YK. Contributed to the writing of the manuscript: KH YK.

yeast and ETEA in humans. Mol Cell Biol 30: 2264–2279. doi:10.1128/MCB.01450-08.

42. Zhao X, Kim Y, Park G, Xu JR (2005) A mitogen-activated protein kinase cascade regulating infection-related morphogenesis in *Magnaporthe grisea*. Plant Cell 17: 1317–1329. doi:doi.org/10.1105/tpc.104.029116.

43. Cherkasova VA, McCully R, Wang Y, Hinnebusch A, Elion EA (2003) A novel functional link between MAP kinase cascades and the Ras/cAMP pathway that regulates survival. Curr Biol 13: 1220–1226. doi:10.1016/S0960-9822(03)00490-1.

44. Kaffarnik F, Müller P, Leibundgut M, Kahmann R, Feldbrügge M (2003) PKA and MAPK phosphorylation of Prf1 allows promoter discrimination in *Ustilago maydis*. EMBO J 22: 5817–5826. doi:10.1093/emboj/cdg554.

45. Lee N, D'Souza CA, Kronstad JW (2003) Of smuts, blasts, mildews, and blights cAMP signaling in phytopathogenic fungi. Ann Rev Phytopathol 41: 399–427. doi:10.1146/annurev.phyto.41.052002.095728.

46. Fortwendel JR, Juvvadi PR, Rogg LE, Asfaw YG, Burns KA, et al. (2012) Plasma membrane localization is required for RasA-mediated polarized morphogenesis and virulence of *Aspergillus fumigatus*. Eukaryot Cell 11: 966–977. doi:10.1128/EC.00091-12.

47. Fujihara N, Sakaguchi A, Tanaka S, Fujii S, Tsuji G, et al. (2010) Peroxisome biogenesis factor PEX13 is required for appressorium-mediated plant infection by the anthracnose fungus *Colletotrichum orbiculare*. Mol Plant Microbe Interact 23: 436–445. doi:10.1094/MPMI-23-4-0436.

48. Gimeno CJ, Ljungdahl PO, Styles CA, Fink GR (1992) Unipolar cell divisions in the yeast *S. cerevisiae* lead to filamentous growth: regulation by starvation and *RAS*. Cell 68: 1077–1090.

49. Pan X and Heitman J (1999) Cyclic AMP-dependent protein kinase regulates pseudohyphal differentiation in *Saccharomyces cerevisiae*. Mol Cell Biol 19: 4847–4887.

50. Ryder LS, Dagdas YF, Mentlak TA, Kershaw MJ, Thornton CR, et al. (2013) NADPH oxidases regulate septin-mediated cytoskeletal remodeling during plant infection by the rice blast fungus. Proc Natl Acad Sci USA 110: 3179–3184. doi:10.1073/pnas.1217470110.

51. Ballou ER, Selvig K, Narloch JL, Nichols CB, Alspaugh JA (2013) Two Rac paralogs regulate polarized growth in the human fungal pathogen *Cryptococcus neoformans*. Fungal Genet Biol 47: 58–75. doi:10.1016/j.fgb.2013.05.006.

52. Tanaka A, Takemoto D, Hyon GS, Park P, Scott B (2008) NoxA activation by the small GTPase RacA is required to maintain a mutualistic symbiotic association between *Epichloë festucae* and perennial ryegrass. Mol Microbiol 68: 1165–1178. doi:10.1111/j.1365-2958.2008.06217.

13

Pathogen and Circadian Controlled 1 (PCC1) Protein Is Anchored to the Plasma Membrane and Interacts with Subunit 5 of COP9 Signalosome in *Arabidopsis*

Ricardo Mir[¤], José León*

Instituto de Biología Molecular y Celular de Plantas, Consejo Superior de Investigaciones Científicas-Universidad Politécnica de Valencia, Valencia, Spain

Abstract

The *Pathogen and Circadian Controlled 1 (PCC1)* gene, previously identified and further characterized as involved in defense to pathogens and stress-induced flowering, codes for an 81-amino acid protein with a cysteine-rich C-terminal domain. This domain is essential for homodimerization and anchoring to the plasma membrane. Transgenic plants with the ß-*glucuronidase (GUS)* reporter gene under the control of 1.1 kb promoter sequence of *PCC1* gene display a dual pattern of expression. At early post-germination, *PCC1* is expressed only in the root vasculature and in the stomata guard cells of cotyledons. During the transition from vegetative to reproductive development, *PCC1* is strongly expressed in the vascular tissue of petioles and basal part of the leaf, and it further spreads to the whole limb in fully expanded leaves. This developmental pattern of expression together with the late flowering phenotype of long-day grown RNA interference (iPCC1) plants with reduced *PCC1* expression pointed to a regulatory role of PCC1 in the photoperiod-dependent flowering pathway. iPCC1 plants are defective in light perception and signaling but are not impaired in the function of the core CO-FT module of the photoperiod-dependent pathway. The regulatory effect exerted by PCC1 on the transition to flowering as well as on other reported phenotypes might be explained by a mechanism involving the interaction with the subunit 5 of the COP9 signalosome (CSN).

Editor: Keqiang Wu, National Taiwan University, Taiwan

Funding: This work was funded by grants BIO2008-00839, BIO2011-27526 and CSD2007-0057 from Ministerio de Ciencia e Innovación of Spain to J.L. A fellowship/contract of the FPU program of the Ministerio de Educación y Ciencia (Spain) funded R.M. work. The funders had no role in study design, data collection and analysis, decision to publish, or preparation of the manuscript.

Competing Interests: The authors have declared that no competing interests exist.

* E-mail: jleon@ibmcp.upv.es

¤ Current address: Department of Botany and Plant Sciences and Center for Plant Cell Biology, University of California, Riverside, California, United States of America

Introduction

The *Pathogen and Circadian Controlled 1 (PCC1)* gene in Arabidopsis was originally identified as a *Pseudomonas syringae*-induced gene with a circadian controlled pattern of expression [1]. Further work revealed *PCC1* as a salicylic acid (SA)-induced gene with a potential function in controlling flowering time under UV-C light stress conditions and also under non-stressed conditions [2]. Based on bioinformatic predictions, PCC1 has been characterized as one of the so-called Cysteine-rich Transmembrane (CYSTM) domain-containing family of proteins [3]. Since plants with reduced *PCC1* expression by means of an RNAi approach (iPCC1 transgenic lines) are late flowering [2] and *PCC1* gene expression is potentially controlled by the circadian clock [1], it seems likely that PCC1 is connected to light signaling. In plants, development from seed germination to the reproductive stage is tightly controlled by light. Light perception and downstream signaling functionally interacts with different hormone signaling pathways in controlling most of the developmental transitions [4]. Gibberellins (GAs) are key hormones in many light-driven transitions [5] including seed germination [6], hypocotyl elongation during skotomorphogenesis [7,8] and flowering time [9]. The control exerted by the combined stimuli of light and GAs on diverse developmental processes is based on multiple regulatory levels from gene transcription [10] to post-transcriptional [11] and post-translational [12] processes. In the model plant *Arabidopsis thaliana*, transcription factors of the bHLH family are key transcriptional regulators in controlling the elongation of hypocotyls in close connection with repressor proteins of the GA-related DELLA family [7,8]. It is also well documented that DELLA proteins are substrates for the post-translational modification based on ubiquitination of lysine residues mediated by the E3 ubiquitin ligase SCF[SLY1] complex [13,14]. The ubiquitinating activity of the SCF complexes is dependent on the function of their four components (RBX1, Skp1-like, Cullin and F-box-proteins). Moreover, cullin must be post-translationally modified by binding the RUB/NEDD8 ubiquitin-like protein for SCF complexes to be assembled and fully active [15]. This ubiquitination machinery is negatively controlled by the function of an eight-subunit protein complex so-called COP9 signalosome (CSN), which has RUB isopeptidase activity that removes RUB modification from cullin proteins [16]. The derubylation reaction is mediated by the subunit CSN5, a zinc metalloprotease, although defective function of any of the subunits make CSN unable in derubylation [17] and cause severe developmental [18] and defense responses [19].

Here we present multiple lines of experimental evidence of the plasma membrane localization for PCC1 protein and its homo-dimerization. We found that the cysteine-rich C-terminus is responsible for both, anchoring to the plasma membrane and dimerization. *PCC1* gene displays a changeable pattern of expression throughout development, which is consistent with potential regulatory roles in both development and defense. Besides, PCC1 interacts with subunits 5a and 5b of CSN at the plasma membrane, which could explain the wide range of altered phenotypes observed in plants with altered *PCC1* transcript levels.

Materials and Methods

Plant Material and Growth Conditions

Arabidopsis thaliana seeds of wild type Col-0, photoreceptor mutants *phyA*, *phyB* (kindly donated by Miguel Blázquez, IBMCP, Valencia, Spain), *cry1*, *cry2* and *cry1cry2* (kindly donated by Jose Jarillo, CBGP, Madrid, Spain), *co-10* mutants and dexametasone-induced overexpression lines oxCO-GR (kindly donated by Federico Valverde, IBVF, Sevilla, Spain), or transgenic lines expressing RNAi constructs for *PCC1* gene previously described [2] were surface sterilized with 30% bleach and 0.01% Tween 20, washed extensively with miliQ sterile water, and sown in Murashige and Skoog medium supplemented with 0.8% agar and 1% sucrose. After 3 d of stratification at 4°C on Petri MS-containing plates, seeds were transferred to a growth chamber under white fluorescent white light (fluence rate of 70 µmol m^{-2} s^{-1}) with a 16 h-light/8 h-dark photoperiod and a controlled temperature of 19–23°C. *Nicotiana benthamina* seeds were sown in soil and grown in green-house under a 16-hlight/ 8-h-dark photoperiod and a controlled temperature of 19–23°C.

Quantitative RT-PCR Analysis and GUS Staining

To quantify transcript levels, total RNAs from wild type and iPCC1 seedlings, harvested at 12 hours after dawn, were extracted, purified with the RNeasy kit (QIAGEN), and analyzed by quantitative RT-PCR techniques as described previously [20]. Primers used for qPCR are included in Table 1. GUS staining was performed in samples harvested at 12 hours after dawn with X-Gluc in the presence of 5 mM ferrycianide/ ferrocyanide redox buffer.

Hormone Treatments and Hypocotyl Elongation Tests

Imbibed seeds from wild type Col-0 and three independent iPCC1 transgenic lines were stratified at 4°C for 4 days, then successively transferred to white light at 21°C for 6 h, and to darkness for additional 18 h before being incubated for 4 additional days under different light qualities supplied by LED lights in a Percival growth chamber. Fluence rates of 70 µmol m^{-2} s^{-1} for white light; 5 µmol m^{-2} s^{-1} for far-red light; 30 µmol m^{-2} s^{-1} for red light and 10 µmol m^{-2} s^{-1} for blue light were used. After different light quality incubation, hypocotyls of seedlings were laid on acetate sheets, scanned and the length measured by using ImageJ software. Around 20 seedlings per genotype and condition were measured in each of the three replicate experiments performed and the mean value calculated. The results are expressed as the mean of three replicates ± SD. When indicated the active gibberellin GA3 or the gibberellins biosynthesis inhibitor paclobutrazol (PAC) were added to the growth media at the indicated concentrations. Treatments with SA were performed by adding the indicated concentrations to liquid media supplemented with 0.02% Tween-20 as surfactant.

Table 1. Oligonucleotides used for qRT-PCR in this work.

Primer name	Sequence (5' to 3')	Reference
qACT2-D	ttgttccagccctcgtttgt	[20]
qACT2-R	tgtctcgtggattccagcag	[20]
qPCC1-2D	tgctccagcctctgtacatca	[2]
qPCC1-2R	cgacttctgtctcatcatgctga	[2]
qFT-D	caaccctcacctccgagaatat	[2]
qFT-R	tgccaaaggttgttccagttgt	[2]
qCO-D	aacgacataggtagtggagagaacaac	[2]
qCO-R	gcagaatctgcatggcaataca	[2]
qGID1a-F	gtgacggttagagaccgcga	[20]
qGID1a-R	tccctcgggtaaaaacgctt	[20]
qGID1b-F	ttacggtcaaggaactcggc	[20]
qGID1b-R	tcgccctgacggttctttc	[20]
qGID1c-F	cggctcaaatcttcgatctgg	[20]
qGID1c-R	ttggcatttgcagggactttc	[20]
qRGA-F	acttcgacgggtacgcagat	[20]
qRGA-R	tgtcgtcaccgtcgttcct	[20]
qGAI-3UTR-F	aatgaattgatctgttgaaccgg	[20]
qGAI-3UTR-R	ggcttcggtcggaaatctatc	[20]

Transient and Stable Transformation of Nicotiana and Arabidopsis with GFP-tagged Versions of PCC1

The PCC1 coding sequence was mobilized from entry Gateway plasmids by recombination to destination binary pGWB5 and pGWB6 vectors [21] to generate constructs for C- and N-terminal PCC1 fusion to GFP. A control GFP-stop-PCC1 construct expressing free GFP and a GFP-Δ177PCC1 construct expressing a truncated version containing the first 177 nucleotides and excluding the 3'-end coding for the C-terminus were also mobilized from entry vectors to the corresponding destination binary vectors using Gateway technology. pGWB5 was also used to generate a GFP-tagged CSN5B subunit of COP9 signalosome. Nicotiana leaves were transiently transformed by infiltration with *Agrobacterium tumefaciens* C58 strain co-transformed (1:1) with the corresponding plasmids and the P19 suppressor of silencing. Plasmids expressing the whole PCC1 protein with C-terminal fusions to GFP were also used to stably transformed *Arabidopsis thaliana* by floral dipping in suspensions of *Agrobacterium tumefaciens* C58 strain transformed with the corresponding plasmids. Primary transformants were selected in kanamycin-supplemented media and homozygous T3 seeds were used throughout this work.

Yeast Two-Hybrid (Y2H) Screening of PCC1 Interactors

A truncated version of PCC1 containing the first 59 amino acids was cloned into pB66 as a C-terminal fusion to Gal4 DNA-binding domain (N-Gal4-PCC1(1-59)-C) and used as a bait to screen a random-primed 7-day old *A. thaliana* seedlings cDNA library containing 98.8 million cDNA clones cloned into pP6 vector. pB66 and pP6 are derived from pAS2ΔΔ and pGADGH plasmids, respectively [22]. Screening was performed by using a mating approach with Y187 (matα) and CG1945 (mata) yeast strains as previously described [22]. A total of 31 His+ colonies were selected on minimal medium lacking tryptophan, leucine and histidine. The prey fragments of the positive clones were amplified by PCR and sequenced at their 5' and 3' junctions. The resulting sequences were used to identify the corresponding interacting proteins in the GenBank database (NCBI).

Protein-Protein Interaction Tests Based on Y2H, BiFC and IP-based Pull-down

Protein interactions were tested in yeast by subcloning baits and preys fused to the DNA binding (DB) and activation (AD) domains of GAL4 and the reverse option in pDBLeu and pPC86 vectors (Invitrogen). Selection was performed by plating serial dilutions of transformed yeasts in minimal medium –Trp-Leu-His supplemented with increasing concentrations of 3-aminotriazole. In planta interactions between proteins were tested by Bifunctional Fluorescence Complementation (BiFC) trough transient co-transformation of Nicotiana leaves by agroinfiltration with pair of proteins each fused to half of the YFP molecule in pYFP43 and pYFN43 vectors [23]. Reconstitution of YFP-based fluorescence was visualized by a TCS SL Leica confocal microscope. Finally, versions of PCC1 fused to GFP, HA and c-myc tags and GFP-tagged versions of GFP-Δ177PCC1 and CSN5B subunit of COP9 signalosome were co-transformed in Nicotiana leaves in the presence of the P19 suppressor of silencing. Total protein extracts were obtained by grinding leaf tissue in liquid nitrogen and extraction with TBS buffer supplemented with protease inhibitor cocktail (Sigma) and detergent as indicated. Immunoprecipitation of total protein extracts was performed with magnetic beads covered with polyclonal rabbit antibodies against HA or GFP tags (Milteny). Pulled-down proteins were further analyzed by Western blot with primary antibodies against HA (polyclonal from Abcam), GFP (monoclonal from Clontech) or c-myc (polyclonal coupled to HRP from Sigma) and secondary anti-rabbit or anti-mouse antibodies coupled to horseradish peroxidase (GE Healthcare). The Enhanced Chemiluniscence (ECL) detection kit (GE Healthcare) was used to expose Fujifilm.

Generation of Double iPCC1/35S::GR-CO Transgenic Plants and Quantification of Flowering Time

iPCC1 plants with extremely reduced levels of PCC1 transcript because of the over-expression of an RNAi construct for PCC1 [2] were crossed to co-2 mutant plants transformed to overexpress CO fused to the Glucocorticoid Receptor (35S::CO-GR/co-2) that makes CO functional in the nucleus only upon treatment with a GR ligand such as dexamethasone (DEX) [24]. A PCR-based genotyping procedure using specific primers for CO (5′-GCA-GAATCTGCATGGCAATGGCAATACA-3′) and PCC1 (5′-CGCTTACTCTGATGTACAGA -3′) and common primers (5′-GCTCCTACAAATGCCATCA-3′) from 35S promoter sequence was used. Flowering time was quantified by counting rosette plus cauline leaves upon bolting as previously reported [25].

Results

PCC1 is a Small Protein Anchored to the Plasma Membrane

Pathogen and Circadian Controlled 1 (PCC1) was originally identified as an early activated gene upon infection with the bacterial pathogen *Pseudomonas syringae AvrRpt2* that shows an expression pattern controlled by the circadian clock [1]. Later on *PCC1* was identified in a comparative transcriptomic analysis of SA-deficient versus wild type plants as a SA- and UV-C light-induced gene that codes for a potential activator of stress-stimulated flowering in Arabidopsis [2]. *PCC1* is a small gene that codes for an 81 amino acids protein with a molecular mass of 8.4 kDa. Despite its low hydrophobicity, estimated by its GRAVY index [26] of –0.129, much lower that the one observed for most of the membrane proteins so far reported, PCC1 has been identified in two previous reports searching for plasma membrane associated proteins in Arabidopsis [27,28]. Further *in silico* analysis of PCC1 sequence with the membrane specific TMPred tool [29] predicts a transmembrane helix in the C-terminus (between amino acids 60 and 78) of the protein (Fig. 1A), which coincides with its most hydrophobic region according to its GRAVY index value, which is slightly over 1.5 between the amino acids 65 and 75. However, the use of different bioinformatic tools for the prediction of the cellular localization of proteins yields unequal results from extracellular to cytoplasmic, nuclear or even chloroplastic localization. This discrepancy in the predictions prompted us to check the subcellular localization of PCC1 through several molecular experimental approaches. First, we generated constructs to transiently express recombinant versions of PCC1 protein fused in its N- and C-terminus to GFP under the 35S promoter in *Nicotiana benthamiana*. Figure 1B shows the levels of PCC1-GFP protein extracted with either TBS buffer or TBS buffer supplemented with different detergents. Only in the presence of detergents, PCC1-GFP protein was efficiently extracted as demonstrated by Western blot using anti-GFP antibody (Fig. 1B). A confocal microscopy analysis of Nicotiana leaves transiently transformed with the 36 kDa fusion proteins PCC1-GFP, GFP-PCC1 and a control GFP-stop-PCC1 construct, which expresses free 28 kDa GFP, pointed to fluorescence associated to the plasma membrane for both PCC1-GFP and GFP-PCC1 fusion proteins, in contrast to the localization of free GFP in the cytosol and nucleus (Fig. 1C). To demonstrate that the membrane localization was dependent on the hydrophobic C-terminus region being anchored to the plasma membrane, we generated a C-terminal fusion to GFP of the truncated version GFP-Δ177PCC1 expressing the first 177 bp of the coding sequence and thus lacking the C-terminus. We first checked that the different constructs expressed proteins of 36 kDa, 33 kDa and 28 kDa corresponding to GFP fusions to the full PCC1 protein, the truncated version and free GFP, respectively (Fig. 1D). The typical fluorescence associated to the plasma membrane of the PCC1-GFP protein was changed to cytoplasmic and nuclear localization for the truncated version lacking the C-terminus (Fig. 1D), thus confirming the essential role for the C-terminus to be anchored to the plasma membrane. These data were further confirmed by co-localization studies of fluorescence associated to GFP and to the membrane-specific staining with fluorophore FM4-64 in Nicotiana leaves transformed with PCC1-GFP and Δ177PCC1-GFP. We found co-localization in the plasma membrane for PCC1-GFP but not for the truncated version (Fig. S1). To check the possibility that PCC1 is associated to the cell wall, protoplasts were isolated from *Nicotiana benthamiana* transiently transformed with *35S::PCC1-GFP* and *35S::GFP-PCC1* constructs as mentioned above. Figure 1E shows that fluorescence associated to GFP was located at the periphery of cell wall-free protoplast, thus confirming that PCC1 is neither located in the cell wall nor in the leaf apoplast. Finally, we also checked whether fluorescence associated to PCC1-GFP expression was still restricted to the plasma membrane in plasmolysed cells of Nicotiana leaves transformed with *35S::PCC1-GFP* construct. Figure S2 shows that GFP fluorescence was detected in the two plasma membranes of adjacent cells and no fluorescence was detected either in the apoplast or cell walls of plasmolysed cells.

To rule out the possibility that transient expression of Arabidopsis *PCC1* in *Nicotiana benthamiana* causes aberrant protein localization on PM, transgenic Arabidopsis lines stably expressing the *35S::PCC1-GFP* construct were generated. Figure 2A shows

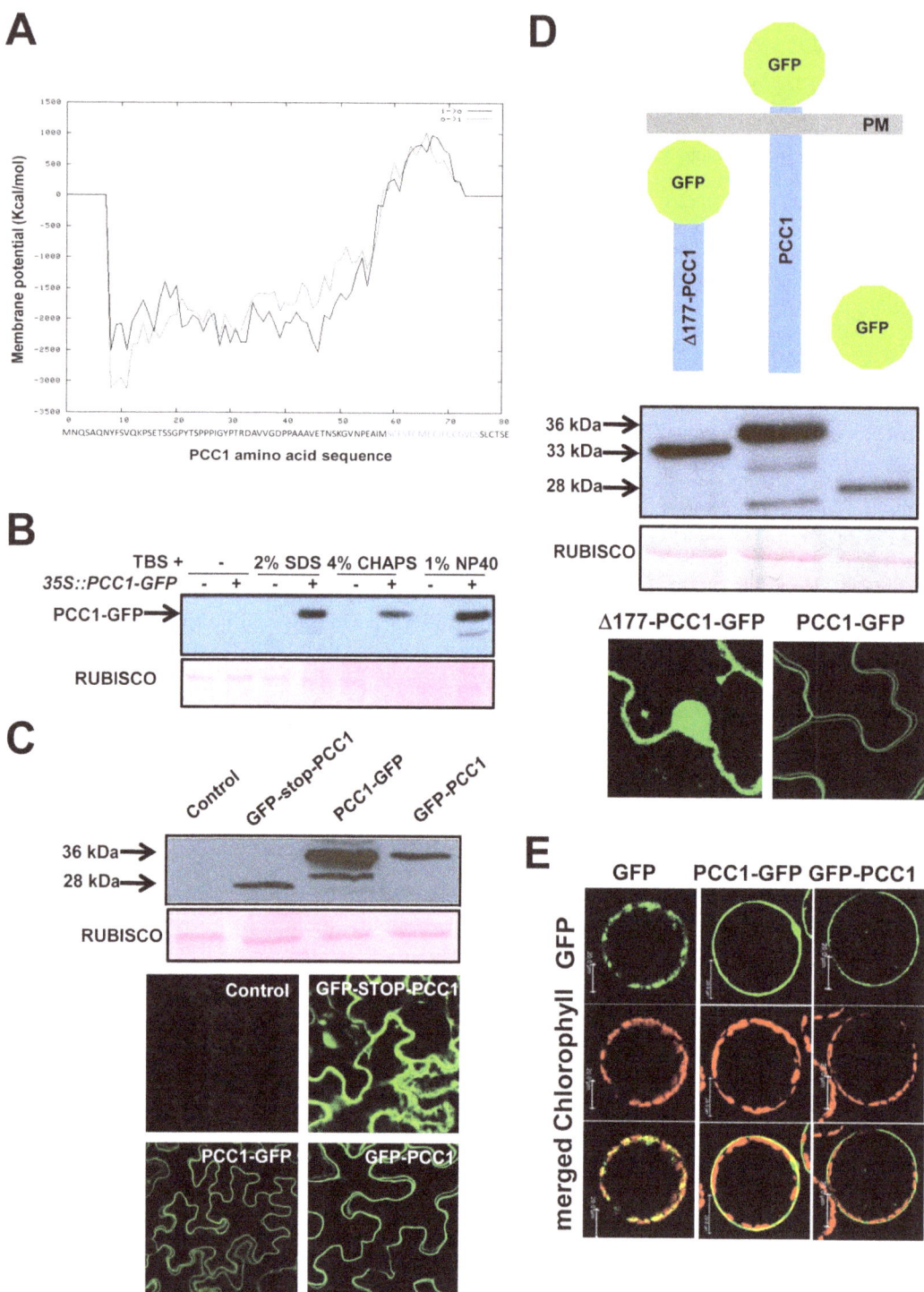

Figure 1. Plasma membrane localization of PCC1 protein. (A) Bioinformatic prediction of transmembrane potential for PCC1 using TMPred tool. The C-terminal domain with positive potential is written in blue in the amino acid sequence below the plot. (B) Extraction of PCC1 protein from *Nicotiana benthamina* leaves transiently transformed with *35S::PCC1-GFP* construct was assessed by using TBS buffer supplemented or not with different detergents as indicated, and further detected by Western blot with anti-GFP antibodies. Ponceau S-stained Rubisco is shown as loading control. (C) Expression of GFP-tagged versions of PCC1 in its C- (PCC1-GFP) and N-terminus (GFP-PCC1) as wells as the free GFP control (GFP-stop-PCC1) was analyzed by Western blot with anti-GFP antibodies and confocal microscopy. (D) Expression of a GFP-tagged truncated PCC1 version (Δ177-PCC1-GFP) without the potential C-terminal membrane-associated domain leads to cytoplasmic localization instead of the membrane localization for the whole GFP-PCC1 protein. (E) Isolation of protoplasts from transiently transformed *Nicotiana benthamina* leaves confirmed the plasma membrane association of PCC1 and allowed to rule out cell wall localization.

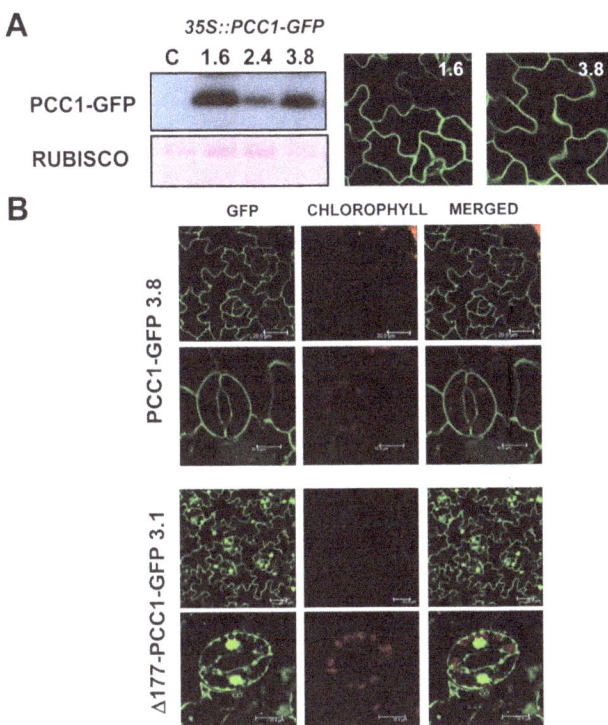

Figure 2. Plasma membrane localization of PCC1 in transgenic 35S::PCC1-GFP Arabidopsis plants. (A) Levels of PCC1-GFP protein in three independent homozygous lines and control C non-transformed plants were analyzed by Western blot with anti-GFP antibodies. PonceauS-stained Rubisco is shown as loading control. Transgenic lines 1.6 and 3.8 with maximal PCC1 expression showed GFP-associated fluorescence by confocal microscopy in the plasma membrane. (B) Membrane-associated localization of PCC1-GFP contrasts with Δ177PCC1-GFP that localizes in both the cytoplasm and nucleus. The second row of images for every genotype shows magnification of stomata guard cells.

that independent transgenic lines overexpressing PCC1-GFP displayed a plasma membrane-associated fluorescence pattern similar to that described above in transient experiments in Nicotiana (Fig. 1). Moreover, transgenic lines expressing 35S::Δ177PCC1-GFP construct showed fluorescence in the cytoplasm and the nuclei (Fig. 2B), thus confirming that PCC1 lacking the C-terminal domain is not anchored to the plasma membrane.

Because some algorithms predicted plastid subcellular localization for PCC1, we focused our attention on these organelles. We found that PCC1 was clearly excluded from chloroplasts in epidermal leaf cells as well as in the guard cells of stomata (Fig. S3), thus allowing to rule out a functional association of PCC1 with plastids.

PCC1 Homodimerization and Anchoring to the Plasma Membrane require its C-terminus

To check whether PCC1 may interact with itself, we first conducted a yeast two hybrid (Y2H) approach. PCC1 protein fused to the activation domain (AD) of GAL4 was expressed in Mav203 strain of Saccharomyces cerevisiae as demonstrated by Western blot with an antibody against the AD (Fig. 3A). Concomitantly, only yeast transformed with both AD-PCC1 and PCC1 fused to the binding domain of GAL4 (BD-PCC1) were able to grow in His-free selective medium (Fig. 3A), thus suggesting that PCC1 interacts with itself to form dimers or higher order

oligomers in yeast. To confirm the interaction in planta we used Bimolecular Fluorescence Complementation (BiFC) assays in Nicotiana benthamiana. Fluorescence was only observed when Nicotiana leaves were transformed with both CYFP-PCC1 and NYFP-PCC1 constructs leading to YFP reconstruction in the plasma membrane (Fig. 3B), thus confirming both the membrane localization and the intermolecular interaction for PCC1. Interestingly, the complemented fluorescence was detected in both plasma membrane and associated vesicle-like formations (Fig. 3B). The interaction was further confirmed by immunoprecipitation (IP)-based pull-down assays followed by Western blot. Nicotiana leaves were co-transformed with PCC1-HA and each of the following constructs expressing PCC1 fused to the indicated tags: PCC1-GFP, GFP-PCC1, GFP-stop-PCC1 and PCC1-myc. After IP with anti-HA, we detected 36 kDa PCC1-GFP and GFP-PCC1 by Western blot with anti-GFP antibody, as well as 24 kDa PCC1-myc in the corresponding immunoprecipitated proteins (Fig. 3C). As a negative control, no GFP-tagged protein was pulled down in leaves co-transformed with PCC1-HA and GFP-stop-PCC1 construct, which expresses free GFP (Fig. 3C), suggesting that interaction based on pull-down techniques was specific. By using a similar pull-down approach with leaves co-transformed with PCC1-HA and the GFP fused to the PCC1 version truncated in its C-terminus (GFP-Δ177PCC1) we further demonstrated that homodimerization of PCC1 required the cysteine-rich C-terminus of the protein. Figure 3D shows that GFP-fused protein was detected in the anti-HA-immunoprecipitated proteins from leaves transformed with PCC1-GFP but not in leaves transformed with GFP-stop-PCC1 or GFP-Δ177PCC1. The reverse IP with anti-GFP antibodies, which pulled down all three GFP, GFP-Δ177PCC1 and full size PCC1-GFP proteins, only allowed detecting the HA-tagged protein in the IP from PCC1-GFP co-transformed leaves (Fig. 3D). Together our findings demonstrate that PCC1 interacts with itself through its C-terminal domain rich in cysteine residues, which, as shown above, is also essential for being anchored to the plasma membrane.

Developmental Pattern of PCC1 Expression

We have previously reported that PCC1 gene expression changes throughout post-germination development, with low levels before the transition to flowering and shifting to high levels during that developmental transition [2]. A 1.1 kb promoter sequence upstream the PCC1 initiation codon was fused to the ß-glucuronidase (GUS) reporter gene and the resulting construct was used to transform Arabidopsis plants. We then selected three independent homozygous p1100PCC1:GUS transgenic lines, which were used to check the PCC1-directed expression in different organs and at different times after germination. In p1100PCC1:GUS lines, the pattern of expression was restricted to the vascular tissue in roots, hypocotyls and cotyledons, as well as to the stomata in cotyledons before the transition from vegetative to reproductive shoot apical meristem (Fig. 4A-C, F). However, PCC1 expression decreased progressively in cotyledons as the transition to reproductive apical meristem approached (Fig. 4O), which under our experimental conditions occurs around 9 days after seed germination [2]. By 8 days after germination, lower expression was detected in stomata and vascular bundles (Fig. 4C). By day 10, GUS staining was strong in the petioles of the first pair of leaves and expanded to the basal part of leaves and subsequently to the rest of the leaf through the vascular tissue and mesophyll (Fig. 4D, G). At longer times after germination, the GUS staining was spread all over the leaf blade and vasculature (Fig. 4E). No expression was detected either in flowers (Fig. 4M), siliques (Fig. 4N) or the elongation zone of the roots (Fig. 4K). GUS staining

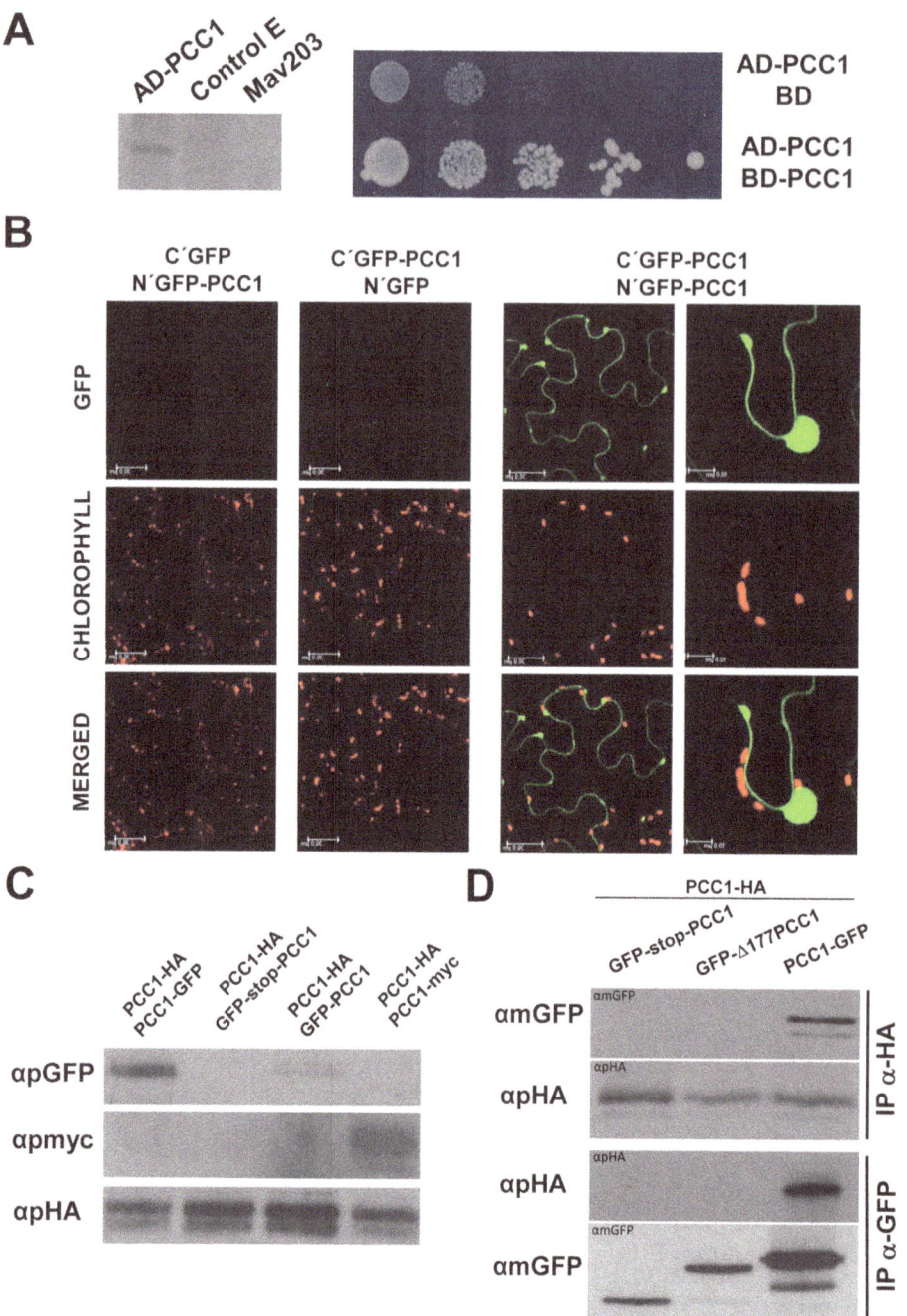

Figure 3. Homodimerization of PCC1. (A) The fusion of PCC1 with the activation domain (AD) of GAL4 is expressed in Mav203 yeast strain as shown by Western blot with anti-AD antibodies. A negative control E and the non-transformed yeast are also shown (top panel). Growth of yeasts co-transformed with AD-PCC1 and PCC1 fused to the DNA binding domain of GAL4 (BD-PCC1) but not with AD-PCC1 and the empty BD vector in minimal media –Leu – Trp –His is indicative of self-interaction of PCC1 in yeast two-hybrid. (B) Bimolecular fluorescence complementation (BiFC)-based demonstration of PCC1-homodimerization and localization of dimers in the plasma membrane of *Nicotiana benthamiana* leaves transiently co-transformed with the indicated constructs. (C) Confirmation of PCC1 homodimerization by pull-down assays in *Nicotiana benthamiana* leaves transiently co-transformed with PCC1-HA and GFP- and c-myc-tagged versions of PCC1 as indicated. Immunoprecipitation (IP) was performed with anti-HA and pulled-down proteins detected by Western blot with the polyclonal antibodies indicated. (D) Homodimerization of PCC1 required the transmembrane C-terminal domain as demonstrated by pull-down assays in *Nicotiana benthamiana* leaves transiently co-transformed with PCC1-HA and GFP-tagged versions of complete and truncated PCC1 molecules. IPs using anti-HA and anti-GFP followed by WB using the indicated antibodies are shown at the left.

was detected in the root cap (Fig. 4K, L). The previously characterized SA-induced pattern of *PCC1* expression [2] was also confirmed in *p1100PCC1:GUS* plants with stronger staining all over SA-treated seedling compared to untreated ones (Fig. 4H-I).

PCC1 and the Photoperiod-Dependent Flowering Pathway

The pattern of expression observed for PCC1 during the events involved in controlling the transition to flowering under inductive

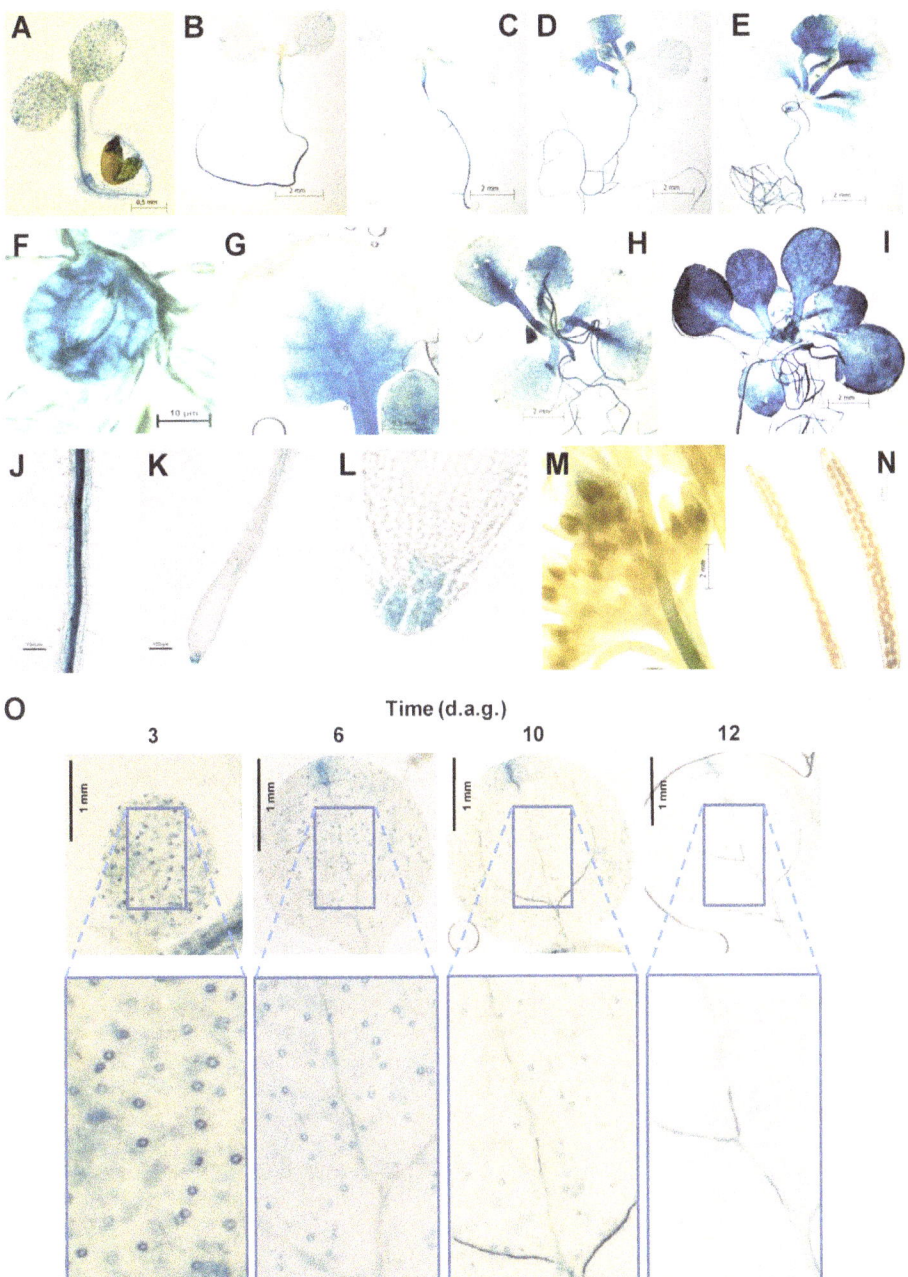

Figure 4. Spatial pattern of *PCC1* expression analyzed with *pPCC1::GUS* transgenic plants. GUS-stained seedlings of (A) 4 days after sowing (d.a.s.); (B) 6 d.a.s.; (C) 8 d.a.s.; (D) 10 d.a.s.; (E) 14 d.a.s. (F) Detail of GUS-stained guard cells of seedling shown in (A). (G) Leaf showing showing staining from the petiole to the distal parts. (H) and (I) Control untreated and 0.1 mM SA-treated 14-day old seedlings, respectively. (J) Vascular tissue stained in the upper part of roots. (K) Absence of GUS staining in the elongation zone and tip of roots. (L) Detail of stained calyptra. (M) and (N) Absence of expression in flower and siliques, respectively. (O) Time-course of GUS staining in cotyledons at different times after germination showing the stomata- and vascular tissue-associated patterns. The generation of *pPCC1::GUS* transgenic lines and the protocols used for GUS staining were previously reported [33].

long days suggests that PCC1 might be connected to or even participate in the signaling pathway controlling the photoperiod-dependent flowering. CONSTANS (CO) and FLOWERING LOCUS T (FT) are essential regulators in the photoperiod-dependent flowering pathway [30] and FT has been characterized as a mobile signal translocated from leaves through the vascular tissue to the shoot apical meristem (SAM) to activate flowering upon interaction with FD transcription factor [31,32]. The fact that *PCC1* expression progresses from SAM to leaves and, in turn,

FT is exported from leaves to SAM, might be related to several alternative hypothesis: PCC1 might interfere with CO activating *FT;* or PCC1 might physically interact with either CO or FT thus modulating their localizations/functions; or might interfere with FT translocation to SAM and its further interaction with FD transcription factor. To test the first hypothesis, iPCC1/*35S::CO-GR/co-2* plants were generated by crossing iPCC1 plants with extremely reduced levels through an RNAi approach [2] to *co-2* mutant plants transformed to overexpress CO fused to the

Glucocorticoid Receptor. Because both constructs contained resistance to kanamycin as selection marker, we developed a PCR-based genotyping procedure using specific primers for *CO* and *PCC1* and common primers from *35S* promoter sequence. We confirmed by qRT-PCR that homozygous double transgenic plants overexpressed CO while showing strongly reduced *PCC1* levels (Fig. 5A). Moreover, *FT* expression was activated upon Dex treatment by inducing CO nuclear translocation (Fig. 5A). We observed that despite the strong down-regulation of *PCC1*, double transgenic plants only flowered earlier in the presence of Dex, this is, when FT was activated (Fig. 5B). These data support that PCC1

is not required for the activation of *FT* by CO and the subsequent floral transition.

To test whether PCC1 might interact *in planta* with CO or FT, BiFC was used in *Nicotiana benthamiana* leaves co-transformed with the corresponding proteins fused to N- and C- terminal parts of YFP. Neither co-transformation with CO and PCC1 or FT and PCC1 were able to reconstruct GFP fluorescence (Fig. 5C). As a positive control, the previously reported interaction of FT and FD was observed in the nuclei (Fig. 5C).

Involvement of PCC1 in Light-Regulated Development

PCC1 regulates flowering time, as demonstrated by the late flowering phenotype observed in iPCC1 plants grown under long day photoperiodic conditions [2]. However, this seems to be not related to interference with the function of the well characterized regulators of the photoperiod-dependent flowering pathway as demonstrated above. Alternatively, PCC1 control of this developmental transition might be related to other key factor such as light perception operating in this pathway. Next, we checked whether iPCC1 plants might display other light-related phenotypes. The photomorphogenic inhibition of hypocotyl elongation, cotyledon opening and acquisition of photosynthetic competence is controlled through light perception and downstream signaling involving Phytochrome Interacting Factors (PIFs) as well as gibberellin-related DELLA proteins [7,8]. We analyzed whether iPCC1 seedlings displayed differential hypocotyl elongation phenotypes compared to wild type seedlings under different qualities of light. Figure 6A shows that iPCC1 hypocotyls grew longer than those of wild type seedlings under blue, red and far-red lights. Whereas iPCC1 hypocotyls were also longer than wild type ones under white light (Fig. 6C and Fig. S4), iPCC1 hypocotyls were not significantly longer than wild type under darkness (Fig. S4). Longer hypocotyls in far red, red, or blue lights might be related to deficiency in the perception of those qualities of lights through phytochromes and cryptochromes, respectively. However, based on comparative transcriptome analysis of iPCC1 vs wild type seedlings [33], no statistically significant change in the transcript levels of the phytochrome and cryptochrome encoding genes was detected (Table S1). Because iPCC1 hypocotyls were not as long as those from *phyA* mutant under far red light, *phyB* mutant under red light or as those from *cry1cry2* double mutant under blue light, we believe that iPCC1 seedlings are somehow partially blind to light in general, or possibly defective in downstream components of light signaling cascades that are common to different light qualities. Regarding this, the levels of *PCC1* transcript are up-regulated in the *phyB* mutant and, in turn, down-regulated in *cry1*, *cry2* and *cry1cry2* mutants when compared to wild type plants (Fig. 6B) suggesting that red and blue light signaling exerts negative and positive modulation, respectively, on *PCC1* gene expression. Since GAs promote hypocotyl elongation, we tested whether iPCC1 and wild type seedlings responded similarly to exogenous GAs under white light. Figure 6C shows that iPCC1 hypocotyls were as responsive as wild types to 4 μM GA$_3$ treatment. By contrast, at 15 μM GA$_3$ wild type hypocotyls were still responsive to GAs but iPCC1 hypocotyls were not responsive anymore (Fig. 6C). We also checked the effect of the GA synthesis inhibitor paclobutrazol (PAC) on hypocotyl elongation under different light qualities. The strong hypocotyl shortening effect exerted by PAC was observed in both wild type and iPCC1 hypocotyls grown under blue, red and far-red lights, but iPCC1 hypocotyls were still significantly longer than wild types in every condition tested (Fig. 6D). Because GA signaling is a key regulatory factor in the control of hypocotyl elongation, we tested by qRT-PCR the transcript levels of GA receptor and DELLA

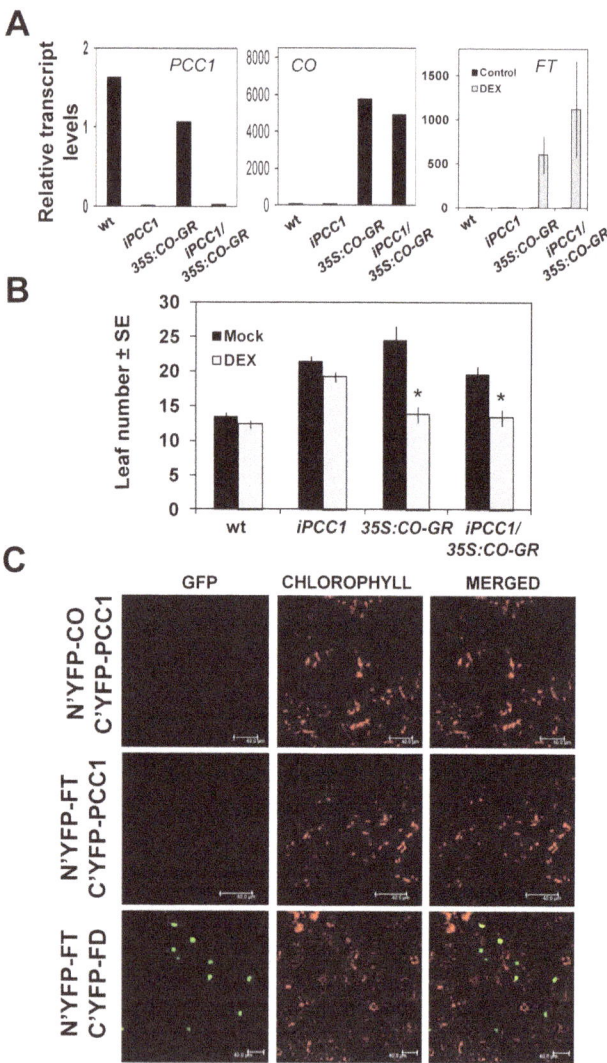

Figure 5. PCC1 does not interfere with photoperiod dependent floral transition pathway nor interacts with its regulatory components. (A) *PCC1*, *CO* and *FT* transcript levels in wild type, double transgenics *iPCC1/35S::CO-GR* and their parental plants in the absence and presence of 10 μM of the GR ligand dexamethasone (DEX). (B) Flowering time quantified by counting total (rosette plus cauline) leaves in long day-grown plants of the genotypes described in (A) treated or not with DEX. * represents statistically significant (p<0.05 in Students t-test) different values in DEX-treated compared to untreated (Mock) seedlings. (C) Analysis of potential interactions between PCC1 and CO or FT by BiFC. The interaction between FT and FD in nuclei is shown as positive control.

encoding genes in wild type and iPCC1 seedlings. Figure 6E shows that genes coding for the GA receptors, *GID1a*, *GID1b* and *GID1c*, were all slightly but significantly down-regulated in iPCC1 seedlings, thus supporting that GA perception might be altered in iPCC1 plants. Neither *RGA* nor *GAI* gene transcripts were significantly altered in iPCC1 seedlings (Fig. 6E). Whether deficient gibberellins perception might be responsible for the delayed flowering observed in iPCC1 plants will require further work. In summary, the partial skotomorphogenic phenotype under different light qualities and the altered sensitivity to GAs in iPCC1 plants as well as the opposite regulation exerted by PHYB and CRY photoreceptors on *PCC1* expression, point to PCC1 as an important regulatory node in photomorphogenic responses.

Y2H-based Screening Reveals the Interaction between PCC1 and the Subunit 5 of the COP9 Signalosome

To clarify the way PCC1 may exert regulation on light-regulated development we conducted a Y2H screening for protein interactors of PCC1. For that purpose, a truncated version of PCC1 containing the first 59 amino acids, thus lacking the C-terminal domain required for membrane anchoring, was fused to the DNA binding domain of *GAL4* (N-Gal4-PCC1(1-59)-C) and used as bait to screen an Arabidopsis library of cDNA clones from 7-day old Arabidopsis seedlings fused to the activation domain of *GAL4*. Among 31 positive clones detected out of 129 million possible interactions tested in the Y2H screening, 7 clones resulted to be either out of frame or antisense sequence clones and were consequently discarded. Another 21 positive clones corresponded to different sequences spanning the locus coding for the CSN5A subunit of the COP9 signalosome (CSN). It has been previously reported that CSN5A binds the DNA binding domain of *GAL4* and is thus frequently considered as a false positive interactor in many different Y2H screenings [34]. The rest 3 identical clones corresponded to the sequence between nucleotides 90 to 1207 of the At1g71230 coding for CSN5B/AJH2, the other subunit 5 of CSN. To confirm the strength of the interaction between PCC1 and CSN5B/AJH2, the clones isolated in the screening were co-transformed in *Saccharomyces cerevisiae* strain AH109 and the transformed yeast was plated in SC minimum medium (-His-Leu-Trp) supplemented with increasing concentrations of 3-aminotriazole (3-AT). Even at 20 mM 3-AT the growth of doubly transformed yeasts was clearly superior relative to the yeasts transformed with the GAL4AD-CSN5B and the empty GAL4BD-plasmids (Fig. 7A), thus suggesting a strong interaction between PCC1 and CSN5B. Because the Y2H system is an heterologous procedure to test plant protein-protein interactions caution is required when interpreting those data. The *in planta* interaction between two proteins must fulfill topological criteria with subcellular localizations being consistent with potential interaction. It has been reported that CSN5A in its monomeric form is located in the cytoplasm and nuclei [35]. We have transiently transformed *Nicotiana benthamina* with a *35S::GFP-CSN5B* construct and found that, similarly to CSN5A, fluorescence was detected in the cytoplasm and nucleus and it can not be ruled out the possibility that is also localized in the plasma membrane (Fig. 7B). By using BiFC, the interaction between PCC1 and both subunits CSN5A and CSN5B of CSN has been confirmed in Nicotiana. Figure 7C shows that the YFP fluorescence was reconstructed in the plasma membrane.

It has been widely characterized that CSN functions by cleaving the ubiquitin-like protein RUB/NEDD8 from cullins [36], which are essential components in the SCF (Skip1-cullin-F-box) complexes involved as E3 ubiquitin ligases in the ubiquitin-proteasome pathway, which is the preponderant protein turnover system in

plants [37,38]. RUB modification of cullins has been characterized to activate the E3 ubiquitin ligase activity [39] and thus CSN should function as a negative regulator of SCF complexes. However, genetic approaches have established that CSN functions *in vivo* promoting E3 ubiquitin ligase activity as a consequence of the protective effect exerted on SCF protein adaptors by limiting their autocatalytic degradation [40]. We have checked whether iPCC1 plants display altered CSN-mediated cleavage of cullin1 (CUL1) when compared to wild type plants. Figure 7D shows that iPCC1 displayed levels of RUB-CUL1 and CUL1 similar to wild type plants, which suggests that PCC1 does not interfere with de-rubylating activity of CSN, at least regarding to CUL1.

Discussion

Despite its small size and the lack of well characterized domains informing about its potential functions, PCC1 protein has a significant impact on a wide array of physiological processes including polar lipid content, ABA-related responses and pathogen defense in Arabidopsis [33]. *PCC1* was initially identified as a gene with a circadian controlled pattern of expression and functionally related to defense against biotrophic pathogens [1]. It was further characterized as a gene with a strictly SA-dependent expression that seems to be involved in controlling both stress-induced and non stressed flowering [2]. Whereas the phenotypic effects of PCC1 gain- and loss-of function have been widely described [1,2,33], much less is known regarding the biochemistry and cell biology of PCC1 protein. Here we have characterized PCC1 as a plasma membrane associated protein that homodimerizes and requires its C-terminal domain to be anchored to the membrane. The C-terminus of PCC1 protein is a cysteine-rich domain that has being used to classify this protein as a member of the so-called Cysteine-rich Trans Membrane (CYSTM) domain-containing proteins [3]. This family of proteins has been proposed to be involved in resistance to stress and particularly against deleterious substances likely through the altered redox potential of the membrane due to the peculiar arrangement of several sulfhydryl groups within the membrane [3]. The alteration of the membrane redox potential might be determinant for quenching radical species or to affect the uptake of metal ions, which has been already reported for CDT1, another plant member of the CYSTM superfamily, involved in the exclusion of heavy metals in *Digitaria ciliaris* and *Oryza sativa* [41]. Besides, it has been also proposed that the N-terminal polar disordered head of the CYSTM properties might under certain conditions assume a certain degree of structure that could be important for alternative functions in the cytoplasm. In this regard, the disordered region upstream of the CYSTM domain has an amino acid composition and organization similar to prion-like proteins that may assume alternative conformations [3]. Although a truncated version of PCC1 lacking its CYSTM-containing C-terminal domain lost the specific plasma membrane localization of the full-length protein, we do not have data supporting whether CYSTM domain might be important for PCC1 function. Further comparative work with plants expressing either the full-length or truncated versions of PCC1 will help to clarify whether the different domains of PCC1 might be important for the regulatory roles exerted by this protein in previously described phenotypes or even in still uncharacterized new processes related to PCC1 function.

Our analysis of the spatial and temporal pattern of *PCC1* expression by using transgenic Arabidopsis plants expressing the *GUS* reporter gene under the control of a 1.1 kb promoter sequence of *PCC1* locus allowed to uncover a dual pattern of expression distinguishable both spatially and temporally. During

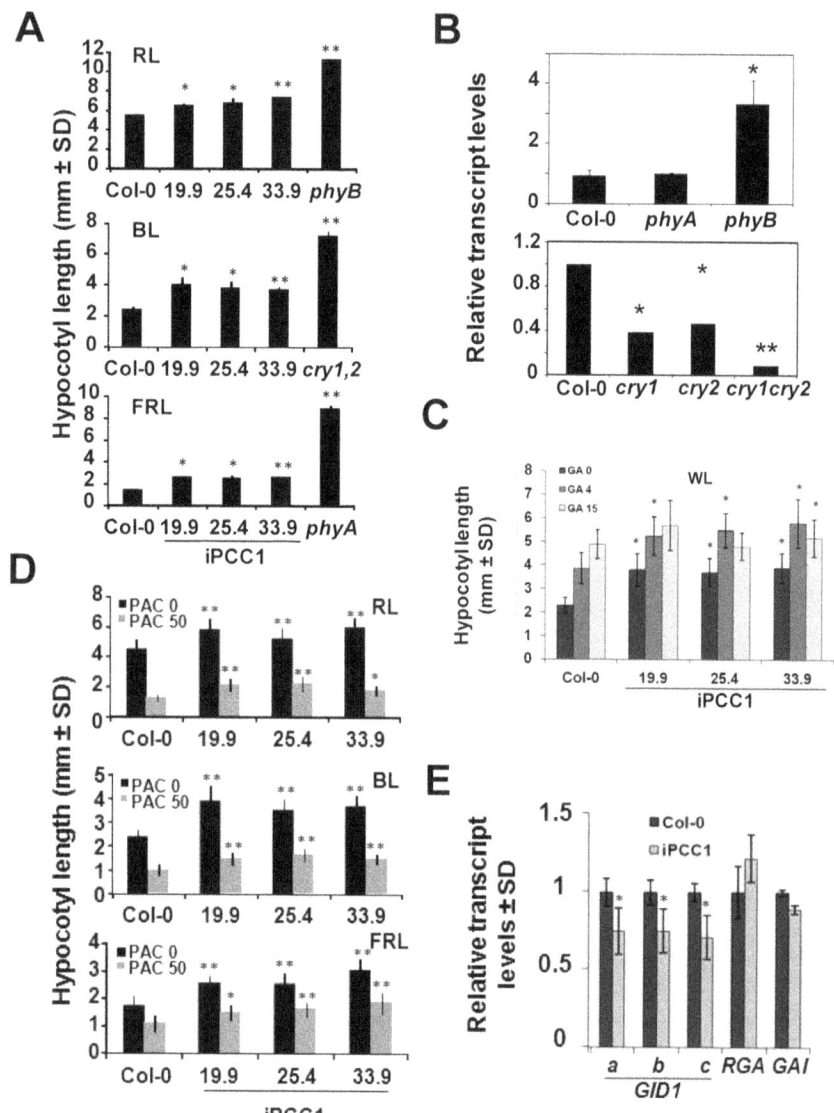

Figure 6. Light- and gibberellin-related hypocotyl phenotypes of iPCC1 plants. Hypocotyl length of seedlings grown under (A) 10 µmole m^{-2} s^{-1} of red light (RL), 30 µmole m^{-2} s^{-1} of blue light (BL), or 5 µmole m^{-2} s^{-1} of far-red light (FRL) for 4 days, and (D) the same light conditions as in (A) but treated as indicated with the gibberellin synthesis inhibitor paclobutrazol (PAC). (B) *PCC1* transcript levels were quantified by qRT-PCR in wild type Col-0 and the indicated photoreceptor mutant seedlings. Values are the mean of three independent biological replicates ± SD and are expressed as relative levels to those detected in wild type seedlings. (C) Hypocotyl length of seedlings grown under 60 µmole m^{-2} s^{-1} of white light (WL), with the indicated µM concentrations of GA3 (GA). Values are the mean of 20 hypocotyls per genotype and condition ± SD. * and ** represents statistically significant (p<0.05 or p<0.01, respectively, in Students t-test) different values in iPCC1 seedlings when compared to wild type seedlings under the same condition. (E) Transcript levels of the indicated genes involved in the perception and signaling of gibberellins were analyzed by qRT-PCR in Col-0 and iPCC1 seedlings grown under 16 h light/8 h darkness photoperiodic white light conditions for 14 days. Values are the mean of three independent biological replicates ± SD and are expressed as relative levels to those detected in wild type seedlings. * represents statistically significant (p<0.05 in Students t-test) different values in iPCC1 seedlings when compared to wild type seedlings.

the first days after seed germination, *PCC1* is expressed in the shoots only in the stomata guard cells and vascular tissues of cotyledons. After that, the transition from vegetative to reproductive shoot apical meristem correlates with the lower *PCC1* expression detected in stomata along with its higher expression in the vascular tissue connecting the apical meristem with the leaves through the vasculature in petioles and basal part of the leaves. Thereafter, *PCC1* expression is expanded all over the leaves. This dual pattern of *PCC1* expression is likely associated to different functions exerted by PCC1. The stomata-associated *PCC1* expression at early stages of development may be related to

defense. In turn, coincidentally with the transition from vegetative to reproductive shoot apical meristem, signaling events originated in the SAM, and translocated to the leaves through the vasculature modulate *PCC1* expression for further developmental control of different responses. Because stomata has been widely characterized as entry sites for different kind of pathogens [42], the expression of *PCC1* in stomata guard cells at early post-germinative development might be connected to defense-related functions of PCC1. This hypothesis is consistent with previously reported phenotypes of enhanced resistance of plants overexpressing *PCC1* gene to the oomycete *Hyaloperonospora parasitica* [1] and

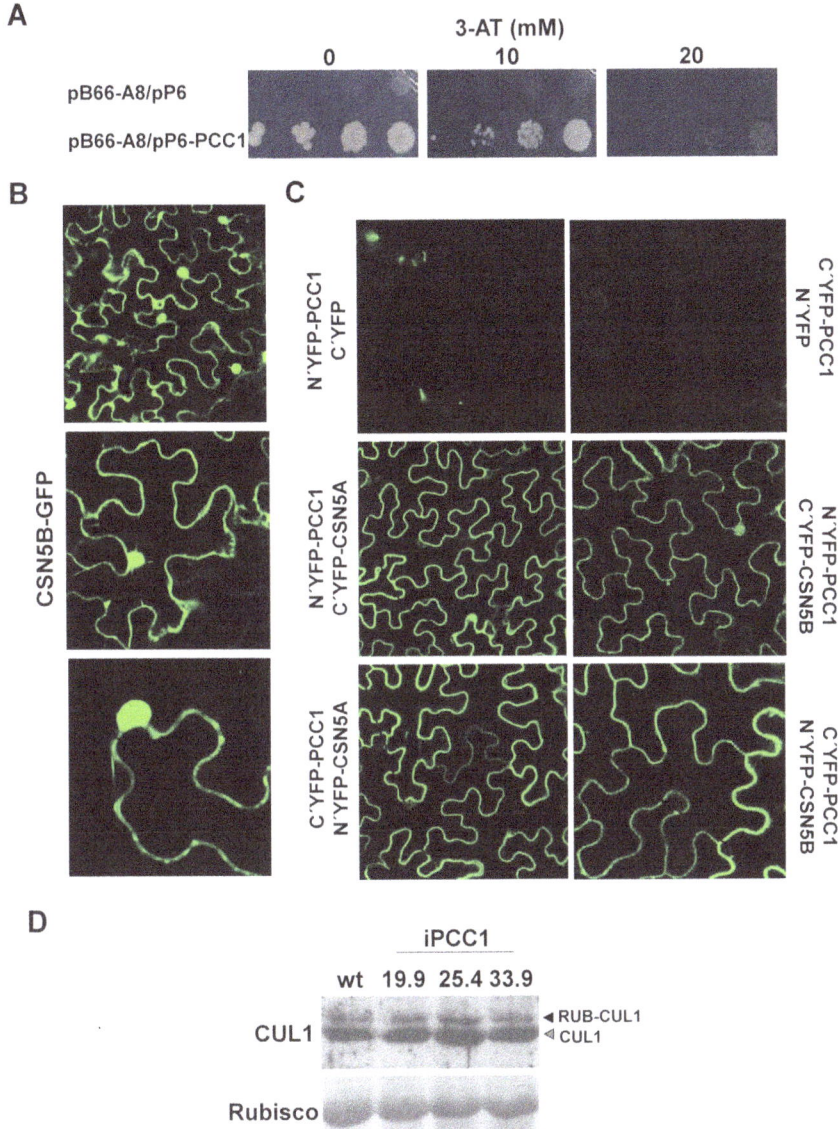

Figure 7. Functional interaction between PCC1 and the subunit 5 of the COP9 signalosome. (A) Growth in -Leu-Trp-His media of yeasts co-transformed with clone A8 corresponding to CSN5B subunit of COP9 signalosome fused to the *GAL4* activation domain (pB66-A8) and either PCC1 fused to the *GAL4* DNA binding domain (pP6-PCC1) or the empty vector (pP6). Growth was tested in increasing concentration of 3-aminotriazol (3-AT). (B) Subcellular localization of CSN5B in cytoplasm and nucleus of *Nicotiana benthamiana* leaves transformed with *35S::GFP-CSN5B*. (C) Interaction of PCC1 with CSN5A and CSN5B in the plasma membrane as demonstrated by BiFC in *Nicotiana benthamiana* leaves transiently co-transformed with the indicated constructs. (D) Levels of free and rubylated (RUB-) forms of CUL 1 in wt and iPCC1 plants as shown by Western blot with anti-CUL1 polyclonal antibodies. Ponceau S-stained Rubisco is shown as loading control.

also with enhanced susceptibility to *Phytophtora brassicae* and enhanced resistance to *Botrytis cinerea* observed for iPCC1 plants [33]. A potential defensive role of PCC1 in stomata might be especially important at early developmental stages of the plant, when physical barrier in cotyledons may not be completely formed. After cotyledons become fully expanded and developed, PCC1 might not be required in stomata for defense and its expression is turned-down in those specialized cells. A second distinguishable function for PCC1 emerges as plants get closer to the developmental transition from vegetative to reproductive shoot apical meristem. A vascular tissue-associated pattern of *PCC1* expression was visualized in *pPCC1::GUS* transgenic lines and it expanded from SAM to the basal part of leaves through leaf petioles. This spatial pattern of expression is interestingly coincident in time with the FT translocation from leaves to SAM [43], which is required to activate the expression of reproductive meristem identity genes such as *AP1* [44] and starting flowering primordia. Despite coincidences suggesting that PCC1 might be involved or interfere with the photoperiod-dependent flowering pathway [2], we have not found evidences demonstrating that PCC1 either interacts with the main players in the photoperiod-dependent flowering pathway, namely CO and FT, or interferes in *FT* activation by CO.

Our results suggest that the potential regulatory role exerted by PCC1 in flowering under inductive conditions should be connected with other factors involved in this developmental transition. We have presented experimental evidence supporting functional interactions between PCC1 and both light- and gibberellins-related signaling pathways (Fig. 6). The involvement of either light or gibberellins in regulating the transition to flowering is widely documented [45]. It has been recently reported that under long days, GA promotes the transition to flowering independently of CONSTANS (CO) and GIGANTEA (GI) by activating the expression of flowering time integrator genes such as FLOWERING LOCUS T (FT) and TWIN SISTER OF FT (TSF) in leaves [46]. This work provides a potential mechanistic explanation of PCC1 being involved in the perception and downstream signaling of stimuli such as light and gibberellins through functional interference with the CSN-mediated role in controlling the activity of multiple E3 ubiquitin ligases. By using BiFC, we have found that interaction between PCC1 and either CSN5A or CSN5B occurs in the plasma membrane (Figure 7). Although CSN5A and CSN5B are localized in cytoplasm and nucleus [35] (Figure 7B) the possibility that they are also partially localized in the plasma membrane cannot be ruled out. This potential plasma membrane localization of CSN5B would allow the interaction with PCC1. However, even if CSN5 subunits are not localized in the plasma membrane in wild type plants, it might be recruited to the membrane upon over-expression of PCC1. In this latter case, the fluorescence that is reconstructed in the plasma membrane in BiFC experiments would correspond to only a fraction of the total CSN5 protein present in the cell. Regarding this, the fact that the yeast 2-hybrid approach we used to identify the interaction between CSN5 and PCC1 was based on the use of a truncated version of PCC1 lacking its C-terminus as bait, suggest that the interaction occurs through the N-terminal domain of PCC1, which is oriented towards the cytoplasm. The interaction of the N-terminus of PCC1 with the cytoplasmic fraction of CSN5 would lead to the recruitment of at least part of the CSN5 subunits to the plasma membrane.

Altered functional interaction between PCC1 and CSN might subsequently alter the ubiquitination status of multiple proteins involved in sensing and transducing multiple stimuli. We have observed that PCC1 interacts with CSN5A and CSN5B subunits, recruiting them to the plasma membrane. Two different scenarios might be considered in the functional interaction between PCC1 and CSN5. This process might scavenge CSN5 subunits preventing the formation of CSN complex and eventually its nuclear translocation. Such a mechanism would imply that iPCC1 plants should have a reduced CSN function in cleaving RUB protein from modified cullins, and such significant alterations have not been detected at least for CUL1 in iPCC1 plants relative to wild type plants. It might be also possible that PCC1-mediated scavenging of CSN5 in the plasma membrane might avoid the assembly of a full CSN complex to be further translocated to the nucleus and function as a negative modulator of COP1-mediated skotomorphogenesis. Accordingly, iPCC1 plants lacking this scavenging mechanism showed partial skotomorphogenic phenotypes under different light conditions (Fig. 6A). However, CSN complex not only modulates COP1 but many other E3 ubiquitin ligases. A model like that would explain that defective CSN function would lead RUB-cullin to accumulate and to keep E3 ubiquitin ligases in their active form. Although we have not observed alterations in CUL1 rubylation we can not rule out the possibility that other cullins may be affected. Alternatively, the function of PCC1 interacting with CSN5 at the plasma membrane might be a mechanism of recruitment of CSN5 to some membrane-associated target acting as potential substrate for CSN5-mediated hydrolytic activity. Whether PCC1-CSN5 interaction could be part of a general regulatory mechanism in

regulating the ubiquitination status of multiple targets or a more specific membrane-associated mechanism affecting particular F-box proteins involved in light- and gibberellin-regulated processes, such as photomorphogenesis or flowering time, or others involved in defense against varied pathogens will require further work.

Despite multiple mechanistic explanations of PCC1 function as regulator of the transition to flowering and photomorphogenesis remain to be clarified, a simple model consistent with the potential functional interaction and the observed phenotypic effects can be drawn. We have previously reported that PCC1 is a positive regulator of the transition to flowering [2]. On the other hand, PHYB seems to promote CO degradation and thus act as a negative regulator of flowering time, whereas, CRY1 and CRY2 stimulate CO accumulation under blue light, and they are thus positive regulators of the transition to flowering [47–49]. These data suggest that a model where PHYB and CRY1,2 regulate PCC1 expression negatively and positively, respectively, is consistent with the flowering time phenotype of photoreceptor mutants and iPCC1 loss-of-function plants. Moreover, the fact that no altered transcript levels of the genes encoding phytochromes and cryptochromes was observed when compared iPCC1 versus wild type plants suggests that PCC1 is not controlling light perception through direct effects on the genes coding for photoreceptors. Alternatively, PCC1 could exert an effect on the inputs of the circadian clock or directly on the function of the oscillator itself. Remarkably, the endogenous content of salicylic acid (SA) levels, which is a potent activator of PCC1 expression, seems to regulate the expression of genes coding for components of the circadian clock such as CCA1 [2]. An increase in the CCA1 transcript levels and a concomitant reduction in CO transcript were detected at dawn in SA-deficient plants, defective in PCC1 expression, when compared with Col-0 plants [2]. We believe that the potential regulatory effect exerted by PCC1 on components of the circadian oscillator would be more important for its role in regulating flowering time than the effect on CO. A still unknown output of the circadian clock would be a potential target for the oscillator-mediated effect triggered by PCC1 in controlling flowering time. Clearly, more work is required to elucidate the mechanism underlying PCC1 control on flowering.

Regarding the photomorphogenesis phenotype, phytochromes and cryptochromes as well as PCC1 are positive regulators of photomorphogenesis and the corresponding mutants are thus skototmorphogenic. However, iPCC1 plants display only partial skotomorphogenic phenotype thus suggesting PCC1 may be involved in light signal transduction downstream photoreceptors together with other functionally-related proteins. CSN accumulates in the nuclei under darkness then promoting the COP1-mediated degradation of positive regulators of photomorphogenesis, such as HY5, and thus promoting skotomorphogenesis. However, under light CSN accumulates in the cytoplasm thus releasing HY5 from degradation and allowing photomoprphogenesis. In this context, PCC1-CSN5 interaction in the plasma membrane would function as a scavenging system to prevent cytoplasm to nucleus CSN translocation and then helping photomorphogenic conditions. Accordingly, plants with defective PCC1 function would lack this scavenging activity thus allowing CSN to be localized in the nuclei promoting skotomorphogenesis. However, because PHYB is a negative regulator of PCC1 expression as described in this work, we believe that PCC1 positive effect on photomorphogenesis should be more related to CRY1,2 function. Regarding this, PCC1 might somehow regulate the interaction between CRY1 and SUPPRESSOR OF PHYA1 (SPA1), which under blue light allows the inactivation of COP1 with the consequent promotion of photomorphogenesis [50,51].

Supporting Information

Figure S1 Co-localization of GFP-PCC1 with stained membranes. Membranes were stained with fluorescent lipid stain FM64 in *Nicotiana benthamiana* leaves transformed with *35S::GFP-PCC1* and *35S::GFP-Δ177-PCC1* constructs expressing the full and truncated versions of tagged PCC1 proteins. Overlays of green and red fluorescence due to GFP and FM64 are shown in the right panels and yellow appeared only when co-localization occurred.

Figure S2 PCC1-GFP is localized only in the plasma membrane in plasmolysed cells of transformed Nicotiana leaves. *Nicotina benthamiana* leaves transiently transformed with *35S::PCC1-GFP* construct were either infiltrated with water (control) or with 0.5 M sorbitol (plasmolysed) and 6 h after infiltration fluorescence was observed under confocal microscopy. Cell contour is marked with red dashed lines.

Figure S3 PCC1 is not localized in plastids. Green fluorescence due to GFP and red fluorescence due to FM64 in membranes and to chlorophyll in plastids (pointed by arrows) are shown together with the corresponding bright field photopgraphs.

Figure S4 Hypocotyl length of Col-0 wild type and iPCC1 seedlings grown under white light or darkness. Values are the mean of 20 hypocotyls per genotype ± SD. * and ** represents statistically significant ($p<0.05$ or $p<0.01$, respectively, in Students t-test) different values in iPCC1 or phyB seedlings when compared to wild type seedlings under the same condition. No statistically significant differences were observed following Students t-test for the seedlings grown under darkness.

Table S1 Comparative transcript levels of phytochrome (PHY) and cryptochrome (CRY) encoding genes in iPCC1 vs Col-0 seedlings.

Acknowledgments

Yeast two-hybrid screening was performed by Hybrigenics, S.A., Paris, France (http://www.hybrigenics.com).

Author Contributions

Conceived and designed the experiments: JL RM. Performed the experiments: RM JL. Analyzed the data: JL RM. Contributed reagents/materials/analysis tools: RM JL. Wrote the paper: JL.

References

1. Sauerbrunn N, Schlaich NL (2004) PCC1: a merging point for pathogen defence and circadian signalling in Arabidopsis. Planta 218: 552–561.
2. Segarra S, Mir R, Martínez C, and León J (2010) Genome-wide analyses of the transcriptomes of salicylic acid-deficient versus wild type plants uncover Pathogen and Circadian Controlled 1 (PCC1) as a regulator of flowering time in Arabidopsis. Plant Cell Environ 33: 11–22.
3. Venancio TM, Aravind L (2010) CYSTM a novel cysteine-rich transmembrane module with a role in stress tolerance across eukaryotes. Bioinformatics 26: 149–152.
4. Lau OS, Deng XW (2010) Plant hormone signaling lightens up: integrators of light and hormones. Curr Opin Plant Biol 13: 571–517.
5. Schwechheimer C (2011) Gibberellin signaling in plants - the extended version. Front. Plant Sci 2: 107.
6. Seo M, Nambara E, Choi G, Yamaguchi S (2009) Interaction of light and hormone signals in germinating seeds. Plant Mol Biol 69: 463–472.
7. De Lucas M, Daviére JM, Rodrigues-Falcon M, Iglesias-Pedraz JM, Lorrain S, et al. (2008) A molecular framework for light and gibberellin control of cell elongation. Nature 451: 480–484.
8. Feng S, Martinez C, Gusmaroli G, Wang Y, Zhou J, et al. (2008) Coordinated regulation of Arabidopsis thaliana development by light and gibberellins. Nature 451: 475–479.
9. Mutasa-Göttgens E, Hedden P (2009) Gibberellin as a factor in floral regulatory networks. J Exp Bot 60: 1979–1989.
10. Bastian R, Dawe A, Meier S, Ludidi N, Bajic VB, et al. (2010) Gibberellic acid and cGMP-dependent transcriptional regulation in Arabidopsis thaliana. Plant Signal Behav 5: 224–232.
11. Yu S, Galvão VC, Zhang YC, Horrer D, Zhang TQ, et al. (2012) Gibberellin regulates the Arabidopsis floral transition through miR156-targeted SQUAMOSA promoter binding-like transcription factors. Plant Cell 24: 3320–3332.
12. Arc E, Galland M, Cueff G, Godin B, Lounifi I, et al. (2011) Reboot the system thanks to protein post-translational modifications and proteome diversity: How quiescent seeds restart their metabolism to prepare seedling establishment. Proteomics 11: 1606–1618.
13. Dill A, Thomas SG, Hu J, Steber CM, Sun TP (2004) The Arabidopsis F-box protein SLEEPY1 targets gibberellin signaling repressors for gibberellin-induced degradation. Plant Cell 16: 1392–1405.
14. Wang F, Deng XW (2011) Plant ubiquitin-proteasome pathway and its role in gibberellin signaling. Cell Res 21: 1286–1294.
15. Hotton SK, Callis J (2008) Regulation of cullin RING ligases. Annu Rev Plant Biol 59: 467–489.
16. Cope GA, Suh GS, Aravind L, Schwarz SE, Zipursky SL, et al. (2002) Role of predicted metalloprotease motif of Jab1/Csn5 in cleavage of Nedd8 from Cul1. Science 298: 608–611.
17. Gusmaroli G, Figueroa P, Serino G, Deng XW (2007) Role of the MPN subunits in COP9 signalosome assembly and activity, and their regulatory interaction with Arabidopsis Cullin3-based E3 ligases. Plant Cell 19: 564–581.
18. Serino G, Deng XW (2003) The COP9 signalosome: regulating plant development through the control of proteolysis. Annu Rev Plant Biol 54: 165–182.
19. Stratmann JW, Gusmaroli G (2012). Many jobs for one good cop - the COP9 signalosome guards development and defense. Plant Sci 185–186: 50–64.
20. Lozano-Juste J, León J (2011) Nitric oxide regulates DELLA content and PIF expression to promote photomorphogenesis in Arabidopsis. Plant Physiol 156: 1410–1423.
21. Nakagawa T, Kurose T, Hino T, Tanaka K, Kawamukai M, et al. (2007) Development of series of gateway binary vectors, pGWBs, for realizing efficient construction of fusion genes for plant transformation. J Biosci Bioeng 104: 34–41.
22. Fromont-Racine M, Rain JC, Legrain P (1997) Toward a functional analysis of the yeast genome through exhaustive two-hybrid screens. Nat Genet 16: 277–282.
23. Belda-Palazón B, Ruiz L, Martí E, Tárraga S, Tiburcio AF, et al. (2012) Aminopropyltransferases involved in polyamine biosynthesis localize preferentially in the nucleus of plant cells. PLoS One 7(10), e46907.
24. Simon R, Igeño MI, Coupland G (1996) Activation of floral meristem identity genes in Arabidopsis. Nature 384: 59–62.
25. Martínez C, Pons E, Prats G, León J (2004) Salicylic acid regulates flowering time and links defence responses and reproductive development. Plant J 37: 209–217.
26. Kyte J, Doolittle RF (1982) A simple method for displaying the hydropathic character of a protein. J Mol Biol 157: 105–132.
27. Marmagne A, Rouet MA, Ferro M, Rolland N, Alcon C, et al. (2004) Identification of new intrinsic proteins in Arabidopsis plasma membrane proteome. Mol Cell Proteomics 3: 675–691.
28. Nühse TS, Stensballe A, Jensen ON, Peck SC (2004) Phosphoproteomics of the Arabidopsis plasma membrane and a new phosphorylation site database. Plant Cell 16: 2394–2405.
29. Hofmann K, Stoffel W (1993) TMbase - A database of membrane spanning proteins segments. Biol Chem Hoppe-Seyler 374: 166.
30. Kobayashi Y, Weigel D (2007) Move on up, it's time for change--mobile signals controlling photoperiod-dependent flowering. Genes Dev 21: 2371–2384.
31. Jaeger KE, Wigge PA (2007) FT protein acts as a long-range signal in Arabidopsis. Curr Biol 17: 1050–1054.
32. Mathieu J, Warthmann N, Küttner F, Schmid M (2007) Export of FT protein from phloem companion cells is sufficient for floral induction in Arabidopsis. Curr Biol 17: 1055–1060.
33. Mir R, Hernández ML, Abou-Mansour E, Martínez-Rivas JM, Mauch F, et al. (2013) Pathogen and Circadian Controlled 1 (PCC1) regulates polar lipid content, ABA-related responses, and pathogen defence in Arabidopsis thaliana. J Exp Bot 64: 3385–3395.
34. Nordgård O, Dahle Ø, Andersen TØ, Gabrielsen OS (2001) JAB1/CSN5 interacts with the GAL4 DNA binding domain: a note of caution about two-hybrid interactions. Biochimie 83: 969–971.

35. Kwok SF, Staub JM, Deng XW (1999) Characterization of two subunits of Arabidopsis 19S proteasome regulatory complex and its possible interaction with the COP9 complex. J Mol Biol 285: 85–95.
36. Nezames CD, Deng XW (2012) The COP9 signalosome: its regulation of cullin-based E3 ubiquitin ligases and role in photomorphogenesis. Plant Physiol 160: 38–46.
37. Moon J, Parry G, Estelle M (2004) The ubiquitin-proteasome pathway and plant development. Plant Cell 16: 3181–3195.
38. Dreher K, Callis J (2007) Ubiquitin, hormones and biotic stress in plants. Ann Bot 99: 787–822.
39. Parry G, Estelle M (2004) Regulation of cullin-based ubiquitin ligases by the Nedd8/RUB ubiquitin-like proteins. Semin Cell Dev Biol 15: 221–229.
40. Wee S, Geyer RK, Toda T, Wolf DA (2005) CSN facilitates Cullin-RING ubiquitin ligase function by counteracting autocatalytic adapter instability. Nat Cell Biol 7: 387–391.
41. Kuramata M, Masuya S, Takahashi Y, Kitagawa E, Inoue C, et al. (2009) Novel cysteine-rich peptides from Digitaria ciliaris and Oryza sativa enhance tolerance to cadmium by limiting its cellular accumulation. Plant Cell Physiol 50: 106–117.
42. Zeng W, Melotto M, He SY (2010) Plant stomata: a checkpoint of host immunity and pathogen virulence. Curr Opin Biotechnol 21: 599–603.
43. Wigge PA (2011) FT, a mobile developmental signal in plants. Curr Biol 21: R374–378.
44. Kardailsky I, Shukla VK, Ahn JH, Dagenais N, Christensen SK, et al. (1999) Activation tagging of the floral inducer FT. Science 286: 1962–1965.
45. Srikanth A, Schmid M (2011) Regulation of flowering time: all roads lead to Rome. Cell Mol Life Sci 68: 2013–2037.
46. Galvão VC, Horrer D, Küttner F, Schmid M (2012) Spatial control of flowering by DELLA proteins in Arabidopsis thaliana. Development 139: 4072–4082.
47. Cerdán PD, Chory J (2003) Regulation of flowering time by light quality. Nature 423:881–885.
48. Guo H, Yang H, Mockler TC, Lin C (1998) Regulation of flowering time by Arabidopsis photoreceptors. Science 279:1360–1363.
49. Mockler T., Guo H, Yang H, Duong H, Lin C (1999) Antagonistic actions of Arabidopsis cryptochromes and phytochrome B in the regulation of floral induction. Development 126: 2073–2082.
50. Liu B, Zuo Z, Liu H, Liu X, Lin C (2011) Arabidopsis cryptochrome 1 interacts with SPA1 to suppress COP1 activity in response to blue light. Genes Dev 25:1029–1034.
51. Weidler G, Zur Oven-Krockhaus S, Heunemann M, Orth C, Schleifenbaum F, et al. (2012) Degradation of Arabidopsis CRY2 is regulated by SPA proteins and phytochrome A. Plant Cell 24:2610–2623.

14

Arabidopsis Plastidial Folylpolyglutamate Synthetase Is Required for Seed Reserve Accumulation and Seedling Establishment in Darkness

Hongyan Meng[1], Ling Jiang[1,2], Bosi Xu[1], Wenzhu Guo[3], Jinglai Li[4], Xiuqing Zhu[4], Xiaoquan Qi[5], Lixin Duan[5], Xianbin Meng[5], Yunliu Fan[1,2], Chunyi Zhang[1,2]*

1 Biotechnology Research Institute, Chinese Academy of Agricultural Sciences, Beijing, People's Republic of China, 2 National Key Facility for Crop Gene Resources and Genetic Improvement (NFCRI), Beijing, People's Republic of China, 3 Huazhong Agricultural University, Wuhan, People's Republic of China, 4 Beijing Institute of Pharmacology and Toxicology, Beijing, People's Republic of China, 5 Institute of Botany, Chinese Academy of Sciences, Beijing, People's Republic of China

Abstract

Interactions among metabolic pathways are important in plant biology. At present, not much is known about how folate metabolism affects other metabolic pathways in plants. Here we report a T-DNA insertion mutant (*atdfb-3*) of the plastidial folylpolyglutamate synthetase gene (*AtDFB*) was defective in seed reserves and skotomorphogenesis. Lower carbon (C) and higher nitrogen (N) content in the mutant seeds than that of the wild type were indicative of an altered C and N partitioning capacity. Higher levels of organic acids and sugars were detected in the mutant seeds compared with the wild type. Further analysis revealed that *atdfb-3* seeds contained less total amino acids and individual Asn and Glu as well as NO_3^-. These results indicate significant changes in seed storage in the mutant. Defects in hypocotyl elongation were observed in *atdfb-3* in darkness under sufficient NO_3^- conditions, and further enhanced under NO_3^- limited conditions. The strong expression of *AtDFB* in cotyledons and hypocotyl during early developmental stage was consistent with the mutant sensitivity to limited NO_3^- during a narrow developmental window. Exogenous 5-formyl-tetrahydrofolate completely restored the hypocotyl length in *atdfb-3* seedlings with NO_3^- as the sole N source. Further study demonstrated that folate profiling and N metabolism were perturbed in *atdfb-3* etiolated seedlings. The activity of enzymes involved in N reduction and assimilation was altered in *atdfb-3*. Taken together, these results indicate that AtDFB is required for seed reserves, hypocotyl elongation and N metabolism in darkness, providing novel insights into potential associations of folate metabolism with seed reserve accumulation, N metabolism and hypocotyl development in Arabidopsis.

Editor: Diane Bassham, Iowa State University, United States of America

Funding: This work was supported by the National Basic Research Program of China (grant no. 2013CB127003 to C.Z.) The funders had no role in study design, data collection and analysis, decision to publish, or preparation of the manuscript.

Competing Interests: The authors have declared that no competing interests exist.

* Email: zhangchunyi@caas.cn

Introduction

The role of seeds is to propagate offspring. In *Arabidopsis thaliana*, seed development can be divided into three stages: cell division or the pre-storage phase, maturation or the storage phase, and the desiccation phase [1,2]. Large quantities of carbon (C) and nitrogen (N) are stored in maturing seeds, mainly in the form of large insoluble compounds [3]. The major storage compounds that accumulate in mature seeds are triacylglycerols (TAGs) and seed storage proteins (SSPs), accounting for 30–45% of the seed dry weight. Small amounts of carbohydrate in the form of sucrose are stored within cotyledons [1,3,4,5,6]. SSPs, including soluble proteins and non-soluble proteins, include two predominant classes, namely, 12S globulin and 2S albumin [1,4,7]. Seed storage accumulation is regulated by many factors, such as hormones, sugars, master regulator genes and transcriptional factors [1]. These seed reserves are used to fuel germination and post-germinative seedling establishment until seedling photosynthesis autotrophy can be efficiently established [4].

Seed germination and post-germinative seedling establishment are metabolically distinct [8,9]. Germination initiates with release from dormancy and seed imbibition and is completed when the radicle emerges through the seed coat [10]. At the beginning of germination, seed reserves other than lipids (TAG) are rapidly converted to soluble metabolites (e.g. glycolysis products, organic acids, and amino acids) that can be transported throughout the seedling to support growth, while the breakdown of seed oil storage TAG is used for subsequent seedling establishment after the radicle has emerged [3,8,9,11,12,13]. Following germination, TAG is broken down to yield free fatty acids (FAs) and glycerol, both of which are ultimately converted to sugars required for post-germinative seedling development [6,12]. The *sdp1* mutant, containing a mutation in sugardependent1 (SDP1), which encodes a patatin domain TAG lipase that initiates TAG breakdown in germinating seeds, displayed slightly delayed seed germination and a much slower post-germinative growth rate than the wild type [14]. Seedlings grown in darkness showed skotomorphogenesis, which is characterized by elongated weak hypocotyls, closed cotyledons, and shortened roots [15]. Seedling establishment and hypocotyl elongation are driven by the catabolism of TAG under dark conditions. Mutants (*icl* and *pck1*) defective in TAG mobilization show shortened hypocotyls in darkness, but hypocotyls could be rescued by providing alternative C sources, such as sucrose [16,17]. N metabolism is also essential for hypocotyl growth. In conifer plants grown in the dark, a portion of N mobilized from the megagametophyte is diverted toward the hypocotyl shortly after germination to produce high levels of Asn, which serves as a reservoir of N to meet subsequent specific developmental demands [18].

Tetrahydrofolate (THF) and its derivatives are collectively called folates. Most cellular folates carry a short poly-γ-Glu tail, which is believed to affect their efficacy and stability. The tail can be removed by γ-glutamyl hydrolase (GGH), a vacuolar enzyme who has an important influence on polyglutamyl tail length and hence on folate stability and cellular folate content [19]. Folylpolyglutamate derivatives are central cofactors for many folate-dependent enzymes [20,21,22,23,24,25]. During the germination process, *de novo* synthesis of THF occurs in pea (*Pisum sativum*) cotyledons, and the inhibition of THF *de novo* synthesis using folate analogs blocks seedling development [26,27,28]. The cotyledonary folate pool contains principally methylated derivatives [28]; the concentration of folylpoly-Glu derivatives increases gradually during germination [29], and the accumulation of folates peaks 3 days after sowing [27].

Plants with defective folate biosynthesis and metabolism showed various aberrant seed and seedling phenotypes. For example, the *globular arrest1* (*gla1*) mutant, which contains a mutation in dihydrofolate synthetase folylpolyglutamate synthetase (DHFS-FPGS) homolog A (DFA), encoding a functional mitochondrial matrix-localized dihydrofolate (DHF) synthetase, exhibited defective embryonic development and did not undergo transition to the heart stage [23,30]. The double knockout (dKO) mutation of 10-formyl-THF deformylase genes, At4g17360 and At5g47435, resulted in defective embryo development, with cells arresting between the heart and early bent cotyledon stages. Mature seeds of dKO were shriveled, accumulated low amounts of lipids, and failed to germinate [31]. A mutation in *AtDFB*, which encodes the plastidial folylpolyglutamate synthetase (FPGS) isoform, displayed short primary roots with a disorganized quiescent center [24,32]. A mutation in *AtDFC*, which encodes the mitochondrial FPGS, was characterized based on its altered N metabolism and enhanced phenotypes to low N stress, providing novel insights into folate biosynthesis and N utilization during early seedling development [33]. To date, the role of folate during skotomorphogenesis in plants remains poorly understood.

In this report, a mutant (*atdfb-3*) carrying a T-DNA insertion in the *AtDFB* gene was characterized for its altered seed reserves and defective seedling establishment with shortened hypocotyls under dark conditions. Early post-germinative growth (before 3 days) in *atdfb-3* required external NO_3^- sufficient conditions, and exogenous application of 5-formyl-tetrahydrofolate (5-F-THF) restored hypocotyl length in *atdfb-3* when NO_3^- was as the sole N source in the medium. The defective hypocotyl elongation could be due to altered seed storage, perturbed folate and N metabolism in *atdfb-3*. This report provides novel insights into a potential associations of folate metabolism with seed reserve accumulation, N metabolism and hypocotyl development elongation in darkness in Arabidopsis.

Results

Reduced seed size and altered C/N partition capacity in mature seeds of *atdfb-3*

A previous report demonstrated that the vegetative phenotype of *atdfb* (*fpgs1*, SALK_133817) did not differ visually from the wild type under light conditions [34]. In this report, SALK_015472 with a T-DNA insertion in the sixth intron of *At5g05980* (Figure S1A in File S1), which encodes the plastidial isoform of FPGS (AtDFB), was obtained from the Arabidopsis Biological Resource Center (The Ohio State University) and named *atdfb-3*, as described previously [32].

First, characteristics of seeds harvested from *atdfb-3* and wild-type plants grown under light conditions, such as seed number per silique and 1000 seed weight, were examined. No significant difference was observed in seed number per silique between *atdfb-3* and the wild type (Figure 1A); however, the width and length of mature *atdfb-3* seeds were slightly but significantly smaller than those of the wild type (Figure 1B). The reduction in dimensions was somewhat reflected by the seed weight, with a significant decrease of 5% in *atdfb-3* compared with the wild type (Figure 1C).

Next, we explored seed reserves in mature *atdfb-3* seeds. We found that C and N levels in *atdfb-3* seeds were 94% and 122%, respectively, compared with wild-type levels (Figure 2A and B). In *AtDFB* complemented plants (Figure S1B and C in File S1), these changes were restored to wild-type levels (Figure 2A and B), indicating they were due to the loss of function of *AtDFB*. These results indicated the altered C and N partitioning capacity observed in *atdfb-3* was due to the loss of function of *AtDFB*.

Altered C and N metabolites in *atdfb-3* seeds

We analyzed metabolites in *atdfb-3* and wild-type seeds using gas chromatography time-of-flight mass spectrometry (GC-TOF-MS). A total of nine metabolites, including two FAs (14:0 and 18:3), three organic acids (oxalic acid, pentanedioic acid, and phosphoric acid), two sugars (galactose and mannose), and two polyols (campesterol and phytol) were higher and three metabolites (20:1, benzoic acid, and lyxose) were lower in *atdfb-3* than in the wild type (Figure 3). In addition, the contents of other metabolites (mainly fatty acids) detected in *atdfb-3* were similar to those in the wild type. These results suggested that the mutation in *AtDFB* altered C-rich metabolites accumulation in mature seeds.

The level of soluble protein was not significantly different from the wild type in *atdfb-3* (Figure 4A), while the total free amino acids were significantly less in *atdfb-3* seeds than in the wild type (Figure 4B). Many individual amino acids accumulated to lower levels in *atdfb-3* than in the wild type, such as Asn, Glu, Asp, Cys, Gly, Pser, Pro, and His. Asn and Glu were both 50% less than the wild type, accounting for the main shortage of total amino acids in *atdfb-3* (Figure 4C). In contrast, some other amino acids accumulated more in *atdfb-3* than in the wild type, such as Gln, Phe, Leu, Ile, Met, β-Aiba, β-Ala, Lys, and γ-Aba. Among them, Leu in *atdfb-3* accumulated to the highest level: 2.5-fold higher than that of the wild type (Figure 4C). As a result, total amino acids in *atdfb-3* seeds were 27% lower than in the wild type, and the Gln/Glu ratio in *atdfb-3* (0.62) was higher than that of the wild type (0.19). Additionally, the NO_3^- content in *atdfb-3* seeds was only 21% of that in the wild type (19.6 µg g^{-1} in *atdfb-3* vs 95.4 µg g^{-1} in the wild type; Figure 4D). These results indicated that the mutation in *AtDFB* reduced the accumulation of N-rich metabolites.

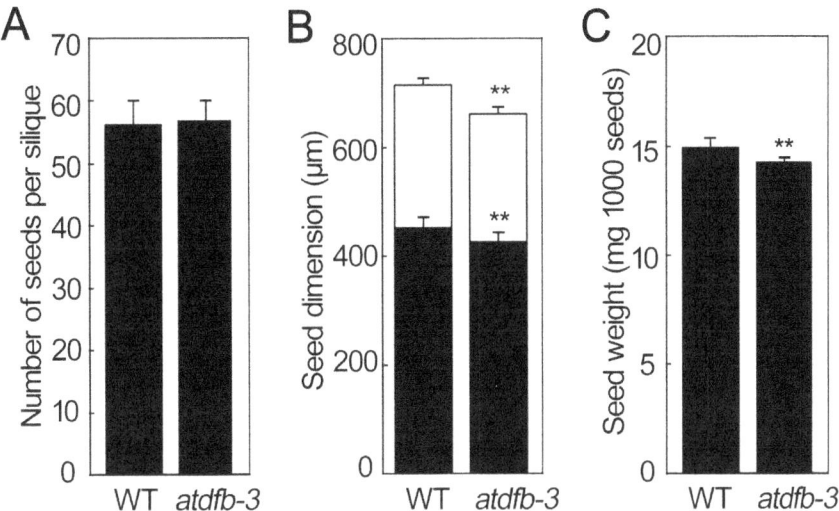

Figure 1. Seed characteristics in WT and *atdfb-3*. (A) Number of seeds per silique. (B) Seed length (black bars) and width (white bars). (C) Seed weight. Data represent means ± SD. A, n = 30; B, n = 3, and each replicate contained 30 seeds. Seeds were viewed using a ZEISS Imager M1 DIC microscope and measured using ImageJ; C, n = 5, and each replicate consisted of a pool from 10 plants. Bars with ** indicate a highly significant difference at P<0.01 (Student's t-test).

Sufficient N supply (but not C) was required for early post-germinative growth of *atdfb-3* in darkness

The post-germinative growth of the mutant was investigated in the dark under various N conditions. After growing on half-strength MS medium (30 N) for 6 days, *atdfb-3* had shortened hypocotyls and primary roots as well as expanded cotyledons and a larger apical hook curvature than the wild type (Figure 5A and B). Similar results were obtained when ammonium (NH_4^+) was omitted and 9.4 mM or 3 mM NO_3^- (9.4 N or 3 N, respectively) was added to the medium (Figure 5A). Interestingly, when the amount of NO_3^- in the medium was decreased further (less than 3 mM), there were no significant changes in the lengths of hypocotyls of the wild-type seedlings, but the mutant displayed even shorter hypocotyls (Figure 5A and B). When the medium was

supplemented with 0.3 mM NO_3^- (0.3 N) or 0 N, these hypocotyl and primary root phenotypes of *atdfb-3* differed significantly from those of the wild type. The cotyledons of *atdfb-3* were folded similarly to those of the wild type; however, the apical hook curvature in *atdfb-3* appeared larger than that in the wild type (Figure 5A). Next, we explored the hypocotyl phenotype further. Interestingly, when NH_4^+ was used as the sole N source in the medium, it could not be used for hypocotyl development in *atdfb-3* (unlike NO_3^-), whether at higher (9.4 and 3 mM) or lower (1 and 0.3 mM) concentrations (Figure S2A and B in File S1). In addition, the mutant could not utilize organic nitrogen Asn or Gln in the medium under dark conditions (Figure S2C in File S1). These results indicated that *atdfb-3* was sensitive to external NO_3^- concentrations during skotomorphogenesis, and the hypocotyl

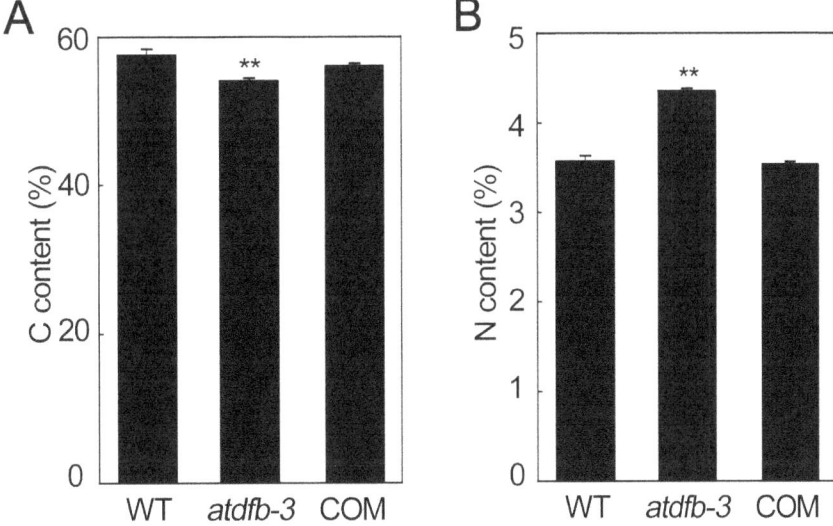

Figure 2. C and N contents in mature WT, *atdfb-3* and *AtDFB* complemented (COM) seeds. (A) C content. (B) N content. Data represent means ± SD. n = 4, and each replicate consisted of 10 mg DW of pooled plant material. Bars with ** indicate a highly significant difference at P<0.01 (Student's t-test).

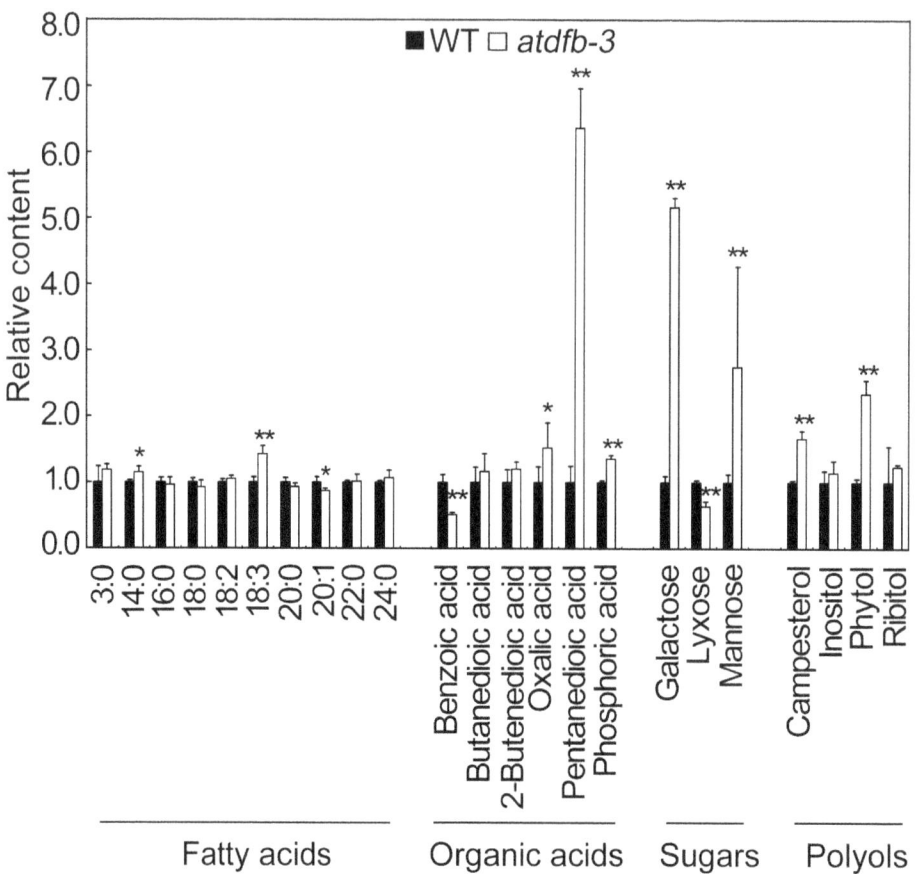

Figure 3. Metabolite profiles in WT and *atdfb-3* seeds. Propanoic acid (3:0), tetradecanoic acid (14:0), hexadecanoic acid (16:0), octadecanoic acid (18:0), 9,12-octadecadienoic acid (18:2), linolenic acid (18:3), eicosanoic acid (20:0), 11-eicosenoic acid (20:1), docosanoic acid (22:0), and tetracosanoic acid (24:0). n = 5, and each replicate consisted of 20 mg of pooled seeds. Data represent means ± SD. White bars with * indicate a significant difference at P<0.05, and ** indicates a highly significant difference at P<0.01 (Student's *t*-test).

elongation in *atdfb-3* required an external, sufficient NO_3^- supply.

Mutant and wild-type seedlings were also grown on media with other nutrient deficiencies. When grown on C-free medium, there was no obvious difference in hypocotyl length between *atdfb-3* and the wild type (Figure 5C). On phosphate (P)-free medium in darkness, the hypocotyl length of *atdfb-3* was 80% compared with that of the wild type (Figure 5C); the ratio was similar to that under N-sufficient conditions (3 mM or higher NO_3^- concentration). These results indicated *atdfb-3* had a specific response to the external NO_3^- supply (but not C or P) during seedling establishment in darkness.

Since hypocotyl elongation in *atdfb-3* was significantly inhibited by 0.3 N (0.3 mM NO_3^-), and the phenotype of the mutant on 9.4 N was similar to that on 1/2 MS, 0.3 N and 9.4 N were used as N-limited and N-sufficient conditions, respectively, in subsequent experiments. Both N conditions were used in our previous report for N limitation analysis in Arabidopsis [33]. The epidermal cell length in *atdfb-3*, as measured using a field emission scanning electronic microscope (FE-SEM), was approximately 86% compared with the wild-type cell length on 9.4 N (396.8±60.6 μm and 462.0±35.8 μm, respectively) and 47% compared with the wild-type cell length on 0.3 N (226.2±36.5 μm and 484.2±35.6 μm, respectively) (Figures 5D and E). These observations demonstrated that *atdfb-3* was defective in hypocotyl cell elongation in darkness.

To explore why *atdfb-3* was sensitive to external NO_3^- concentrations, the stage at which N-sufficient conditions were required for *atdfb-3* hypocotyl development was investigated by removing NO_3^- from the medium at various time points after sowing. Seedlings first grown on 9.4 N for 0 to 6 days were transferred to 0.3 N for the remaining days, for a total growth time of 6 days. A significant difference in hypocotyl length between *atdfb-3* and the wild type was observed when seedlings grown on 9.4 N for 2 days before transferring to 0.3 N (Figure 5F). Hypocotyl length of *atdfb-3* first grown on 9.4 N for 3 days or longer time and then transferred to 0.3 N was similar to that of the mutant grown on 9.4 N for 6 days (Figure 5F). These results indicated that N-sufficient conditions were important for *atdfb-3* during the first 3 days. In further time-course experiments, the hypocotyl length of seedlings grown under N-limited conditions and then transferred to N-sufficient conditions was shorter than those continuously grown on 9.4 N for 6 days and longer than those continuously grown on 0.3 N for 6 days (Figure 5G). The less time *atdfb-3* was grown on 0.3 N before transferring to N-sufficient conditions, the longer the hypocotyls (Figure 5G). The hypocotyl in 6-day-old wild-type seedlings grown on 0.3 N for 2 to 5 days before transferring to 9.4 N was longer than those continuously grown on 0.3 N or 9.4 N for 6 days (Figure 5G); however, this phenomenon was not observed when wild-type seedlings were transferred from N-sufficient to N-limited conditions (Figure 5F). It is possible that the transferring from N-limited

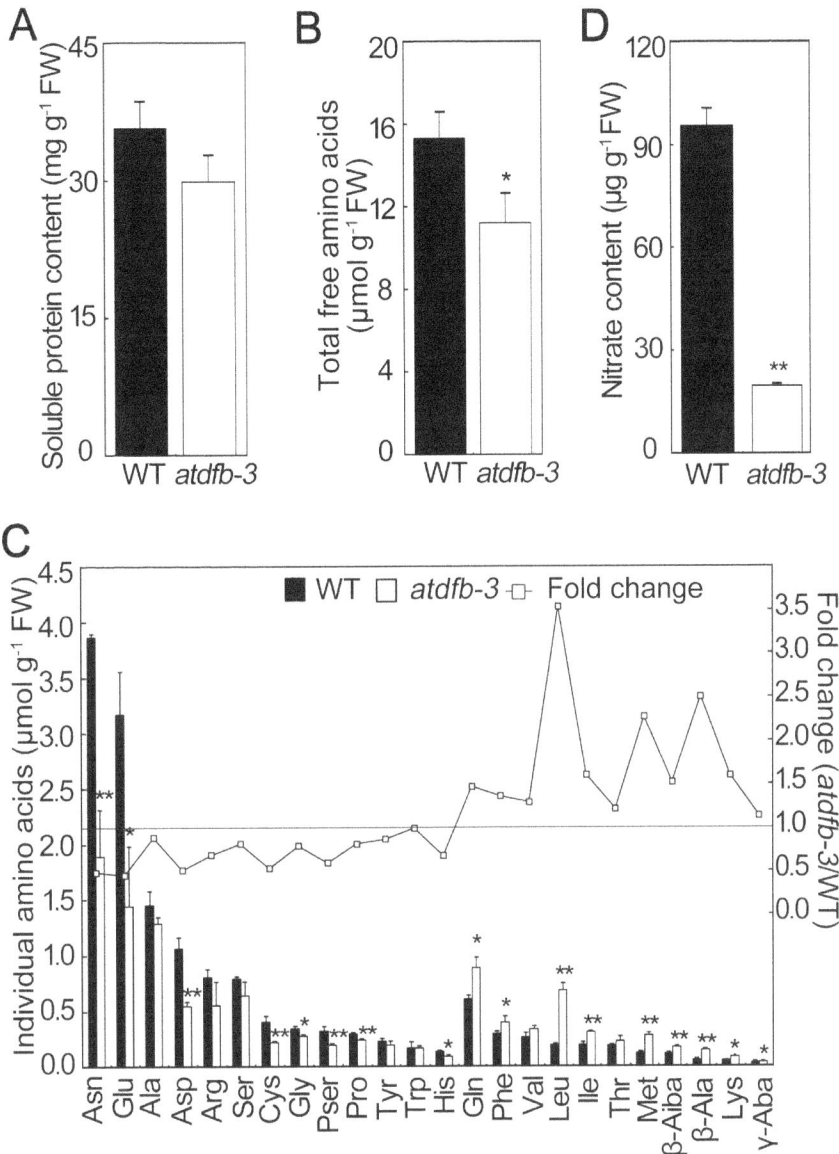

Figure 4. Contents of N-rich metabolites in WT and *atdfb-3* seeds. (A) Soluble protein. (B) Total amino acids. (C) Individual amino acids. (D) NO_3^- content. Data represent means ± SD. n = 3. A, each replicate consisted of 50 mg pooled plant material; B and C, each replicate consisted of 300 mg pooled plant material; D, each replicate consisted of 1 g pooled plant material. Bars with * indicate a significant difference at $P < 0.05$, and ** indicates a highly significant difference at $P < 0.01$ (Student's t-test).

to N-sufficient conditions stimulated hypocotyl elongation in the wild type. These results indicated that N-sufficient conditions were required for early growth of *atdfb-3*, and that the response of *atdfb-3* to low N stress occurred within a narrow developmental window (3 days or earlier).

The endosperm in *atdfb-3* was removed under both N conditions to explore whether the storage in embryo or endosperm in *atdfb-3* was altered, which would affect hypocotyl development under dark conditions. When the endosperm was removed, wild-type hypocotyls were slightly shorter than those with endosperm present under both 9.4 N and 0.3 N conditions (Figure 5H). However, the hypocotyls in *atdfb-3* without endosperm under 9.4 N were only 30% of those in *atdfb-3* with the endosperm, and even shorter than those in *atdfb-3* with endosperm under 0.3 N (Figure 5H). Defects in hypocotyl elongation were more significant in *atdfb-3* without endosperm than that with endosperm under

both N conditions (Figure 5H). These results indicated that, unlike the wild type, the embryo alone could not satisfy hypocotyl development in *atdfb-3* under N-sufficient conditions, and that the endosperm was vital for *atdfb-3* hypocotyl development. Under N-limited condition, external N could not satisfy hypocotyl development of *atdfb-3* even with the endosperm. The requirement of sufficient NO_3^- during early hypocotyl development in *atdfb-3* could be due to altered seed storage in embryo.

AtDFB was expressed in early developmental stage in Arabidopsis

Since *AtDFB* is important for early seedling establishment under dark conditions, we investigated the expression pattern of *AtDFB* during the early stage in etiolated seedlings to illustrate its importance in hypocotyl development. A plasmid containing an *AtDFB* promoter-driven GUS fragment was introduced into the

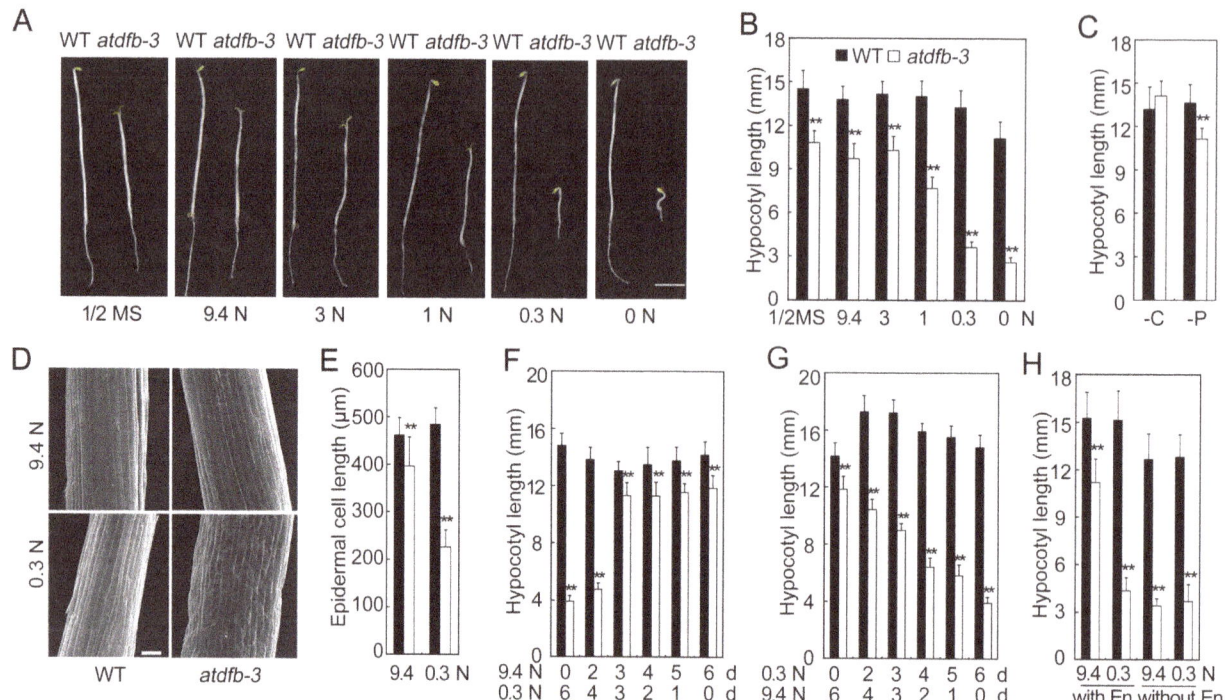

Figure 5. Maladjustment time window in *atdfb-3* etiolated seedlings. (A) Images of 6-day-old etiolated WT and *atdfb-3* seedlings grown on different amounts of nitrogen (N). Scale bar, 3 mm. (B) Hypocotyl lengths of 6-day-old etiolated WT and *atdfb-3* seedlings grown on different amounts of nitrogen (N). (C) Effects of carbon (C)-free or phosphate (P)-free media on the hypocotyl lengths of 6-day-old WT and *atdfb-3* etiolated seedlings. (D) FE-SEM image of hypocotyl cells of 6-day-old etiolated WT and *atdfb-3* seedlings grown on 9.4 N and 0.3 N medium. Scale bar, 100 μm. (E) Hypocotyl cell length in 6-day-old etiolated WT and *atdfb-3* seedlings on 9.4 N and 0.3 N medium. (F) Hypocotyl length of 6-day-old WT and *atdfb-3* etiolated seedlings. Both genotypes seedlings were grown on 9.4 N medium for 0 to 6 days and then transferred to 0.3 N medium for the remaining days (0 to 6 days). (G) Hypocotyl length of 6-day-old WT and *atdfb-3* etiolated seedlings. Seedlings of both genotypes were grown on 0.3 N medium for 0 to 6 days and then transferred onto 9.4 N medium for the remaining days (0 to 6 days). (H) Hypocotyl length of 6-day-old WT and *atdfb-3* etiolated seedlings with or without endosperm in 9.4 N and 0.3 N media. Data represent the means ± SD (in panel B, C, E, F, G and H, n = 15). White bars with * indicate a significant difference at $P<0.05$, and ** indicates a highly significant difference at $P<0.01$ (Student's *t*-test).

wild-type plants (Figure 6A). Histochemical GUS staining showed that *AtDFB* was strongly expressed in cotyledons and hypocotyls in 2-day-old germinating seeds, while it was strongly expressed only in cotyledons in 3-day-old etiolated seedlings (Figure 6B). The expression pattern of *AtDFB* under light was similar to that in the dark (Figure S3 in File S1). These results indicated that *AtDFB* was expressed in early seedling developmental stage.

Rescue of *atdfb-3* seedling establishment by exogenous 5-F-THF depended on the nitrate supply

5-F-THF is able to rescue the *atdfb* defects in primary root development in light [27]. We added various concentrations of 5-F-THF or 5-methyl-tetrahydrofolate (5-M-THF) to 0.3 N medium to determine whether folate derivatives could rescue defects in hypocotyl elongation in *atdfb-3* in dark. The difference in hypocotyl length between the mutant and wild-type seedlings was reduced when grown on 0.3 N with 0.5 μM or 5 μM 5-F-THF, while treatment with 50 or 500 μM 5-F-THF rescued hypocotyl elongation in *atdfb-3* seedlings (Figure S4A in File S1). 5-M-THF stimulated hypocotyl elongation in both *atdfb-3* and wild-type seedlings. Interestingly, under N-limited conditions, disparities between the mutant and wild type decreased with increasing 5-M-THF concentrations, but the hypocotyl length in *atdfb-3* was still only 67% of the wild type when grown with 500 μM 5-M-THF (Figure S4B in File S1). These results indicated that 5-M-THF can only partially rescue the hypocotyl elongation

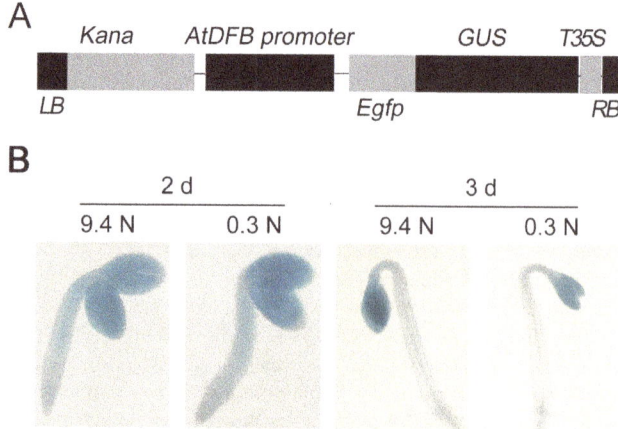

Figure 6. Histochemical localization of *AtDFB* promoter activity and *AtDFB* expression patterns. (A) Schematic diagram of the *AtDFB: GUS* construct. LB and RB indicate the left and right borders, respectively, and Kana indicates the kanamycin resistance gene. (B) GUS staining of 2- and 3-day-old etiolated seedlings under 9.4 N or 0.3 N conditions.

in *atdfb-3*. Thus, 50 μM 5-F-THF was chosen to rescue the hypocotyl elongation defect in *atdfb-3* under dark conditions.

The hypocotyl length of 6-day-old *atdfb-3* seedlings was restored to the wild-type level under both N-sufficient and N-limited conditions with 50 μM 5-F-THF (Figure 7A). Further analysis indicated that 5-F-THF could rescue the hypocotyl elongation defects in *atdfb-3* seedlings under N-limited conditions at various experimental time points (Figure 7B). We next examined the stages at which folate was vital for hypocotyl elongation. Both *atdfb-3* and the wild type seeds were grown on N-limited conditions with 5-F-THF for 0 to 6 days and then moved to conditions without 5-F-THF for the remaining time, for a total growth time of 6 days. We found that the mutant seedlings grown on medium with 5-F-THF for only 2 days and then transferred to conditions without 5-F-THF could adapt to low-N conditions, demonstrating the same hypocotyl length as the wild type at day 6 (Figure 7C). However, *atdfb-3* seedlings grown on medium with 5-F-THF for 1 day before transferring to conditions without 5-F-THF could not adapt to N-limited conditions, similar to those without 5-F-THF treatment under N-limited conditions (Figure S4C in File S1). Additionally, the mutant seedlings grown on medium without 5-F-THF for 2 or more days before transferring to medium with 5-F-THF showed shorter hypocotyls than did the wild-type seedlings; the longer time the mutant was grown on medium without 5-F-THF before transferring to medium with 5-F-THF, the shorter the hypocotyls (Figure 7D). These results suggested that intact folate metabolism was necessary for early (2 days or earlier) developmental stages in Arabidopsis.

When no N was applied to the medium, 5-F-THF could not rescue hypocotyl defects in *atdfb-3*. Meanwhile, 5-F-THF could not restore hypocotyl defects in *atdfb-3* when NH_4^+ was the sole N source in the medium (Figure S5A and B in File S1). These results indicated that the recovery of hypocotyl development in *atdfb-3* by 5-F-THF depended on exogenous NO_3^- supply.

Folate metabolism was altered in *atdfb-3* germinating seeds and etiolated seedlings

To increase our understanding of how folate metabolism was perturbed in the mutant, liquid chromatography/mass spectroscopy (LC-MS) was employed to profile various folate derivatives in early developmental stage of 2-day-old germinating seeds. We found that 5-M-THF was the major folate derivative, accounting

for 70% of the total folates (Figure S6 and Table S1 in File S1). Under N-sufficient conditions, the mutant contained less 5-F-THF, 5-M-THF, and total folates than the wild type (approximately 30%, 80%, and 75% of the wild type, respectively). Under N-limited conditions, the contents of most folate derivatives decreased in the wild type, e.g. 67% reduction for 5-F-THF and 35% for 5-M-THF, respectively, but remained unchanged in *atdfb-3* (Figure S6 and Table S1 in File S1).

To further determine how the *AtDFB* mutation interferes with folate metabolism in dark-grown seedlings, various folate derivatives and poly-glutamylated 5-M-THF and 5-F-THF were examined in 6-day-old etiolated seedlings (Figure 8). First, we found that there was a difference in folate derivative contents between the mutant and wild type. Under N-sufficient conditions, the mutant contained less 5-F-THF, 5-M-THF, and total folates than the wild type (approximately 70%, 36%, and 51% of the wild type, respectively). 5-M-THF, the major folate derivative, constituted the major deficiency in total folates in *atdfb-3* seedlings. Under N-limited conditions, higher accumulation of folate derivatives including 5-F-THF and DHF was observed in *atdfb-3* than that of the wild type, which was opposite to that under N-sufficient conditions; however, 5-M-THF remained less in *atdfb-3* than in the wild type (Figure 8A and Table S2 in File S1). Second, we found that the folate derivatives profiling of the mutant and wild type responded differentially to N limitation. For example, N limitation led to a 50% decrease in 5-F-THF in the wild type, but had no effects on *atdfb-3*, resulting in a 1.6-fold accumulation of 5-F-THF in *atdfb-3* as compared to the wild type. N limitation led to no significant decrease in total folates in the wild type, but a 44% increase in *atdfb-3* seedlings, resulting in a drastic reduction of the difference from 49% to 16% between the mutant and wild type (Figure 8A and Table S2 in File S1).

The levels of polyglutamated folates with 5-, 6-, 7-, and 8-Glu tails were compared between the mutant and wild type based on relative peak areas due to a lack of standards. There was a significant difference between the two genotypes. Under N-sufficient conditions, most striking difference was observed for both 5-M-THF-Glu7 and 5-F-THF-Glu7, i.e. higher accumulation in *atdfb-3* than the wild type. In addition, 5-M-THF-Glu6 was less and 5-F-THF-Glu8 was higher in *atdfb-3* than that of the wild type, respectively (Figure 8B). Under N-limited conditions, both 5-M-THF-Glu7 and 5-F-THF-Glu7 remained higher in the mutant than in the wild type as observed under N-sufficient

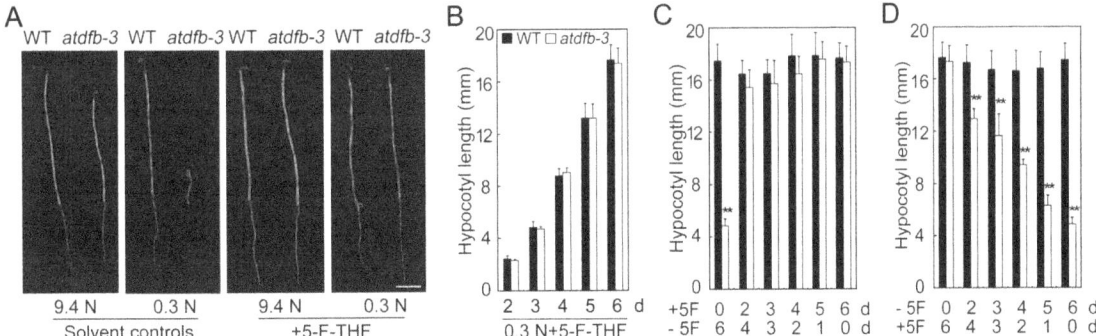

Figure 7. Exogenous 5-F-THF restored the wild-type hypocotyl phenotype in etiolated *atdfb-3* seedlings. (A) Images and hypocotyl of 6-day-old etiolated WT and *atdfb-3* seedlings germinated on 9.4 N or 0.3 N medium with or without 50 μM 5-F-THF. (B) Hypocotyl length of WT and *atdfb-3* under N-limited conditions with 5-F-THF for various days in darkness. (C) Hypocotyl length of 6-day-old WT and *atdfb-3* etiolated seedlings. Seedlings of both genotypes were grown on 0.3 N medium with 50 μM 5-F-THF for 0 to 6 days and then transferred to 0.3 N medium without 5-F-THF for the remaining time (0 to 6 days). (D) Hypocotyl length of 6-day-old WT and *atdfb-3* etiolated seedlings. Seedlings of both genotypes were grown on 0.3 N medium for 0 to 6 days and then syntferred to 0.3 N medium with 50 μM 5-F-THF for the remaining time (0 to 6 days). Data represent means ± SD (n = 15). White bars with ** indicate a highly significant difference at $P<0.01$ (Student's *t*-test).

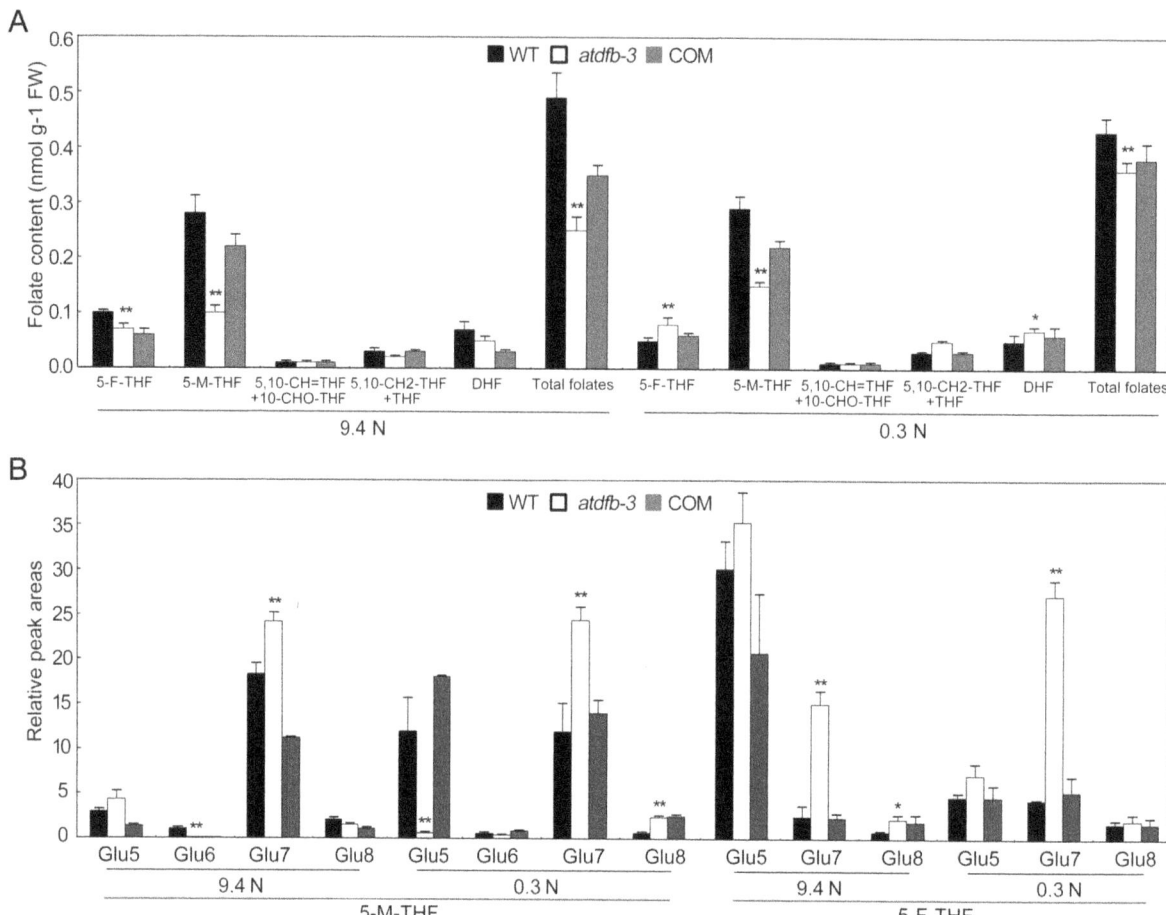

Figure 8. Folate profiles in 6-day-old WT, *atdfb-3*, and *AtDFB* complemented (COM) seedlings grown on 9.4 N or 0.3 N medium. (A) Levels of individual folates and total folates in seedlings. The folate species detected were: 5-formyl-THF (5-F-THF), 5-methyl-THF (5-M-THF), 5,10-methenyl-THF (5,10-CH=THF), 10-formyl-THF (10-CHO-THF), 5,10-methylene THF (5,10-CH$_2$-THF), tetrahydrofolate (THF), and dihydrofolate (DHF). Note that 10-formyl-THF (10-CHO-THF) and 5,10-CH=THF are grouped together and THF and 5,10-methylene THF (5,10-CH$_2$-THF) are grouped together because the procedure used for folate analysis results in inter-conversion of these pairs of folate species. (B) Relative LC-MS peak areas of folylpolyglutamates (5-M-THF-Glu$_n$ and 5-F-THF-Glu$_n$, n = 5, 6, 7, or 8). Data are means ± SD (n = 5). Each replicate consisted of 100 mg of pooled plant material. A significant difference at $P<0.05$ is indicated by *, and a highly significant difference at $P<0.01$ is indicated by ** (Student's *t*-test).

conditions. 5-M-THF-Glu5 and 5-M-THF-Glu8 was around 19 folds less and 2.9 folds higher in *atdfb-3* than in the wild type, respectively (Figure 8B). Moreover, the folate derivatives with polyglutamates in the mutant and wild type differed in responding to N limitation. For example, N limitation led to a 3.1-fold increase of 5-M-THF-Glu5 in the wild type, but a 6.1-fold decrease in *atdfb-3*. The pattern of the folylpolyglutamation profile in the complemented transformants was similar to that in the wild type (Figures 8B).

We also analyzed the expression of genes involved in folate biosynthesis and C1 metabolism (Figure S7 in File S1). The expression of *AtDFA* and *AtDFC* were enhanced in 2-day-old *atdfb-3* seedlings due to loss function of *AtDFB* especially under 0.3 N, but not in 6-day-old seedlings (Figure S7A and B in File S1). Most of these genes, including *AMINODEOXYCHORISMATE LYASE* (*ADCL*), *10-FORMYL-THF DEFORMYLASE 2* (*FDF2*), *5-FORMYL-THF CYCLOLIGASE* (*5-FCL*), *10-FORMYL-THF SYNTHETASE* (*THFS*), *γ-GLUTAMYL HYDROLASE 1* (*GGH1*) and *GGH2*, showed higher expression in *atdfb-3* than in the wild type under both N conditions. Low N stimulated the expression of *ADCL*, *PDF2* and *MTHFR2* in both genotypes, but

only that of *5-FCL* in the mutant and *THFS* in the wild type, respectively (Figure S7C in File S1).

N metabolism was affected in *atdfb-3* germinating seeds and etiolated seedlings

Under N-sufficient conditions, the C and N contents in *atdfb-3* were unchanged compared with the wild type, while the N content in *atdfb-3* increased by 9% under N-limited conditions (Figure 9A and B). In addition, the mutant accumulated 23% more soluble protein under 9.4 N and 32% more under 0.3 N than did the wild type (Figure 9C). These results indicated that N metabolism in germinating *atdfb-3* seeds was altered under N-limited conditions.

Because *atdfb-3* showed a failure of seedling establishment when grown on low N, we sought to understand how N metabolism was affected by the *AtDFB* mutation by analyzing N-relating metabolites and enzyme activities. Under N-sufficient conditions, there was no significant difference in NO$_3^-$ and NO$_2^-$ contents between *atdfb-3* and the wild type; however, under N-limited conditions lower level of NO$_3^-$ and higher level of NO$_2^-$ were detected in *atdfb-3* than in the wild type (Figure 10A and B), whereas no significant difference in NH$_4^+$ contents was observed (Figure S8 in File S1). Subsequently, activities of the enzymes

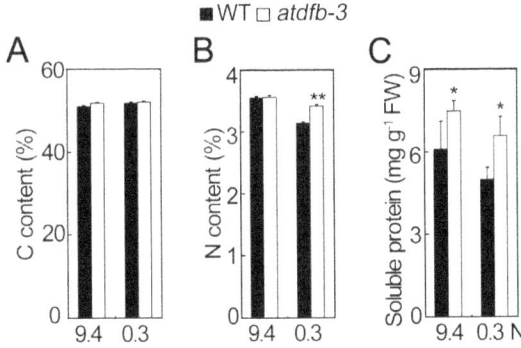

Figure 9. N Metabolites in 2-day-old WT and *atdfb-3* germinating seeds under 9.4 N or 0.3 N. (A) C content. (B) N content. (C) Soluble protein content. Data represent means ± SD. A–B, n = 4, and each replicate consisted of 10 mg DW pooled plant material; C, n = 3, and each replicate consisted of 50 mg pooled plant material. A significant difference at $P<0.05$ is indicated by *, and a highly significant difference at $P<0.01$ is indicated by ** (Student's *t*-test).

involved in N metabolism, such as nitrite reductase (NiR), glutamine synthetase (GS), and glutamine 2-oxoglutarate amino transferase (GOGAT) were investigated under both N conditions. NiR activity was lower in *atdfb-3* than in the wild type under both 9.4 N and 0.3 N, about 51% and 61% of the wild type, respectively (Figure 10C). Under 9.4 N or 0.3 N, GS activity in *atdfb-3* was 83% or 85% of the wild type, respectively (Figure 10D). The GOGAT activity did not differ significantly between *atdfb-3* and the wild type under 9.4 N, while that in *atdfb-3* was only half of the wild type under 0.3 N (Figure 10E). These results implied that activity of the enzymes involved in N reduction and assimilation was altered in *atdfb-3* etiolated seedlings.

Next, transcripts of genes involved in NO_3^- transport, NO_3^- reduction, and N assimilation were also examined in *atdfb-3*. When N was sufficient, *atdfb-3* had lower level of expression of *NITRATE TRANSPORTER 1.1* (*NRT1.1*) than the wild type; however, transcripts of *GS1:1* and *GS2* were more abundant in *atdfb-3* than in the wild type. Low-N stress increased the expression level of *NRT1.1* and *GS1:4* in both genotypes (Figure 10F). Taken together, these results indicated that N metabolism was perturbed in *atdfb-3*.

Discussion

Under light conditions, mutation in the *AtDFB* resulted in a short primary root due to perturbed folate profile [32]. In this report, *atdfb-3* was characterized for its defects in seed reserves and hypocotyl elongation in the dark due to loss of function of *AtDFB*, providing novel insights into a potential link among folate metabolism, seed reserves, and hypocotyl development in Arabidopsis.

AtDFB mutation altered seed storage

Seed storage compound synthesis and accumulation in maturing seeds of Arabidopsis are under the control of many factors, such as hormones, sugars, master regulator genes, and transcription factors [4]. Folates also play an important role in seed development, since *gla1* and *fpgs1fpgs2* exhibited defective embryo development [30,34]. The double knockout (dKO) mutation of 10-formyl-THF deformylase resulted in shriveled seeds and low amounts of lipids, such as 20:1 [31]. A slightly but significantly lower level of 20:1, one of the markers for storage oil

in Arabidopsis [35], was also detected in *atdfb-3* than in the wild type (Figure 3), probably indicative of a low oil storage in the mutant.

Arabidopsis mutants that have defective seed storage mobilization had shorter hypocotyls in the dark [16,17], and *atdfb-3* also showed shortened hypocotyls (Figure 5A). There is a possibility that high level of mannose in *atdfb-3* seeds reduced storage mobilization rate, as exogenous mannose did greatly reduce the rate of storage lipid mobilization in germinating Arabidopsis seeds [36]. It was also reported that galactose that accumulated during seed maturation could provide easily available energy and also be an important component of the sugar signaling pathway during germination of pea seeds [9,37]. Thus, high content of galactose accumulated in *atdfb-3* seeds might promote germination while mannose inhibits storage mobilization. Besides, mannose and galactose are intermediates of ascorbic acid biosynthesis [38]. Therefore, it needs further investigation that whether the accumuation of mannose and galactose is due to deficient ascorbic acid biosynthesis in *atdfb-3*.

It was also observed that oxalic acid was accumulated in *atdfb-3* seeds (Figure 3). Oxalate could be produced by glycolate or glyoxylate during photorespiration, or by the breakdown of ascorbic acid [39,40]. Previously we reported that the mitochondrial AtDFC is involved in regulation of N metabolism in Arabidopsis by linking folate metabolism with photorespiration [33]. However, it seems unlikely that the oxalate accumulation is due to photorespiration alteration given the fact that Gly/Ser ratio, an indicator for photorespiration, was not changed in the mutant (Figure 4C). Oxalate accumulation was observed both in *atdfb-3* and the mutant of oxalyl-CoA synthetase, an enzyme that catalyzes the first step of oxalate catabolism [41]. However, it remains unclear that whether the oxalate accumulation in *atdfb-3* seeds is attributable to decreased oxalate catabolism. Pentanedioic acid was dramatically increased in the *atdfb-3* seeds (Figure 3), but it is unknown yet how folates affect its biosynthesis in plants to date. In addition, altered levels of many individual amino acids in *atdfb-3* verified the role of folates in amino acids metabolism [31,32,33,42,43].

Folate biosynthesis and polyglutamylation were responsive to low N stress

Low N stress enhanced the expression of folate synthesis and metabolism genes (such as *AtDFA* and *THFS*) in the mutant (Figure S7A in File S1). As a result, similar contents of total folates in the mutant and wild type were achieved (Figure S6 in File S1). This is different from our previous report that the total folates level in *atdfc* remained lower than in the wild type under N limitation [33]. Under N-sufficient conditions, hypocotyls of 6-day-old etiolated *atdfb-3* seedlings were only slightly shorter than the wild type; however, under N-limited conditions, the mutant had significantly reduced hypocotyls (Figure 5A and B). In association with this, 5-M-THF-Glu5 and 5-F-THF-Glu7 in *atdfb-3* was 0.05- and 6-fold of that in the wild type under N-limited conditions, respectively (Figure 8B). Considering that 5-M-THF-Glu5 and 5-F-THF-Glu7 were the most changed folate derivative and responded in an opposite manner when the mutant was subjected to low N, we hypothesize that 5-M-THF-Glu5 and 5-F-THF-Glu7 may, at least in part, play an important but contrasting role in regulation of hypocotyl development. Meanwhile, we could also speculate that there is not a causal relationship between total folate level and shortened hypocotyl length in the mutant under N limitation because the mutant and wild type had a similar level of total folates (Figure 8A and S6).

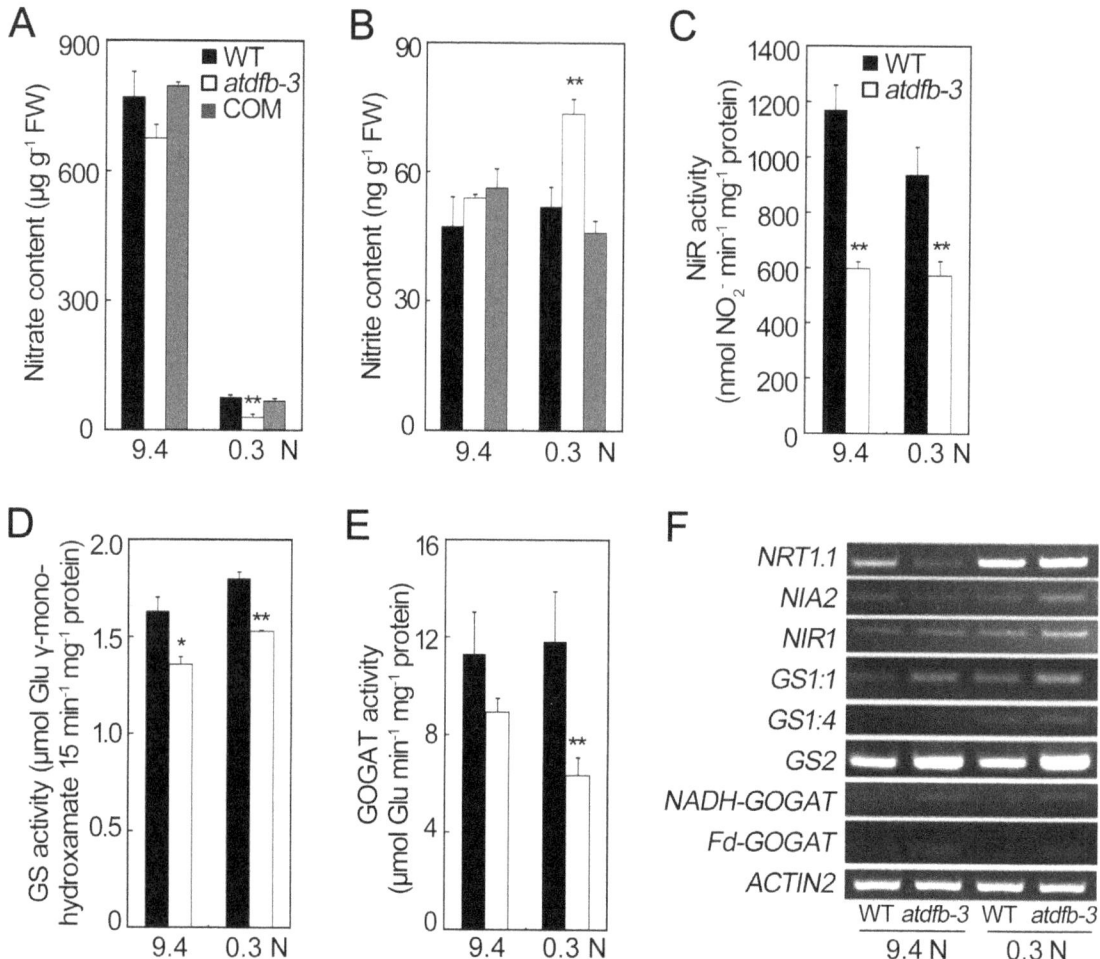

Figure 10. Altered biochemical characteristics of N reduction and assimilation in 6-day-old etiolated WT, *atdfb-3* **and** *AtDFB* **complemented (COM) seedlings on 9.4 N or 0.3 N medium.** (A) NO_3^- content. (B) NO_2^- content. (C) NiR activity. (D) GS activity. (E) GOGAT activity. (F) Altered transcript levels of genes involved in nitrate transport, reduction, and N assimilation. Data are means ± SD (n = 3). In panel (A) and (B), each replicate consisted of 1 g of pooled plant material. In panel (C) and (D), each replicate consisted of 200 mg of pooled plant material. In panel (E), each replicate consisted of 500 mg of pooled plant material. A significant difference at $P<0.05$ is indicated by *, and a highly significant difference at $P<0.01$ is indicated by ** (Student's *t*-test).

Exogenous nitrate was required for the rescue of *atdfb-3* mutant by 5-F-THF application

Seed reserves are mobilized to fuel seedlings until autotrophic growth. Given the massive reduction in nitrate level in *atdfb-3* (Figure 4D), it was reasonable that the mutant requires sufficient external N. 5-F-THF could rescue the defects in hypocotyl elongation in *atdfb-3* only in the presence of NO_3^-, but not NH_4^+ in the medium (Figure S5 in File S1). Owing to the observation of reduction of GS and GOGAT activity under N-limited conditions (Figure 10), NH_4^+ assimilation is probably deficient in *atdfb-3*. Thus, the mutant could not utilize the sole N source NH_4^+ to fuel seedling development. Additionally, exogenous 5-F-THF failed to rescue the mutant when NO_3^- was absent (Figure S5 in File S1), demonstrating an absolute necessity of NO_3^- for seedling establishment in *atdfb-3*.

The exogenously supplied 5-F-THF was probably absorbed and converted into other active forms of folate by 5-FCL, MTHFR, or other enzymes that convert folate derivatives [44,45]. The recovery of hypocotyl elongation in *atdfb-3* treated with 5-F-THF was probably due to the excess of folates in the monoglutamylated form, which might be able to complement

the mutation by accomplishing the same functions as a small amount of polyglutamylated folates. Given the fact that microtubule cytoskeleton plays a crucial role during hypocotyl elongation [46], and abnormal actin cytoskeleton was observed in *atdfb* primary root [32], it's conceivable that 5-F-THF may promote hypocotyl cell elongation in *atdfb-3* through regulating cytoskeleton stabilization. Taken together, these observations suggest that folate-regulated N metabolism is important for *atdfb-3* seedling development in the dark.

Perturbation of N metabolism was caused by *atdfb* mutation

Defective hypocotyl elongation in *atdfb-3* is accompanied by perturbed N metabolism due to loss function of *AtDFB*. Under N-sufficient conditions, activities of NiR and GS were both lower in *atdfb-3* than those of the wild type, indicative of an impaired N reduction and assimilation, although the contents of NO_3^-, NO_2^- and NH_4^+ in *atdfb-3* were similar to those in the wild type (Figure 10 and Figure S8 in File S1). Under N-limited conditions, lower content of NO_3^- may reflect a decreased NO_3^- uptake and/or reduction in *atdfb-3* (Figure 10A), and the lower NiR activity could partly explain why high content of NO_2^- accumulated in *atdfb-3* (Figure 10B and C). Apart from this, low activities of GS and GOGAT were also observed under 0.3 N (Figure 10D and E), suggesting a reduced N assimilation ability in *atdfb-3* as compared to the wild type. Taken together, the ability of N reduction and assimilation in *atdfb-3* was significantly lower than in the wild type under N limitation, resulting in defective hypocotyl development. Furthermore, it was reported that folates can be oxidized by NO_2^- to several pterin products [47], therefore it is possible that insufficient folate derivatives could not effectively protect the etiolated seedling from toxicity of significantly accumulated NO_2^- under N-limited conditions. Thus, the drastically shortened hypocotyl of *atdfb-3* under 0.3 N could be partly due to the NO_2^- toxicity. However, how the altered folates profiling or polyglutamylation affects N metabolism in *atdfb-3* awaits further investigation. One of the possibilities is that polyglutamylated 5-M-THF or 5-F-THF may act as a regulator of the N metabolism enzymes.

Taking together, we provide genetic evidence that the plastidial isoform FPGS is required for normal seed reserve accumulation and hypocotyl elongation under dark conditions. The *AtDFB* mutation results in altered seed storage, perturbed folate profile, altered N metabolism and shorter hypocotyls in etiolated seedlings, and exogenous 5-F-THF recovered shortened hypocotyls of the mutant to the wild-type level when NO_3^- was present in the growth conditions. However, the underlying mechanism through which folate regulates seed reserve accumulation and hypocotyl development during skotomorphogenesis as well as the relationship among folate metabolism, N metabolism and hypocotyl development require further investigation.

Materials and Methods

Plant materials and growth conditions

Arabidopsis wild-type (*Arabidopsis thaliana*, ecotype Columbia), the T-DNA insertion mutant of *AtDFB* (SALK_015472, called *atdfb-3* in this report), and the *AtDFB* complemented line are grown in identical growth chambers under a 16-h photoperiod (photosynthetic photon flux density 60 $\mu E\ m^{-2}\ s^{-1}$) and a day/night temperature of 22/16°C before being harvested. For biochemical analysis, and metabolite measurement assays, seeds from various genetic backgrounds were harvested at the same time and were after-ripened for 3 months.

For the Petri dish-based N limitation experiments, when NO_3^- was used as the sole N source, NH_4^+ was removed from the half-strength MS medium [48]. The K^+ level was balanced with KCl to maintain 9.4 mM K^+. In this report, 9.4 mM NO_3^- (9.4 N) was used as the N-sufficient condition and 0.3 mM NO_3^- (0.3 N) as the N-limited condition. When NH_4^+ was used as the sole N source, NO_3^- was removed from the half-strength MS medium, and NH_4Cl was then added to the desired N concentration. When Asn and Gln were used as the sole N sources, NO_3^- and NH_4^+ were removed from the half-strength MS medium, and the K^+ level was balanced with KCl to maintain 9.4 mM K^+. For C- or P-deficiency experiments, sucrose or KH_2PO_4 was not added to half-strength MS medium. Endosperm/seed coat tissues were removed from the embryo using a dissecting microscope, after allowing seeds to soften in water for 6 h at 4°C [17]. For all experiments mentioned above, the wild-type and *atdfb-3* seeds were sterilized, grown on the same plate, treated at 4°C in the dark for 2 days, and then moved to a growth chamber at 22°C under continuous dark conditions. Digital photographs of hypocotyls at various stages of etiolated seedling growth were acquired using a Nikon 700 camera, and their lengths were measured using ImageJ.

For expression pattern analysis, 1,406 bp of the *AtDFB* promoter was amplified and cloned into the binary vector pKGWFS 7.0. The construct was introduced into wild-type plants using the floral dipping method. The homozygous *AtDFB: GUS*-transformed seedling were stained according to Francisco [49], and observed under the stereoscope from Nikon DIGITAL CAMERA Dxm 1200F.

For the 5-F-THF and 5-M-THF [(6R, S)-5-formyl-5,6,7,8-tetrahydrofolic acid and (6R, S)-5-methyl-5,6,7,8-tetrahydrofolic acid, calcium salt; Schircks Laboratories, Switzerland] supplementation experiments, a stock solution was added to the growth medium to achieve the desired working concentration. A stock solution of 5 mM 5-F-THF or 5-M-THF was prepared in deionized water. Seeds were planted directly on the medium with or without the abovementioned folate derivatives. The hypocotyl length assays were performed as described above.

Microscopic analysis

The hypocotyls of 6-day-old etiolated wild-type and *atdfb-3* seedlings grown on 9.4 N or 0.3 N were observed according to the method by Cowling *et al.* under a Hitachi S1-4800 high-resolution FE-SEM [50]. The cells of the midportions of hypocotyls were observed using a ZEISS Imager M1 DIC microscope and a 10× objective lens [51].

Biochemical analysis

Biochemical analysis procedure was according to Jiang *et al.* [33]. C and N contents were analyzed using a Perkin Elmer 2400 Series II CHNS/O Elemental Analyzer (www. perkinelmer.com), and the value indicated the percentage of C or N in total dry weight (mg/100 mg DW). Free amino acids were analyzed using an ASI.KAUNER amino acid analyzer A200 (www.knauer.net). Soluble proteins were extracted from the frozen seedling powder using 100 mM HEPES-KOH (pH 7.5) and 0.1% Triton X-100 and assayed using a commercial protein assay kit (Bio-Rad). NO_3^- and NO_2^- were measured as described by Oliveira [39]. NH_4^+ was measured according to Andrew *et al.* [52].

Seed metabolite profile analysis using GC-TOF-MS

Seed metabolite analysis using GC-TOF-MS was performed using a method modified from that described previously [9,53]. Seeds (approximately 20 mg) were homogenized using a pre-cooled mortar and pestle with liquid nitrogen and extracted in 1.5 ml of a methanol: chloroform: water extraction solution (2.5:1:1, v/v/v). Internal standards (50 μl 1 mg ml^{-1} ribitol in water and 20 μl 2 mg ml^{-1} C^{13}-nonadecanoic acid in chloroform) were subsequently added. The mixture was extracted for 2 h at 37°C with shaking at 1,500 rpm. After 10 min of centrifugation at 12,000 rpm, 400 μl water and chloroform were added to the supernatant, respectively. Following vortexing and a 5-min centrifugation at 12,000 rpm, 200 μl methanol-water phase was isolated and reduced to dryness in a vacuum. Meanwhile, 400 μl chloroform-lipid phase was obtained and concentrated to dryness using nitrogen gas. Residues were re-dissolved and derivatized for 2 h at 37°C (in 25 μl 20 mg ml^{-1} methoxyamine hydrochloride in pyridine) followed by a 30-min treatment with 50 μl N-methyl-N-(trimethylsilyl) trifluoroacetamide at 37°C. Each 1 μl aliquot of the derivatives was injected in a splitless mode using an autosampler into an Agilent 6890 GC system coupled to a LECO Pegasus IV time-of-flight mass spectrometer system (LECO Corporation, USA). A DB-5MS capillary column (30 m×0.25 mm i.d., 0.25-μm film thickness, Agilent J&W Scientific, USA) was used to separate the samples. The injector temperature was 280°C. The

Helium gas flow rate through the column was 1.0 ml min^{-1}. The column temperature was held at 80°C for 1 min and then increased by 10°C min^{-1} to 310 °C and held there for 10 min. The column effluent was introduced into the ion source of a Pegasus IV TOF-MS. The transfer line and the ion source temperatures were 280 and 200°C, respectively. The electron energy was 70 eV, and mass data were collected in a full-scan mode (m/z 50–650). The detector voltage was set at 1,650 V. All samples were randomized, and five biological replicates were analyzed within 24 h of chemical derivatization. Raw data were processed using LECO ChromaTOF v3.32. Information, including the peak area and retention time, for each detected metabolite was obtained. According to the retention time, the name of each metabolite was obtained by searching the NIST MS Search 2.0 database. When the matching score was higher than 800, the result was considered credible, and the metabolite was further analyzed. The relative contents of the metabolites are shown, and those of the wild type were normalized to values of 1.

Folate profile analysis using LC-MS

The following folates were purchased from Schircks Laboratories: 5-M-THF, THF, 5-F-THF, 5,10-methenyltetrahydrofolate, and DHF. The 2-day-old etiolated seedlings grown on 9.4 N and 0.3 N medium plates were used for identification of folate profiles. The 6-day-old etiolated seedlings grown on 9.4 N and 0.3 N medium plates were used for identification of folate profiles and folylpolyglutamates of 5-M-THF-Glu$_n$ and 5-F-THF-Glu$_n$ (n = 5, 6, 7, and 8). The methods for sample preparation and metabolite measurement were described previously [33]. The experiments included five biological replicates.

NiR, GS and GOGAT enzyme activity analysis

The method for enzyme activity analysis was similar to our previous report [33]. The 200 mg powdered tissues of 6-day-old etiolated seedlings for NiR analysis were added to 0.6 ml of extraction buffer containing 50 mM potassium phosphate buffer (pH 7.5), 1 mM EDTA, 10 mM 2-mercaptoethanol, 100 mM phenylmethanesulfonyl fluoride, and 5 mg PVP and then homogenized. The homogenate was centrifuged, and the supernatant (crude enzyme solution) was used for the NiR activity analysis. A blank sample, in which sulfanilamide was added prior to the extract, was used for background reading. NiR activity was assayed following Takahashi et al. [54], with modifications, to measure the decrease of NO_2^- in the assay mixture. A 45 µl sample of the crude enzyme solution was transferred to a 1.5 ml centrifuge tube, and 195 µl of the assay solution containing 50 mM potassium phosphate buffer (pH 7.5), 1 mM NaNO$_2$, and 1 mM methyl viologen was added. The reaction was started by adding 60 µl of 57.4 mM Na$_2$S$_2$O$_4$ in 290 mM NaHCO$_3$ (final Na$_2$S$_2$O$_4$ concentration in the assay solution, 11.5 mM), and the reaction was run for 5 min at 30°C. A 0.3 ml aliquot was transferred to a new tube containing 0.7 ml water and mixed vigorously to stop the reaction, after which 1 ml 1% (w/v) sulfanilamide in 3 N HCl and 1 ml 0.02% (w/v) N-1-naphthylethylenediameine dihydrochloride were added. The absorbance of this mixture at 520 nm was measured. NiR enzyme activity was expressed as nmole NO_2^- used per min per mg protein.

For assessment of total GS activities, freshly harvested samples (500 mg) were ground on ice with 1 ml extraction buffer consisting of 100 mM Tris-HCl (pH 7.6), 1 mM MgCl$_2$, 1 mM EDTA, and 10 mM 2-mercaptoethanol. Semi-synthetase GS activity was assayed, with NH$_2$OH used as an artificial substrate, by quantifying the formation of glutamic acid γ-monohydroxamate.

The homogenates were centrifuged at 12,000 g for 30 min at 4°C, and the supernatant was analyzed for total GS activities. Total GS activity was measured in a preincubation assay buffer (30°C) consisting of 37.5 mM imidazole buffer (pH 7.0), 30 mM sodium glutamate, 25 mM MgSO$_4$, 50 mM NH$_2$OH, and 3 mM ATP. The reaction was terminated after 15 min at 30°C by addition of acidic FeCl$_3$ solution (88 mM FeCl$_3$, 670 mM HCl, and 200 mM trichloroacetic acid). After allowing 10 min for the color development, the reaction mixture was centrifuged at 4,000 g at room temperature for 10 min, and 2 ml of supernatant was then transferred from each well into a new tube. The A$_{540}$ was measured in a spectrophotometer quantification reader [55]. GS enzyme activity was expressed as µmol Glu γ-monohydroxamate formed per 15 min per mg protein.

For assessment of GOGAT activities, freshly harvested samples (200 mg) were ground on ice with 0.6 ml extraction buffer consisting of 100 mM potassium phosphate buffer (pH 7.4), 1.28 mM EDTA, and 10 mM 2-mercaptoethanol. GOGAT activity was assayed by quantifying the formation of Glu and using NADH used as the substrate. The reaction mixture consisted of 100 mM potassium phosphate buffer (pH 7.4), 10 mM Gln, 10 mM 2-oxoglutarate, 0.05 mM NADH, and extract. After a 5-min pre-incubation at 30°C, the reaction was started by adding the reductant solution (1.68 mg Na$_2$S$_2$O$_4$ and 3.48 mg NaHCO$_3$ in 1 ml of reaction solution). After a 15 min of incubation at 30°C, the reaction was terminated by heating to 98°C for 5 min. The Glu concentration was then determined using the ninhydrin reaction [56]. GOGAT activity was expressed as µmol Glu formed per min per mg protein.

Accession numbers

Sequence data from this article can be found in the Arabidopsis Genome Initiative or GenBank/EMBL databases under the following accession numbers: AT5G05980 (AtDFB), AT3G18780 (ACTIN2), AT1G12110 (NRT1.1), AT1G37130 (NIA2), AT2G15620 (NIR1), AT5G37600 (GS1:1), AT5G16570 (GS1:4), AT5G35630 (GS2), AT5G53460 (NADH-GOGAT), AT5G04140 (Fd-GOGAT), AT5G41480 (AtDFA), AT3G10160 (AtDFC), AT3G55630 (AtDFD), AT5G57850 (ADCL), AT5G47435 (FDF2), AT5G13050 (5-FCL), AT1G50480 (THFS), AT1G78660 (GGH1), AT1G78680 (GGH2), AT2G44160 (MTHFR2), and AT3G07270 (GTPCHI).

Supporting Information

File S1 Contains the following files: Figure S1. Identification of atdfb-3 and AtDFB complemented (COM) line. (A) Gene map of AtDFB (At5g05980). Boxes indicate exons and lines indicate introns. T-DNA insertion site for the mutant is indicated. Arrows indicate the positions of the primers (F and R) used for RT-PCR. (B) Schematic diagram of the ProAtDFB:AtDFB-HWG complemented construct. LB and RB indicate the left and right borders, respectively, and Hyg indicates the hygromycin resistance gene. (C) AtDFB transcripts in wild-type (WT), atdfb-3, and one representative COM plant. Total RNA was prepared from 14-day-old seedlings grown in light. ACTIN2 transcripts were used as a loading control. **Figure S2. Hypocotyl phenotypes of 6-day-old etiolated WT and atdfb-3 at various concentrations of NH$_4^+$ (A, B) or organic nitrogen (C). Figure S3. GUS staining of 2- and 3-day-old light-grown AtDFB: GUS seedlings under 9.4 N or 0.3 N conditions. Figure S4. Hypocotyl length of WT and atdfb-3 under N-limited conditions with 5-F-THF or 5-M-THF treatment.** (A) Hypocotyl length of 7-day-old etiolated WT and *atdfb-3* seedlings after application of various concentrations of 5-F-THF under N-limited conditions. (B) Hypocotyl length of 6-day-old etiolated WT and *atdfb-3* seedlings after application of various concentrations of 5-M-THF under N-limited conditions. (C) Hypocotyl length of 6-day-old WT and *atdfb-3* etiolated seedlings grown on 0.3 N medium with 50 μM 5-F-THF and then transferred to 0.3 N medium without 5-F-THF for the remaining days. **Figure S5. Hypocotyl phenotype of 6-day-old etiolated WT and *atdfb-3* under 0 N or NH₄⁺ with 5-F-THF treatment.** (A) Image of hypocotyl phenotype of 6-day-old WT and *atdfb-3* under 0 N or 3 mM NH_4^+ with 5-F-THF. (B) Hypocotyl length of 6-day-old WT and *atdfb-3* etiolated seedlings grown on 0 N (upper panel) or 3 mM NH_4^+ with 50 μM 5-F-THF (lower panel). **Figure S6. Folate profiles in 2-day-old WT and *atdfb-3* germinating seeds under 9.4 N or 0.3 N. Figure S7. Transcript levels of genes involved in folate biosynthesis and metabolism. Figure S8. Ammonium content in 6-day-old WT and *atdfb-3* seedlings in the dark. Table S1. Profiles of total folates and various folate species in 2-day-old WT and *atdfb-3* germinating seeds grown on 9.4 N or 0.3 N medium in the dark.**

Table S2. Profiles of total folates and various folate species in 6-day-old WT, *atdfb-3*, and *AtDFB* complemented (COM) etiolated seedlings grown on 9.4 N or 0.3 N medium.

Acknowledgments

We are grateful to Flanders Interuniversity Institute for Biotechnology (Belgium) for providing the pHWG plasmid. We thank Jin Si at the Animal Nutrition and Feed Research Institute, Chinese Academy of Agricultural Sciences (CAAS) for help with the free amino acid analyses. We thank Chengjun Ji in the Department of Ecology, Peking University for help with the C and N content analyses. We thank Zhen Xue at the Metabolic Analysis Platform of the Institute of Botany, Chinese Academy of Sciences for help with the metabolite analyses.

Author Contributions

Conceived and designed the experiments: CYZ YLF. Performed the experiments: HYM BSX WZG JLL XQZ. Analyzed the data: HYM LJ CYZ. Contributed reagents/materials/analysis tools: XQQ LXD XBM. Wrote the paper: HYM LJ CYZ.

References

1. Baud S, Boutin J-P, Miquel M, Lepiniec L, Rochat C (2002) An integrated overview of seed development in *Arabidopsis thaliana* ecotype WS. Plant Physiology and Biochemistry 40: 151–160.
2. Weber H, Borisjuk L, Wobus U (2005) Molecular physiology of legume seed development. Annu Rev Plant Biol 56: 253–279.
3. Graham IA (2008) Seed storage oil mobilization. Annu Rev Plant Biol 59: 115–142.
4. Baud S, Dubreucq B, Miquel M, Rochat C, Lepiniec L (2008) Storage reserve accumulation in Arabidopsis: metabolic and developmental control of seed filling. Arabidopsis Book 6: e0113.
5. Fait A, Nesi AN, Angelovici R, Lehmann M, Pham PA, et al. (2011) Targeted enhancement of glutamate-to-gamma-aminobutyrate conversion in Arabidopsis seeds affects carbon-nitrogen balance and storage reserves in a development-dependent manner. Plant Physiol 157: 1026–1042.
6. O'Neill CM, Gill S, Hobbs D, Morgan C, Bancroft I (2003) Natural variation for seed oil composition in Arabidopsis thaliana. Phytochemistry 64: 1077–1090.
7. Heath J, Weldon R, Monnot C, Meinke D (1986) Analysis of storage proteins in normal and aborted seeds from embryo-lethal mutants of Arabidopsis thaliana. Planta 169: 304–312.
8. Cernac A, Andre C, Hoffmann-Benning S, Benning C (2006) WRI1 is required for seed germination and seedling establishment. Plant physiology 141: 745–757.
9. Fait A, Angelovici R, Less H, Ohad I, Urbanczyk-Wochniak E, et al. (2006) Arabidopsis seed development and germination is associated with temporally distinct metabolic switches. Plant Physiol 142: 839–854.
10. Bewley JD (1997) Seed Germination and Dormancy. Plant Cell 9: 1055–1066.
11. Andre C, Benning C (2007) Arabidopsis seedlings deficient in a plastidic pyruvate kinase are unable to utilize seed storage compounds for germination and establishment. Plant Physiol 145: 1670–1680.
12. Chen M, Thelen JJ (2010) The plastid isoform of triose phosphate isomerase is required for the postgerminative transition from heterotrophic to autotrophic growth in Arabidopsis. Plant Cell 22: 77–90.
13. Penfield S, Graham S, Graham IA (2005) Storage reserve mobilization in germinating oilseeds: Arabidopsis as a model system. Biochem Soc Trans 33: 380–383.
14. Eastmond PJ (2006) SUGAR-DEPENDENT1 encodes a patatin domain triacylglycerol lipase that initiates storage oil breakdown in germinating Arabidopsis seeds. Plant Cell 18: 665–675.
15. Josse EM, Halliday KJ (2008) Skotomorphogenesis: the dark side of light signalling. Curr Biol 18: R1144–1146.
16. Eastmond PJ, Germain V, Lange PR, Bryce JH, Smith SM, et al. (2000) Postgerminative growth and lipid catabolism in oilseeds lacking the glyoxylate cycle. Proc Natl Acad Sci U S A 97: 5669–5674.
17. Penfield S, Rylott EL, Gilday AD, Graham S, Larson TR, et al. (2004) Reserve mobilization in the Arabidopsis endosperm fuels hypocotyl elongation in the dark, is independent of abscisic acid, and requires PHOSPHOENOLPYRUVATE CARBOXYKINASE1. Plant Cell 16: 2705–2718.
18. Canas RA, de la Torre F, Canovas FM, Canton FR (2006) High levels of asparagine synthetase in hypocotyls of pine seedlings suggest a role of the enzyme in re-allocation of seed-stored nitrogen. Planta 224: 83–95.
19. Akhtar TA, Orsomando G, Mehrshahi P, Lara-Nunez A, Bennett MJ, et al. (2010) A central role for gamma-glutamyl hydrolases in plant folate homeostasis. Plant J 64: 256–266.
20. Hanson AD, Gregory JF 3rd (2002) Synthesis and turnover of folates in plants. Curr Opin Plant Biol 5: 244–249.
21. Hanson AD, Gregory JF 3rd (2011) Folate biosynthesis, turnover, and transport in plants. Annu Rev Plant Biol 62: 105–125.
22. Hanson AD, Roje S (2001) One-Carbon Metabolism in Higher Plants. Annu Rev Plant Physiol Plant Mol Biol 52: 119–137.
23. Ravanel S, Cherest H, Jabrin S, Grunwald D, Surdin-Kerjan Y, et al. (2001) Tetrahydrofolate biosynthesis in plants: molecular and functional characterization of dihydrofolate synthetase and three isoforms of folylpolyglutamate synthetase in Arabidopsis thaliana. Proc Natl Acad Sci U S A 98: 15360–15365.
24. Sahr T, Ravanel S, Rebeille F (2005) Tetrahydrofolate biosynthesis and distribution in higher plants. Biochem Soc Trans 33: 758–762.
25. Van Wilder V, De Brouwer V, Loizeau K, Gambonnet B, Albrieux C, et al. (2009) C1 metabolism and chlorophyll synthesis: the Mg-protoporphyrin IX methyltransferase activity is dependent on the folate status. New Phytol 182: 137–145.
26. Gambonnet B, Jabrin S, Ravanel S, Karan M, Douce R, et al. (2001) Folate distribution during higher plant development. Journal of the Science of Food and Agriculture 81: 835–841.
27. Jabrin S, Ravanel S, Gambonnet B, Douce R, Rébeillé F (2003) One-carbon metabolism in plants. Regulation of tetrahydrofolate synthesis during germination and seedling development. Plant physiology 131: 1431–1439.
28. Roos A, Cossins E (1971) Pteroylglutamate derivatives in *Pisum sativum* L. Biosynthesis of cotyledonary tetrahydropteroylglutamates during germination. Biochem J 125: 17–26.
29. Chan PY, Coffin JW, Cossins EA (1986) In Vitro synthesis of pteroylpoly-γ-glutamates by cotyledon extracts of Pisum sativum L. Plant and cell physiology 27: 431–441.
30. Ishikawa T, Machida C, Yoshioka Y, Kitano H, Machida Y (2003) The GLOBULAR ARREST1 gene, which is involved in the biosynthesis of folates, is essential for embryogenesis in Arabidopsis thaliana. Plant J 33: 235–244.
31. Collakova E, Goyer A, Naponelli V, Krassovskaya I, Gregory JF 3rd, et al. (2008) Arabidopsis 10-formyl tetrahydrofolate deformylases are essential for photorespiration. Plant Cell 20: 1818–1832.
32. Srivastava AC, Ramos-Parra PA, Bedair M, Robledo-Hernandez AL, Tang Y, et al. (2011) The folylpolyglutamate synthetase plastidial isoform is required for postembryonic root development in Arabidopsis. Plant Physiol 155: 1237–1251.
33. Jiang L, Liu Y, Sun H, Han Y, Li J, et al. (2013) The Mitochondrial Folylpolyglutamate Synthetase Gene Is Required for Nitrogen Utilization during Early Seedling Development in Arabidopsis. Plant Physiol 161: 971–989.
34. Mehrshahi P, Gonzalez-Jorge S, Akhtar TA, Ward JL, Santoyo-Castelazo A, et al. (2010) Functional analysis of folate polyglutamylation and its essential role in plant metabolism and development. Plant J 64: 267–279.
35. Lemieux B, Miquel M, Somerville C, Browse J (1990) Mutants of Arabidopsis with alterations in seed lipid fatty acid composition. Theor Appl Genet 80: 234–240.
36. To JP, Reiter WD, Gibson SI (2002) Mobilization of seed storage lipid by Arabidopsis seedlings is retarded in the presence of exogenous sugars. BMC Plant Biol 2: 4.
37. Blochl A, Peterbauer T, Richter A (2007) Inhibition of raffinose oligosaccharide breakdown delays germination of pea seeds. J Plant Physiol 164: 1093–1096.
38. Conklin PL, DePaolo D, Wintle B, Schatz C, Buckenmeyer G (2013) Identification of Arabidopsis VTC3 as a putative and unique dual function

protein kinase::protein phosphatase involved in the regulation of the ascorbic acid pool in plants. J Exp Bot 64: 2793–2804.

39. Franceschi VR, Nakata PA (2005) Calcium oxalate in plants: formation and function. Annu Rev Plant Biol 56: 41–71.

40. Yu L, Jiang J, Zhang C, Jiang L, Ye N, et al. (2010) Glyoxylate rather than ascorbate is an efficient precursor for oxalate biosynthesis in rice. Journal of Experimental Botany 61: 1625–1634.

41. Foster J, Kim HU, Nakata PA (2012) A previously unknown oxalyl-CoA synthetase is important for oxalate catabolism in Arabidopsis. The Plant Cell Online 24: 1217–1229.

42. Ravanel S, Block MA, Rippert P, Jabrin S, Curien G, et al. (2004) Methionine metabolism in plants: chloroplasts are autonomous for de novo methionine synthesis and can import S-adenosylmethionine from the cytosol. J Biol Chem 279: 22548–22557.

43. Wei Z, Sun K, Sandoval FJ, Cross JM, Gordon C, et al. (2013) Folate polyglutamylation eliminates dependence of activity on enzyme concentration in mitochondrial serine hydroxymethyltransferases from Arabidopsis thaliana. Arch Biochem Biophys 536: 87–96.

44. Anguera MC, Suh JR, Ghandour H, Nasrallah IM, Selhub J, et al. (2003) Methenyltetrahydrofolate synthetase regulates folate turnover and accumulation. J Biol Chem 278: 29856–29862.

45. Roje S, Janave MT, Ziemak MJ, Hanson AD (2002) Cloning and characterization of mitochondrial 5-formyltetrahydrofolate cycloligase from higher plants. J Biol Chem 277: 42748–42754.

46. Li J, Wang X, Qin T, Zhang Y, Liu X, et al. (2011) MDP25, a novel calcium regulatory protein, mediates hypocotyl cell elongation by destabilizing cortical microtubules in Arabidopsis. Plant Cell 23: 4411–4427.

47. Reed LS, Archer MC (1979) Action of sodium nitrite on folic acid and tetrahydrofolic acid. Journal of Agricultural and Food Chemistry 27: 995–999.

48. Murashige T, Skoog F (1962) A revised medium for rapid growth and bio assays with tobacco tissue cultures. Physiologia plantarum 15: 473–497.

49. Francisco P, Li J, Smith SM (2010) The gene encoding the catalytically inactive beta-amylase BAM4 involved in starch breakdown in Arabidopsis leaves is expressed preferentially in vascular tissues in source and sink organs. J Plant Physiol 167: 890–895.

50. Cowling RJ, Harberd NP (1999) Gibberellins control Arabidopsis hypocotyl growth via regulation of cellular elongation. Journal of Experimental Botany 50: 1351–1357.

51. Renault H, El Amrani A, Palanivelu R, Updegraff EP, Yu A, et al. (2011) GABA accumulation causes cell elongation defects and a decrease in expression of genes encoding secreted and cell wall-related proteins in Arabidopsis thaliana. Plant Cell Physiol 52: 894–908.

52. Andrew KN, Worsfold PJ, Comber M (1995) On-line flow injection monitoring of ammonia in industrial liquid effluents. Analytica Chimica Acta 314: 33–43.

53. Roessner U, Luedemann A, Brust D, Fiehn O, Linke T, et al. (2001) Metabolic profiling allows comprehensive phenotyping of genetically or environmentally modified plant systems. Plant Cell 13: 11–29.

54. Takahashi M, Sasaki Y, Ida S, Morikawa H (2001) Nitrite reductase gene enrichment improves assimilation of NO(2) in Arabidopsis. Plant physiology 126: 731–741.

55. O'Neal D, Joy KW (1973) Glutamine synthetase of pea leaves. I. Purification, stabilization, and pH optima. Arch Biochem Biophys 159: 113–122.

56. Lancien M, Martin M, Hsieh MH, Leustek T, Goodman H, et al. (2002) Arabidopsis glt1-T mutant defines a role for NADH-GOGAT in the non-photorespiratory ammonium assimilatory pathway. Plant J 29: 347–358.

15

Insect Attraction versus Plant Defense: Young Leaves High in Glucosinolates Stimulate Oviposition by a Specialist Herbivore despite Poor Larval Survival due to High Saponin Content

Francisco R. Badenes-Perez[1,2]*, Jonathan Gershenzon[3], David G. Heckel[1]

1 Department of Entomology, Max Planck Institute for Chemical Ecology, Jena, Germany, **2** Instituto de Ciencias Agrarias, Consejo Superior de Investigaciones Científicas, Madrid, Spain, **3** Department of Biochemistry, Max Planck Institute for Chemical Ecology, Jena, Germany

Abstract

Glucosinolates are plant secondary metabolites used in plant defense. For insects specialized on Brassicaceae, such as the diamondback moth, *Plutella xylostella* L. (Lepidoptera: Plutellidae), glucosinolates act as "fingerprints" that are essential in host plant recognition. Some plants in the genus *Barbarea* (Brassicaceae) contain, besides glucosinolates, saponins that act as feeding deterrents for *P. xylostella* larvae, preventing their survival on the plant. Two-choice oviposition tests were conducted to study the preference of *P. xylostella* among *Barbarea* leaves of different size within the same plant. *P. xylostella* laid more eggs per leaf area on younger leaves compared to older ones. Higher concentrations of glucosinolates and saponins were found in younger leaves than in older ones. In 4-week-old plants, saponins were present in true leaves, while cotyledons contained little or no saponins. When analyzing the whole foliage of the plant, the content of glucosinolates and saponins also varied significantly in comparisons among plants that were 4, 8, and 12 weeks old. In *Barbarea* plants and leaves of different ages, there was a positive correlation between glucosinolate and saponin levels. This research shows that, in *Barbarea* plants, ontogenetical changes in glucosinolate and saponin content affect both attraction and resistance to *P. xylostella*. Co-occurrence of a high content of glucosinolates and saponins in the *Barbarea* leaves that are most valuable for the plant, but are also the most attractive to *P. xylostella*, provides protection against this specialist herbivore, which oviposition behavior on *Barbarea* seems to be an evolutionary mistake.

Editor: William J. Etges, University of Arkansas, United States of America

Funding: This research was supported by the Max Planck Society (http://www.mpg.de/en) and the Spanish Ministry of Science and Innovation (grant#AGL2010-18151) (http://www.idi.mineco.gob.es/). The funders had no role in study design, data collection and analysis, decision to publish, or preparation of the manuscript.

Competing Interests: The authors have declared that no competing interests exist.

* E-mail: frbadenes@ica.csic.es

Introduction

According to the optimal defense theory, the most valuable parts of a plant should also be the most protected against herbivores [1,2]. Young leaves are supposed to be more valuable than older ones because they can make a higher contribution to the fitness of the plant as a result of having relatively higher photosynthetic potential [3]. In agreement with this theory, it has been found that, within a plant, different organs and leaves can contain different concentrations of defense compounds [4,5]. This the case for glucosinolates, plant secondary metabolites used for defense and found mainly in plants of the order Brassicales [6,7], which have been found in higher concentrations in younger compared to older leaves within the same plant [4,8–11]. At the whole plant level, glucosinolate content also changes over time, but not in a linear manner [4,8,12]. Like glucosinolates, saponins are plant secondary metabolites used in plant defense [13–15]. With the exception of insects specialized on saponin-rich plants [16], saponins act as feeding-deterrents and are toxic [17–19]. We have not found any studies addressing changes in saponin content with leaf age in Brassicaceae, but saponin content in leaves has

been shown to decrease with leaf age in American holly, *Ilex opaca* Aiton (Aquifoliaceae) [20]. Saponin content also changes over time at the whole plant level, often increasing with plant age, although decreases with plant age and seasonal fluctuations have also been recorded [21–23]. In the genus *Barbarea* R. Br. (Brassicaceae), the only one in Brassicaceae where saponins have been found so far [24–28], seasonal fluctuations in saponin content seem to occur as inferred by changes in resistance to the flea beetle *Phyllotreta nemorum* L. (Coleoptera: Chrysomelidae) [29,30].

The diamondback moth, *Plutella xylostella* L. (Lepidoptera: Plutellidae), is an insect herbivore specialized on glucosinolate-containing crucifers [31–33]. Specialist insects like *P. xylostella* have evolved mechanisms to avoid the toxicity of glucosinolates, which are used in host plant recognition and act as feeding and oviposition stimulants [7,33–38]. Larvae of *P. xylostella* cannot survive on some varieties and types of *B. vulgaris* despite these plants being highly preferred for oviposition by *P. xylostella* adults [24,25,28,39–42]. This oviposition mistake of *P. xylostella* has been investigated in pest management to use *Barbarea* plants as a dead-end trap crop for *P. xylostella*, which is considered one of the most damaging insect pests of cruciferous crops throughout the world

[28,32,40,43]. Among the varieties and types of *B. vulgaris* on which *P. xylostella* cannot survive are *B. vulgaris* var. *variegata* and G-type (glabrous) *B. vulgaris* var. *arcuata*, while P-type (pubescent) *B. vulgaris* var. *arcuata* allows survival of *P. xylostella* larvae [24,25]. The resistance of *B. vulgaris* to *P. xylostella* is caused by the triterpenoid saponins 3-*0*-β-cellobiosylhederagenin (saponin 1) and 3-*0*-β-cellobiosyloleanolic acid (saponin 2), which act as feeding-deterrents or are correlated with deterrency in *P. xylostella* larvae [24,25]. Other *B. vulgaris* types and other *Barbarea* spp. containing saponins 1 and 2 are also resistant to *P. xylostella* [28].

Adult oviposition and larval feeding preference for younger over older leaves of a particular host plant is a common trend among many herbivorous insects, especially in specialists, including *P. xyllostella* [9,44,45]. Within the same plant, two-choice oviposition preference tests have shown that *P. xylostella* prefers to oviposit on younger leaves of <3.0 maximum leaf diameter compared to older leaves of >6.0 maximum leaf diameter of *B. vulgaris* [44]. Given the type of rosette growth of *Barbarea* plants, at the age of the plants used in the study by Badenes-Perez et al (2005) and here, leaf size was correlated with leaf age. Although containing lower content of toxic metabolites than younger leaves, older leaves may be less nutritious for insects than younger leaves [46,47]. Feeding on older leaves can also increase insect mandibular wear more than feeding on younger leaves because of the increased toughness of older leaves [48]. At the whole plant level, the plant phenological age hypothesis also predicts that herbivores prefer and perform better on younger compared to older plants [49]. However, there are many cases in which insects prefer older plants over younger ones [50,51]. Among *Brassica oleracea* L. and *B. vulgaris* plants aged between 6 and 14 weeks old, *P. xylostella* also preferred to oviposit on older versus younger plants, even though survival and development of *P. xylostella* larvae can be negatively affected by plant age [52].

In relation to the preference of *P. xylostella* for younger leaves and older plants of *B. vulgaris* [44,52], it is not known whether there could be an association between this preference and plant concentrations of glucosinolates and saponins, the former being oviposition and feeding stimulants, and the latter preventing the survival of the insect on the plant. We hypothesize that, given the known attraction of *P. xylostella* to glucosinolates, if glucosinolate content in *Barbarea* leaves is higher in younger compared to older leaves as it happens in other Brassicaceae, *P. xylostella* would preferentially oviposit on young leaves. It is not known how saponins vary with leaf and plant age in *Barbarea* spp. and whether they are correlated with changes in glucosinolate content. A correlation between plant content of glucosinolates and saponins could protect *Barbarea* plants from specialist insects adapted to glucosinolates. Furthermore, as both glucosinolates and saponins are plant defense compounds, their co-occurrence would have implications for the protection of *Barbarea* plants, not only against *P. xylostella*, but against other herbivores as well. To test our hypotheses, we conducted two-choice oviposition preference tests and measured glucosinolate and saponin concentrations in *Barbarea* leaves of different size to test the association between leaf size, oviposition preference, and glucosinolate and saponin concentrations. We also measured glucosinolate and saponin content in plants of different age. Besides analyzing true leaves, we analyzed the glucosinolate and saponin content of cotyledons. Larval survival and oviposition preference tests were also conducted with cotyledons and true leaves within the same plant.

Results

Analysis of Glucosinolates and Saponins in Barbarea spp

A significant negative relationship was found between leaf size and content of glucosinolates for both G-type ($y = 13.01 - 0.95x$; $n = 100$; $r = 0.39$; $F_{1,98} = 17.01$; $P \leq 0.001$) and P-type *B. vulgaris* ($y = 14.81 - 0.86x$; $n = 20$; $r = 0.44$; $F_{1,19} = 4.20$; $P = 0.050$) (Figs. 1A and 1B). In G-type *B. vulgaris*, the glucosinolate that decreased the most with increasing leaf size was the dominant glucosinolate (*S*)-2-hydroxy-2-phenylethylglucosinolate (S2OH2PE) ($y = 11.95 - 0.89x$; $n = 100$; $r = 0.38$; $F_{1,98} = 16.13$; $P \leq 0.001$), but concentrations of (*R*)-2-hydroxy-2-phenylethylglucosinolate (R2OH2PE) ($y = 0.26 - 0.17x$; $n = 100$; $r = 0.31$; $F_{1,98} = 10.34$; $P = 0.002$), indol-3-ylmethylglucosinolate (I3M) ($y = 0.64 - 0.04x$; $n = 100$; $r = 0.42$; $F_{1,98} = 21.01$; $P \leq 0.001$), and 4-methoxyindol-3-ylmethylglucosinolate (4MOI3M) ($y = 0.09 - 0.01x$; $n = 100$; $r = 0.29$; $F_{1,98} = 8.93$; $P = 0.003$) also decreased with leaf size. In P-type *B. vulgaris*, only the dominant glucosinolate R2OH2PE ($y = 4.25 - 1.07x$; $n = 20$; $r = 0.46$; $F_{1,19} = 4.69$; $P = 0.044$) decreased significantly with leaf size; concentrations of the other glucosinolates found did not vary significantly with leaf size.

In G-type *B. vulgaris*, a significant negative relationship was also found between leaf size and content of saponin 1 ($y = 7.50 - 0.61x$; $n = 100$; $r = 0.51$; $F_{1,98} = 34.89$; $P \leq 0.001$) and saponin 2 ($y = 2.50 - 0.23x$; $n = 100$; $r = 0.47$; $F_{1,98} = 27.99$; $P \leq 0.001$) (Fig. 2). In these same leaves of different size, there was a significant positive relationship between saponin and glucosinolate content for both saponin 1 (ln (y+1) = $1.36 + 0.26$ln (x+1); $n = 100$; $r = 0.21$; $F_{1,98} = 4.56$; $P = 0.035$) and saponin 2 (ln (y+1) = $1.47 + 0.43$ln (x+1); $n = 100$; $r = 0.25$; $F_{1,98} = 6.44$; $P = 0.013$).

When leaves of different sizes were grouped into three groups according to maximum leaf width (large, >50 mm; medium, 20–50 mm; and small, <20 mm), there were significant differences in the content of glucosinolates ($F_{2,108} = 224.31$; $P \leq 0.001$) and saponins 1 ($F_{2,81} = 19.36$; $P \leq 0.001$) and 2 ($F_{2,81} = 9.06$; $P \leq 0.001$) among the three different leaf sizes (Tables 1 and 2). Small and large leaves had, respectively, the highest and the lowest concentrations of both glucosinolates and saponins. For these leaves of different size, there was a significant positive relationship between saponin and glucosinolate content for both saponin 1 ($y = 2.10 + 1.63*10^{-5}x$; $n = 30$; $r = 0.40$; $F_{1,28} = 5.41$; $P = 0.027$) and saponin 2 ($y = 1.97 + 1.02*10^{-5}x$; $n = 30$; $r = 0.45$; $F_{1,28} = 7.22$; $P = 0.012$) in G-type *B. vulgaris*; for saponin 1 ($y = 1.81 + 9.91*10^{-5}x$; $n = 30$; $r = 0.79$; $F_{1,28} = 54.55$; $P \leq 0.001$) in *B. verna* (the relationship was not statistically significant for saponin 2); and for saponin 1 ($y = 6.61 + 1.80*10^{-4}x$; $n = 30$; $r = 0.48$; $F_{1,28} = 8.29$; $P = 0.008$) in *B. rupicola*, which did not contain any saponin 2.

There were significant differences in the content of glucosinolates in cotyledons and true leaves of plants of G-type *B. vulgaris*, P-type *B. vulgaris*, NAS-type *B. vulgaris*, *B. vulgaris variegata*, *B. rupicola*, and *B. verna* ($F_{1,134} = 261.79$; $P \leq 0.001$) (Table 3). True leaves contained approximately 2.3, 7.4, 2.9, 3.0, 3.5, and 4.2 times more glucosinolates than cotyledons in G-type *B. vulgaris*, P-type *B. vulgaris*, NAS-type *B. vulgaris*, *B. vulgaris variegata*, *B. rupicola*, and *B. verna*, respectively. Present in higher concentrations in true leaves than in cotyledons were the individual glucosinolates S2OH2PE ($F_{1,134} = 51.45$; $P \leq 0.001$); R2OH2PE ($F_{1,134} = 44.22$; $P \leq 0.001$); I3M ($F_{1,134} = 18.63$; $P \leq 0.001$); and 2-phenylethylglucosinolate (2PE) ($F_{1,134} = 399.78$; $P \leq 0.001$). Concentrations of S2OH2PE were 2.6 and 3.3 times higher in true leaves than in cotyledons in G-type *B. vulgaris* and *B. vulgaris variegata*, respectively. Concentrations of R2OH2PE were up to 18.8 times higher in true leaves than in cotyledons in P-type *B. vulgaris*. In *B. verna*, concentrations of I3M were up to 4.8 times higher in true leaves than in

A

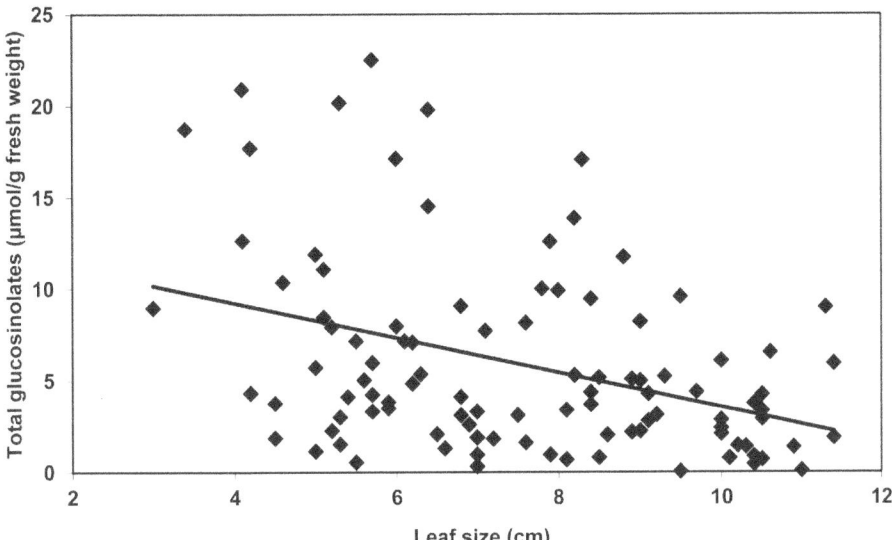

B

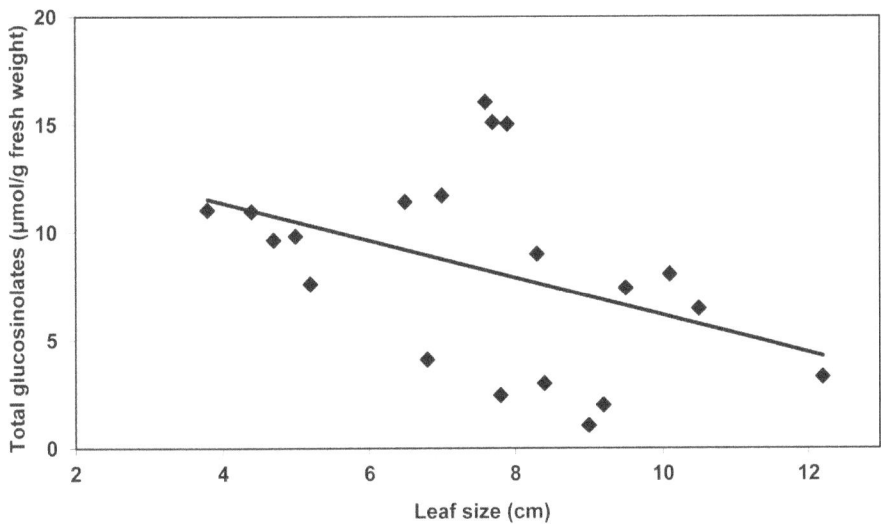

Figure 1. Total glucosinolates (μmol/g of plant fresh weight) in leaves of different size in G-type *Barbarea vulgaris* **var.** *arcuata* **(A) and P-type** *B. vulgaris* **var.** *arcuata* **(B).** For G- and P-type *B. vulgaris* leaves, maximum leaf width ranged from 3.0 to 11.4 cm (n = 100) and from 4.4 to 12.2 cm (n = 20), respectively.

cotyledons. In G-type *B. vulgaris*, concentrations of 2PE were up to 36.0 times higher in true leaves than in cotyledons. Concentrations of 4MOI3M, however, were lower in true leaves than in cotyledons ($F_{1,134} = 113.77$; $P \leq 0.001$). In *B. verna*, concentrations of 4MOI3M were up to 24.0 times lower in true leaves than in cotyledons.

There were significant differences in the content of saponins 1 ($F_{1,35} = 32.48$; $P \leq 0.001$) and 2 ($F_{1,35} = 5.49$; $P = 0.025$) in true leaves and cotyledons of plants of G-type *B. vulgaris* and *B. verna* (Table 4). No saponins were found in cotyledons. Similarly, when comparing true leaves and cotyledons of plants of NAS-type *B. vulgaris*, *B. vulgaris variegata*, and *B. rupicola*, we found significant differences in the content of saponins 1 ($F_{1,40} = 48.54$; $P \leq 0.001$)

and 2 ($F_{1,40} = 1400.14$; $P \leq 0.001$) (Table 4). No saponins were found in true leaves and cotyledons of *B. rupicola*. In NAS-type *B. vulgaris*, saponin 1 was found in all true leaves and in 3 out of 13 cotyledons analyzed (the peak areas of the signal of [M-H]⁻ were on average 9.7 times smaller for cotyledons than for true leaves in the plants which cotyledons contained saponins). In *B. vulgaris variegata*, saponin 1 was found in all true leaves and in 1 out of 5 cotyledons analyzed (the peak areas of the signal of [M-H]⁻ were 7.1 times smaller for cotyledons than for true leaves in the plant of *B. vulgaris variegata* which cotyledon contained saponins). Saponin 2 was not detected in cotyledons. In plants of G- and P-type *B. vulgaris*, *B. rupicola*, and *B. verna*, 5 h after cutting all leaves except either one cotyledon or one true leaf, there were no differences in

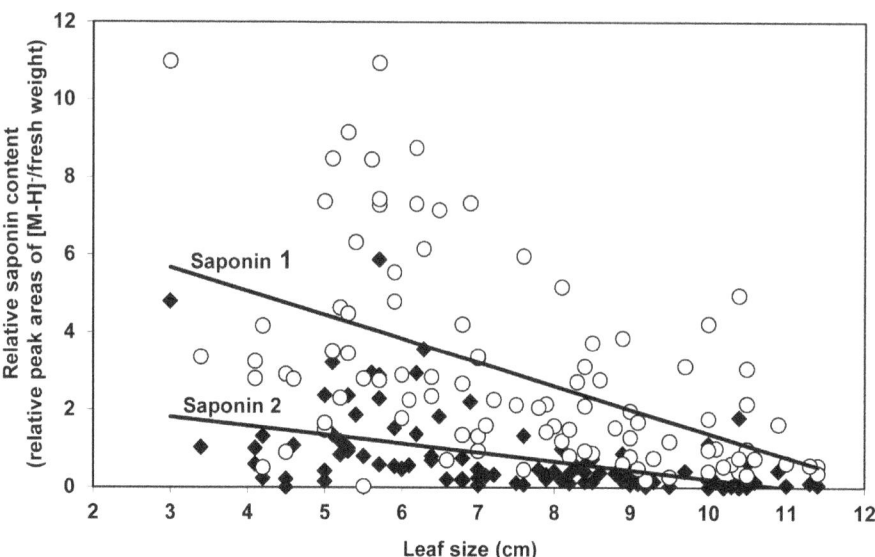

Figure 2. Relative content of 3-*O*-*β*-cellobiosylhederagenin (saponin 1) and 3-*O*-*β*-cellobiosyloleanolic acid (saponin 2) in leaves of different size in G-type *Barbarea vulgaris* var. *arcuata*. Maximum leaf width ranged from 3.0 to 11.4 cm (n = 100). Units of peak areas for the signal of the molecular ion in the negative-ion mass spectrum [M-H]⁻ divided by 100,000/mg of leaf fresh weight.

the content of glucosinolates ($F_{1,64}$ = 0.11; P = 0.743) (Table S1) and saponins 1 ($F_{1,48}$ = 0.42; P = 0.436) and 2 ($F_{1,48}$ = 0.52; P = 0.473) (Table S2) compared to the same type of leaves in intact plants with all their leaves remaining (Table S1). The glucosinolate content of *Barbarea* seeds is shown on Table 5. *Barbarea* seeds did not contain saponins 1 and 2.

When analyzing the whole plant foliage, there were significant differences in glucosinolate content among plants of different age in both G-type ($F_{2,27}$ = 10.70; P≤0.001) and P-type plants ($F_{2,27}$ = 56.29; P≤0.001) (Fig. 3). In both G- and P-type *B. vulgaris* plants, total glucosinolate content was highest in 8-week-old plants and lowest in 4-week-old plants. Among the individual glucosinolates in G-type *B. vulgaris*, those that varied the most with plant age were I3M ($F_{2,27}$ = 43.87; P≤0.001), S2OH2PE ($F_{2,27}$ = 7.58; P = 0.002) and 4MOI3M ($F_{2,27}$ = 4.49; P = 0.021), while 2PE ($F_{2,27}$ = 0.15; P = 0.865) and R2OH2PE ($F_{2,27}$ = 0.30; P = 0.741) did not show significant variation (Table 6). Among the individual glucosinolates that varied the most with plant age in P-type *B. vulgaris* were R2OH2PE ($F_{2,27}$ = 66.32; P≤0.001), I3M ($F_{2,27}$ = 30.78; P≤0.001), and 4MOI3M ($F_{2,27}$ = 4.63; P = 0.018), while 2PE ($F_{2,27}$ = 3.33; P = 0.051) and S2OH2PE ($F_{2,27}$ = 2.14; P = 0.138) did not vary significantly (Table 6). There were also significant differences in the content of saponins 1 ($F_{2,27}$ = 8.51; P = 0.001) and 2 ($F_{2,27}$ = 3.86; P = 0.034) among the G-type *B. vulgaris* plants of different age (Fig. 4). As in the case of total glucosinolate content, the content of saponins 1 and 2 was highest in 8-week-old plants and lowest in 4-week-old plants. For G-type *B. vulgaris* plants of different ages, there was also a significant positive relationship between saponin and glucosinolate content for both saponin 1 (y = 1.14+30.35x; n = 28; r = 0.65; $F_{1,26}$ = 19.05; P≤0.001) and saponin 2 (y = 1.90+62.90x; n = 28; r = 0.68; $F_{1,26}$ = 22.12; P≤0.001).

Oviposition Preference Tests between Leaves of Different Size within the Same Plant

There was a significant negative relationship between leaf size and number of eggs laid by *P. xylostella* per leaf area (y = 1.84–0.17x; n = 84; r = 0.58; $F_{1,83}$ = 42.15; P≤0.001) (Fig. 5A). The

number of eggs laid per leaf area was also positively correlated with leaf glucosinolate content (y = 0.32+0.04x; n = 84; r = 0.34; $F_{1,83}$ = 10.40; P = 0.002) (Fig. 5B). In the case of *P. xylostella* oviposition preference between true leaves and cotyledons within the same plant, significantly more eggs were laid on true leaves than on cotyledons ($F_{1,18}$ = 12.62; P = 0.002). When considering the numbers of eggs laid by *P. xylostella* per leaf area, however, these differences were not statistically significant ($F_{1,18}$ = 0.47; P = 0.502) (Table 7).

Survival of Larvae on True Leaves and Cotyledons within the Same Plant

On true leaves of plants, 100% and 80% of *P. xylostella* larvae survived after 5 days on P-type *B. vulgaris* and *B. rupicola*, respectively (for each *Barbarea* tested, n = 5 plants, each with 5 larvae). No larvae survived the 5-day period on true leaves of G-type *B. vulgaris*, *B. vulgaris variegata*, NAS-type *B. vulgaris*, and *B. verna*. On cotyledons, however, survival of *P. xylostella* larvae after 5 days was high for all *Barbarea* plants tested: 100%, 100%, 80%, 100%, 100% and 100% on G-type *B. vulgaris*, P-type *B. vulgaris*, *B. vulgaris variegata*, NAS-type *B. vulgaris*, *B. rupicola*, and *B. verna*, respectively.

Discussion

Our research shows that in *Barbarea*, the only genus of the Brassicaceae family known to simultaneously contain glucosinolates and saponins, content of these two plant defense compounds are negatively correlated with leaf size. Oviposition preference by *P. xylostella* was also negatively correlated with leaf size because *P. xylostella* laid more eggs per leaf area on smaller leaves than on larger ones. *P. xylostella* and many other herbivores use plant secondary metabolites as "fingerprints" to recognize hosts and oviposit on them [34,53]. In our study, attraction to glucosinolates seemed to be more important for ovipositing *P. xylostella* than avoidance of saponins, which is consistent with the presence and absence of glucosinolates and saponins, respectively, on the leaf surface of *Barbarea* in concentrations perceivable by *P. xylostella*

Table 1. Mean ± SE glucosinolates (μmol/g of leaf fresh weight) concentrations in large, medium and small leaves of *B. rupicola*, *B. verna*, and G- and P- type *B. vulgaris* var. *arcuata*.

	Leaf size	Total glucosinolates	R2OH2PE	S2OH2PE	I3M	4MOI3M	2PE
B. rupicola	large	1.44±0.10	0.00±0.00	0.00±0.00	0.02±0.00	0.01±0.00	1.40±0.10
B. rupicola	medium	9.04±0.86	0.00±0.00	0.00±0.00	0.43±0.04	0.02±0.00	8.59±0.82
B. rupicola	small	14.04±0.95	0.00±0.00	0.00±0.00	0.93±0.06	0.03±0.01	13.07±0.91
B. verna	large	2.46±0.15	0.00±0.00	0.00±0.00	0.03±0.00	0.01±0.00	2.42±0.15
B. verna	medium	5.54±0.66	0.00±0.00	0.00±0.00	0.10±0.02	0.01±0.00	5.43±0.65
B. verna	small	13.02±1.07	0.00±0.00	0.00±0.00	0.43±0.04	0.03±0.01	12.57±1.03
G-type *B.vulgaris*	large	0.36±0.05	0.00±0.00	0.31±0.05	0.05±0.01	0.00±0.00	0.00±0.00
G-type *B.vulgaris*	medium	2.83±0.44	0.00±0.00	2.40±0.37	0.42±0.08	0.00±0.00	0.01±0.00
G-type *B.vulgaris*	small	6.09±0.62	0.00±0.00	4.40±0.43	1.34±0.22	0.01±0.00	0.34±0.07
P-type *B.vulgaris*	large	0.74±0.16	0.63±0.16	0.00±0.00	0.10±0.01	0.00±0.00	0.00±0.00
P-type *B.vulgaris*	medium	4.98±0.96	4.53±0.88	0.00±0.00	0.43±0.09	0.00±0.00	0.02±0.00
P-type *B.vulgaris*	small	11.20±0.64	9.92±0.55	0.00±0.00	1.21±0.11	0.01±0.00	0.06±0.01

Abbreviations for glucosinolates are: (*R*)-2-hydroxy-2-phenylethylglucosinolate (R2OH2PE), (*S*)-2-hydroxy-2-phenylethylglucosinolate (S2OH2PE), indol-3-ylmethylglucosinolate (I3M), 4-methoxyindol-3-ylmethylglucosinolate (4MOI3M), and 2-phenylethylglucosinolate (2PE).
Large, medium and small leaves had a maximum leaf width of >50 mm, 20–50 mm, and <20 mm, respectively. For each plant and leaf size n = 10.

Table 2. Mean $\pm$ SE 3-0-β-cellobiosylhederagenin (saponin 1) and 3-0-β-cellobiosyloleanolic acid (saponin 2) in large, medium, and small leaves within the same plant.

	Type of leaf	Saponin 1	Saponin 2
B. rupicola	large	$(1.1\pm1.1)*10^2$	0.00 ± 0.00
B. rupicola	medium	$(7.4\pm3.3)*10^3$	0.00 ± 0.00
B. rupicola	small	$(1.9\pm0.7)*10^4$	0.00 ± 0.00
B. verna	large	$(1.5\pm0.3)*10^4$	$(4.5\pm3.2)*10^2$
B. verna	medium	$(5.3\pm1.1)*10^4$	$(2.1\pm1.1)*10^3$
B. verna	small	$(9.0\pm0.9)*10^4$	0.00 ± 0.00
G-type *B.vulgaris*	large	$(1.2\pm0.3)*10^4$	$(4.4\pm1.4)*10^3$
G-type *B.vulgaris*	medium	$(1.2\pm0.3)*10^5$	$(7.4\pm1.8)*10^4$
G-type *B.vulgaris*	small	$(2.0\pm0.4)*10^5$	$(1.1\pm0.2)*10^5$

Large, medium and small leaves had a maximum leaf width of >50 mm, 20–50 mm, and <20 mm, respectively. For each plant and leaf type n = 10. Saponin concentrations given as peak areas for the signal of the molecular ion in the negative-ion mass spectrum [M-H]$^{-}$/mg of leaf fresh weight.

[34]. In *Barbarea* plants with saponins, these were found in true leaves of all sizes, while no saponins (or very small amounts of them) were found in cotyledons. Larvae of *P. xylostella* could survive on cotyledons, even in those *Barbarea* plants whose true leaves contained enough saponin concentrations to prevent their survival.

In Lepidoptera, survival is greatly determined by the oviposition behavior of adult females, as immature stages have limited mobility [54]. Consequently, most ovipositing Lepidoptera prefer to oviposit on hosts where their larvae are able to survive, but there are cases in which the correlation between oviposition preference and larval performance is poor, and several hypotheses have been put forward to interpret this apparently non-adaptive behavior [55–57]. With few exceptions, such as ovipositing on *Barbarea* [28,39,40], *P. xylostella* oviposition preference and larval performance are positively correlated [58]. However, the oviposition preference for smaller *Barbarea* leaves over larger ones demonstrated here for *P. xylostella* seems to be a non-adaptive mechanical response to cues given by plant secondary metabolites (glucosinolates) specific from their cruciferous host plants. Given that no *P. xylostella* larvae survive on resistant *Barbarea*, and that survival of larvae is less likely on small leaves that contain high concentrations of saponins, there cannot be any selective advantage in the oviposition behavior of *P. xylostella* on *Barbarea*. The relatively low content of saponins in larger leaves of *Barbarea* would make *P. xylostella* more likely to survive on the plant, yet larger leaves also have relatively low concentrations of glucosinolates, which make them less stimulatory for *P. xylostella* larvae [35]. The preference of *P. xylostella* moths for younger leaves within a *Barbarea* plant represents a second "oviposition mistake", on top of the known "oviposition mistake" of *P. xylostella* preferring resistant *Barbarea* plants over other host plants that allow survival of its larvae [28,40–42].

Cotyledons serve as a storage of nutrients for the growing plant and they are the first photosynthetic tissue appearing above-ground when germination occurs [59]. Cotyledons of brassicaceous plants usually contain variable amounts of glucosinolates [8,60]. In *Barbarea* plants, glucosinolates, which can defend the plants against generalist herbivores, were present in cotyledons, but saponins, which could also protect the plant against specialist herbivores like *P. xylostella*, were not (or were present in very low concentrations). No saponin 2 was detected in cotyledons, indicating that synthesis of saponin 2 could be subsequent to that

of saponin 1 (assuming that saponins are not translocated from cotyledons to other parts of the plant). Unlike glucosinolates, saponins were not found in the seeds of *Barbarea*, indicating that saponins may start being produced in cotyledons and true leaves after some time once true leaves appear. Lack of saponins in seeds and cotyledons indicates that, for some time, at the seedling stage, the plant may not be protected against *P. xylostella* and other herbivores. However, given the small size of cotyledons, they do not provide sufficient food for a *P. xylostella* larva to develop from first instar to pupa (Badenes-Perez, personal observation). Even though *P. xylostella* is known to oviposit on cotyledons of crucifer seedlings in the field, upon egg hatch, larvae move to true leaves, where they prefer to feed [61]. With the exception of 4MOI3M, we found that cotyledons contained lower concentrations of glucosinolates than true leaves in the same plant. This, together with the low frequency of cotyledons with saponins, the low concentrations of saponins found in those cotyledons with saponins, and the ensuing survival of *P. xylostella* larvae on cotyledons, indicates that cotyledons are not as protected from herbivory as true leaves. However, as *Barbarea* spp. are early successional biennial plants that appear early in the season [62], cotyledons might be important for the plant only for a relatively short time, when the presence of herbivores and the visibility of the plant as a seedling may be relatively low. This would also be in agreement with the plant apparency hypothesis [63].

Our analyses show that total glucosinolate content in 4-, 8-, and 12-week-old plants varies, but not in the same linear manner as oviposition preference by *P. xylostella* varies among plants of similar ages as described by Badenes-Perez et al (2005). Non-linear ontogenetic changes in the content of defense compounds have been interpreted as part of a dynamic pattern, also affected by the development of herbivore tolerance and resource allocation constraints in the plant [59]. Besides plant glucosinolate content, the increase in leaf area and leaf number that occurs with plant age may affect oviposition preference by *P. xylostella* [52]. In *B. vulgaris*, the increase in number of leaves and total leaf area when comparing 6- and 12-week-old plants was 11.6 and 42.2 fold, respectively [52].

The simultaneous presence of high content of glucosinolates and saponins in small/young leaves of *Barbarea*, which are the most valuable for the plant, but also the most attractive to ovipositing *P. xylostella*, provides protection against this specialist herbivore. The association between glucosinolates and saponins could indicate

Table 3. Mean ± SE glucosinolates (μmol/g of leaf fresh weight) concentrations in cotyledons and true leaves in plants of *B. rupicola*, *B. verna*, G- and P-type *B. vulgaris* var. *arcuata*, NAS-type *B. vulgaris*, and *B. vulgaris* var. *variegata*.

	Type of leaf	Total glucosinolates	S2OH2PE	R2OH2PE	I3M	4MOI3M	2PE
G-type *B.vulgaris*	cotyledon	1.439±0.147	1.147±0.120	0.010±0.005	0.262±0.036	0.018±0.004	0.001±0.001
G-type *B.vulgaris*	true leaf	3.320±0.410	3.046±0.384	0.004±0.003	0.231±0.029	0.002±0.001	0.036±0.014
P-type *B.vulgaris*	cotyledon	0.649±0.112	0.000±0.000	0.352±0.071	0.270±0.050	0.014±0.002	0.013±0.006
P-type *B.vulgaris*	true leaf	4.860±0.643	0.000±0.000	4.444±0.620	0.399±0.061	0.001±0.000	0.016±0.005
NAS-type *B.vulgaris*	cotyledon	0.329±0.066	0.000±0.000	0.016±0.002	0.164±0.041	0.008±0.001	0.141±0.028
NAS-type *B.vulgaris*	true leaf	0.956±0.135	0.000±0.000	0.023±0.005	0.181±0.020	0.002±0.001	0.750±0.125
B. rupicola	cotyledon	1.985±0.092	0.000±0.000	0.010±0.002	0.066±0.009	0.019±0.002	1.889±0.087
B. rupicola	true leaf	7.042±0.415	0.000±0.000	0.012±0.003	0.199±0.017	0.003±0.001	6.827±0.405
B.vulgaris variegata	cotyledon	1.052±0.385	0.857±0.336	0.000±0.000	0.147±0.064	0.007±0.002	0.042±0.020
B.vulgaris variegata	true leaf	3.188±0.773	2.841±0.676	0.002±0.002	0.256±0.097	0.002±0.000	0.087±0.023
B. verna	cotyledon	2.061±0.088	0.000±0.000	0.005±0.002	0.053±0.006	0.024±0.002	1.979±0.083
B. verna	true leaf	8.662±0.509	0.000±0.000	0.012±0.003	0.255±0.017	0.001±0.000	8.385±0.501

Abbreviations for glucosinolates are: (*R*)-2-hydroxy-2-phenylethylglucosinolate (R2OH2PE), (*S*)-2-hydroxy-2-phenylethylglucosinolate (S2OH2PE), indol-3-ylmethylglucosinolate (I3M), 4-methoxyindol-3-ylmethylglucosinolate (4MOI3M), and 2-phenylethylglucosinolate (2PE).

For true leaves, the largest true leaf of each plant was taken. For each plant and leaf type, replication was: n = 15 for G- and P-type *B. vulgaris* and *B. rupicola*, n = 13 for NAS-type *B. vulgaris*, n = 10 for *B. verna*, and n = 5 for *B. vulgaris* var. *variegata*.

Table 4. Mean ± SE 3–0-β-cellobiosylhederagenin (saponin 1) and 3–0-β-cellobiosyloleanolic acid (saponin 2) in cotyledons and true leaves in plants of *B. rupicola*, *B. verna*, G-type *B. vulgaris* var. *arcuata*, NAS-type *B. vulgaris*, and *B. vulgaris* var. *arcuata*.

	Type of leaf	Saponin 1	Saponin 2
G-type *B.vulgaris*	cotyledon	0.00±0.00	0.00±0.00
G-type *B.vulgaris*	true leaf	0.15±0.04	0.04±0.06
B. rupicola	cotyledon	0.00±0.00	0.00±0.00
B. rupicola	true leaf	0.00±0.00	0.00±0.00
B. verna	cotyledon	0.00±0.00	0.00±0.00
B. verna	true leaf	0.05±0.01	0.00±0.00
NAS-type *B.vulgaris*	cotyledon	$(1.3±0.8)*10^4$	0.00±0.00
NAS-type *B.vulgaris*	true leaf	$(4.0±0.6)*10^5$	$(1.2±0.2)*10^5$
B. vulgaris variegata	cotyledon	$(2.0±2.0)*10^4$	0.00±0.00
B. vulgaris variegata	true leaf	$(4.9±0.9)*10^5$	$(1.3±0.5)*10^5$

Traces of saponin 2 (less than 0.01 μmol/g plant fresh weight) were also detected in *B. verna*.
For true leaves, the largest true leaf of each plant was taken. For each plant and leaf type and treatment: n = 10 for G-type *B.vulgaris* and *B. verna*, n = 13 for NAS-type *B. vulgaris*, and n = 5 for *B. rupicola* and *B. vulgaris variegata*. Saponin concentrations given as μmol/g of leaf fresh weight for G-type *B. vulgaris* var. *arcuata* and *B. verna* and as peak areas for the signal of the molecular ion in the negative-ion mass spectrum $[M-H]^{-}$/mg of leaf fresh weight for *B. rupicola*, NAS-type *B. vulgaris*, and *B. vulgaris* var. *arcuata*. Analyses were also conducted with P-type *B. vulgaris* var. *arcuata*, which did not have any saponins 1 and 2.

that, from an evolutionary point of view, in *Barbarea*, saponins might have appeared after glucosinolates, enabling plants to be defended against insects that had adapted to glucosinolate-defended plants. Saponins would then be what has been called a "second line of defense", appearing as a response to herbivores that have overcome the "first line of defense" provided by glucosinolates [25,64].

Materials and Methods

Ethics Statement

Insects collected in Kenya were collected at Athi River, 40 km southeast of Nairobi, Kenya, in 2005 by Dr. Bernhard Löhr, and sent by him in July 2005 to MPICE in Jena under EU permit number EG-D-TH1-390390 AG39/2005. Insects collected in Spain were collected in 2013 in Arganda del Rey, Madrid, at the experimental farm "La Poveda", which belongs to the Institute of Agricultural Sciences (CSIC). A permit was not required for the collection of insects at the collecting site in Spain.

Plant Growth and Insect Culture

Experiments were conducted in the laboratory at the Max Planck Institute for Chemical Ecology in Jena, Germany. *Barbarea rupicola* Moris, *B. verna* (Mill.) Asch., and *B. vulgaris* var. *variegata* seeds were purchased from B & T World Seeds (Aigues-Vives, France). *Barbarea vulgaris* var. *arcuata* G-type seeds were purchased from Rieger-Hofmann GmbH (Blaufelden-Raboldshausen, Germany) and P-type seeds were collected in Tissø (Denmark) and donated to us by Dr. Jens K. Nielsen. Seeds of NAS-type *B. vulgaris* were collected in The Netherlands and donated to us by Dr. Hanneke van Leur. The NAS-type of *B. vulgaris* was not classified varietally, although morphologically they would belong to what botanists consider var. *arcuata* or var. *vulgaris* [65]. Additional G-type *Barbarea vulgaris* var. *arcuata* seeds from Jena (Germany) were provided by Dr. Tamara Krügel. All plants used in the experiments were grown in pots in the greenhouse using a substrate of peat moss with clay. In the experiments testing the effect of different leaf size on glucosinolate and saponin content, plants were approximately 10 weeks old at the time when the experiments were conducted and they were grown in 20-cm-diameter pots. To compare differences in glucosinolate and saponin content between true leaves and cotyledons, 4-week-old

Table 5. Mean ± SE glucosinolates (μmol/g of seed) in seeds of *B.verna* (n = 4), G-type *B. vulgaris* var. *arcuata* (n = 2), P-type *B. vulgaris* var. *arcuata* (n = 4), and *B. vulgaris* var. *variegata* (n = 2).

Glucosinolates	*B. verna*	*B. vulgaris* G-type	*B. vulgaris* P-type	*B. vulgaris* variegata
Total	76.73±3.91	31.65±0.59	60.32±7.94	53.64±0.08
R2OH2PE	0.02±0.01	0.56±0.01	53.51±4.78	1.18±0.01
S2OH2PE	0.00±0.00	30.69±0.53	0.00±0.00	43.21±0.18
I3M	0.09±0.01	0.15±0.01	0.22±0.01	0.18±0.01
2PE	76.62±3.90	0.25±0.04	6.59±3.46	9.08±0.08

Abbreviations for glucosinolates are: (R)-2-hydroxy-2-phenylethylglucosinolate (R2OH2PE), (S)-2-hydroxy-2-phenylethylglucosinolate (S2OH2PE), indol-3-ylmethylglucosinolate (I3M), and 2-phenylethylglucosinolate (2PE).
Saponins 1 and 2 were not found in the seeds.

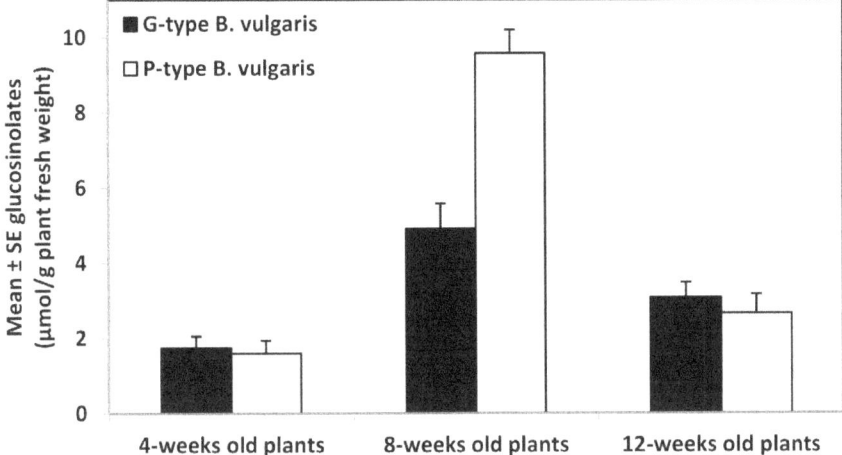

Figure 3. Mean ± SE glucosinolates (μmol/g of plant fresh weight) in the whole foliage of G- and P-type *Barbarea vulgaris* var. *arcuata* plants of 4-, 8-, and 12-weeks old plants (n = 10).

plants were used and they were grown in 8-cm-diameter pots. Glucosinolate and saponin content was also compared among plants that were 4, 8, and 12 weeks old and grown in 15-cm-diameter pots. All plants used in the experiments were grown in the glasshouse at 22–28°C under 16 h supplemental light from Master Sun-T PIA Agro 400 or Master Sun-T PIA Plus 600 W Na lights (Philips, Turnhout, Belgium). *P. xylostella* used in the experiments were either collected in Kenya (provided by Dr. Bernhard Löhr) or collected in Spain. Insects were later reared on cabbage plants in a climate-controlled chamber (16:8 h light:dark, 21±2°C and 55±5 RH).

Analysis of Glucosinolates and Saponins in Barbarea spp

Glucosinolate and saponin content in individual cotyledons, in individual true leaves, in foliage of whole plants, and in seeds was determined as in Badenes-Perez et al. (2010). Cotyledons and leaves were cut approximately from the middle of their petiole. In the experiment comparing cotyledons and true leaves, the first (largest) true leaf in 4-week-old plants was used in the analyses.

Foliage of whole plants was harvested by cutting approximately half of the plant from the crown in the case of the comparison of foliage among plants 4, 8, and 12 weeks old. In the experiment testing whether changes in glucosinolate and saponin content occurred when only one cotyledon or one true leaf was left per plant, all the leaves of the plant, except either one cotyledon or the largest true leaf, were cut with scissors. The content of glucosinolates and saponins in the remaining cotyledon or true leaf was compared to a control of similar leaves in intact plants (having no leaves cut) 5 hours after removal of leaves. This experiment was arranged to assess whether an associated experiment conducted to test *P. xylostella* oviposition preference between true leaves and cotyledons could be influenced by glucosinolate and saponin induction at the time as a result of the mechanical removal of leaves. The time period of 5 hours to study changes in glucosinolate and saponin content in the plants was set to coincide with the time during which *P. xylostella* had laid most of its eggs in the course of the experiment [31,66]. For glucosinolate and saponin analysis in *Barbarea* seeds, 20 mg of seeds were

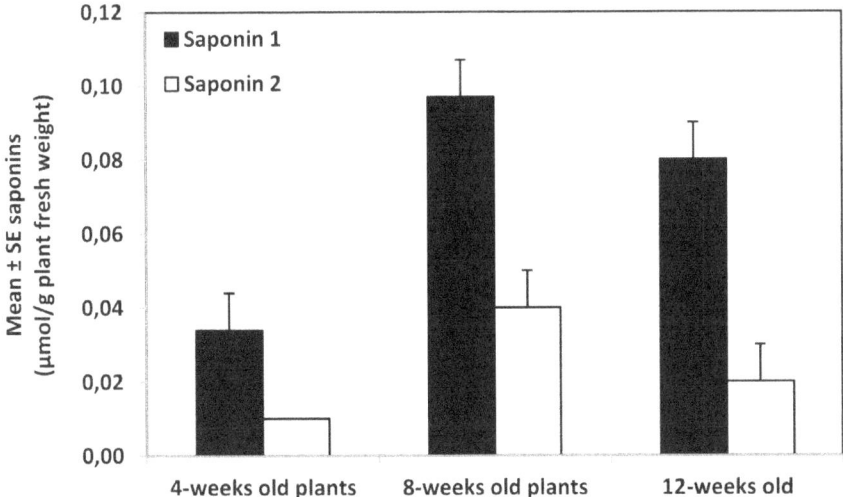

Figure 4. Mean ± SE 3-*O*-β-cellobiosylhederagenin (saponin 1) and 3-*O*-β-cellobiosyloleanolic acid (saponin 2) in the whole foliage of G-type *Barbarea vulgaris* var. *arcuata* plants of 4-, 8-, and 12-weeks old plants (n = 10).

Table 6. Mean ± SE glucosinolates (μmol/g of plant fresh weight) in foliage of G- and P- type *B. vulgaris* var. *arcuata* plants 4-weeks old, 8-weeks old, and 12-weeks old (n = 10).

Glucosinolates	*B. vulgaris* var. *arcuata* G-type			*B. vulgaris* var. *arcuata* P-type		
	4-weeks old	8-weeks old	12-weeks old	4-weeks old	8-weeks old	12-weeks old
Total	1.73±0.32	4.89±0.67	3.06±0.40	1.58±0.34	9.57±0.62	2.66±0.50
R2OH2PE	0.22±0.11	0.19±0.04	0.14±0.03	1.02±0.24	8.97±0.64	2.45±0.47
S2OH2PE	1.46±0.35	4.20±0.67	2.78±0.37	0.50±0.38	0.00±0.00	0.00±0.00
I3M	0.04±0.01	0.47±0.06	0.13±0.02	0.04±0.01	0.56±0.08	0.18±0.04
4MOI3M	0.00±0.00	0.02±0.01	0.00±0.00	0.00±0.00	0.01±0.00	0.02±0.00
2PE	0.01±0.00	0.01±0.00	0.01±0.00	0.02±0.01	0.03±0.01	0.01±0.00

Abbreviations for glucosinolates are: (*R*)-2-hydroxy-2-phenylethylglucosinolate (R2OH2PE), (*S*)-2-hydroxy-2-phenylethylglucosinolate (S2OH2PE), indol-3-ylmethylglucosinolate (I3M), 4-methoxyindol-3-ylmethylglucosinolate (4MOI3M), and 2-phenylethylglucosinolate (2PE). Traces of 4MOI3M (less than 0.01 μmol g^{-1} of plant fresh weight) were found in 4- and 12-weeks old G-type *B. vulgaris* var. *arcuata* plants and in 4-weeks old P-type *B. vulgaris* var. *arcuata* plants. Traces of S2OH2PE (less than 0.01 μmol g^{-1} of plant fresh weight) were found in 12-weeks old P-type *B. vulgaris* var. *arcuata* plants.

analyzed for each plant type and replicate. Glucosinolates and saponins were extracted with 80% aqueous methanol (methanol:water 80:20, v:v). For glucosinolate determination, 4-hydroxybenzylglucosinolate was added as an internal standard. The methanolic extract was loaded onto DEAE Sephadex columns, followed by washing steps and by sulfatase treatment and elution of desulfoglucosinolates. Desulfoglucosinolates were separated on reversed-phase chromatography and quantified with a diode array detector at 229 nm (Agilent 1100 HPLC system, Agilent Technologies, Waldbronn, Germany), using a relative response factor of 2.0 for aliphatic and 0.5 for indole glucosinolates. The response factors we used were based on Brown et al. (2002). For saponin determination, the HPLC system indicated above was coupled to an ESI ion-trap mass spectrometer (Esquire 6000, Bruker Daltonics, Bremen, Germany) operated in negative mode in the range m/z 250–1700, with skimmer voltage, −40 eV; capillary exit voltage, −150.6 eV; capillary voltage, 4,000 V; nebulizer pressure, 35 psi; drying gas, 10 l min-1; and gas temperature, 330°C. Saponins were quantified by the peak areas for the signal of the molecular ion in the negative-ion mass spectrum [M-H]$^{-}$ and in some of the analyses we used a standard curve created with an isolated standard of saponin 2.

Oviposition Preference Tests between Leaves of Different Size within the Same Plant

Oviposition preference between leaves of different type within the same plant was assessed in plexiglass tubes 3.0 cm (inner diameter) by 10.0 cm (length) with plants of G-type *B. vulgaris* var. *arcuata*. Each tube had a 0.5-cm-diameter hole in the middle, through which a piece of dental wick soaked with a 10% sugar solution was inserted into the tube as a food source for the moth inside. One mated female moth was placed in each tube, where it was offered two 7.1 cm^2 circular disks of the abaxial side of *B. vulgaris* leaves. For each tube, the ends of a single tube were attached to two different leaves in the same *B. vulgaris* plant with the help of rubber bands and parafilm. The leaves compared had a difference in maximum leaf diameter ranging from 0 to 58 mm. A total of 42 comparisons involving 84 leaves were conducted (besides these 84 leaves, 16 additional leaves in which *P. xylostella* had not laid any eggs were taken to have more data points to analyze the relationship between leaf size and glucosinolate and saponin content). After one day, the number of eggs on each plant was counted in the laboratory using a dissecting microscope. The

leaves used in the oviposition preference experiments were photographed with a digital camera and leaf areas were determined using WinFOLIA leaf area analysis software (Regent Instruments Inc., Quebec, Canada).

Oviposition preference experiments were also conducted with cotyledons and true leaves of the same plant for which all other leaves had been removed by cutting them with scissors. Immediately after cutting the leaves, oviposition preference tests were conducted in 32.5×32.5×32.5 cm polyester cages with 96×26 mesh (MegaView Science Education Services Co., Ltd., Taichung, Taiwan). Multiple cages were used, each of which was considered a replicate. One mated female moth was released in each experimental arena containing one *Barbarea* plant with only one true leaf (the largest) and one cotyledon. The experiment was replicated four times for each comparison. A small plastic cup with a 10% sugar solution on cotton was placed in the middle of the cage to provide a food source for the moths. Moths were allowed to oviposit overnight in the darkness from 19:00 to 7:00 h. *P. xylostella* lays most of its eggs during the first 3 h of scotophase and the peak oviposition occurs between 19:00 and 20:00 h [31,66]. The number of eggs on each plant was counted in the laboratory using a dissecting microscope.

Survival of Larvae on True Leaves and Cotyledons within the Same Plant

Survival of first-instar larvae of *P. xylostella* was monitored over a period of 5 days. Using a brush, one *P. xylostella* larva was placed individually on a plant containing either one true leaf or one cotyledon (five plants and five larvae were used in total per treatment). The plants tested were *B. rupicola*, *B. verna*, G- and P-type *B. vulgaris* var. *arcuata*, *B. vulgaris* var. *variegata*, and NAS-type *B. vulgaris*.

4.4. Statistical Analysis

Differences in eggs laid by *P. xylostella* per leaf area and in glucosinolate and saponin content among *Barbarea* leaves of different size were analyzed using analysis of variance (ANOVA) and simple regressions with SPSS. When significant treatment differences were indicated by a significant F-test at $P \leq 0.05$, means were separated by Fisher's Protected least significant difference (LSD). Differences in *P. xylostella* oviposition preference between cotyledons and true leaves were analyzed with a paired t-test with SPSS. In order to normalize the residuals, data were transformed

A

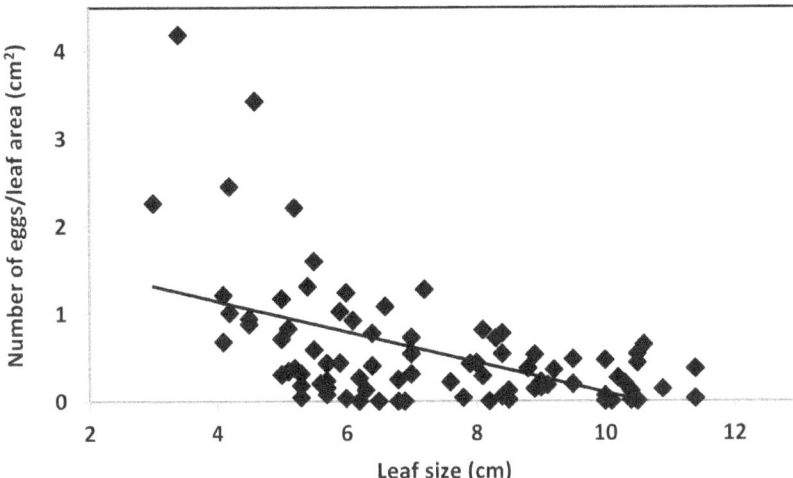

B

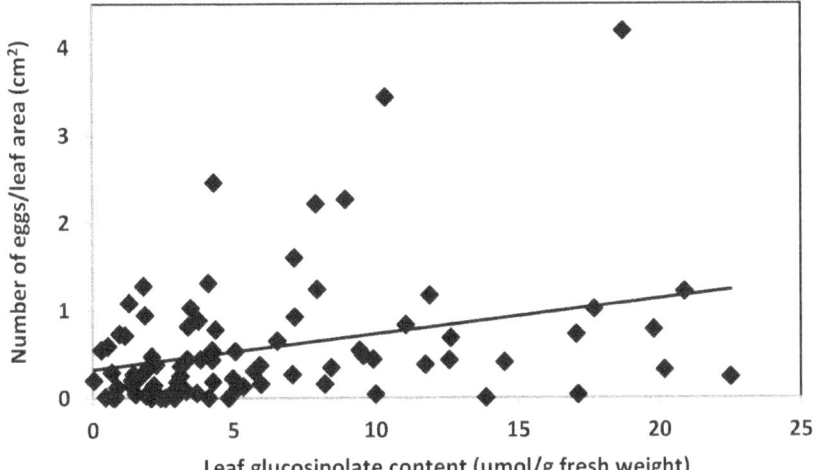

Figure 5. Correlation between leaf size and number of eggs laid by *P. xylostella* **per leaf area (A) and between leaf glucosinolate content and number of eggs laid by** *P. xylostella* **per leaf area (B). Leaves of G-type** *Barbarea vulgaris* **var.** *arcuata* **plants were used (n = 84).**

Table 7. Leaf areas of cotyledons and true leaves of *B. rupicola*, *B. verna*, and G-type *B. vulgaris*, and numbers of eggs laid by *P. xylostella* on cotyledons and true leaves in two-choice preference tests (n = 4 for each plant and leaf type).

	Type of leaf	Leaf area (cm²)	Eggs per leaf	Eggs/cm² of leaf area
B. rupicola	cotyledon	0.48±0.09	1.25±0.95	2.73±2.29
B. rupicola	true leaf	4.70±1.07	9.25±3.15	2.23±0.77
B. verna	cotyledon	1.31±0.26	1.75±1.11	1.17±0.53
B. verna	true leaf	5.41±1.58	8.75±4.31	2.20±1.02
G-type *B.vulgaris*	cotyledon	0.91±0.14	2.25±1.31	2.76±1.82
G-type *B.vulgaris*	true leaf	5.19±0.80	9.00±4.38	2.48±1.66

Data shown as mean ± SE.

prior to analysis using a natural log (x+1) function. Although all tests of significance were based on the transformed data, only untransformed data are presented.

Supporting Information

Table S1 Mean ± SE glucosinolates (μmol/g of leaf fresh weight) concentrations in cotyledons and true leaves in *Barbarea* plants six hours after removing the rest of the leaves in the plant or leaving them intact. As true leaf, the largest true leaf of the plant was taken. For each plant and leaf type and treatment n = 5.

Table S2 Mean ± SE 3-*0-β*-cellobiosylhederagenin (saponin 1) and 3-*0-β*-cellobiosyloleanolic acid (saponin 2) in cotyledons and true leaves of *Barbarea* plants six hours after removing the rest of the leaves in the plant or leaving them intact. As true leaf, the largest true leaf of the

plant was taken. For each plant and leaf type and treatment n = 5. Saponin concentrations given as μmol/g of leaf fresh weight.

Acknowledgments

We thank Dr. Michael Reichelt for help with HPLC and LC-MS analysis of glucosinolates and saponins. Thanks to Jutta Steffen and Christin Heinrich for insect rearing and technical assistance during the experiments. Thanks to Drs. Jens K. Nielsen, Niels Agerbirk, Hanneke van Leur, and Tamara Krügel for providing *B. vulgaris* seeds. Thanks to Drs. Niels Agerbirk and Tetsuro Shinoda for providing saponin standards. Thanks to Dr. Bernhard Löhr for providing insects. Thanks to Birgit Hohmann and Beatriz Parrado Márquez for help growing plants.

Author Contributions

Conceived and designed the experiments: FRBP JG DGH. Performed the experiments: FRBP. Analyzed the data: FRBP. Contributed reagents/materials/analysis tools: FRBP JG DGH. Wrote the paper: FRBP. Provided comments on the manuscript: JG DGH.

References

1. Rhoades DF (1979) Evolution of plant chemical defense against herbivores. In: Rosenthal GA, Janzen DH, editors. Herbivores: Their Interactions with Secondary Plant Metabolites. New York, USA: Academic Press. 3–54.
2. McKey D (1974) Adaptive patterns in alkaloid physiology. American Naturalist 108: 305–320.
3. Harper JL (1989) The value of a leaf. Oecologia 80: 53–58.
4. Brown PD, Tokuhisa JG, Reichelt M, Gershenzon J (2003) Variation of glucosinolate accumulation among different organs and developmental stages of *Arabidopsis thaliana*. Phytochemistry 62: 471–481.
5. van Dam NM, de Jong TJ, Iwasa Y, Kubo T (1996) Optimal distribution of defences: are plants smart investors? Functional Ecology 10: 128–136.
6. Halkier BA, Gershenzon J (2006) Biology and biochemistry of glucosinolates. Annual Review of Plant Biology 57: 303–333.
7. Hopkins RJ, van Dam NM, van Loon JJA (2009) Role of glucosinolates in insect-plant relationships and multitrophic interactions. Annual Review of Entomology 54: 57–83.
8. Petersen B, Chen S, Hansen C, Olsen C, Halkier B (2002) Composition and content of glucosinolates in developing *Arabidopsis thaliana*. Planta 214: 562–571.
9. Gutbrodt B, Dorn S, Unsicker S, Mody K (2012) Species-specific responses of herbivores to within-plant and environmentally mediated between-plant variability in plant chemistry. Chemoecology 22: 101–111.
10. Lambdon PW, Hassall M, Boar RR, Mithen R (2003) Asynchrony in the nitrogen and glucosinolate leaf-age profiles of *Brassica*: is this a defensive strategy against generalist herbivores? Agriculture, Ecosystems & Environment 97: 205–214.
11. Shelton AL (2005) Within-plant variation in glucosinolate concentrations of *Raphanus sativus* across multiple scales. Journal of Chemical Ecology 31: 1711–1732.
12. Wentzell AM, Kliebenstein DJ (2008) Genotype, age, tissue, and environment regulate the structural outcome of glucosinolate activation. Plant Physiology 147: 415–428.
13. Osbourn A (1996) Saponins and plant defence – a soap story. Trends in Plant Science 1: 4–9.
14. Francis G, Kerem Z, Makkar HPS, Becker K (2002) The biological action of saponins in animal systems: a review. British Journal of Nutrition 88: 587–605.
15. Sparg SG, Light ME, van Staden J (2004) Biological activities and distribution of plant saponins. Journal of Ethnopharmacology 94: 219–243.
16. Matsuda K, Kaneko M, Kusaka K, Shishido T, Tamaki Y (1999) Soyasaponins as feeding stimulants to the oriental clouded yellow larva, *Colias erate poliographus* (Lepidoptera: Pieridae). Applied Entomology and Zoology 33: 255–258.
17. Adel M, Sehnal F, Jurzysta M (2000) Effects of alfalfa saponins on the moth *Spodoptera littoralis*. Journal of Chemical Ecology 26: 1065–1078.
18. Jain DC, Tripathi AK (1991) Insect feeding-deterrent activity of some saponin glycosides. Phytotherapy Research 5: 139–141.
19. De Geyter E, Swevers L, Caccia S, Geelen D, Smagghe G (2012) Saponins show high entomotoxicity by cell membrane permeation in Lepidoptera. Pest Management Science 68: 1199–1205.
20. Potter D, Kimmerer T (1989) Inhibition of herbivory on young holly leaves: evidence for the defensive role of saponins. Oecologia 78: 322–329.
21. Pecetti L, Biazzi E, Tava A (2010) Variation in saponin content during the growing season of spotted medic *Medicago arabica* (L.) Huds. Journal of the Science of Food and Agriculture 90: 2405–2410.
22. Szakiel A, Pączkowski C, Henry M (2011) Influence of environmental abiotic factors on the content of saponins in plants. Phytochemistry Reviews 10: 471–491.
23. Teng H-M, Fang M-F, Cai X, Hu Z-H (2009) Localization and dynamic change of saponin in vegetative organs of *Polygala tenuifolia*. Journal of Integrative Plant Biology 51: 529–536.
24. Agerbirk N, Olsen CE, Bibby BM, Frandsen HO, Brown LD, et al. (2003) A saponin correlated with variable resistance of *Barbarea vulgaris* to the diamondback moth *Plutella xylostella*. Journal of Chemical Ecology 29: 1417–1433.
25. Shinoda T, Nagao T, Nakayama M, Serizawa H, Koshioka M, et al. (2002) Identification of a triterpenoid saponin from a crucifer, *Barbarea vulgaris*, as a feeding deterrent to the diamondback moth, *Plutella xylostella*. Journal of Chemical Ecology 28: 587–599.
26. Nielsen JK, Nagao T, Okabe H, Shinoda T (2010) Resistance in the plant, *Barbarea vulgaris*, and counter-adaptations in flea beetles mediated by saponins. Journal of Chemical Ecology 36: 277–285.
27. Nielsen NJ, Nielsen J, Staerk D (2010) New resistance-correlated saponins from the insect-resistant crucifer *Barbarea vulgaris*. Journal of Agricultural and Food Chemistry 58: 5509–5514.
28. Badenes-Perez FR, Reichelt M, Gershenzon J, Heckel DG (2014) Using plant chemistry and insect preference to study the potential of *Barbarea* (Brassicaceae) as a dead-end trap crop for diamondback moth (Lepidoptera: Plutellidae). Phytochemistry: 137–144.
29. Nielsen JK, de Jong PW (2005) Temporal and host-related variation in frequencies of genes that enable *Phyllotreta nemorum* to utilize a novel host plant, *Barbarea vulgaris*. Entomologia Experimentalis Et Applicata 115: 265–270.
30. Agerbirk N, Olsen CE, Nielsen JK (2001) Seasonal variation in leaf glucosinolates and insect resistance in two types of *Barbarea vulgaris* ssp. *arcuata*. Phytochemistry 58: 91–100.
31. Talekar NS, Shelton AM (1993) Biology, ecology, and management of the diamondback moth. Annual Review of Entomology 38: 275–301.
32. Furlong MJ, Wright DJ, Dosdall LM (2013) Diamondback moth ecology and management: problems, progress, and prospects. Annual Review of Entomology 58: 517–541.
33. Ratzka A, Vogel H, Kliebenstein DJ, Mitchell-Olds T, Kroymann J (2002) Disarming the mustard oil bomb. Proceedings of the National Academy of Sciences of the United States of America 99: 11223–11228.
34. Badenes-Perez FR, Reichelt M, Gershenzon J, Heckel DG (2011) Phylloplane location of glucosinolates in *Barbarea* spp. (Brassicaceae) and misleading assessment of host suitability by a specialist herbivore. New Phytologist 189: 549–556.
35. van Loon JJA, Wang CZ, Nielsen JK, Gols R, Qiu YT (2002) Flavonoids from cabbage are feeding stimulants for diamondback moth larvae additional to glucosinolates: chemoreception and behaviour. Entomologia Experimentalis et Applicata 104: 27–34.
36. Spencer JL, Pillai S, Bernays EA (1999) Synergism in the oviposition behavior of *Plutella xylostella*: sinigrin and wax compounds. Journal of Insect Behavior 12: 483–500.
37. Renwick JAA, Haribal M, Gouinguené S, Stadler E (2006) Isothiocyanates stimulating oviposition by the diamondback moth, *Plutella xylostella*. Journal of Chemical Ecology 32: 755–766.
38. Badenes-Perez FR, Reichelt M, Heckel DG (2010) Can sulfur fertilisation increase the effectiveness of trap crops for diamondback moth, *Plutella xylostella* (L.) (Lepidoptera: Plutellidae)? Pest Management Science 66: 832–838.
39. Idris AB, Grafius E (1996) Effects of wild and cultivated host plants on oviposition, survival, and development of diamondback moth (Lepidoptera: Plutellidae) and its parasitoid *Diadegma insulare* (Hymenoptera: Ichneumonidae). Environmental Entomology 25: 825–833.

40. Shelton AM, Nault BA (2004) Dead-end trap cropping: a technique to improve management of the diamondback moth, *Plutella xylostella* (Lepidoptera: Plutellidae). Crop Protection 23: 497–503.

41. Lu JH, Liu SS, Shelton AM (2004) Laboratory evaluations of a wild crucifer *Barbarea vulgaris* as a management tool for the diamondback moth *Plutella xylostella* (Lepidoptera : Plutellidae). Bulletin of Entomological Research 94: 509–516.

42. Badenes-Perez FR, Shelton AM, Nault BA (2004) Evaluating trap crops for diamondback moth, *Plutella xylostella* (Lepidoptera : Plutellidae). Journal of Economic Entomology 97: 1365–1372.

43. Shelton AM, Badenes-Perez FR (2006) Concepts and applications of trap cropping in pest management. Annual Review of Entomology 51: 285–308.

44. Badenes-Perez FR, Nault BA, Shelton AM (2006) Dynamics of diamondback moth oviposition in the presence of a highly preferred non-suitable host. Entomologia Experimentalis et Applicata 120: 23–31.

45. Raupp MJ, Denno RF (1983) Leaf age as a predictor of herbivore distribution and abundance. In: Denno RF, McClure MS, editors. Variable plants and herbivores in natural and managed systems. New York, NY: Academic Press. 91–124.

46. Raupp MJ, Werren JH, Sadof CS (1988) Effects of short-term phenological changes in leaf suitability on the survivorship, growth, and development of gypsy moth (Lepidoptera: Lymantriidae) larvae. Environmental Entomology 17: 316–319.

47. Kause A, Ossipov V, Haukioja E, Lempa K, Hanhimäki S, et al. (1999) Multiplicity of biochemical factors determining quality of growing birch leaves. Oecologia 120: 102–112.

48. King BH, Crowe ML, Blackmore MD (1998) Effects of leaf age on oviposition and on offspring fitness in the imported willow leaf beetle *Plagiodera versicolora* (Coleoptera: Chrysomelidae). Journal of Insect Behavior 11: 23–36.

49. Karban R (1990) Herbivore outbreaks on only young trees: testing hypotheses about aging and induced resistance. Oikos 59: 27–32.

50. Heisswolf A, Obermaier E, Poethke HJ (2005) Selection of large host plants for oviposition by a monophagous leaf beetle: nutritional quality or enemy-free space? Ecological Entomology 30: 299–306.

51. Spangler SM, Calvin DD (2000) Influence of sweet corn growth stages on European corn borer (Lepidoptera: Crambidae) oviposition. Environmental Entomology 29: 1226–1235.

52. Badenes-Perez FR, Nault BA, Shelton AM (2005) Manipulating the attractiveness and suitability of hosts for diamondback moth (Lepidoptera : Plutellidae). Journal of Economic Entomology 98: 836–844.

53. Städler E, Baur R, De Jong R (2002) Sensory basis of host-plant selection: in earch of the "fingerprints" related to oviposition of the cabbage root fly. Acta Zoologica Academiae Scientiarum Hungaricae 48 (Suppl. 1): 265–280.

54. Renwick JAA (1989) Chemical ecology of oviposition in phytophagous insects. Experientia 45: 223–228.

55. Janz N (2002) Evolutionary ecology and oviposition strategies. In: Hilker M, Meiners T, editors. Chemoecology of Insect Eggs and Egg Deposition. Oxford, UK: Blackwell Publishing Ltd. 349–376.

56. Larsson S, Ekbom B (1995) Oviposition mistakes in herbivorous insects: confusion or a step towards a new host plant? Oikos 72: 155–160.

57. Thompson JN (1999) What we know and do not know about coevolution: insect herbivores and plants as a test case. In: Olff H, Brown VK, Drent RH, editors. Herbivores: Between Plants and Predators. Oxford, UK: Blackwell Science Ltd. 7–30.

58. Zhang P-J, Lu Y-b, Zalucki M, Liu S-S (2012) Relationship between adult oviposition preference and larval performance of the diamondback moth, *Plutella xylostella*. Journal of Pest Science 85: 247–252.

59. Boege K, Marquis RJ (2005) Facing herbivory as you grow up: the ontogeny of resistance in plants. Trends in Ecology & Evolution 20: 441–448.

60. Wallace SK, Eigenbrode SD (2002) Changes in the glucosinolate–myrosinase defense system in *Brassica juncea* cotyledons during seedling development. Journal of Chemical Ecology 28: 243–256.

61. Uematsu H (1996) Inter-leaf movement of larvae of diamondback moth, *Plutella xylostella* L. (Lepidoptera: Yponomeutidae) on rape (*Brassica napus*) seedlings. Japanese Journal of Applied Entomology and Zoology 40: 35–38.

62. Root RB, Tahvanainen J (1969) Role of winter cress, *Barbarea vulgaris*, as a temporary host in seasonal development of crucifer fauna. Annals of the Entomological Society of America 62: 852–855.

63. Feeny P (1976) Plant apparency and chemical defense. In: Wallace J, Mansell R, editors. Biochemical Interaction Between Plants and Insects: Springer US. 1–40.

64. Feeny P (1977) Defensive Ecology of the Cruciferae. Annals of the Missouri Botanical Garden 64: 221–234.

65. van Leur H, Raaijmakers CE, van Dam NM (2006) A heritable glucosinolate polymorphism within natural populations of *Barbarea vulgaris*. Phytochemistry 67: 1214–1223.

66. Pivnick KA, Jarvis BJ, Gillott C, Slater GP, Underhill EW (1990) Daily patterns of reproductive activity and the influence of adult density and exposure to host plants on reproduction in the diamondback moth (Lepidoptera: Plutellidae). Environmental Entomology 19: 587–593.

16

Epistatic Association Mapping for Alkaline and Salinity Tolerance Traits in the Soybean Germination Stage

Wen-Jie Zhang[1,2][♪], **Yuan Niu[1]**[♪], **Su-Hong Bu[1]**[♪], **Meng Li[1]**[♪], **Jian-Ying Feng[1]**, **Jin Zhang[1]**, **Sheng-Xian Yang[1]**, **Medrine Mmayi Odinga[1]**, **Shi-Ping Wei[1]**, **Xiao-Feng Liu[1]**, **Yuan-Ming Zhang[1]***

1 Section on Statistical Genomics, State Key Laboratory of Crop Genetics and Germplasm Enhancement, Department of Crop Genetics and Breeding, Nanjing Agricultural University, Nanjing, Jiangsu, China, 2 Institute of Crop Research, Ningxia Academy of Agriculture and Forestry Sciences, Yinchuan, Ningxia, China

Abstract

Soil salinity and alkalinity are important abiotic components that frequently have critical effects on crop growth, productivity and quality. Developing soybean cultivars with high salt tolerance is recognized as an efficient way to maintain sustainable soybean production in a salt stress environment. However, the genetic mechanism of the tolerance must first be elucidated. In this study, 257 soybean cultivars with 135 SSR markers were used to perform epistatic association mapping for salt tolerance. Tolerance was evaluated by assessing the main root length (RL), the fresh and dry weights of roots (FWR and DWR), the biomass of seedlings (BS) and the length of hypocotyls (LH) of healthy seedlings after treatments with control, 100 mM NaCl or 10 mM Na_2CO_3 solutions for approximately one week under greenhouse conditions. A total of 83 QTL-by-environment (QE) interactions for salt tolerance index were detected: 24 for LR, 12 for FWR, 11 for DWR, 15 for LH and 21 for BS, as well as one epistatic QTL for FWR. Furthermore, 86 QE interactions for alkaline tolerance index were found: 17 for LR, 16 for FWR, 17 for DWR, 18 for LH and 18 for BS. A total of 77 QE interactions for the original trait indicator were detected: 17 for LR, 14 for FWR, 4 for DWR, 21 for LH and 21 for BS, as well as 3 epistatic QTL for BS. Small-effect QTL were frequently observed. Several soybean genes with homology to *Arabidopsis thaliana* and soybean salt tolerance genes were found in close proximity to the above QTL. Using the novel alleles of the QTL detected above, some elite parental combinations were designed, although these QTL need to be further confirmed. The above results provide a valuable foundation for fine mapping, cloning and molecular breeding by design for soybean alkaline and salt tolerance.

Editor: Boris Alexander Vinatzer, Virginia Tech, United States of America

Funding: This work was supported by grant 2011CB109306 from the National Basic Research Program of China; grants 30971848, 31301004, and 31301229 from the National Natural Science Foundation of China; grants KYT201002 and KYZ201202-9 from the Fundamental Research Funds for the Central Universities; grants 20100097110035 and 20120097110023 from Specialized Research Fund for the Doctoral Program of Higher Education; 111 project (B08025); and a Project Funded by the Priority Academic Program Development of Jiangsu Higher Education Institutions. The funders had no role in study design, data collection and analysis, decision to publish, or preparation of the manuscript.

Competing Interests: The authors have declared that no competing interests exist.

* E-mail: soyzhang@njau.edu.cn

♪ These authors contributed equally to this work.

Introduction

Soil salinity and alkalinity are important abiotic stresses that adversely affect crop productivity and quality [1]. Approximately 20% of irrigated agricultural land is adversely affected by salinity and alkalinity [2], and salt-affected agricultural areas are continuously increasing. The salinity threat to agriculture exists in more than 100 countries [3]. Salinity inhibits seed germination and seedling growth; reduces nodulation; causes severe leaf chlorosis, bleaching and necrosis; and decreases total biomass and seed yield [4–6]. In China, there are 6.7 million ha of saline irrigated land, and 52.5–61.0% of soybean production loss is due to alkaline and salinity stresses [7]. The development of soybean cultivars with high salt tolerance is recognized as an efficient way to maintain sustainable soybean production in a salt stress environment [8–10]. However, the genetic architecture of the tolerance must first be elucidated. Therefore, the importance of alkaline and salt tolerance necessitates association mapping for these traits in soybean.

During the past decade, many attempts have been made to dissect the genetic mechanisms of alkaline and salt tolerances in *Arabidopsis* [11,12], rice [13,14] and tomato [15], with the greatest progress achieved in *Arabidopsis* and rice. First, studies showed that this type of tolerance is controlled by polygenes [12,14–16]. Then, many quantitative trait loci (QTL) for the tolerance were identified [11,13,17,18]. Finally, a number of candidate genes and candidate pathways for the evolution of salinity tolerance have been reported [19] (http://www.arabidopsis.org/), e.g., the Na^+/H^+ ion antiporter gene *AtNHX1* [20], the salt overly sensitive pathway gene *SOS1–SOS3* [21–23], the K^+/Na^+ homeostasis genes *SKC1* [24] and *Saltol* [25], the stomatal aperture control gene *DST* [26] and the ABA signaling or synthesis gene *RAS1* [27]. However, very little is known about the genetics of alkaline and salt tolerances in soybean.

Although there have been some classic inheritance studies of salt tolerance in soybean [28,29], the molecular inheritance of this tolerance needs to be addressed. Lee et al. [30] identified one major QTL for salt tolerance, which was associated with markers sat_091, satt237 and satt339 on linkage group N. Based on the assumption that this QTL was identical to the *Ncl* locus identified by Abel [28], this locus was further confirmed in F_2 and recombinant inbred lines (RIL) [10,31]. Additional loci have been reported; e.g., Chen et al. [16] detected 11 QTL in RIL; Lee et al. [32] found a different single dominant gene in F_2; and Tuyen et al. [33] mapped a major QTL on linkage group D2 in F_2 and F_6. In addition, Cho et al. [34] detected two markers, satt285 and satt079, that were significantly associated with foliar TRG accumulation, which is correlated with NaCl stress in soybean. It should be noted that all the above results in soybean were obtained from biparental populations. However, biparental population mapping strategies have several drawbacks [35]. For example, if the two parental lines do not segregate at a particular QTL, the QTL cannot be detected regardless of how many offspring are sampled in the mapping population. To overcome such shortcomings, association mapping strategies are recommended [36,37], and many such studies have been conducted [38–47]. However, very little is known about the detection of both QTL-by-environment (QE) and QTL-by-QTL (QQ) interactions.

The objective of this study was to mine novel QTL for alkaline and salt tolerances in soybean using epistatic association mapping (EAM) [36]. Elite alleles from the detected QTL were used to design parental combinations for cultivar improvement.

Results

Evaluation of phenotypic values for STI and ATI

Alkaline and salt tolerances were measured in the LR, FWR, DWR, BS and LH of 257 soybean cultivars in 2009 and 2010, and the mean values, standard deviations, ranges, skewness and kurtosis were calculated (Tables 1 and S1). All the traits exhibited continuous distribution, and most showed a normal distribution. It should be noted that 2 germplasm accessions, Fengzitianandou and Baiqiu 1, were found to be highly resistant to salt; furthermore, 8 germplasm accessions, Linanbayuebai, Shengxiantiangengdou, Zunyizongzidou, Beichuanwuyanwo, Fengzitianandou, Guangxidalidou, Hedou 6 and Jidou 13, were highly resistant to alkaline condition.

The analysis of variance showed significant differences among all the cultivars for all the traits (Table 2), indicating that genetic variation exists among all the cultivars. Most of the other studied factors were also significant (Table 2), suggesting that treatment, year and interactions should be considered in joint association mapping.

Mapping QTL for STI traits

A total of 83 QE interactions (24 for LR-STI, 12 for FWR-STI, 11 for DWR-STI, 15 for LH-STI and 21 for BS-STI) and one epistatic QTL of FWR-STI were detected using EAM implemented with an empirical Bayes algorithm. Among these QTL, 19 were confirmed using an enriched compressed mixed linear model (E-cMLM) method. Most of the detected QTL showed small effects on these traits, except for one LR-STI QTL associated with satt022 and one FWR-STI epistatic QTL between markers satt656 and sat_256. A summary of all the detected QTL, including the marker name, linkage group, position, variance, r^2 value and QTL type, is presented in Table 3. If two linked QTL were separated by less than 5 cM, they were considered a single QTL.

A total of 24 LR-STI QTL, with total phenotypic variance explained (PVE) values of 0.06–7.43%, were identified and mapped to linkage groups A1, A2, B1, C1, C2, D1b, E-H and J-N. Among these QTL, 6 were further identified by E-cMLM. It should be noted that the QTL associated with satt022 had a PVE

greater than 5%. Furthermore, the two QTL associated with markers satt615 and sat_354 should be considered a single QTL because the genetic distance between the two markers is 2.5 cM. In addition, the QTL associated with markers satt226, satt615 and sat_354 were simultaneously identified in 2009 and 2010 using E-cMLM.

A total of 13 FWR-STI QE, with 0.17–3.10% PVE, were identified and mapped to linkage groups C2, D1b, E, G, I and M-O. One epistatic QTL with 15.80% PVE was identified between markers satt656 and sat_256, and sat_256 was also found to exhibit environmental interaction. Among these QTL, 3 were also identified using E-cMLM.

A total of 11 DWR-STI QE, with 0.13-2.05% PVE, were found and mapped to linkage groups E-G, I, K, M and N. Of these QTL, 3 were also identified using E-cMLM.

A total of 15 LH-STI QE, with 0.23–4.50% PVE, were identified and mapped to linkage groups A1, A2, B2, C2, D1a, E-G, J, N and O. Of these QTL, 4 were also identified using E-cMLM. Several closely linked QTL, e.g., sat_280 and sat_266 (3.82 cM), were considered a single QTL.

A total of 21 BS-STI QE, with 0.20–3.56% PVE, were identified and mapped to linkage groups A1-G, I, J, M and N. Of these QTL, 4 were also identified using E-cMLM. Closely linked QTL, e.g., satt452 and satt045 (1.56 cM), were considered a single QTL.

Mapping QTL for ATI traits

A total of 86 QE interactions (17 for LR-ATI, 16 for FWR-ATI, 17 for DWR-ATI, 18 for LH-ATI and 18 for BS-ATI) were identified using EAM implemented with an empirical Bayes algorithm. Among these QTL, 29 were confirmed using an E-cMLM method. Most of the detected QTL showed small effects on these traits. A summary of all the detected QTL is presented in Table 3.

A total of 17 LR-ATI QTL, with 0.41–4.20% PVE, were detected and mapped to linkage groups A2, B2, C2, D2, F and J-N. Of these QTL, 8 were also identified using E-cMLM. It should be noted that several markers were associated with both LR-ATI and LR-STI, i.e., satt615-sat_354, satt160, satt247, satt245 and satt022. Fourteen closely linked marker pairs were found to be associated with both LR-STI and LR-ATI, e.g., satt289 (LR-STI) and satt277 (LR-ATI). In addition, using E-cMLM, two markers, satt683 and sat_280, were found to be associated with LR-ATI in 2009 and 2010.

A total of 16 FWR-ATI QE, with 0.31–4.16% PVE, were detected and mapped to linkage groups A1, A2, B2, C1, D2, E–H, J–M and O. Of these QTL, 6 were also identified using E-cMLM. It should be noted that several markers were associated with both FWR-ATI and FWR-STI, i.e., sat_354, satt352 and satt463. Three closely linked marker pairs were associated with both FWR-STI and FWR-ATI, e.g., satt262 (FWR-STI) and satt094 (FWR-ATI).

A total of 17 DWR-ATI QE, with 0.10–3.12% PVE, were identified and mapped to linkage groups A1, B1, B2, C2, D1b, D2, E, F, H, J, K, M and O. Of these QTL, 6 were also detected using E-cMLM. Several closely linked QTL were considered a single QTL, e.g., satt509 and sat_261 (0.45 cM). It should be noted that two markers were associated with both DWR-ATI and DWR-STI, i.e., sct_190 and satt220, and nine closely linked marker pairs were associated with both DWR-STI and DWR-ATI, e.g., satt441 (DWR-STI) and satt417 (DWR-ATI). In addition, using E-cMLM, marker satt132 was found to be associated with DWR-ATI in 2009 and 2010.

Table 1. Phenotypic variation in alkaline and salt tolerance indices measured in 257 soybean cultivars in 2009 and 2010.

Year	Indicator	Trait	Mean	Std Dev	Minimum	Maximum	Skewness	Kurtosis
2009	STI	LR	44.58	12.53	−2.22	67.15	−0.88	0.68
		LH	51.03	14.67	−0.26	82.48	−0.50	0.16
		FWR	63.92	13.51	9.81	95.71	−0.84	1.58
		DWR	51.56	15.67	−8.00	87.37	−0.53	0.31
		BS	47.22	9.64	21.52	74.89	−0.18	−0.03
	ATI	LR	32.24	18.38	−19.36	89.95	−0.04	0.87
		LH	19.24	16.86	−42.05	72.88	−0.12	1.13
		FWR	62.67	16.96	−8.96	95.06	−1.52	4.08
		DWR	49.14	18.98	−30.00	91.04	−0.85	2.13
		BS	34.03	11.68	−10.41	64.11	−0.82	1.79
2010	STI	LR	45.26	9.23	1.70	69.00	−0.72	1.99
		LH	46.80	15.25	5.47	79.24	−0.34	−0.20
		FWR	50.48	12.55	0.31	82.69	−0.43	1.03
		DWR	32.54	18.18	−28.57	71.43	−0.70	0.80
		BS	40.46	10.71	−30.65	66.94	−1.12	7.47
	ATI	LR	21.71	16.48	−20.57	66.01	0.11	−0.08
		LH	11.54	12.01	−34.91	42.86	−0.46	0.73
		FWR	40.98	22.26	−17.29	82.16	−0.52	−0.38
		DWR	30.90	21.49	−23.53	78.57	−0.22	−0.52
		BS	23.52	13.74	−24.13	50.46	−0.47	−0.05

STI: salt tolerance index; ATI: alkaline tolerance index; LR: Length of main root; FWR: fresh weights of roots; DWR: dry weights of roots; BS: biomass of seedlings; LH: length of hypocotyls.

A total of 18 LH-ATI QE, with 0.10–3.12% PVE, were found and mapped to linkage groups A1, B2, C2, D1b, D2, E-I, K and N. Of these QTL, 5 were also identified using E-cMLM. It should be noted that three markers were associated with both LH-ATI and LH-STI, i.e., staga001, satt309 and sat_372. Three closely linked marker pairs were associated with both LH-STI and LH-ATI, e.g., satt045 (LH-STI) and satt452 (LH-ATI). In addition,

using E-cMLM, marker satt452 was found to be associated with LH-ATI in 2009 and 2010.

A total of 18 BS-ATI QE, with 0.06–4.96% PVE, were detected and mapped to linkage groups A1, B2, C1, C2, D2-G, K, M and N. Of these QTL, 4 were also identified using E-cMLM. Seven markers were associated with both BS-ATI and BS-STI, i.e., satt382, AW277661, satt422, sat_354, satt160, sat_256 and satt022. Five closely linked marker pairs were associated with

Table 2. Analysis of variance in the length of the main root (LR), fresh and dry weights of roots (FWR and DWR), biomass of seedlings (BS) and length of hypocotyls (LH).

Source of variation	DF	Length of main root		Length of hypocotyls		Fresh weight of roots		Dry weight of roots		Biomass of seedlings	
		MS	F	MS	F	MS	F	MS	F	MS	F
Year	1	3268.810	2689.26**	287.580	341.04**	3.625	1977.39**	0.006	4.17*	12.547	1190.54**
Treat	2	9080.705	7470.72**	5287.965	6271.05**	13.870	7565.39**	0.044	30.98**	84.024	7972.70**
Year × Treat	2	179.925	148.02**	17.630	20.91**	0.206	112.27**	0.010	6.60**	0.793	75.23**
Variety	256	34.002	27.97**	11.707	13.88**	0.070	38.12**	0.002	1.24*	0.749	71.04**
Year × Variety	236	15.650	12.88**	4.400	5.22**	0.019	10.54**	0.002	1.15	0.130	12.36**
Treat × Variety	512	6.760	5.56**	3.724	4.42**	0.012	6.72**	0.001	1.00	0.060	5.69**
Year × Treat × Variety	461	4.579	3.77**	2.079	2.47**	0.008	4.38**	0.002	1.06	0.031	2.92**
Residual	1447	1.216		0.843		0.002		0.001		0.011	

*and **: significance at the 0.05 and 0.01 levels, respectively. DF: degree of freedom; MS: mean square.

Table 3. Association mapping for alkaline and salt tolerance indices using epistasis association mapping.

Trait	Marker	Linkage group	Position (cM)	LOD	Var	r² (%)	E-cMLM	Previous study
LR-STI	sat_267	A1	78.44	6.51	1.81	1.51		
	satt200	A1	92.88	4.21	0.49	0.41		
	satt390	A2	9.14	11.09	0.07	0.06		
	satt632	A2	51.5	4.64	0.76	0.63		
	satt509	B1	32.5	12.36	0.28	0.24		Chen et al (2008)
	satt453	B1	123.95	3.09	2.12	1.76		
	satt565	C1	0.00	3.23	2.94	2.45		
	satt289	C2	112.34	4.19	0.70	0.59	√(10⁶)	Cho et al (2002)
	sat_254	D1b	46.91	6.73	2.09	1.74		Chen et al (2008)
	satt226	D2	85.15	4.84	2.79	2.32	sat_354☆(09), satt615(10)	
	satt615-sat_354	D2	91.2–93.7	4.59–7.29	1.05–5.94	0.87–4.94	satt615(10), sat_354(09)	
	satt045	E	46.65	4.39	1.74	1.45	√(09)	
	sat_381	E	64.18	3.05	0.48	0.40	satt045(09)	
	AW186493	F	21.04	3.41	0.42	0.35		
	satt160	F	33.18	4.90	2.55	2.12		
	satt309	G	4.53	4.92	2.21	1.83		Cho et al (2002)
	AF162283	G	87.94	5.41	0.47	0.39		Chen et al (2008)
	satt222	H	68.08	3.85	0.46	0.39		
	sat_228	J	23.91	3.24	0.48	0.40		Chen et al (2008)
LR-ATI	sat_280	N	43.45	5.78	9.71	2.94	satt683(09, 10)	
	satt237	N	74.98	4.48	1.28	0.39		
	satt022	N	102.05	10.29	13.89	4.20		
FWR-STI	satt289	C2	112.34	7.28	1.13	0.53		Cho et al (2002)
	satt296	D1b	52.61	7.08	0.99	0.46	satt488 (10)	Chen et al (2008)
	sat_354	D2	93.7	5.99	6.65	3.10		
	satt672	D2	114.97	3.20	0.36	0.17		
	satt649	F	5.36	4.57	3.71	1.73		

Trait	Marker	Linkage group	Position (cM)	LOD	Var	r² (%)	E-cMLM	Previous study
LR-STI	satt215	J	44.08	7.55	0.63	0.52		Chen et al (2008)
	satt247-satt417	K	43.95–46.2	3.08–12.64	0.29–3.20	0.24–2.66	satt102 (10)	Chen et al (2008)
	satt463-satt245	M	50.09–53.54	3.03–4.38	0.83–1.57	0.69–1.30		Chen et al (2008)
	satt323	M	60.04	4.48	0.39	0.32		Chen et al (2008)
	satt022	N	102.05	15.56	8.94	7.43		Lee et al (2004); Chen et al (2008)
LR-ATI	satt329	A2	110.94	4.95	0.74	0.22		
	satt070	B2	72.8	3.62	4.46	1.35		
	satt277	C2	107.58	3.10	7.06	2.13		
	satt514	D2	85.69	3.22	2.41	0.73		Tuyen et al (2010)
	satt615-sat_354	D2	91.2–93.7	3.29–4.13	2.07–9.08	0.63–2.75		Tuyen et al (2010)
	satt659	F	26.7	6.34	9.40	2.84	√(09)	
	satt160	F	33.18	11.43	8.71	2.64	√(09)	
	satt132	J	39.18	3.48	7.76	2.35	√(10)	
	satt431	J	78.57	3.85	1.35	0.41	√(09)	
	satt247	K	43.95	6.06	6.02	1.82		
	satt238	L	19.93	3.53	3.94	1.19	√(10)	
	satt652	L	30.87	5.63	0.86	0.26	√(09)	
	satt245	M	53.54	3.92	3.99	1.21	sat_256 (09)	
	satt683	N	34.52	4.66	5.44	1.64	√(09, 10)	
FWR-ATI	satt142	H	86.48	3.86	3.43	0.67	√(10)	
	satt215	J	44.08	5.37	2.40	0.47	satt414 (10)	
	satt417	K	46.2	10.28	7.33	1.44		
	sct_190	K	77.37	16.51	1.85	0.36		
	satt238	L	19.93	4.61	3.13	0.62	√(09)	
	satt527	L	70.36	7.19	1.56	0.31		
	satt463	M	50.09	4.49	15.36	3.02	√(09)	
	satt094	O	56.58	4.55	7.44	1.46		

Table 3. Cont.

Left portion

Trait	Marker	Linkage group	Position (cM)	LOD	Var	r² (%)	E-cMLM	Previous study
	satt659	F	26.7	3.28	3.11	1.45		
	satt352	G	50.52	3.20	4.04	1.89	satt138 (09)	Chen et al (2008)
	satt440	I	112.69	3.49	4.06	1.90		Chen et al (2008)
	satt463	M	50.09	4.46	1.41	0.66		Chen et al (2008)
	sat_256	M	74.52	11.55	6.12	2.86		Chen et al (2008)
	satt022	N	102.05	9.30	4.20	1.96		Li et al (2004); Chen et al (2008)
	satt262	O	57.02	3.92	0.44	0.21		
	satt656× sat_256	F×M	135.11× 74.52	3.40	33.83	15.80	satt656 (10)	Chen et al (2008)
FWR-ATI	sat_344	A1	19.37	3.66	15.00	2.95		
	satt632	A2	51.5	3.38	5.11	1.00		
	satt577	B2	6.05	3.09	4.53	0.89	sat_342 (10)	
	AW277661	C1	74.79	4.93	5.61	1.10		
	sat_354	D2	93.7	7.23	15.26	3.00	sat_365 (10)	Tuyen et al (2010)
	satt452	E	45.09	3.09	2.77	0.54		
	satt160	F	33.18	6.28	21.18	4.16		
	satt352	G	50.52	4.70	5.43	1.07		
DWR-ATI	satt307	C2	121.26	4.28	0.47	0.10		
	satt282	D1b	76.09	3.81	6.55	1.33		
	sat_354	D2	93.7	6.19	13.17	2.67		Tuyen et al (2010)
	satt263	E	45.4	4.10	4.93	1.00		
	satt149	F	18.12	3.81	1.26	0.26		
	satt142	H	86.48	7.14	1.88	0.38	√(2010)	
	satt132	J	39.18	3.08	6.92	1.40	satt215(09,10)	
	satt417	K	46.2	4.11	4.51	0.92		
	sct_190-satt260	K	77.37–80.12	4.10–3.64	1.99–2.39	0.40–0.49		
	satt463	M	50.09	4.23	15.36	3.12	satt323(10)	
	satt220-satt626	M	56.29–58.59	3.49–3.60	1.12–5.19	0.23–1.05	satt323 (10)	

Right portion

Trait	Marker	Linkage group	Position (cM)	LOD	Var	r² (%)	E-cMLM	Previous study
DWR-STI	satt045	E	46.65	3.85	2.41	0.64		
	AW186493	F	21.04	3.72	0.50	0.13		Cho et al (2002)
	satt160	F	33.18	5.56	7.73	2.05		
	satt688	G	12.54	4.20	1.19	0.32		Chen et al (2008)
	satt440	I	112.69	4.20	2.78	0.74		
	satt441	K	46.2	6.17	1.78	0.47		Chen et al (2008)
	sct_190	K	77.37	3.23	1.51	0.40	√(09)	
	satt220	M	56.29	4.39	4.19	1.11		Chen et al (2008)
	satt250	M	107.69	3.61	2.39	0.63		Chen et al (2008)
	sat_266	N	47.27	3.08	6.85	1.81	satt631(09)	Lee et al (2004); Chen et al (2008)
	satt022	N	102.05	3.08	6.02	1.59		
DWR-ATI	satt382	A1	26.42	9.90	14.04	2.85		
	satt509-sat_261	B1	32.5–32.95	3.61–5.89	1.39–1.92	0.28–0.39		
	satt070	B2	72.8	5.12	9.98	2.03		
	staga001	C2	119.84	5.35	3.82	0.78		
LH-STI	satt309	G	4.53	15.41	7.39	3.25		Cho et al(2002)
	sat_372	G	107.75	3.46	4.58	2.01		
	satt132	J	39.18	4.34	4.95	2.18		Chen et al (2008)
LH-ATI	satt683	N	34.52	3.34	1.66	0.73		
	sat_280-sat_266	N	43.45–47.27	3.78–4.11	2.26–5.84	0.99–2.56		
	satt262	O	57.02	4.89	3.61	1.59		
	satt382	A1	26.42	3.59	8.02	3.53		
	sat_342	B2	20.3	3.31	7.76	3.42		
	staga001	C2	119.84	11.30	8.28	3.65		
	satt357	C2	151.91	3.40	1.04	0.46		
	sat_254	D1b	46.91	4.90	1.88	0.83	√(09)	

Table 3. Cont.

Trait	Marker	Linkage group	Position (cM)	LOD	Var	r² (%)	E-cMLM	Previous study
	sat_256	M	74.52	3.97	10.68	2.17	satt323 (10)	
	satt592	O	100.37	3.31	9.23	1.87	sat_274 (10)	
LH-STI	satt200	A1	92.88	5.52	0.53	0.23		
	satt632	A2	51.5	4.88	4.06	1.79	aw132402(09)	
	aw132402	A2	67.86	12.53	3.20	1.41	√(09)	
	sat_355	B2	66.23	4.58	1.83	0.81		Chen et al (2008)
	staga001	C2	119.84	8.20	2.64	1.16	sat_252 (09)	Cho et al (2002)
	sat_160	D1a	104.27	5.15	3.68	1.61		
	satt411	E	12.92	8.64	1.33	0.58		
	satt045	E	46.65	13.30	5.73	2.51	√(09)	
satt160		F	33.18; 6.15	10.25	4.50			
LH-ATI	satt255	N	76.48	3.60	0.24	0.10		
BS-STI	satt382	A1	26.42	3.50	2.79	2.42		
	aw132402	A2	67.86	3.56	1.06	0.92		
	satt453	B1	123.95	4.13	1.03	0.90		
	satt070	B2	72.8	5.28	2.11	1.84	√(09)	Chen et al (2008)
	satt534	B2	87.58	4.21	3.08	2.68	satt070(09)	
	AW277661	C1	74.79	4.05	0.81	0.71		
	satt640	C2	30.46	3.20	0.37	0.32		
	satt422	C2	44.66	3.73	0.73	0.63		
	satt289	C2	112.34	3.93	0.62	0.54		Cho et al (2002)
	satt296	D1b	52.61	4.02	0.40	0.35		Chen et al (2008)
	sat_354	D2	93.7	5.39	3.82	3.32		
	satt452-satt045	E	45.09–46.65	3.50–5.63	0.49–1.36	0.42–1.18	√(09)	
	satt659	F	26.7	5.81	1.56	1.35		
	satt160	F	33.18	8.59	4.09	3.56		
	satt309	G	4.53	6.91	2.24	1.95		Cho et al (2002)
	satt672	D2	114.97	3.90	0.59	0.26	satt413(09)	
	satt452	E	45.09	5.51	4.65	2.05	satt685(09); satt263(10)	
	AW186493	F	21.04	5.16	0.68	0.30		
	satt656	F	135.11	6.18	3.39	1.49	satt160 (10)	
	satt309	G	4.53	4.37	5.20	2.29		
	satt688	G	12.54	4.86	0.15	0.07		
	sat_372	G	107.75	3.04	5.81	2.56		
	satt302	H	81.04	7.24	0.15	0.07		
	satt354	I	46.22	4.84	2.26	0.99	satt419 (09)	
	sat_419	I	98.11	3.46	3.68	1.62		
satt440			112.69; 7.93	2.18	0.96			
	satt260	K	80.12	4.25	1.59	0.70		
BS-STI	satt237	N	74.98	5.84	1.45	1.26		Lee et al (2004); Chen et al (2008)
BS-ATI	satt022	N	102.05	4.21	3.05	2.66		Lee et al (2004); Chen et al (2008)
	sat_344	A1	19.37	4.72	3.88	2.04		
	satt382	A1	26.42	5.72	5.51	2.90		
	satt577	B2	6.05	4.84	3.19	1.68	sat_342 (10)	
	AW277661	C1	74.79	5.63	2.69	1.42		
	satt422	C2	44.66	4.22	0.11	0.06		
	staga001	C2	119.84	10.08	4.52	2.38	√(09)	
	satt669	D2	67.7	4.25	1.14	0.60	satt226(10)	Tuyen et al (2010)
	sat_354	D2	93.7	6.24	5.66	2.98	satt226(10)	Tuyen et al (2010)
	satt263	E	45.4	5.13	1.71	0.90		
	satt160	F	33.18	9.66	9.42	4.96		
	satt138	G	55.99	6.76	1.18	0.62		
	sat_372	G	107.75	4.10	6.20	3.27		
	satt417	K	46.2	9.64	2.71	1.43		
	sct_190	K	77.37	3.81	0.84	0.44		

Table 3. Cont.

Trait	Epistatic association mapping						Similar results	
	Marker	Linkage group	Position (cM)	LOD	Var	r² (%)	E-cMLM	Previous study
	satt626	M	58.59	3.06	0.87	0.46		
	sat_256	M	74.52	8.91	6.80	3.58		
	satt631	N	26.13	3.01	1.57	0.83		
	satt022	N	102.05	3.37	5.70	3.00		

Epistatic association mapping						Similar results	
Marker	Linkage group	Position (cM)	LOD	Var	r² (%)	Previous study	E-cMLM
satt270	I	50.11	6.90	0.88	0.77		
sat_228	J	23.91	4.02	0.23	0.20	Chen et al (2008)	
sat_256	M	74.52	5.28	3.80	3.30	Chen et al (2008)	satt245(09)
satt250	M	107.69	3.51	0.44	0.39	Chen et al (2008)	

*: the same associated marker detected by different approaches;
§: year, 09:2009; 10: 2010;
#: marker in similar result column is linked to the associated marker in this study; LR: Length of main root; FWR: Fresh weight of root; DWR: Dried weight of root; LH; Length of hypocotyl; BS: Biomass.

both BS-STI and BS-ATI, e.g., satt289 (BS-STI) and staga001 (BS-ATI).

In summary, the above mapping results produced three particularly noteworthy observations. First, almost all the detected QTL were small (1–5%). Second, almost all the detected QTL were found to exhibit environmental interactions. Finally, some markers were found to be associated with the STI and ATI indicators of multiple traits. For example, marker sat_354 was associated with LR-ATI, LR-STI, FWR-ATI, FWR-STI, BS-ATI and BS-STI, and markers satt160 and satt022 were associated with LR-STI, LR-ATI, BS-STI and BS-ATI. These data indicate that correlations exist among the above five traits (Table 2).

Mapping QTL from original traits

Original trait observations and marker information were used to perform EAM. A total of 78 QE (17 for LR, 14 for FWR, 4 for DWR, 21 for LH and 21 for BS) and 3 epistatic QTL for BS were detected and are listed in Table 4. Among these QTL, 74 were also identified using E-cMLM. Most of the detected QTL showed small effects on these traits, except for two epistatic QTL: sat_390 × sat_254 and satt244 × sat_160.

A total of 17 QE interactions for LR, with 0.07–2.38% PVE, were detected and mapped to linkage groups C2, D1a-F, H, J–L, N and O. Of these QTL, 16 were also identified using E-cMLM. Note that seven markers, satt289, sat_252, satt672, satt263, satt149, sat_224 and satt244, were found to be associated with LR in more than two environments. When all the LR datasets were jointly analyzed by E-cMLM, 15 QTL-by-year interactions were detected, which was consistent with the identification of 15 QE interactions using EAM.

A total of 14 FWR QE interactions, with 0.06–1.06% PVE, were identified and mapped to linkage groups A1-B1, C2, D2-F and J-N. Of these QTL, 13 were also identified using E-cMLM. Note that nine QTL were found to be associated with LR in more than two environments. When all the FWR datasets were jointly analyzed using E-cMLM, 13 main-effect QTL, one QTL-by-treatment interaction and QTL-by-year interaction were detected, which was consistent with the identification of 13 QE interactions using EAM.

Four DWR QE interactions, with 0.08-4.38% PVE, were identified and mapped to linkage groups B1, D2, H and K. Of these QTL, one was also identified using E-cMLM.

A total of 21 LH QE interactions, with 0.07-2.00% PVE, were identified and mapped to linkage groups B1, B2, C2-L and N. All of these QTL were also identified using E-cMLM. Note that nine QTL were found to be associated with LH in more than two environments. When all the LH datasets were jointly analyzed using E-cMLM, 20 main-effect QTL and five QTL-by-year interactions were detected. Among these QTL, twenty were found to be consistent with the QE interactions for LH index identified using EAM.

A total of 21 QE and 3 QQ interactions for BS, with 0.08-11.41% PVE, were detected and mapped to linkage groups A2, B1, C2-F, H and J-O. Of these QTL, 23 were also identified using E-cMLM. Several closely linked QTL were considered a single QTL, e.g., satt413 and satt672 (1.36 cM), and thirteen QTL were found to be associated with BS in more than two environments. When all the BS datasets were jointly analyzed using E-cMLM, 26 main-effect QTL and two QTL-by-year interactions were detected. Among these QTL, 23- were found to be consistent with the QE and QQ interactions for BS identified using EAM.

A comparison of the QTL for ATI and STI indicators revealed 56 common QTL: 14 for LR, 11 for FWR, 3 for DWR, 15 for LH and 13 for BS. For example, markers satt289, satt222 and satt022

Table 4. Association mapping for alkaline and salt tolerance traits using epistasis association mapping and an enriched compressed mixed linear model (E-cMLM).

Trait	Epistatic association mapping						Similar result		
	Marker	Linkage group	Position (cM)	LOD	Var	r^2(%)	E-cMLM	Jointly E-cMLM	Previous study
LR	satt289	C2	112.34	24.49	0.03	0.18	sat_252⁎(09CKˢ,09AL)	staga001(M×Y)	Cho et al(2002)
	sat_252	C2	127	4.69	0.10	0.73	√(09CK, 09AL)	√(M×Y)	Cho et al(2002)
	satt580	D1a	62.36	6.55	0.02	0.18		satt077(M×Y)	
	satt296	D1b	52.61	4.70	0.03	0.19	√(09CK)	satt282(M×Y)	Chen et al (2008)
	satt459	D1b	118.62	5.37	0.08	0.56	satt274(10AL)	√(M×Y)	
	satt672	D2	114.97	4.32	0.03	0.21	satt413(10CK,10SA)	satt413(M×Y)	
	satt263	E	45.4	3.33	0.06	0.45	satt452(09AL), sat_381(10AL)	√(M×Y)	
	satt149	F	18.12	4.31	0.03	0.25	√(10CK), satt659(09CK), BE806387(10AL)	AW186493(M×Y)	
	satt222	H	68.08	3.76	0.07	0.48	satt142(09SA)	√(M×Y)	
	sat_224	J	75.12	3.15	0.13	0.98	√(09CK), satt431(09AL)		
	satt244	J	65.04	3.79	0.32	2.38	sat_224(09CK), satt431(09AL)	sat_165(M×Y)	Chen et al (2008)
	sct_190	K	77.37	7.81	0.04	0.30		√(M×Y)	
	satt166	L	66.51	27.34	0.04	0.28			
	satt683	N	34.52	7.86	0.11	0.81	sat_266(10AL)	satt631(M×Y)	
	satt234	N	84.59	3.47	0.01	0.07	satt237(10SA)	satt255(M×Y)	Li et al (2004);Chen et al (2008)
	satt022	N	102.05	3.86	0.31	2.31		√(M×Y)	Li et al (2004);Chen et al (2008)
	satt331	O	93.37	6.07	0.04	0.29	satt592(09SA)	√(M×Y)	
FWR	sat_267	A1	78.44	3.40	0.00	0.86	satt200(09CK,10CK)	√(M)	
	satt390	A2	9.14	3.14	0.00	0.25		√(M,M×T)	
	satt509	B1	32.5	6.39	0.00	0.37			
	satt289	C2	112.34	18.14	0.00	0.27	satt277(09SA,09AL,10SA)	√(M)	Chen et al (2008)
	sat_252	C2	127	4.17	0.00	0.99	satt277(09SA, 09AL,10SA)	√(M)	Cho et al (2002)
	satt672	D2	114.97	8.78	0.00	0.23	satt256(09CK,09SA,10CK)	√(M)	Cho et al (2002)
	sat_381	E	64.18	6.12	0.00	0.78		√(M)	
	satt269	F	11.37	6.21	0.00	0.21	BE806387(09CK,09AL),satt149(10SA)	satt149(M)	
	sat_390	F	1.79	3.40	0.00	1.06	BE806387(09CK,09AL),satt149(10SA)	√(M)	
	satt215	J	44.08	3.81	0.00	0.35	√(09SA),satt132(10AL)	satt244(M)	Chen et al (2008)
	sct_190	K	77.37	40.07	0.00	0.19	satt260(10SA)	√(M)	
	satt238	L	19.93	3.86	0.00	0.30		satt652(M)	
	satt463	M	50.09	5.31	0.00	0.87	√(09AL),satt626(09CK),satt245(09SA)	√(M,M×Y)	Chen et al (2008)
	satt234	N	84.59	87.87	0.00	0.06	satt022(09CK),satt237(09SA)	satt237(M)	Li et al (2004);Chen et al (2008)
DWR	satt453	B1	123.95	3.83	0.00	4.38	√(10SA)		
	satt669	D2	67.7	3.11	0.00	0.13			

Table 4. Cont.

Trait	Epistatic association mapping						Similar result		Previous study
	Marker	Linkage group	Position (cM)	LOD	Var	r²(%)	E-cMLM	Jointly E-cMLM	
LH	satt302	H	81.04	3.33	0.00	0.32			
	sct_190	K	77.37	10.32	0.00	0.08			
	satt453	B1	123.95	5.65	0.05	0.75	√(09SA)	√(M)	
	sat_355	B2	66.23	3.78	0.01	0.22	satt534(10SA)	√(M)	Chen et al (2008)
	satt289	C2	112.34	24.06	0.02	0.25	sat_252(09CK),staga001(09SA,10SA,10AL),satt202(10CK)	satt277(M)	Cho et al (2002)
	satt254	D1a	56.43	3.49	0.03	0.42	satt580(09SA)	AW781285(M)	
	satt580	D1a	62.36	11.47	0.00	0.01	satt580(09-NaCl)	AW781285(M)	
	satt296	D1b	52.61	10.20	0.03	0.48	√(09AL,10SA,10AL),sat_254(09CK)	√(M)	Chen et al (2008)
	sat_254	D1b	46.91	4.36	0.06	1.00	√(09CK),satt296(09AL,10SA,10AL)	satt296(M)	Chen et al (2008)
	satt669	D2	67.7	3.55	0.02	0.33	satt226(10CK)	satt226(M)	Tuyen et al (2010)
	sat_354	D2	93.7	4.75	0.12	2.00	√(09SA),satt514(10CK)	√(M,M×Y)	Tuyen et al (2010)
	sat_381	E	64.18	7.56	0.03	0.46	√(09AL),satt045(09SA),satt263(10AL)	√(M)	
	satt149	F	18.12	16.35	0.00	0.07	BE806387(09CK,09AL,10CK,10AL),AW186493(09SA),satt659(10SA)	BE806387(M)	
	satt160	F	33.18	3.74	0.11	1.82	BE806387(09CK,09AL,10CK),AW186493(09SA),satt659(10SA,10AL)	√(M)	
	satt309	G	4.53	4.25	0.10	1.55	satt688(10CK)	√(M,M×Y)	Cho et al(2002)
	satt222	H	68.08	4.03	0.02	0.36		√(M,M×Y)	
	satt354	I	46.22	3.73	0.06	0.98	√(10CK,10AL),satt270(09CK,09AL)	√(M,M×Y)	
	sat_228	J	23.91	4.41	0.02	0.39	sat_165(09AL,10SA,10AL)	sat_165(M)	Chen et al (2008)
	satt441	K	46.2	4.61	0.03	0.42		√(M)	Chen et al (2008)
	satt166	L	66.51	5.46	0.01	0.17	satt527(10SA)		
	satt683	N	34.52	18.58	0.03	0.43		√(M)	
	satt234	N	84.59	85.46	0.00	0.07		satt237(M,M×Y)	Li et al (2004);Chen et al (2008)
	sat_266	N	47.27	5.49	0.06	0.97		√(M)	
BS	satt390	A2	9.14	5.67	0.00	0.22	√(09SA,09AL)	√(M)	
	satt233	A2	100.08	6.10	0.00	0.55	√(09CK,10SA,10AL)	√(M,M×Y)	
	satt453	B1	123.95	4.13	0.00	0.46		√(M)	
	satt565	C1	0	5.65	0.00	0.53			
	sat_252	C2	127	3.84	0.00	0.74	satt277(09SA,09AL)	√(M)	Cho et al (2002)
	satt580	D1a	62.36	3.41	0.00	0.21	AW781285(10CK)	√(M×Y)	
	satt274	D1b	116.34	9.04	0.00	0.58	√(09AL,10AL)	√(M)	
	satt256	D2	124.3	3.28	0.00	0.51	√(09CK,09SA,09AL)	satt672(M)	
	satt413-satt672	D2	113.61-114.97	5.91-13.60	0.00	0.27-0.41	satt256(09CK,09SA,09AL)	√(M)	
	sat_381	E	64.18	5.37	0.00	0.74		√(M)	

Table 4. Cont.

Trait	Epistatic association mapping						Similar result		Previous study
	Marker	Linkage group	Position (cM)	LOD	Var	r^2(%)	E-cMLM	Jointly E-cMLM	
	satt269	F	11.37	6.44	0.00	0.27	BE806387(09CK,09AL)	ˇ(M)	
	satt656	F	135.11	4.82	0.00	0.95		ˇ(M)	Chen et al (2008)
	satt659×sat_354	F×D2	26.7×93.7	3.04	0.01	3.29	BE806387(09CK,09AL),sat_354(09SA)	satt659(M),sat_362(M)	Tuyen et al (2010)
	sat_390×sat_254	F×D1b	1.79×46.91	3.23	0.01	9.00	BE806387(09CK,09AL),sat_254(09CK)	sat_390(M),satt296(M)	
	satt222	H	68.08	7.70	0.00	0.29	ˇ(09SA,09AL)	ˇ(M)	
	satt132	J	39.18	4.79	0.00	0.46	satt215(09AL)	ˇ(M)	Chen et al (2008)
	satt244	J	65.04	4.18	0.00	1.60	satt215(09AL)		Chen et al (2008)
	satt244×sat_160	J×D1a	65.04×104.27	3.15	0.02	11.41	satt215(09AL)	satt244(M),sat_160(M)	
	sct_190	K	77.37	42.52	0.00	0.11	ˇ(09CK,09AL),satt260(10CK,10SA)	ˇ(M)	
	satt238	L	19.93	4.23	0.00	0.18		satt652(M)	
	sat_391	M	1.02	5.57	0.00	1.36	ˇ(10CK)	ˇ(M)	
	satt234	N	84.59	20.49	0.00	0.08	satt022(09CK,09AL)	ˇ(M)	Li et al (2004);Chen et al (2008)
	satt022	N	102.05	3.97	0.00	1.42	ˇ(09CK,09AL)	ˇ(M)	Li et al (2004);Chen et al (2008)
	satt331	O	93.37	5.54	0.00	0.18	sat_274(09AL)	ˇ(M)	

ˇ: the same association marker detected by different approach;

§, year, i.e., 2009, and 2010; SA: salt; AL: alkaline.

*: marker in similar result column is linked to the associated marker in this study; M: main-effect QTL; Y: year; T: treatment. LR: Length of main root; FWR: Fresh weight of root; DWR: Dried weight of root; LH: Length of hypocotyl; BS: Biomass.

were associated with both LR and LR-STI, and markers satt683 and satt022 were associated with both LR and LR-ATI. Among these markers, marker satt022 is common; in other words, marker satt022 was associated with LR, LR-STI and LR-ATI.

Predictions for novel parental combinations

The best way to improve a trait is to pyramid all the desirable elite alleles into one cultivar, if possible. By maximizing the number of elite alleles using a Monte Carlo simulation experiment, the ideal novel cultivar combination can be designed. Using this method, 21 elite alleles of 27 QTL for LR-STI were pyramided by combining cultivars Guangxibayuehuang, 0804, Shangqiu 832012, Qingyuanxiaoqingdou, Daheidou, Ganjiangnan, Shuangliuliuyuehuang, Baiqiu 1, Xu 0701 and Wenfeng 5.

Discussion

The mapping results from this study are reliable for three reasons. First, 19 QTL for STI, 29 QTL for ATI, and 74 QTL for original traits, which were detected using EAM, were confirmed by E-cMLM (Tables 3 and 4). Furthermore, 36 QTL for STI (11 for LR-STI, 7 for FWR-STI, 5 for DWR-STI, 4 for LH-STI and 9 for BS-STI), 6 QTL for ATI (2 for LR-ATI, 1 for FWR-ATI, 1 for DWR-ATI and 2 for BS-ATI) and 29 QTL for original traits (6 for LR, 6 for FWR, 10 for LH and 7 for BS) found in this study have also been identified by other researchers (Tables 3 and 4). For example, one major QTL on linkage group N plays an essential role in enhancing soybean salt tolerance in different genetic backgrounds [10,28,30], and the present study also confirmed several associations with salt tolerance, e.g., satt237 for BS-STI and satt234 for LR, LH, FWR and BS. Chen et al. [16] identified 11 QTL significantly associated with salt tolerance, and Cho et al. [34] detected 15 QTL on linkage groups B2, C2, D2, G, J and K for foliar TRG accumulation, which is postulated to function as a compatible solute and/or osmoprotectant under adverse NaCl stress conditions. These 26 QTL were all confirmed in this study. Finally, several salt tolerance genes from *Arabidopsis thaliana* and soybean (http://www.ncbi.nlm.nih.gov/gene?term=salt%20tolerance) were found to be located in close proximity to tolerance-associated markers in this study.

The potential candidate genes for salt tolerance are summarized in Tables S2 and S3. Among these candidate genes, four soybean genes, Glyma13g41980, Glyma15g03400, Glyma17g37430 and Glyma17g35340, which have homologs in *Arabidopsis*, were associated with five markers examined in this study, with physical distances of 135.76–820.04 kb. Six soybean salt tolerance genes, *LOC100808889, LOC100807827, LOC100800981, LOC100795117, LOC100814727* and *LOC100797515*, were closely linked to seven markers analyzed in this study, with physical distances of 135.41–619.74 kb. Although these consistent results were observed, one might raise doubts about our conclusions because we did not consider cultivar relatedness. In our opinion, the inclusion of all the main and epistatic QTL in the genetic model reduced the importance of controlling for genetic background.

Tuyen et al. [33] reported that tolerance to NaCl may not always accompany tolerance to alkaline stress. Their evidence was based on a discrepancy between the QTL for alkaline tolerance, located between markers satt669 and sat_300 on linkage group D2, and the QTL for saline tolerance, located on linkage group N, in wild soybean JWS156-1. However, the opposite phenomenon was observed in the present study. For example, satt237 on linkage group N was associated with both LR-ATI and BS-STI; satt615 on linkage group D2 was associated with both LR-STI and LR-ATI; and sat_354 on linkage group D2 was associated with LR-STI, LR-ATI, FWR-STI, FWR-ATI, BS-STI and BS-ATI.

Similar results for other linkage groups were also found in this study. More importantly, the cultivar Fengzitianandou was found to be highly resistant to NaCl and alkaline stresses. In addition, tolerance to salt and alkaline stresses is related to tolerance to Al in soybean.

Sharma et al. [48] identified two major QTL for tolerance to Al. These two QTL were also detected in the present study and were located in the marker intervals satt160-satt252 (F) and satt202-satt371 (C2), respectively. Note that satt160 was simultaneously associated with LR-STI, LR-ATI, FWR-ATI, DWR-STI, LH-STI, BS-STI, BS-ATI and LH. Bianchi-Hall et al. [49] identified six QTL associated with tolerance to Al stress. Among these QTL, *Al tol1-1, Al tol1-3, Al tol1-5* and *Al tol1-6* were found to be located in close proximity to the associated markers in this study. For example, satt329 (associated with LR-ATI) and satt233 (associated with BS) on linkage group A2 were close to *Al tol1-1*; satt509 (associated with LR-STI, DWR-ATI and FWR) on linkage group B1 was close to *Al tol1-3*; satt215 (associated with LR-STI, FWR-ATI and FWR), satt431 (associated with LR-ATI), satt132 (associated with DWR-ATI, LH-STI and BS), satt244 (associated with LR and BS) and sat_224 (associated with LR) on linkage group J were close to *Al tol1-5*; and satt238 (associated with LR-ATI, FWR-ATI, FWR and BS) on linkage group L was linked to *Al tol1-6*. These results demonstrate that the mechanisms of tolerance to alkaline/salt and Al conditions are similar.

Most of the detected QTL showed small effects, except for four QTL: one main-effect QTL associated with satt022 and three epistatic QTL: satt656 × sat_256 for LR-STI, sat_390 × sat_254 for BS and satt244 × sat_160 for BS. The results of this study demonstrate that large differences in tolerance are caused not only by a few large-effect QTL but also by the cumulative effect of numerous small-effect QTL. This small-effect QTL phenomenon has been observed in *Arabidopsis* [50], rice [51] and maize [52], although the high PVE of the epistatic QTL is surprising because alkaline and salt tolerances in plants are derived from interactive molecular pathways [53]. In addition, almost all QTL exhibit significant environmental interactions, indicating that the genetic architecture across various salt and alkaline stresses is sensitive or variable.

Research into tolerance to NaCl and alkaline stresses usually employs two types of indicators: a salt and alkaline tolerance index and an original trait indicator. These two types of indicators were also analyzed in this study, producing complementary results: tolerance genes were found proximal to both types of associated markers.

The utilization of varieties with multiple resistance genes is an effective way to reduce the effects of adversity. As the numbers of both cloned genes and detected QTL increase, a Monte Carlo simulation experiment becomes a valuable means of parental combination prediction and selection strategy design [36,54]. In this study, several parental combinations were predicted. Certain cultivars are encountered repeatedly in these combinations and may be used to improve multiple traits. For an STI indicator, e.g., Sudou 1 may be chosen for improving FWR and BS; Daheidou for LR and LH; Qingyuanxiaoheidou for LR, LH and BS; Qinyan 1 for DWR and BS; and Wuhuabayuehuang for DWR and BS. These predictions may be valuable for several reasons. First, two of the selected parents, Zhejiangsiyuebai (DWR-STI) and Caoqing (BS-STI), are on the list of 348 ancestors of 651 soybean cultivars released during 1923–1995 in China [55]. Youbian 31, Sudou 1, Fengjiao 66–12, Ludou 1, Wenfeng 5, and Jindou 2 are on the list of 171 varieties that were bred earlier and were derived from 519 soybean cultivars [55]. Second, some accessions exhibit beneficial agronomic traits and strong resistance to adversity, i.e., Sudou 1,

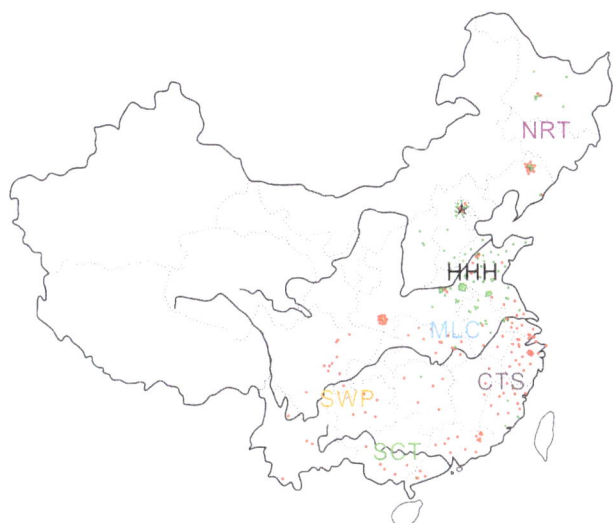

Figure 1. Distribution of 257 soybean cultivars.

Ludou 1, Kexi 8, Daheidou, Edou 2, Niumaohuang, Wenfeng 5, Dianbaiheidou and Baiqiu 1. Finally, some selected cultivars have been widely grown in certain areas, i.e., Sudou 1, Ludou 1 and Wenfeng 5. In addition, we should consider the sizes of novel alleles when predicting parental combinations. Thus, the tolerant variety Fengzitianandou may be included in a parental combination because of its large effect. Of course, a prerequisite for the above prediction is that all the above QTL will be further confirmed.

The phenomenon of QTL clusters has been reported in soybean [56] and cotton [57]. Previous work has indicated that numerous disease resistance loci are clustered in various regions of the soybean genome, e.g., on chromosomes D1b and F. This phenomenon was also evident in the current results. For example, QTL linked to markers satt226, satt615, sat_354 and satt514 on chromosome D2 were clustered in an 8.55 cM interval region. Three salt tolerance genes, Glyma17g37430, Glyma17g35340 and LOC100807827, were found to be located near the tolerance-associated QTL clusters of satt256, satt413 and satt672 (Tables S2 and S3). The phenomenon of QTL clusters may be used to explain trait correlation [57].

Materials and Methods

Plant materials and DNA marker analysis

All the 257 soybean cultivars were obtained, by stratified random sampling, from 6 geographic ecotypes in China (Fig. 1), were planted in three-row plots in a completely randomized design, and were evaluated at the Jiangpu experimental station at Nanjing Agricultural University in 2009 and 2010. The plots were 1.5 m wide and 2 m long. These seeds were used to conduct plastic container experiments.

Approximately 0.3 g of fresh leaves obtained from each cultivar in 2009 was used to extract genomic DNA using the cetyltri-methylammonium bromide method, as described by Lipp et al. [58]. To screen for polymorphisms among all the cultivars, PCR was performed with 135 simple sequence repeat (SSR) primer pairs. The primer sequences were obtained from the soybean database Soybase (http://www.ncbi.nlm.nih.gov). PCR was performed as described by Xu et al. [56].

Evaluation of alkaline and salt tolerances

A salt-water flooding method [59] was used to evaluate the alkaline and salt tolerances of all the soybean cultivars. In brief, twelve soybean seeds for each cultivar were sown in a $30 \times 20 \times 15$ cm plastic container with sand added to a height of $3.5 \times$ cm. The seeds were then treated with control (CK, pH: 7.0), $100 \times$ mM NaCl (pH: 7.0) and $10 \times$ mM Na_2CO_3 (pH: 11.1) solutions, with two replications each. A 350-ml aliquot of the appropriate solution for each treatment was applied to each plastic container filled with sand. Twelve soybean seeds for each treatment were grown in a growth chamber under white fluorescent light (600 μmol m^{-2} s^{-1}; 14 h light/10 h dark) at 25 ± 1°C. The length of the main root (LR), fresh and dry weights of roots (FWR and DWR), biomass of seedlings (BS) and length of hypocotyls (LH) of healthy seedlings from 5 plants in plastic containers under simulated alkaline and salt conditions were measured 7 days after sowing. The units used were centimeters for length and grams for weight. To measure the degree of salt and alkaline tolerances, the original trait observations may be transformed into salt and alkaline tolerance indices for each trait using the below equations:

$$\text{salt tolerance index (STI)} = (x_{CK} - x_{NaCl})/x_{CK} \times 100\%$$

$$\text{alkaline tolerance index(ATI)} = \left(x_{CK} - x_{Na_2CO_3}\right)\Big/x_{CK} \times 100\%$$

where x_{CK}, x_{NaCl} and $x_{Na_2CO_3}$ stand for the phenotypic values exhibited following control, saline and alkaline treatments, respectively.

Population structure

Population structure plays an important role in association mapping. To investigate the population structures of all the selected cultivars, the STRUCTURE program [60] and the approach of Evanno et al. [61] were employed. The number of subpopulations (K) was set from 2 to 10. The number of replicates for each K was 20, and the total average of the mean log-likelihood at a fixed value of K was used. Using the ad hoc statistic ΔK, which is based on the rate of change in the log-

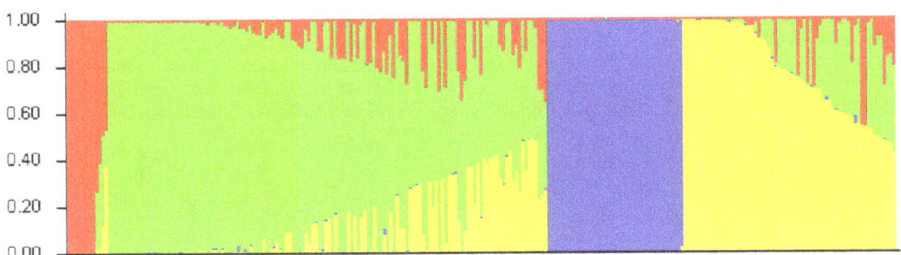

Figure 2. Plot of posterior probabilities (*y*-axis) against four subgroups on each cultivar (*x*-axis) using STRUCTURE software.

probability of data between successive K values, the ΔK value was much higher for the model parameter $K = 4$ than for other values of K (Fig. 2). By combining this high ΔK value with knowledge of the breeding history of these cultivars, a K value of 4 was chosen. The Q matrix was calculated based on information from 135 SSR markers and was incorporated into the association mapping.

Epistatic association mapping

The EAM approach suggested by Lü et al. [36] was also used to analyze all the five-trait datasets in this study; for technical details, the reader is referred to the original study by Lü et al. [36]. All the above analyses were performed using the SAS program.

The enriched compression mixed linear model (E-cMLM) approach, suggested by Li [37] and expanded by Zhang et al. [62], was used to confirm the results of the epistatic association mapping. In this analysis, the dataset for each year and each treatment was analyzed, and all the datasets for all the two-year and three-treatment datasets were jointly analyzed.

Supporting Information

Table S1 Phenotypic variation in the length of the main root (LR), fresh and dry weights of roots (FWR and DWR), biomass of seedlings (BS) and length of hypocotyls (LH) among healthy seedlings measured in 257 soybean cultivars in 2009 and 2010.

Table S2 SSR markers located near soybean genes with homology to salt tolerance genes in *Arabidopsis thaliana*.

Table S3 SSR markers located near salt tolerance genes in soybean.

Author Contributions

Conceived and designed the experiments: YMZ. Performed the experiments: WJZ YN SPW XFL SXY YMZ. Analyzed the data: YN SHB ML WJZ JYF JZ YMZ. Contributed reagents/materials/analysis tools: SHB ML. Wrote the paper: YMZ YN MMO.

References

1. Zhu JK (2001) Plant salt tolerance. Trends in Plant Science 6:66–71.
2. Yamaguchi T, Blumwald E (2005) Developing salt-tolerant crop plants: challenges and opportunities. Trends Plant Sci 10:615–620.
3. Kaman H, Cetin M, Kirda C (2011) Effects of Lower Seyhan Plain irrigation on groundwater depth and salinity. Journal of Food, Agriculture & Environment 9:648–652.
4. Abel GH, Mackenzie AJ (1964) Salt tolerance of soybean varieties (*Glycine max* L. Merrill) during germination and later growth. Crop Sci 14:157–161.
5. Abel G (1969) Inheritance of the capacity for chloride inclusion and chloride exclusion by soybeans. Crop Sci 9:697-698.
6. Parker MB, Gains TP, Hook JE, Gascho GJ, Maw BW (1987) Chloride and water stress effects on soybean in pot culture. Journal of Plant Nutrition 10:517–538.
7. Lam HM, Chang RZ, Shao GH, Liu ZT (ed.) (2009) Research on tolerance to stresses in Chinese soybean. Beijing: China Agriculture Press.
8. Chinnusamy V, Jagendorf A, Zhu JK (2005) Understanding and improving salt tolerance in plants. Crop Sci 45:437–448.
9. Genc Y, Oldach K, Verbyla AP, Lott G, Hassan M, et al. (2010) Sodium exclusion QTL associated with improved seedling growth in bread wheat under salinity stress. Theor Appl Genet 121: 877–894.
10. Hamwieh A, Tuyen DD, Cong H, Benitez ER, Takahashi R, et al. (2011) Identification and validation of a major QTL for salt tolerance in soybean. Euphytica 179:451–459.
11. Quesada V (2002) Genetic architecture of NaCl tolerance in *Arabidopsis*. Plant Physiology 130:951–963.
12. Katori T, Ikeda A, Iuchi S, Kobayashi M, Shinozaki K, et al. (2010) Dissecting the genetic control of natural variation in salt tolerance of Arabidopsis thaliana accessions. Journal of Experimental Botany 61:1125–1138.
13. Lin HX, Zhu MZ, Yano M, Gao JP, Liang ZW, et al. (2004) QTLs for Na$^+$ and K$^+$ uptake of the shoots and roots controlling rice salt tolerance. Theor Appl Genet 108:253–260.
14. Wang ZF, Wang JF, Bao YM, Wu YY, Zhang HS (2011) Quantitative trait loci controlling rice seed germination under salt stress. Euphytica 178:297–307.
15. Foolad MR, Stoltz T, Dervinis C, Rodriguez RL, Jones RA (1997) Mapping QTLs conferring salt tolerance during germination in tomato by selective genotyping. Molecular Breeding 3:269–277.
16. Chen HT, Cui SY, Fu SX, Gai JY, Yu DY (2008) Identification of quantitative trait loci associated with salt tolerance during seedling growth in soybean (*Glycine max* L.). Australian Journal of Agricultural Research 59:1086–1091.
17. Clerkx EJM, El-Lithy ME, Vierling E, Ruys GJ, Blankestijn-De Vries H, et al. (2004) Analysis of natural allelic variation of Arabidopsis seed germination and seed longevity traits between the accessions Landsberg erecta and Shakdara, using a new recombinant inbred line population. Plant Physiology 135:432–443.
18. DeRose-Wilson L, Gaut BS (2011) Mapping salinity tolerance during *Arabidopsis thaliana* germination and seedling growth. PLoS ONE 6:e22832.
19. Munns R (2005) Genes and salt tolerance: bringing them together. New Phytologist 167:645–663.
20. Apse MP, Aharon GS, Snedden WS, Blumwald E (1999) Salt tolerance conferred by overexpression of a vacuolar Na$^+$/H$^+$ antiport in *Arabidopsis*. Science 285:1256-1258.
21. Liu J, Ishitani M, Halfter U, Kim C-S, Zhu JK (2000) The *Arabidopsis thaliana* SOS2 gene encodes a protein kinase that is required for salt tolerance. Proc Natl Acad Sci USA 97:3730–3734.
22. Shi H, Ishitani M, Kim C, Zhu JK (2010) The arabidopsis thaliana salt tolerance gene SOS1 encodes a putative Na$^+$/H$^+$ antiporter. Proc Natl Acad Sci USA 97(12):6896–6901.
23. Qiu QS, Guo Y, Dietrich MA, Schumaker KS, Zhu JK (2002) Regulation of SOS1, a plasma membrane Na$^+$/H$^+$ exchanger in *Arabidopsis thaliana*, by SOS2 and SOS3. Proc Natl Acad Sci USA 99:8436–8441.
24. Ren ZH, Gao JP, Li LG, Cai XL, Huang W, et al. (2005) A rice quantitative trait locus for salt tolerance encodes a sodium transporter. Nat Genet 37:1141–1146.
25. Thomson MJ, De Ocampo M, Egdane J, Rahman MA, Sajise AG, et al. (2010) Characterizing the saltol quantitative trait locus for salinity tolerance in rice. Rice 3: 148–160.
26. Huang XY, Chao DY, Gao JP, Zhu MZ, Shi M, et al. (2009) A previously unknown zinc finger protein, DST, regulates drought and salt tolerance in rice via stomatal aperture control. Genes & Development 23: 1805–1817.
27. Ren ZH, Zheng ZM, Chinnusamy V, Zhu JH, Cui XP, et al. (2010) RAS1, a quantitative trait locus for salt tolerance and ABA sensitivity in *Arabidopsis*. Proc Natl Acad Sci USA 107:5669–5674.
28. Abel G (1969) Inheritance of the capacity for chloride inclusion and chloride exclusion by soybeans. Crop Sci 9:697–698.
29. Shao G, Chang R, Chen Y (1994) Study on Inhedtanee of Salt Tolerance in Soybean. Acta Agronomica Sinica 20: 721–726.
30. Lee GJ, Carter TE Jr, Villagarcia MR, Li Z, Zhou X, et al. (2004) A major QTL conditioning salt tolerance in S-100 soybean and descendent cultivars. Theor Appl Genet 109:1610–1619.
31. Hamwieh A, Xu DH (2008) Conserved salt tolerance quantitative trait locus (QTL) in wild and cultivated soybeans. Breeding Sci 58:355–359.
32. Lee JD, Shannon JG, Vuong TD, Nguyen HT (2009) Inheritance of salt tolerance in wild soybean (*Glycine soja* Sieb. and Zucc.) Accession PI483463. Journal of Heredity 100:798–801.
33. Tuyen DD, Lal SK, Xu DH (2010) Identification of a major QTL allele from wild soybean (*Glycine soja* Sieb. & Zucc.) for increasing alkaline salt tolerance in soybean. Theor Appl Genet 121:229–236.
34. Cho Y, Njiti VN, Chen X, Triwatayakorn K, Kassem MA, et al. (2002) Quantitative trait loci associated with foliar trigonelline accumulation in *Glycine Max* L. Journal of Biomedicine and Biotechnology2(3):151–157.
35. He X-H, Qin H, Hu Z, Zhang Y-M (2011) Mapping of epistatic quantitative trait loci in four-way crosses. Theor Appl Genet 122:33–48
36. Lü H-Y, Liu X-F, Wei S-P, Zhang Y-M (2011) Epistatic association mapping in homozygous crop cultivars. PLoS ONE 6(3):e17773.
37. Li M (2011) Methodologies for functional mapping of quantitative trait loci and genome-wide association study (Ph D dissertation). Nanjing Agricultural University.
38. Hansen M, Kraft T, Ganestam S, Säll T, Nilsson NO (2001) Linkage disequilibrium mapping of the bolting gene in sea beet using AFLP markers. Genet Res 77: 61–66.
39. Hauser MT, Harr B, Schlötterer C (2001) Trichome distribution in *Arabidopsis thaliana* and its close relative Arabidopsis lyrata: molecular analysis of the candidate gene *GLABROUS1*. Mol Biol Evol 18:1754–1763.

40. Thornsberry JM, Goodman MM, Doebley J, Kresovich S, Nielsen D, et al. (2001) *Dwarf8* polymorphisms associate with variation in flowering time. Nat Genet 28:286–289.

41. Palaisa KA, Morgante MM, Williams M, Rafalski A (2003) Contrasting effects of selection on sequence diversity and linkage disequilibrium at two phytoene synthase loci. Plant Cell 15:1795–1806.

42. Wilson LM, Whitt SR, Ibáñez AM, Rocheford TR, Goodman MM, et al. (2004) Dissection of maize kernel composition and starch production by candidate associations. Plant Cell 16:2719–2733.

43. Jun TH, Van K, Kim MY, Lee SH (2008) Association analysis using SSR markers to find QTL for seed protein content in soybean. Euphytica 162:179–191.

44. Singh RK, Bhat KV, Bhatia VS, Mohapatra T, Singh NK (2008) Association mapping for photoperiod insensitivity ait in soybean. National Academy Science Letters 31:281–284.

45. Wang J, McClean PE, Lee R, Goos RJ, Helms T (2008) Association mapping of iron deficiency chlorosis loci in soybean (*Glycinemax* L. Merr.) advanced breeding lines. Theor Appl Genet 116:777–787.

46. Li YH, Smulders MJM, Chang RZ, Qiu LJ (2011) Genetic diversity and association mapping in a collection of selected Chinese soybean accessions based on SSR marker analysis. Conservation Genetics 12:1145–1157.

47. Niu Y, Xu Y, Liu XF, Yang SX, Wei SP, et al. (2013) Association mapping for seed size and shape traits in soybean cultivars. Molecular Breeding 31:785–794.

48. Sharma AD, Sharma H, Lightfoot DA (2011) The genetic control of tolerance to aluminum toxicity in the 'Essex' by 'Forrest' recombinant inbred line population. Theor Appl Genet 122:687–694.

49. Bianchi-Hall CM, Carter TE Jr, Bailey MA, Mian MAR, Rufty TW, et al. (2000) Aluminum tolerance associated with quantitative trait loci derived from soybean PI 416937 in hydroponics. Crop Sci 40:538–545.

50. El-Lithy ME, Bentsink L, Hanhart CJ, Ruys GJ, Rovito D, et al. (2006) New *Arabidopsis* Recombinant Inbred Line populations genotyped using SNPWave and their use for mapping flowering-time quantitative trait loci. Genetics 172:1867–1876.

51. Uwatoko N, Onishi A, Ikeda Y, Kontani M, Sasaki A, et al. (2008) Epistasis among the three major flowering time genes in rice: coordinate changes of photoperiod sensitivity, basic vegetative growth and optimum photoperiod. Euphytica163(2):167–175.

52. Buckler ES, Holland JB, Bradbury PJ, Acharya CB, Brown PJ, et al. (2009) The genetic architecture of maize flowering time. Science 325:714–718.

53. Komeda Y (2004) Genetic Regulation of Time to Flower in Arabidopsis thaliana. Annu Rev Plant Biol 55, 521—535.

54. Zhang Y-M, Mao Y, Xie C, Smith H, Luo L, et al. (2005) Mapping QTL using naturally occurring genetic variance among commercial inbred lines among maize (*Zea mays* L.). Genetics 169:2267–2275.

55. Cui ZL, Gai JY, Carter TE Jr, Qiu JX, Zhao TJ (1999) The released Chinese soybean cultivars and their pedigree analysis (1923–1995). Beijing, China, China Agriculture Publishing House.

56. Xu Y, Li H-N, Li G, Wang X, Cheng L, et al. (2011) Mapping quantitative trait loci for seed size traits in soybean (*Glycine max* L. Merr.). Theor Appl Genet 122:581–594.

57. Qin H, Guo W, Zhang Y-M, Zhang TZ (2008) QTL Mapping of yield and fiber traits based on a four-way cross population in *Gossypium hirsutum* L. Theor Appl Genet117:883–894.

58. Lipp M, Brodmann P, Pietsch K, Pauwels J, Anklam E, et al. (1999) IUPAC collaborative trail study of a method to detect genetically modified soybeans and maize in dried powder. Journal of AOAC International 82:923–928

59. Sobhanian H, Razavizadeh R, Nanjo Y, Ehsanpour AA, Jazii FR, et al. (2010) Proteome analysis of soybean leaves, hypocotyls and roots under salt stress. Proteome Science 8:19.

60. Pritchard JK, Stephens M, Donnelly P (2000) Inference of population structure using multilocus genotype data. Genetics 155:945–959.

61. Evanno G, Regnaut S, Goudet J (2005) Detecting the number of clusters of individuals using the software STRUCTURE: a simulation study. Molecular Ecology 14:2611–2620.

62. Zhang Z, Ersoz E, Lai CQ, Todhunter RJ, Tiwari HK, et al. (2010) Mixed linear model approach adapted for genome-wide association studies. Nat Genet 42:355–360.

17

Whole Transcriptome Profiling of Maize during Early Somatic Embryogenesis Reveals Altered Expression of Stress Factors and Embryogenesis-Related Genes

Stella A. G. D. Salvo[1], Candice N. Hirsch[2], C. Robin Buell[3,4], Shawn M. Kaeppler[1], Heidi F. Kaeppler[1]*

1 Department of Agronomy, University of Wisconsin, Madison, Wisconsin, United States of America, 2 Department of Agronomy and Plant Genetics, University of Minnesota, St. Paul, Minnesota, United States of America, 3 Department of Plant Biology, Michigan State University, East Lansing, Michigan, United States of America, 4 DOE Great Lakes Bioenergy Research Center, Michigan State University, East Lansing, Michigan, United States of America

Abstract

Embryogenic tissue culture systems are utilized in propagation and genetic engineering of crop plants, but applications are limited by genotype-dependent culture response. To date, few genes necessary for embryogenic callus formation have been identified or characterized. The goal of this research was to enhance our understanding of gene expression during maize embryogenic tissue culture initiation. In this study, we highlight the expression of candidate genes that have been previously regarded in the literature as having important roles in somatic embryogenesis. We utilized RNA based sequencing (RNA-seq) to characterize the transcriptome of immature embryo explants of the highly embryogenic and regenerable maize genotype A188 at 0, 24, 36, 48, and 72 hours after placement of explants on tissue culture initiation medium. Genes annotated as functioning in stress response, such as glutathione-S-transferases and germin-like proteins, and genes involved with hormone transport, such as PINFORMED, increased in expression over 8-fold in the study. Maize genes with high sequence similarity to genes previously described in the initiation of embryogenic cultures, such as transcription factors BABY BOOM, LEAFY COTYLEDON, and AGAMOUS, and important receptor-like kinases such as SOMATIC EMBRYOGENESIS RECEPTOR LIKE KINASES and CLAVATA, were also expressed in this time course study. By combining results from whole genome transcriptome analysis with an in depth review of key genes that play a role in the onset of embryogenesis, we propose a model of coordinated expression of somatic embryogenesis-related genes, providing an improved understanding of genomic factors involved in the early steps of embryogenic culture initiation in maize and other plant species.

Editor: Xianlong Zhang, National Key Laboratory of Crop Genetic Improvement, China

Funding: Funding for this research was provided by National Institute of Food and Agriculture, United States Department of Agriculture Hatch funding (WIS01226). The funders had no role in study design, data collection and analysis, decision to publish, or preparation of the manuscript.

Competing Interests: The authors have declared that no competing interests exist.

* Email: hfkaeppl@wisc.edu

Introduction

In order to meet global food, feed, and fiber needs in the face of climate change and predicted population growth, current and future crop improvement efforts will likely include the utilization of biotechnology-based approaches [1,2]. This includes the discovery and functional analysis of agriculturally important genes for crop research and product development. Currently, most of the crop genetic engineering systems utilize embryogenic, regenerable tissue cultures as a critical part of the transformation process [3]. Totipotent, embryogenic cultures are also desirable for efficient somatic embryo production for other agricultural biotechnology applications such as clonal propagation, production of synthetic seed [4], and the proposed utilization of somatic embryos for gamete cycling in rapid breeding [5].

At the molecular level, it is widely accepted that the induction of somatic embryogenesis involves massive cellular reprogramming and activation of various signaling cascades [6,7]. The necessary triggers that induce somatic embryogenesis in tissue culture are tantamount with stress response [8,9]. As the accumulation of near-damaging cellular signals trigger change, only specific genotypes are capable of efficient cellular adaptation, pluripotency, and embryogenic competence in tissue culture [10].

The majority of crop genotypes within species display low embryogenic growth response in culture. This genotype-dependent culture response decreases the efficiency and significantly limits the application of clonal propagation schemes and current transformation systems in the genetic study and improvement of crop plants [11]. In maize, the inbred line A188, which displays a high embryogenic culture response, has been utilized in investigations on the inheritance and genetic control of the genotype-dependent culture response [12–14] and in improving embryogenic response efficiency and regeneration ability in tissue culture [15,16].

Despite the agronomic importance, few genes with a direct role in the induction of somatic embryogenesis in tissue culture have been identified, and their role in embryogenic culture response in maize and other crops is not understood. In *Brassica napus*,

Table 1. Highly expressed genes in immature zygotic embryo explants in tissue culture.[a]

Gene	Function	FPKM						Cluster Number
		0 h	24 h	36 h	48 h	72 h		
GRMZM2G020940	unknown	889.44	1548.51	1168.15	1203.65	1282.68		6
GRMZM2G080603	grp1 (glycine-rich protein1)	1005.03	1816.68	2345.42	2365.11	2412.29		6
GRMZM2G480954	unknown	22.07	1217.33	993.33	844.36	571.95		6
GRMZM2G153292	tua2 (alpha tubulin2)	1292.61	1065.69	1384.88	1349.02	1313.01		6
GRMZM2G080274	ARATH HON1 Group	1205.96	164.49	251.87	293.59	288.91		6
GRMZM2G337229	ole1 (oleosin1)	1129.77	1341.73	1377.95	1155.93	812.47		6
GRMZM2G051943	chitinase A1	1.01	1636.10	2150.28	1676.96	1331.93		3
GRMZM2G332838	Histone H4	1295.95	267.26	649.95	642.20	566.99		6
GRMZM2G011523	unknown	6.47	1952.69	1191.21	825.81	451.70		3
GRMZM2G057823	ald1 (aldolase1)	1421.28	802.19	931.20	729.20	695.52		6
GRMZM2G088511	unknown	998.56	928.75	1323.82	1023.23	1005.62		6
GRMZM2G084195	Histone H4	1273.38	358.28	845.45	883.28	770.60		6
GRMZM2G091715	unknown	1207.21	495.31	568.79	534.97	466.14		6
GRMZM2G303374	unknown	954.01	888.49	1058.47	1105.93	1267.46		6
GRMZM2G152466	tua4 (alpha tubulin4)	1504.67	548.22	923.32	1070.86	1147.94		6
GRMZM2G165901	rab15 (responsive to abscisic acid15)	1063.15	2870.95	2945.72	2222.61	1919.31		6
GRMZM2G072855	Histone H4	1242.98	248.54	541.83	557.07	509.32		6
AC233865.1_FG001	Histone H4	2001.74	427.13	765.30	812.44	745.99		6
GRMZM2G031545	unknown	801.56	714.73	982.54	1193.72	1366.95		6
GRMZM2G156632	WIP1 (wound induced protein1)	2.36	7158.54	2770.89	788.17	254.05		3
GRMZM2G028393	sci1 (subtilisin-chymotrypsin inhibitor homolog1)	18.32	4544.08	2155.63	1779.17	1152.94		3
GRMZM2G126900	unknown	1.15	1308.76	571.75	490.23	435.86		3

[a]Fragments per kilobase of exon per million fragments mapped (FPKM) at 0, 24, 36, 48, and 72 h after placement on tissue culture initiation medium and the assigned gene cluster number determined by k-means analysis.

Arabidopsis, and Chinese white poplar, the transcription factor, BABY BOOM, when ectopically expressed in recalcitrant lines in tissue culture, was shown to induce somatic embryogenesis [6,7]. LEAFY COTYLEDON and PINFORMED genes have been thoroughly studied in zygotic embryogenesis in normal seed development with some studies suggesting that these genes may also be important to somatic embryogenesis in tissue culture [7]. In addition, regulatory genes such as AGAMOUS, WUSCHEL and CLAVATA have been studied in Arabidopsis for their role in meristem formation, somatic embryo formation, and callus maintenance [6,7], yet their role in maize tissue culture is not well understood.

To improve the understanding of genomic factors involved in early somatic embryogenesis in maize, we examined the transcriptome of the highly embryogenic maize inbred line A188 at 0 to 72 hours (h) after placement of immature embryo explant tissues onto culture initiation medium. Some of the first embryogenesis-related alterations in cell processes and cell division that are necessary for efficient embryogenic response occur during the early initiation stages. Based on our findings, we propose a coordinated expression model for somatic embryogenesis-related genes and describe an overview of global expression trends highlighting genes that are up- and down-regulated during the time course of the study. Genes related to somatic embryogenesis in other species and the relative expression of maize genes with high sequence similarity is also discussed. This research provides important information relating to the improvement of crop tissue culture and genetic engineering systems.

Results

Global analysis

RNA-seq reads were generated for nine immature embryo samples consisting of 25 embryos per sample of the maize inbred line A188. Embryo samples were placed on culture initiation medium from 0 to 72 hours. In total, the number of reads per sample ranged from 11 million (M) to 36 M (Table S1). Reads were aligned to the B73 maize reference genome sequence [17] and a set of representative transcript assemblies (RTAs) missing in the B73 reference genome sequence that were identified in transcriptome analyses of 503 maize inbred lines including B73 [18]. Expression values were determined using fragments per kb exon model per million mapped reads (FPKMs) using Cufflinks [19]. Biological replicates at each time point were correlated to assess data reproducibility. Pair-wise Pearson's correlations of expression values between embryos from two different donor plants collected at the same time point ranged from 0.9643 to 0.9927 (Table S2), indicating a high degree of reproducibility. Based on this analysis, average expression values from the two replicates were used for downstream analyses. A total of 28,992 annotated B73 reference genes and 6,405 RTAs were expressed in at least one time point (FPKM>0); while 10,464 reference genes and 2,276 RTAs were not expressed in any sample (Table S3).

The highest expressed reference genes across time points included genes that function in stress response, RNA binding, DNA synthesis and chromatin structure (Table 1). For example, GRMZM2G156632 is a highly expressed gene which is annotated as wound induced protein 1 (WIP1). Another gene related to plant defense that was among the highest expressed genes was GRMZM2G051943, which encodes for chitinase A1. The RTA with the highest expression at 0 h was joint_Locus_12721 with an FPKM value of 399.89 which decreased over 8-fold to 49.13 at 72 h. The highest RTA expressed at 36 and 48 h was joint_Locus_33043 with an FPKM value of 13.07 at 0 h and FPKM

values of 309.83 and 200.60 at 36 and 48 h, respectively. This RTA was annotated as encoding IN2-1, which based on sequence similarity, is a glutathione S-transferase (GST) protein. The highest expressed RTA at 72 h was joint_Locus_83 (204.13 FPKM) which did not match any known gene annotations.

Characterization of genes with 8-fold or greater expression change. In order to gain an understanding of genes expressed in this time course study, we selected genes differentially expressed by at least 8-fold compared to the control time point (0 h). Comparison of gene expression patterns across the surveyed time points indicated that the largest number of genes with a change in expression profile was from 0 to 24 h (Figure 1). This is supported by the observation that 1,856 genes were expressed at (or greater than) an 8-fold change when comparing 0 vs 24 h, 1,559 genes at an 8-fold change when comparing 0 vs 36 h, 1,496 genes at an 8-fold change when comparing 0 vs 48 h, and 1,488 genes at an 8-fold change when comparing 0 vs 72 h. Similarly, comparisons at other time points revealed 177, 45, and 41 genes differentially expressed 8-fold in comparisons of 24 vs 36, 36 vs 48, and 48 vs 72 h, respectively. Most genes differentially expressed at 8-fold were up-regulated. For example, 72%, 67%, 66%, and 72% of the genes differentially expressed when compared to 0 h were up-regulated at 24, 36, 48, and 72 h time points, respectively (Table S4). When considering a 2-fold change in expression, 8,174 genes were differentially expressed when comparing 0 vs 24 h, 6,737 genes when comparing 0 vs 36 h, 6,444 genes when comparing 0 vs 48 h, and 6,580 when comparing 0 vs 72 h.

The most abundant genes with large expression changes were enriched for biological processes such as oxidation-reduction processes, metabolic processes, protein phosphorylation, and transmembrane transport (Table S5). For example, genes with an 8-fold expression change or greater at 24 h were enriched for antiporter and transmembrane transport such as GRMZM2G006894, a hydrogen-exporting ATPase and GRMZM2G479906, GRMZM2G415529, and GRMZM2G366146 are ABC transporters. Other genes up-regulated at least 8-fold at 24 h were involved in transport of amino acids, sugars, and peptides or were specific to transmembrane transport of important nutrients. There were fewer genes that were down-regulated at 8-fold or greater. These genes were involved in membrane transport of amino acids and metals. For example, GRMZM2G140328 and GRMZM5G892495 are down-regulated 8-fold at 24 h and are both involved in calcium signaling. Genes that were up-regulated 8-fold or greater at 72 h revealed glycosyl-related genes such as GRMZM2G179063, a UDP-glucosyltransferase, and iron ion binding genes such as GRMZM2G103773, a BRASSINOSTEROID-6-OXIDASE 2. Genes down-regulated greater or equal to 8-fold between 0 h and 72 h were also enriched for genes involved in stress response such as ATP binding heat shock proteins GRMZM2G360681 and GRMZM2G310431, genes involved in nutrient assimilation such as GRMZM2G087254 and AC189750.4_FG004 both adenylyl-sulfate reductases, and genes involved in regulation of transcription such as GRMZM2G011789, a CCAAT box binding transcription factor.

Characterization of genes grouped by k-means analysis. The induction of somatic embryogenesis involves a complex coordination of multiple pathways [20,21]. Genes involved in hormone response, signal transduction, stress response, transcriptional regulation and cellular reorganization have been described previously [7,9,20]. We sought to determine if our maize transcriptome data supported concepts and models regarding these major biological functions during the very early stages of embryogenic tissue culture initiation. Using k-means clustering

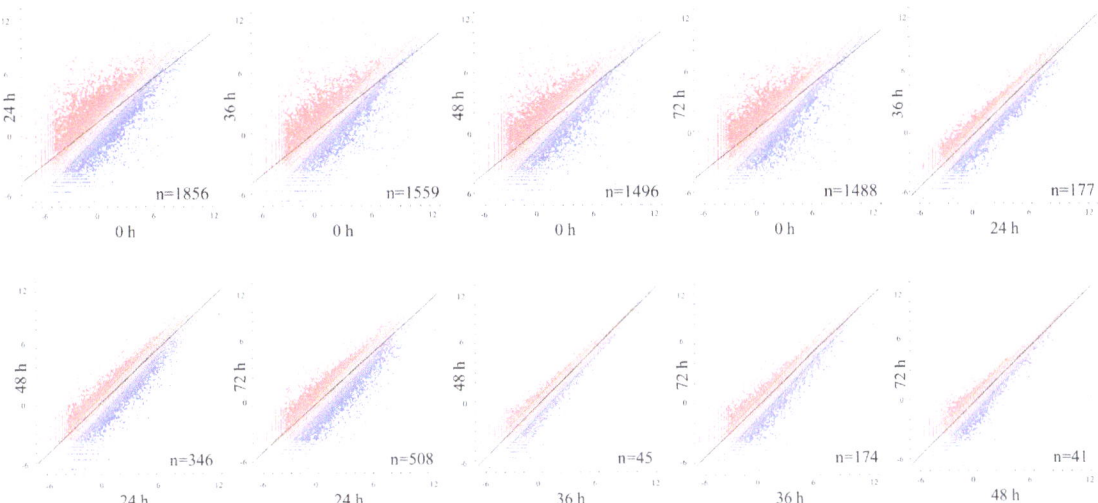

Figure 1. Gene expression changes in early somatic embryogenesis. Scatter plots of gene expression changes as log2 values of fragments per kilobase of exon model per million fragments mapped (FPKM) in immature zygotic embryo explants of maize inbred line A188 after placement on culture initiation medium for each time point comparison at 0, 24, 36, 48, and 72 h where n is the number of genes differentially expressed greater than 8-fold for each time point comparison. Red dots represent genes that are up-regulated, blue dots represent genes that are down-regulated, the middle green line indicates no fold change in expression, the two outer green lines indicate a 2-fold change in expression, and the solid black line is the best fit linear correlation.

with six clusters, we identified groups of genes with similar expression patterns including: (1) up-regulated and then down-regulated during the developmental window highlighted in this study, (2) both up- and down-regulated during the time course, (3) genes with an expression trend towards increased up-regulation from 0 to 24 h, (4) genes with a higher up-regulation later in the time course at 36, 48 and 72 h compared to all other genes expressed in the developmental window, (5) genes with an expression trend towards large-scale down-regulation from 0 to 24 h, and (6) genes with constitutive expression throughout the study (Figure 2 and Table S6).

Gene ontology enrichment was significant for clusters 1, 2, 3, and 6 (Table S7). Genes in cluster 1 were enriched for protein kinase and phosphorylation activity. Specifically, these genes were enriched for functions involving DUF26 signaling receptor kinases and post-translational modification receptor like kinases, as well as UDP glucosyl and glucoronyl transferases. Gene expression values in cluster 2, enriched for apoptotic processes, ranged from a minimum FPKM of 0.006 to a maximum of 4.414. Since the expression of these genes in cluster 2 was very low, these FPKM values could be inaccurate and attributed to noise. Genes in cluster 3 (initially up-regulated) were involved in numerous functions including transmembrane transport activity, oxidation-reduction

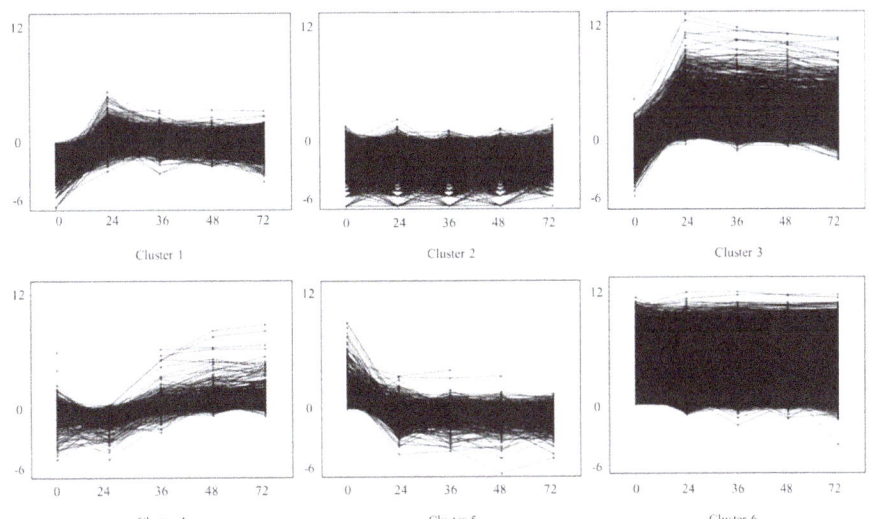

Figure 2. K-means clustering of genes expressed during early somatic embryogenesis. Log2 values of fragments per kilobase of exon model per million fragments mapped (FPKM) in genes with greater than zero FPKM expressed in immature zygotic embryo explants of maize inbred line A188 at 0, 24, 36, 48, and 72 h after placement on culture initiation medium grouped by expression trends as uncentered Pearson's correlation coefficient in six k-means clusters.

Table 2. Somatic embryogenesis-related genes, National Center for Biotechnology Information (NCBI) accessions, and plant species where accession was previously characterized.

Gene name	NCBI gene accession number	NCBI protein accession number	Species	Reference
AGL15	U22528	AAA65653	Arabidopsis	[65]
AtLEC2	AF400124	AAL12005	Arabidopsis	[48,76]
BnBBM1	AF317904	AAM33800	B. napus	[33,34,36,37]
CLV1	U96879	AAB58929	Arabidopsis	[63]
ZAG1	L18924	AAA02933	Maize	[77]
ZAG2	L18925	AAA03024	Maize	[77]
ZmLEC1	AF410176	AAK95562	Maize	[47,78]
ZMM2	L81162	AAB81103	Maize	[79]
ZmPIN1a	DQ836239	ABH09242	Maize	[54,55]
ZmPIN1b	DQ836240	ABH09243	Maize	[54,55]
ZmPIN1c	EU570251	ACB55418	Maize	[54]
ZmSERK1	AJ400868	CAC37640	Maize	[80,81]
ZmSERK2	AJ400869	CAC37641	Maize	[80,81]
ZmSERK3	AJ400870	CAC37642	Maize	[80]
ZmWUS1	AM234744	CAJ84136	Maize	[60]
ZmWUS2	AM234745	CAJ84137	Maize	[60]

processes, and heme binding or iron ion binding such as cytochrome P450 related genes. Finally, genes in cluster 6 were enriched for intracellular functions such as chromatin structure and DNA synthesis, ribosomal proteins synthesis, transcription factors, cell transport and RNA processing. A total of 2,704 RTAs, the largest proportion of RTAs grouped into a k-means cluster, were grouped into cluster 6 (Table S6).

Since gene ontology enrichment was not significant for clusters 4 and 5, MapMan B73 5b gene annotations [22] were used to describe genes in these clusters. Genes in cluster 4 were related to protein degradation, signaling receptor kinases, transcription factors, and genes involved in hormone metabolism and secondary metabolism. Genes in cluster 5 were involved in similar functions to cluster 4, but in addition, some cluster 5 genes were annotated for functions in amino acid and lipid metabolism (Table S6). RTAs with annotations relating to transcription factors that promote embryo development which could be involved in somatic embryogenesis are RTA joint_Locus_9393 annotated as an ethylene responsive transcription factor and expressed from 32.70 to 22.46 at 0 h and 72 h, respectively. Similarly, the RTA joint_Locus_7247 which was annotated as encoding an AP2 domain transcription factor was expressed from 10.01 to 14.33 at 0 and 72 h, respectively. Both RTAs were grouped into cluster 6 by k-means analysis. RTAs with interesting annotations and expression trends related to stress factors are joint_Locus_19099 annotated as encoding a GST-30 which showed a decrease in expression from 60.98 at 0 h to 5.41 at 72 h, and joint_Locus_9459, annotated as a cytochrome P450 in maize, which showed an increase in expression from 1.82 at 0 h to 22.04 at 24 h.

Candidate genes previously described in somatic embryogenesis

We performed an in-depth review of the literature to identify major candidate genes previously reported or suggested to be important for somatic embryogenesis in maize and other species,

and using sequence similarity we identified orthologs in maize for genes identified in other species (Table 2).

Genes involved in stress responses previously suggested to be important in somatic embryogenesis include GST and germin like-proteins (GLP). Using gene accessions [23] and protein sequence similarity, we identified 15 maize GST genes, of which several showed an 8-fold or greater increase from 0 to 24 h (Table 3). In addition, one maize GLP gene GRMZM2G045809, annotated as ZmGLP2-1 [24], was up-regulated greater than 8-fold from 1.44 at 0 h to 251.27 FPKM at 72 h (Table 3). These stress response genes exhibiting a large fold change and increased expression from 0 to 24 h were grouped into k-means cluster 3.

Genes involved in embryogenic pathway initiations include BABY BOOM (BBM) and LEAFY COTYLEDON (LEC) genes [7]. In this study, we highlight three maize genes that showed high sequence similarity to the highly conserved AP2 binding domain of *Brassica napus* BBM (BnBBM1, accession number AF317904). GRMZM2G366434, GRMZM2G141638, and GRMZM2G139082 are 91.2%, 92.5%, and 93.2% similar to the translated amino acid sequence of BnBBM1, respectively (Figure S1). GRMZM2G366434 showed a 4-fold up-regulation relative to 0 h at 36, 48 and 72 h (Figure 3A), GRMZM2G141638 also increased during this time course (Figure 3B), and GRMZM2G139082 increased over 4-fold from 0 to 72 h (Figure 3C). These maize BBM-like genes were grouped into cluster 3.

GRMZM2G011789, the maize ortholog to LEAFY COTYLEDON1 (ZmLEC1), was grouped into cluster 5 by k-means analysis. GRMZM2G011789 was expressed early initially (62.07 FPKM at 0 h) and then decreased dramatically to 2.90 at 24 h and 0.56 at 72 h (Figure 3D). Using sequence similarity, we found that maize gene GRMZM2G405699 is 47.4% similar in protein sequence to the Arabidopsis LEAFY COTYLEDON2 (AtLEC2, Figure S2) and 99.8% similar to the maize VIVIPARIOUS1 (VP1) gene GRMZM2G133398 (GenBank accession M60214). GRMZM2G405699 showed a moderate increase in expression during the time course of this study from 23.05 FPKM at 0 h to 30.50 FPKM at 72 h (Figure 3E), and GRMZM2G133398 (VP1)

Table 3. Expression of maize glutathione-S-transferase genes (ZmGST) and maize germin-like proteins (ZmGLP) in tissue culture.[a]

Gene name	Annotated name	FPKM					Fold change	Cluster number
		0 h	24 h	36 h	48 h	72 h		
ZmGST 8	GRMZM2G156877	0.63	25.66	32.87	33.53	24.51	8-fold	3
ZmGST 9	GRMZM2G126763	1.14	0.14	0.34	0.54	0.45	8-fold	2
ZmGST 10	GRMZM2G096153	21.22	17.45	18.86	23.83	24.42		6
ZmGST 11	GRMZM2G119499	1.02	0.92	2.81	7.00	7.15		4
ZmGST 12	GRMZM2G096269	1.04	0.84	2.43	4.76	2.32		4
ZmGST 13	GRMZM2G126781	0	0	0.22	0.65	0		
ZmGST 14	GRMZM2G175134	1.14	2.90	6.02	12.58	15.26	8-fold	3
ZmGST 15	GRMZM2G150474	0.18	3.25	3.85	3.66	2.64	8-fold	3
ZmGST 16	GRMZM5G895383	0	0.13	0.12	0.26	0.43		
ZmGST 18	GRMZM2G019090	7.05	121.54	206.88	175.17	105.79	8-fold	6
ZmGST 19	GRMZM2G335618	0.73	52.51	45.53	50.03	49.92	8-fold	3
ZmGST 20	GRMZM2G434541	1.93	16.64	10.82	8.91	4.71	8-fold	3
ZmGST 21	GRMZM2G428168	15.61	151.54	198.39	219.32	194.54	8-fold	6
ZmGST 22	GRMZM2G330635	59.24	208.94	140.95	136.98	118.47	8-fold	6
ZmGST 23	GRMZM2G416632	5.56	211.88	200.78	209.11	143.91	8-fold	3
ZmGST 24	GRMZM2G032856	0.06	9.28	4.88	3.05	1.12	8-fold	1
ZmGST 25	GRMZM2G161905	0.91	51.28	15.15	6.36	1.75	8-fold	3
ZmGST 26	GRMZM2G363540	0.59	0.07	0	0	0	8-fold	
ZmGST 27	GRMZM2G077206	0.08	0.15	0.09	0.06	0		
ZmGST 28	GRMZM2G146475	7.88	16.79	13.13	15.44	11.86		6
ZmGST 29	GRMZM2G127789	0.45	0.70	0.59	0.61	0.44		2
ZmGST 30	GRMZM2G044383	41.73	49.57	30.16	15.25	3.37	8-fold	6
ZmGST 31	GRMZM2G475059	10.05	52.94	40.30	31.53	21.71		6
ZmGST 32	GRMZM2G041685	0	0.38	0.07	0.31	0.45		
ZmGST 33	GRMZM2G028821	0	0	1.28	6.88	11.82		
ZmGST 34	GRMZM2G145069	0	0	0	0	0		
ZmGST 34	GRMZM2G149182	0	0	0	0	0		
ZmGST 35	GRMZM2G161891	0	0.07	0.46	0.87	2.06		
ZmGST 37	GRMZM2G178079	4.09	15.62	17.86	18.05	20.88		6
ZmGST 38	GRMZM2G066369	0.21	0	0.26	0.54	0.29		
ZmGST 40	GRMZM2G054653	0.08	0.37	0.52	0.40	0.68	8-fold	2
ZmGST 41	GRMZM2G097989	26.40	26.00	22.70	32.69	34.13	8-fold	6
ZmGST 42	GRMZM2G025190	2.10	197.29	133.85	90.71	66.98	8-fold	3
ZmGLP2-1	GRMZM2G045809	1.44	25.16	373.23	412.11	251.27	8-fold	3

Table 3. Cont.

Gene name	Annotated name	FPKM					Fold change	Cluster number
		0 h	24 h	36 h	48 h	72 h		
ZmGLP3-1	AC190772.4_FG011	0	0.11	7.97	4.56	2.48		
ZmGLP3-2	GRMZM2G030772	0	0	6.04	5.85	1.73		
ZmGLP3-3	GRMZM2G149714	0	0.07	2.77	2.45	1.23		
ZmGLP3-16	GRMZM2G072965	0	0.15	0.24	0.49	0.17		
ZmGLP10-1	GRMZM2G178817	0	0.62	1.49	0.45	0.18		
ZmGLP10-2	GRMZM2G071390	0	0.64	0.59	0.34	0		
ZmGLP10-3	GRMZM2G049930	0	1.39	2.88	0.79	0.32		

[a]Fragments per kilobase of exon per million fragments mapped (FPKM) at 0, 24, 36, 48, and 72 h after placement on tissue culture initiation medium, genes with an 8-fold change in expression or greater as compared to the 0 h time point, and the assigned gene cluster designated by k-means analysis.

showed a different expression pattern with a moderate decrease from 59.04 FPKM at 0 h to 46.03 FPKM at 72 h (Figure 3F). GRMZM2G405699 and GRMZM2G133398 were grouped into k-means cluster 6.

SOMATIC EMBRYOGENESIS RECEPTOR-LIKE KINASE (SERK) genes are also important for embryogenic pathway initiation. In this study, expression of SERK1 (ZmSERK1, GRMZM5G870959) was minimal, ranging from 4.23 to 5.86 FPKM. Similarly, the orthologs to maize SERK2 (ZmSERK2, GRMZM2G115420) and the ortholog to maize SERK3 (ZmSERK3, GRMZM2G150024) showed very similar magnitudes in expression and trend increasing from about 15 to 20 FPKM. In our study, both ZmSERK2 and ZmSERK3 increased nearly 2-fold from 0 h to 24 h. Maize SERK genes were grouped into cluster 6.

PIN1 is involved in auxin transport [25]. The maize PINFORMED1 (PIN1) gene, (ZmPIN1a, GRMZM2G098643) displayed up-regulation with FPKM values of 11.02 at 0 h to 153.59 at 72 h (Figure 4A), and additional orthologs to maize PINFORMED1 (ZmPIN1b, GRMZM2G074267) and (ZmPIN1c, GRMZM2G149184) also increased in expression. ZmPIN1a and ZmPIN1b were grouped into cluster 6; ZmPIN1c was grouped into cluster 3 (Figure 2).

Known genes involved in embryo formation and development include WUSCHEL, CLAVATA, AGAMOUS and WOX genes. ZmWUS1 (GRMZM2G010929) was minimally expressed, not exceeding 1 FPKM during this time course (Figure 4B), and ZmWUS2 (GRMZM2G028622) was not expressed in any sample. GRMZM2G14151 has high sequence similarity to the CLAVATA (CLV1) gene in Arabidopsis (Figure S3) and increased in expression from 13.14 FPKM at 0 h to 20.06 FPKM at 72 h (Figure 4C). GRMZM2G14151 was grouped into k-means cluster 6. Maize genes that are orthologs to AGAMOUS, which include ZMM2 (GRMZM2G359952), ZAG1 (GRMZM2G052890), and ZAG2 (GRMZM2G160687), showed minimal expression during the time course of this study (Figure 4D and Table 2). A BLAST search for AGL15 revealed a number of maize genes with high sequence similarity. For example, ZmMADS69 (GRMZM2G171650), ZmMADS52 (GRMZM2G446426), and ZmMADS73 (GRMZM2G046885) show 67.74%, 64.41%, and 42.11% sequence similarity to the AGL15 amino acid sequence in Arabidopsis. These MADS box transcription factors were grouped into k-means cluster 6 and were moderately expressed throughout the time course, with FPKM values greater than 10 at every time point (Figure S4A–C). One maize gene, ZmMADS11 (GRMZM2G139073), has 45.75% sequence similarity to AGL15. ZmMADS11 was grouped into k-means cluster 3 and was shown to have an 8-fold expression change at each time point compared to 0 h (Figure S4D). We also examined the expression of maize WUSCHEL-related homeobox domain (WOX) genes and found ZmWOX2A (GRMZM2G108933), ZmWOX5A (GRMZM2G478396), ZmWOX5B (GRMZM2G116063), and ZmWOX11/12B (GRMZM2G314064) showed an 8-fold increase in expression after placement of immature embryos into the tissue culture environment and grouped into k-means clusters 1, 2, and 3 while other maize WOX genes were grouped into clusters 5 and 6 (Table 4).

Discussion

Somatic embryogenesis-related genes have been extensively characterized in Arabidopsis; however, relatively few have been evaluated in maize. Using transcriptome data of maize in embryogenic tissue culture initiation, this study provides an in-

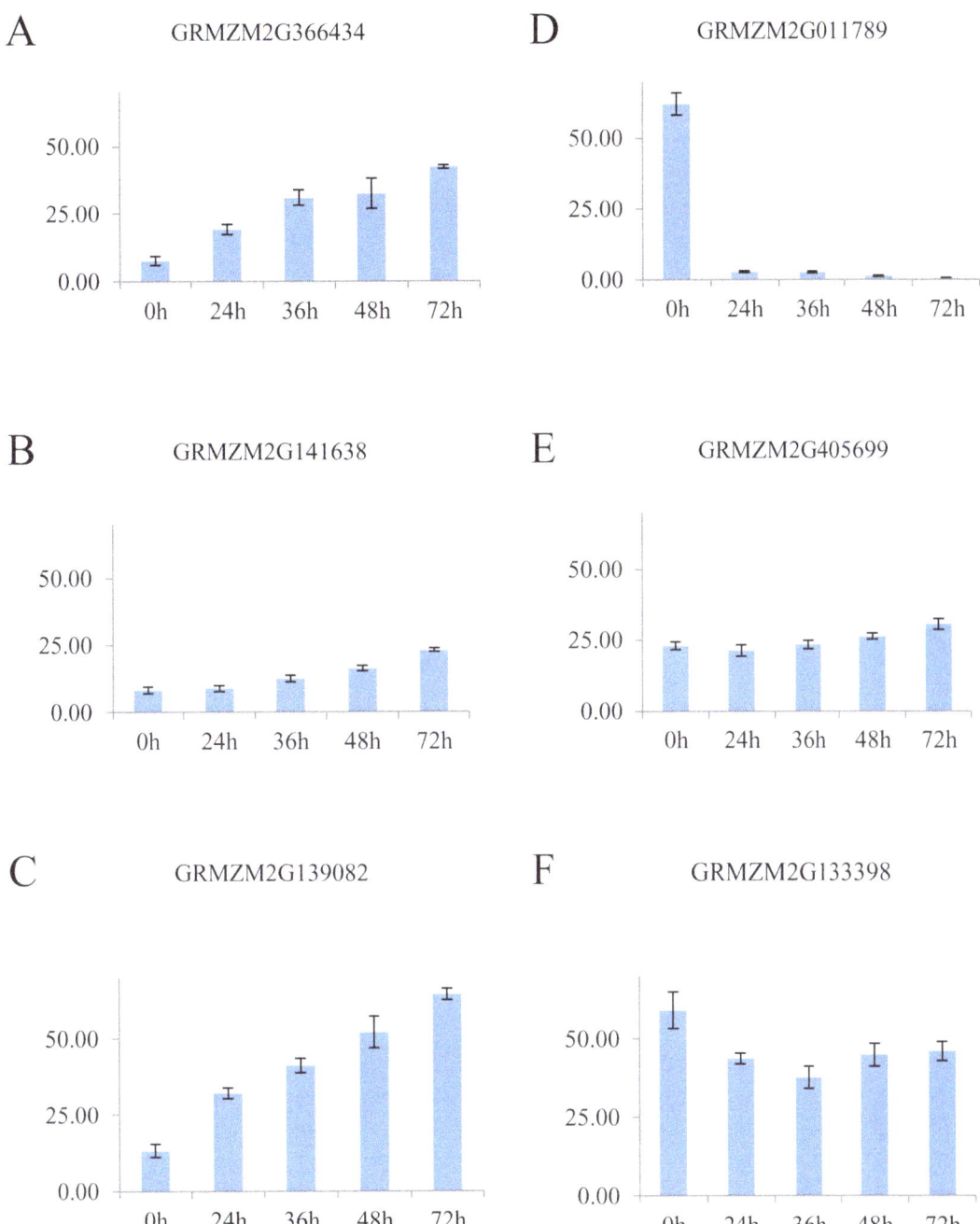

Figure 3. Gene expression of somatic embryogenesis genes involved in induction. Average values of fragments per kilobase of exon model per million fragments mapped (FPKM) of genes involved in the induction of somatic embryogenesis expressed in immature zygotic embryo explants of maize inbred line A188 at 0, 24, 36, 48, and 72 h after placement on culture initiation medium where (A), (B), and (C) are BBM-like maize genes with high sequence similarity to *Brassica napus* BABY BOOM (BnBBM1), (D) is maize LEAFY COTYLEDON1 (ZmLEC1), (E) is a maize gene with high sequence similarity to *Arabidopsis thaliana* LEAFY COTYLEDON2 (AtLEC2), and (F) is maize VIVIPARIOUS1 (VP1). Bars indicate average mean ± SE (n = 4 for 0, 24, 36, and 48 h include technical and biological replicates; n = 2 for 72 h include only technical replicates).

depth look at the major candidate genes discussed in previous reviews and research studies on somatic embryogenesis. Moreover, we propose a model (Figure 5) based on coordinated expression (Table S8) of somatic embryogenesis-related genes highlighted in this study and their relative expression in early embryogenic tissue culture response.

Genes associated with stress response in tissue culture

Our observations support previous reviews on the transition to somatic embryogenesis, with our whole transcriptome data showing a large number of genes expressed during the early stages of somatic embryogenesis from 0 h to 24 h in tissue culture. Gene enrichment analysis of genes clustered based on k-means, and of genes grouped by large fold changes when compared to the control time point 0 h, supported major biological functions

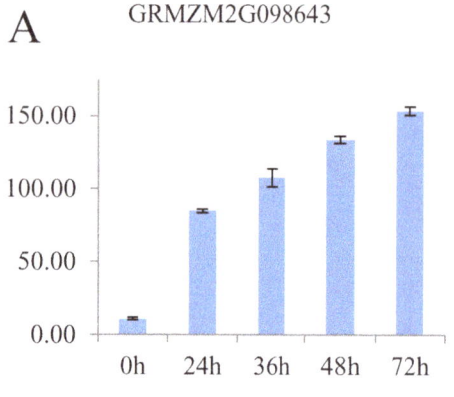

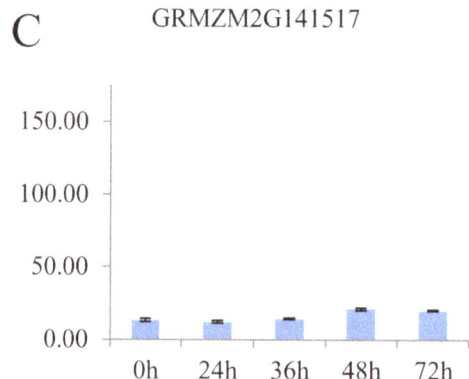

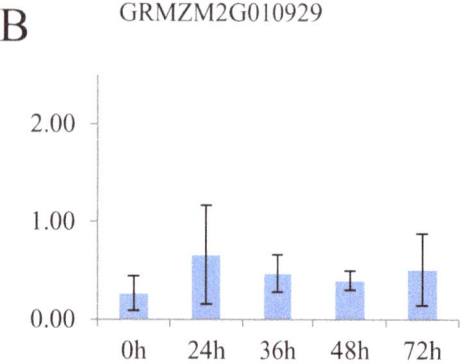

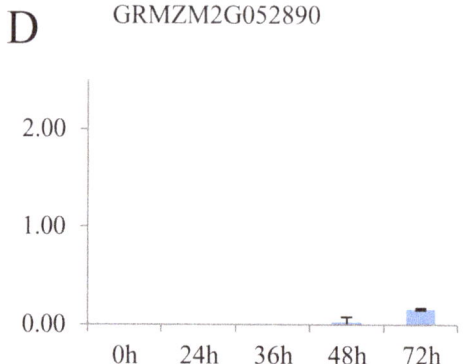

Figure 4. Gene expression of somatic embryogenesis genes involved in callus initiation and maintenance. Fragments per kilobase of exon model per million fragments mapped (FPKM) of maize genes associated with embryogenic callus induction expressed in immature zygotic embryos of maize inbred line A188 at 0, 24, 36, 48, and 72 h after placement on culture initiation medium where (A) is the maize ortholog to PINFORMED1 (ZmPIN1a), (B) is the maize ortholog to WUSCHEL (ZmWUS1), (C) is maize gene with high sequence similarity to CLAVATA (CLV1) and (D) is a maize ortholog to AGAMOUS (ZAG1). Bars indicate average mean ± SE (n = 4 for 0, 24, 36, and 48 h include technical and biological replicates; n = 2 for 72 h include only technical replicates).

suggested in previous studies as important for somatic embryogenesis such as stress response, transmembrane transport, and hormone metabolism [6,8,9,21]. Genes with a large fold change and genes grouped in cluster 3 in this study include cytochrome P450, UDP-glucosyl, and glucoronyl transferases. In another study involving an embryogenic maize line from China, genes differentially expressed in the early stages of embryogenesis were also related to stress where metabolism of xenobiotics by cytochrome P450 was identified as one of the most significant pathways by enrichment analysis of differentially expressed genes in samples grown 1–5 days after tissue culture [26].

In this study, we identified two maize genes, WIP1 (GRMZM2G156632) and chitinase A1 (GRMZM2G051943) which were up-regulated over 1500-fold from 0 h to 24 h. These genes have been previously described in plant defense and stress response [27,28], but have not, until now, been associated with tissue culture response in maize. WIP1 has previously been characterized as a defense gene based on its involvement in hypersensitive defense response [28]. Chitinase proteins have been suggested to promote somatic embryogenesis [9] since one study in carrot showed that a non-embryogenic mutant line was triggered

to produce somatic embryos after the addition of chitinase proteins in the tissue culture medium [29].

GSTs are a family of genes also involved in plant defense [30] and we observed 15 out of the 33 maize GST genes with large fold expression changes during early somatic embryogenesis (Table 3). It has been suggested that some GSTs may function in tissue dedifferentiation by affecting the cell's redox status by changing endogenous levels of important plant growth hormones such as auxin [6]. GSTs were also detected in chicory during somatic embryogenesis in callus cultures initiated with leaf tissue [30], and GSTs were also expressed in response to auxin treatment in *Cyclamen persiucum* tissue culture with an initial up-regulation during the first 4 hours followed by down-regulation at 72 h [31]. In this study, GST genes were found to be coexpressed with BBM, WUS, PIN, and SERK genes (Figure 5, Table S8). We also detected one maize GLP gene (GRMZM2G045809) with a large fold change in expression at 72 h (Table 3). Moreover, this GLP gene was shown to be coexpressed with the BBM transcription factor (Figure 5). GLPs are proteins that also affect the plant redox status and are involved in developmental regulation. In wheat embryogenic callus cultures, GLPs were detected as early as 2 to 72 hours after plating explant tissues in culture [32]. GLPs are

Whole Transcriptome Profiling of Maize during Early Somatic Embryogenesis Reveals Altered Expression...

221

Table 4. Expression of WUSCHEL-related maize WOX genes in tissue culture.[a]

Gene name	Annotated name	Sequence similarity (%)	FPKM					Fold change	Cluster number
			0 h	24 h	36 h	48 h	72 h		
ZmWOX2A	GRMZM2G108933	100	0.70	0.34	0.08	0.04	0.11	8-fold	2
ZmWOX2B	GRMZM2G339751	100	0.18	0.68	0.15	0	0		
ZmWOX3A	GRMZM2G122537	100	0.23	0.98	0.32	0.53	0.39		2
ZmWOX3B	GRMZM2G069028	85.71	1.50	1.10	0.52	0.22	0.79		5
ZmWOX3B	GRMZM2G140083	84.81	0	0	0	0	0		
ZmWOX5A	GRMZM2G478396	96.72	0	4.51	9.06	6.20	6.61	8-fold	
ZmWOX5B	GRMZM2G116063	100	0.06	2.17	1.91	4.53	6.27	8-fold	1
ZmWOX9A	GRMZM2G133972	100	0.49	0.55	0.19	0.07	0		
ZmWOX9B	GRMZM2G031882	100	5.95	3.07	2.91	2.08	1.57		6
ZmWOX9C	GRMZM2G409881	100	5.53	3.01	4.86	3.42	2.85		6
ZmWOX11/12B	GRMZM2G314064	98.46	2.51	11.29	15.50	17.96	14.80	8-fold	3
ZmWOX13A	GRMZM2G038252	100	0	0	0	0	0		
ZmWOX13A	GRMZM2G069274	100	0	0	0	0	0		
ZmWOX13B	GRMZM5G805026	100	5.86	5.46	6.21	7.46	6.05		6
KNOTTED1	GRMZM2G017087	100	20.67	15.33	11.52	10.66	10.94		6

[a]Fragments per kilobase of exon per million fragments mapped (FPKM) at 0, 24, 36, 48, and 72 h after placement on tissue culture initiation medium, genes with an 8-fold change in expression or greater as compared to the 0 h time point, and the assigned gene cluster designated by k-means analysis.

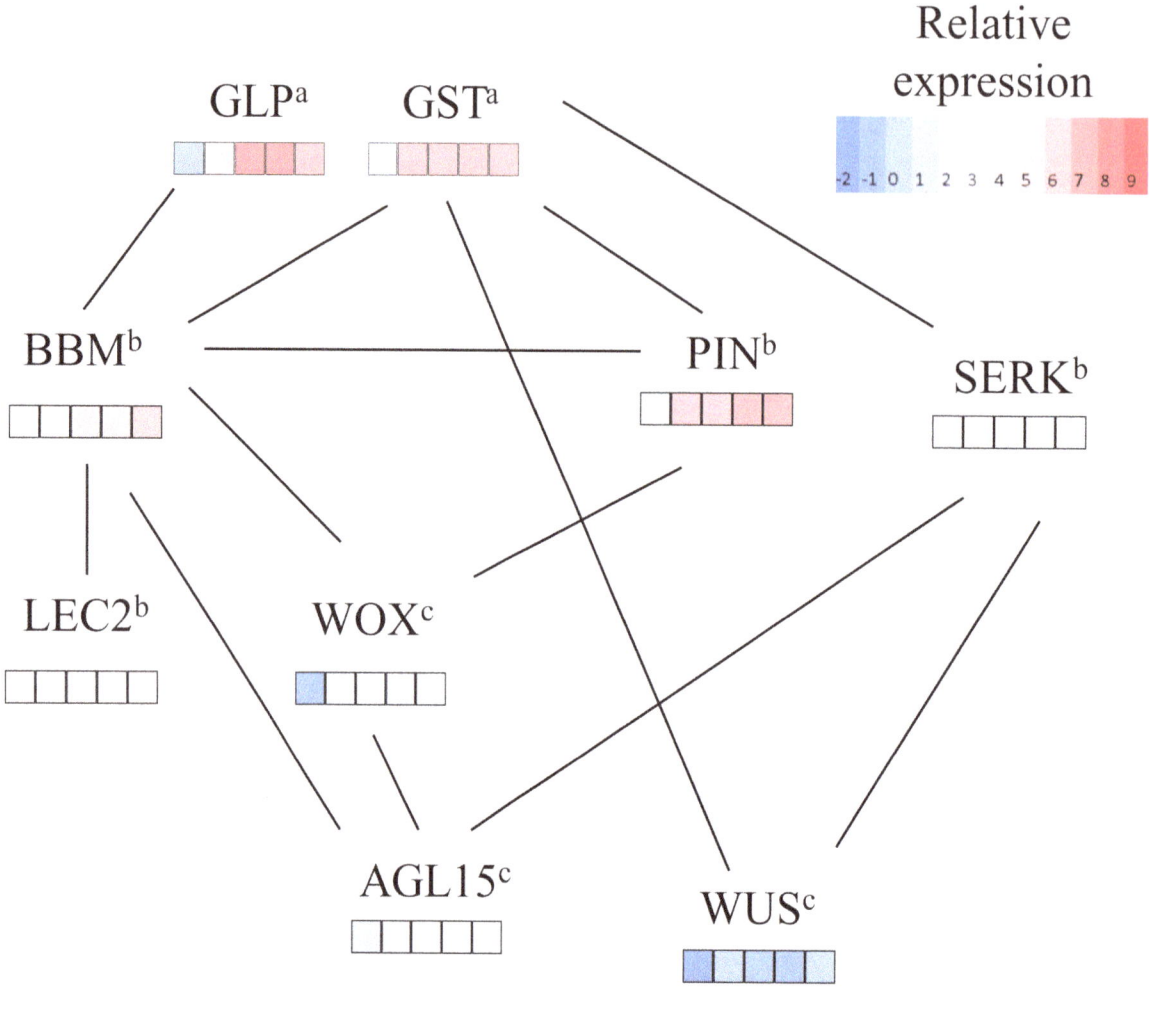

aGenes most commonly associated with stress response.

bGenes most commonly described as being involved in embryogenic pathway initiation.

cGenes most commonly described as being involved in somatic embryo formation and development.

Figure 5. Relative expression of coexpressed somatic embryogenesis-related genes. A proposed model of somatic embryogenesis-related gene expression networks as determined by coexpression (solid lines) with a correlation coefficient greater or equal to 0.9 between genes expressed during the early stages of somatic embryogenesis. Relative expression of transcripts detected in immature zygotic embryo explant tissues in tissue culture were detected in the inbred line A188 and reported as the average log2 expression displayed by color coded values as depicted by the figure legend for each gene. The first left most box under each gene name is the average log2 transformation of fragments per kilobase of exon model per million fragments mapped (FPKM) at 0 h, the second box at 24 h, the third at 36 h, the fourth at 48 h, and the fifth right most box is the average FPKM 72 h. Glutathione-S-transferases (GSTs) and germin-like proteins (GLPs) are stress response genes that are triggered in early somatic embryogenesis. BABY BOOM (BBM), an APETALA-like ethylene-responsive element transcription factor, and LEAFY COTYLEDON2 (LEC2), a B3 domain transcription factor, promote somatic embryogenesis. PINFORMED (PIN) genes mediate auxin transport and establish essential endogenous auxin concentrations in the cell. SOMATIC EMBRYOGENESIS RECEPTOR LIKE KINASES (SERK) genes are also involved in somatic embryogenesis and hormone metabolism. WUSCHEL (WUS), a homeodomain transcription factor, regulates stem cell fate during embryo formation and development. AGAMOUS like-15 (AGL15), a MADS box transcription factor, also promotes somatic embryo formation and is also involved in meristem development. WUSCHEL-related homeobox domain (WOX) genes have also been detected during somatic embryogenesis in embryogenic genotypes but not in non-embryogenic genotypes.

typically detected in embryogenic tissues, but not in non-embryogenic tissues. GLPs with superoxide dismutase activity promote the production of hydrogen peroxide (H_2O_2), a type of oxidative stress. It has been suggested that the H_2O_2 produced may serve as a secondary signaling molecule acting to promote somatic embryogenesis [8,9].

Genes involved in embryogenic pathway initiation

Embryogenic pathway initiation is marked when somatic cells acquire embryogenic competence and proliferate as embryogenic cells capable of forming somatic embryos [8]. One gene that has been attributed to initiation of somatic embryogenesis across plant species is BBM. BBM genes highlighted in this study were found to be coexpressed with GLP, GST, PIN, WOX, LEC2, and AGL15 genes (Figure 5). BBM was first discovered in investigations of *Brassica napus* microsporogenesis by subtractive hybridization [33]. The gene was consistently expressed only in embryogenic microspore cultures. Sequence analysis showed that BBM has two unique binding domains: an APETALA-like AP2 binding domain and an ethylene-responsive element binding factor, both characteristic of functioning in plant hormone signaling and regulation [33]. Overexpression of BBM in Arabidopsis and *B. napus* led to the induction of somatic embryogenesis and regeneration ability without the addition of exogenous plant hormones [33]. This observation suggested that BBM acts as a stimulator of plant hormone production, triggering signaling pathways important for somatic embryogenesis [33,34]. Overexpression of BBM-induced embryo formation enhanced regeneration ability in Chinese white poplar [35] and tobacco, [36] and improved transformation efficiency in sweet pepper [37]. In another study focused on transforming artificial chromosomes into maize, the shuttle vector used contained a BBM homolog called ZmODP2 to promote cell division and callus growth after transformation [38]. Researchers suggested that the presence of this construct improved transformation efficiency in maize tissue culture by 20–50%. In our study, we identified three maize genes GRMZM2G366434, GRMZM2G141638, and GRMZM2G139082 with high sequence similarity to BnBBM1 and which contain the conserved and unique AP2 binding domain (Figure S1). These maize genes were also shown to increase in expression during early somatic embryogenesis (Figure 3) in this study. When we compared these maize gene expression trends to transcripts detected in the maize B73 gene atlas [39], expression was not at all detected (0 FPKM) in whole seeds or endosperm at 10, 12, 14, and 16 days after pollination. Expression was, however, detected in zygotic embryos 16 days after pollination, in germinating seed, in the primary root and in V3 stem and shoot apical meristem [40].

Another important group of genes involved in embryogenic pathway initiation are the LEC genes. LECs are transcription factors identified in studies of zygotic embryogenesis in plants that have been proposed to be important for somatic embryogenesis [7,41]. Mutational analysis of LEC genes showed their function in early zygotic embryogenesis, specifically, to maintain suspensor cell fate and specify cotyledon identity [42]. LEC genes play an important regulatory role directly interacting with hormone response genes [43,44]. AtLEC1 was cloned and ectopically expressed in transgenic Arabidopsis seeds, showing its essential role in germination and embryonic organ identity [45]. AtLEC1 in Arabidopsis in tissue culture was also shown to be differentially expressed in embryogenic compared to non-embryogenic samples [46]. One study on the highly embryogenic maize hybrid HiII, ZmLEC1 transcription in somatic embryos showed a high initial expression and then a decrease in expression during early development of [47]. We found a similar result in our study, where ZmLEC1 (GRMZM2G011789) decreased in expression over 20 fold during the first 24 hours in tissue culture. In contrast, the expression of a maize gene similar to AtLEC2 (GRMZM2G405699) based on high sequence similarity (Figure S2) increased steadily in this study (Figure 3), was grouped into k-means cluster 6 (Table S6), and is coexpressed with BBM (Figure 5). The role for AtLEC2 in Arabidopsis zygotic embryogenesis to induce somatic embryos by activating auxin responsive genes was proven by ectopic expression [48,49]. AtLEC2 is nearly identical to VP1 in Arabidopsis [7,49] and in this study, we found that the maize gene GRMZM2G405699, which is most similar to AtLEC2, is also highly similar to maize VP1 (GRMZM2G133398). Both AtLEC2 and VP1 genes share the

same class of unique B3 domains. One study involving gene expression analysis on T-DNA insertion lines in Arabidopsis suggested a role for VP1-like genes in recruiting chromatin-remodeling factors that can either activate or repress LEC1-like activity during seed development [50]. Moreover, it has been suggested that this complex network involving LEC1 and LEC2 genes in seed development can up-regulate important transcription factors such as BBM during early zygotic embryogenesis in Arabidopsis [44]. Our study showed GRMZM2G405699 coexpressed with maize BBM-like genes during early somatic embryogenesis (Figure 5).

PIN1 genes encode influx and efflux carrier proteins that mediate auxin transport in early zygotic embryo development [51]. In this study, PIN1 was coexpressed with GST, BBM, and WOX genes (Figure 5). In Arabidopsis, PIN genes, are essential for embryonic stem cell growth [52] and are expressed in early proembryonic development [53]. In maize, PIN1 genes also play a role in auxin transport and tissue differentiation during zygotic embryogenesis [54]. The ZmPIN1a gene was highly expressed in this study (Figure 4A), and increased dramatically from the 24 to 72 h time point. A study using *in situ* hybridization of ZmPIN1a and ZmPIN1b showed transcript abundance and protein localization of PIN proteins in maize kernels, endosperm and embryo [54,55]. The authors also suggested a role for PIN1 in maize development during zygotic embryogenesis in mediating polar auxin transport and patterning during development [25,54]. In tissue culture, an important step in establishing embryogenic patterning in embryos is apical-basal rearrangement [51]. Our observations also show that PIN1 genes are expressed during tissue culture response in early somatic embryogenesis.

Genes involved in somatic embryo formation and development

There is evidence that genes involved in meristem formation are also important in somatic embryo formation. For example, WUS is a homeodomain transcription factor involved in shoot and floral meristem development specifically as a regulator of stem cell fate and organ identity [56]. WUS expression has been detected in a small group of cells described as the organizing center of meristematic tissue. This organizing center is localized underneath a larger mass of stem cells [52,57,58]. WUS has an important role in regulating and activating pluripotent stem cells by promoting proliferation genes and repressing developmental regulators [59]. While, it has been shown in Arabidopsis that PIN1-mediated auxin transport directly induces WUS expression in early somatic embryogenesis [52], in our study, ZmWUS genes were minimally expressed, however, were coexpressed with GST and SERK genes (Figure 5). We hypothesize that the developmental window highlighted in this study may have captured a time when the organizing center was just initiating in development and that transcripts detected represented few cells showing WUS activity during the early stages of stem cell development. In addition, it is plausible that suitable endogenous auxin concentrations were just beginning to establish. Over time, more cells either localized in or on the organizing center would also display WUS transcriptional activity.

Detailed analysis of the expression pattern of WUS orthologs in maize and rice showed that WUS genes in higher plants did not mimic expression localized in the organizing center as it did in Arabidopsis implying a major modification in plant evolution [60]. Our findings showed that WUS genes had minimal to no expression in early embryogenesis in tissue culture but some maize WOX genes increased in expression over 8-fold. WOX genes were coexpressed with BBM, PIN and AGL15 (Figure 5).

WOX expression was detected in somatic embryogenesis in other plants where efficient embryogenic callus cultures are also genotype-dependent [61]. In our study, we highlight a number of maize WOX genes with differential expression compared to time point 0 h (Table 4) of which ZmWOX2A, ZmWOX5A/5B, and ZmWOX11/12B showed an 8-fold change in expression. Some examples of WOX genes in tissue culture in other plants include, WOX2 associated with somatic embryogenesis in conifer tissue culture [62] and WOX11 in grapes detected in embryogenic versus non-embryogenic cultivars *in vitro* [61].

Expression of a CLV1-like gene (GRMZM2G141517), however, showed a steady increase in this study (Figure 4C). CLV1 is a receptor-like kinase also involved in shoot and floral meristem development [63] and acts upstream of WUS. CLV1 represses WUS activity by interacting in a regulatory loop with WUS to promote callus initiation and maintenance [64]. Capturing expression of these genes at later time points in tissue culture would provide insight on transcriptional activity between the regulatory loop between WUS and CLV1 in maize.

Another meristem-related gene discovered in Arabidopsis is AGAMOUS, a MADS box transcription factor involved in flower development and organ differentiation [65]. AGAMOUS had been shown to interact directly with WUS also by repressing its expression in floral stem cells [58,66]. Similar to ZmWUS, genes highly similar to AGAMOUS, such as ZAG1, ZAG2 and ZMM2 were minimally expressed in this study. However, we did observe differential expression relative to 0 h for an AGL15-like gene (GRMZM2G139073) (Figure S4D). In addition, GLP, PIN, WOX and BBM genes were coexpressed with GRMZM2G139073 (Figure 5, Table S8). AGL15 in *B. napus* and Arabidopsis embryos [65] have been shown to be preferentially localized in embryonic tissues [67]. Additionally, AGL15 was shown to promote somatic embryo development in Arabidopsis and soybeans [68]. Studies also suggest that AGL15 in Arabidopsis interacts with LEC2 directly [43,69] and, immunoprecipitation and time-of-flight mass spectrometry revealed AGL15 was included in the SERK1 complex *in vivo* [70]. To date, there have been no studies of AGL15–like genes expressed in maize somatic embryos that have been reported. From studies on AGL15 in Arabidopsis in promoting somatic embryogenesis and interacting with LEC2 and SERK1, we hypothesize that maize AGL15-like genes may also be important for callus initiation and maintenance.

Conclusion

Deciphering the underlying genetic mechanisms controlling somatic embryogenesis in tissue culture is important for improving our understanding of the basic processes involved in somatic embryo formation, and in the development of embryogenic tissue culture systems that are less genotype dependent. Although few major genes that promote somatic embryogenesis in Arabidopsis and other plants species have been described, even fewer genes have been studied and their expression revealed in the context of the whole transcriptome in tissue cultures of maize. In this study, we highlighted the expression of maize genes with high sequence similarity to BBM, LEC2, CLV1, and AGL15, and maize SERK and PIN genes, and discussed their potential role in somatic embryogenesis. Many of the somatic embryogenesis related genes analyzed in this study fall into a k-means clusters 3 with an expression trend towards an initial large up-regulation and a second cluster number 6, with genes that are moderately to highly expressed throughout the targeted developmental. However, clusters 4 and 5 also show interesting expression trends that could

be important for further studies due to their large up- and down-regulation expression trends, respectively. In this investigation, we also highlighted maize gene families, mainly GST, GLP, and WOX genes and identified specific genes within gene families with altered expression. A number of specific genes discussed in this study could be potential candidates for further testing regarding their importance and contribution to embryogenesis in tissue culture in maize.

Whole transcriptome profiling during the very early stages in the initiation of somatic embryogenesis in culture of the highly embryogenic, regenerable maize genotype, A188, now provides new information on the expression of somatic embryogenesis-related genes in maize. By studying the whole transcriptome during a specific developmental window, we were able to provide data on transcripts detected for major genes previously described with a role in embryogenesis. This information can be utilized to help us better understand major gene functions and expression networks involved in the induction of somatic embryogenesis in culture. Investigations involving fine-mapping and identification of specific genes in maize that confer regeneration ability could build on the findings reported here to further enhance our understanding of which many genes expressed in concert are possible key factors underlying the genotype dependent nature of tissue culture phenotypes. In the same way, a study involving the analysis of the whole transcriptome of isogenic lines differing in their ability to produce embryogenic, regenerable cultures, and their representative transcripts that are not mapped to the reference genome, could also add to identifying causal genes, providing a deeper understanding of the somatic embryogenesis-related genes we described here, and allow determination of their level of significance in the process. Improving our understanding of the biological processes and the genetic mechanisms that confer efficient tissue culture response such as somatic embryogenesis *in vitro* will help crop improvement strategies and functional genomics testing that is necessary to increase agricultural productivity in a changing global agricultural landscape.

Materials and Methods

Plant material and tissue culture initiation

Field grown donor plants were grown at the West Madison Agricultural Research Station (Madison, WI). Immature maize embryos from two plants of the maize inbred line A188 were isolated and cultured as previously described [71] with minor modifications. Briefly, ten days after pollination, 125 immature embryos (1.0–1.2 mm from scutellar tip to base) from each of two maize ears were harvested, aseptically dissected from kernels, and then placed onto culture initiation medium by placing embryos axis side down (scutellum side up) on modified N6 tissue culture medium [72]. The medium was prepared with N6-basal salts [72] at 3.98 g/L (PhytoTechnolgies Lab, product number M524), 2 mL/L of 1 mg/mL 2, 4-D stock, 2.875 g/L L-proline, 30 g/L sucrose, 3.5 g/L gelzan, pH to 5.8. After autoclaving, filter sterilized N6 vitamins stock (1,000x solution) and silver nitrate stock solution prepared as per protocol [71] were added. Prior to embryo isolation, ears were surface sterilized in a 50% commercial bleach (8.25% sodium hypochlorite) solution with a drop of Tween 20, and then rinsed 3 times in sterile, deionized water. A total of 210 embryos from each of two donor plants, or two biological replicates, were used for this study. Ten to 25 embryos were harvested for the each of 0, 24, 36, and 48 h time points. Ten embryos from only one plant, or one biological sample, were harvested for the 72 h time point. For the first time point (0 h), the embryos were aseptically dissected from kernels and immediately placed into liquid nitrogen without placement on culture medium. For subsequent time points at 24 h, 36 h, 48 h, and 72 h after plating, embryos were aseptically isolated and placed onto culture initiation medium.

RNA-seq Library Construction and Sequencing

RNA was extracted using the Invitrogen TRIzol reagent according to the manufacturer's instructions (Invitrogen, http://www.invitrogen.com). Samples were processed using the RNeasy MinElute Cleanup kit (Qiagen, http://www.qiagen.com). RNA quality was assessed using the Agilent RNA 6000 Pico Kit Bioanalyzer prior to preparation of the sequence library. Approximately 5 µg of total RNA was processed for mRNA isolation, fragmented, converted to cDNA, and PCR amplified according to the Illumina TruSeq RNA Sample Prep Kit as per the provided protocol, and sequenced on an Illumina HiSeq 2000 (San Diego, CA) at the University of Wisconsin Biotechnology Center (Madison, WI). Two technical sequencing replicates were conducted for each of the two biological sample collections for the 0, 24, 36, and 48 h time points and one biological sample for the 72 h time point, each with 101 nucleotide single-end reads. Sequences are available in the Sequence Read Archive at the National Center for Biotechnology Information (BioProject accession number PRJNA242658). Sequence quality for each sample was evaluated using the FastQC software (http://www.bioinformatics.bbsrc.ac.uk/projects/fastqc) and all samples passed quality control analysis. For subsequent analyses, FPKM values from the two technical sequencing replicates were averaged to represent transcript abundance for each time point 0, 23, 36, and 48 h. FPKM values from two technical sequencing replicates were averaged from one biological sample for time point 72 h.

Data Analysis

To quantify transcript abundance, sequence reads for each sample were mapped to the maize v2 pseudomolecules (AGPv2; http://ftp.maizesequence.org) [17] and 8,681 non-RTAs that were assembled using RNA-seq reads from 503 diverse maize inbred lines [18]. Mapping was performed using Bowtie version 0.12.7 [73] and TopHat version 1.4.1 [74] with a minimum and maximum intron length of 5 bp and 60,000 bp respectively and the no-novel-indels option. All other parameters were set to the default values. Normalized gene expression levels were determined using Cufflinks version 1.3.0 [19] setting a maximum intron size of 60,000 bp, the version 5b annotation (http://ftp.maizesequence.org) as the reference annotation, and the AGPv2 fasta sequences for the bias detection and correction algorithm. All other parameters were set to the default values. Pearson's correlation of transcript abundance estimates were measured between biological replicates. Transcripts for samples for 0, 24, 36, and 48 h time points were averaged between the two biological replicate samples while transcripts for the 72 h time point represented transcripts detected in only one biological sample.

K-means clusters were determined using uncentered Pearson's correlation coefficients in DNA Star ArrayStar version 5.1.0 build 114 allowing 6 clusters and 100 iterations. Only genes with an FPKM value greater than zero at any given time point were included. For an analysis of differential gene expression, each time point was compared to the control time at 0 h. A threshold for differential expression of greater than 8-fold for raw FPKM values was used. In order to include genes that may have not been expressed at any given time point but then showed expression at other time points, we included genes with a sum of 2 FPKM or greater in the differential gene expression analysis. Raw values were log2 transformed and visualized on a scatter plot in DNA

Star ArrayStar version 5.1.0 build 114. In order to determine coexpression of selected genes, an analysis was done in the R using the xtable statistical computing package version 3.0.2 to calculate the Pearson correlation where the minimum coefficient was set to a threshold of 0.75. In the discussion highlighting the coexpression of specific somatic embryogenesis-related genes, the threshold was set to 0.90.

Gene ontology enrichment analysis was conducted in the PlantGSEA database (http://structuralbiology.cau.edu.cn/PlantGSEA/index.php) [75] to describe groups of genes in specific clusters or groups of genes differentially expressed with large fold expression changes in different time point comparisons. Enrichment analysis determined maize gene sets that characterized each group as determined by statistical analysis. Fisher's exact test took into account the number of genes in the group query, the total number of genes in a gene set, and the number of overlapping genes. A multiple test false discovery rate correction using the Yekutieli method was set to a cutoff P-value at 0.05. Additional annotations were determined by MapMan genome release for *Zea mays* based on B73 5b filtered gene sets (http://mapman.gabipd.org/) [22].

Maize sequence similarity

Maize genes with high sequence similarity to somatic embryogenesis related genes were determined by comparing the maize 5b.60 protein sequences to the protein sequence of previously cloned and characterized genes using BLASTP in the MaizeGDB BLAST POPcorn Project Portal (http://popcorn.maizegdb.org/main/index.php). Input parameters were set to an e-value cutoff of 1e-4 and the maximum number of hits was set to 500. Maize genes with a percent identity greater than or equal to 50% were analyzed for the presence of the conserved binding domain or other features specific to the gene of interest in the National Center for Biotechnology Information (NCBI) Batch Web CD-Search Tool (http://www.ncbi.nlm.nih.gov/Structure/bwrpsb/bwrpsb.cgi) and NCBI Conserved Domains CD-Search tool (http://www.ncbi.nlm.nih.gov/Structure/cdd/wrpsb.cgi) to determine genes with the best match. Sequence similarity reported in this study by pairwise alignment was done in LALIGN (http://embnet.vital-it.ch/software/LALIGN_form.html) as the percent identity by local or global alignment.

Supporting Information

Figure S1 Sequence alignment to determine sequence similarity of BABY BOOM in maize. Multiple sequence alignment of the conserved 147 amino acid sequence of BABY BOOM1 (gene accession AF317904) and three maize annotated proteins with high sequence similarity: GRMZM2G366434_P01, GRMZM2G141638_P01, and GRMZM2G139082_P02.

Figure S2 Sequence alignment to determine sequence similarity of LEAFY COTYELODN2 in maize. Sequence alignment between LEAFY COTYELODN2 in Arabidopsis (gene accession AF400124) and maize protein GRMZM2G405699_P01.

Figure S3 Sequence alignment to determine sequence similarity of CLAVATA in maize. Sequence alignment of the 191 amino acid translated sequence representing the catalytic domain of protein kinases superfamily of CLAVATA (CLV1) in Arabidopsis (gene accession U96879) and the maize protein GRMZM2G141517_P01.

Figure S4 Gene expression trends in early somatic embryogenesis of maize genes that are similar to AGL15. Fragments per kilobase of exon model per million fragments mapped (FPKM) of maize genes with high sequence similarity to Arabidopsis gene AGL15 (gene accession U22528) associated with embryogenic callus induction expressed in immature zygotic embryos of maize inbred line A188 at 0, 24, 36, 48, and 72 h after placement on culture initiation medium. Genes shown include (A) ZmMADS69 (GRMZM2G171650), (B) ZmMADS52 (GRMZM2G446426), (C) ZmMADS73 (GRMZM2G046885), and (D) SILKY1 (GRMZM2G139073). (n = 4 for 0, 24, 36, and 48 h include technical and biological replicates; n = 2 for 72 h include only technical replicates).

Table S1 Summary data on transcripts detected. A summary of the number of reads mapped for each sample of immature zygotic embryos from two ears of maize inbred line A188 collected at 0, 24, 36, 48 h and one ear collected at 72 h after placement on tissue culture initiation medium.

Table S2 Correlation of transcript abundance between biological replicates across time. Pearson's correlation of transcript abundance estimates measured as fragments per kilobase of exon model per million fragments mapped (FPKM) between samples of immature zygotic embryo explants from two maize ears of inbred line A188 collected at 0, 24, 36, 48, and 72 h after placement on tissue culture initiation medium.

Table S3 Gene expression of reference genes and representative transcript assemblies (RTA) detected in early somatic embryogenesis. Fragments per kilobase of exon model per million fragments mapped (FPKM) of gene transcripts detected in immature zygotic embryo explants of maize in inbred line A188 at 0, 24, 36, 48, and 72 h after placement on tissue culture initiation medium averaged across two biological replicates and gene annotations based on MapMan 5b filtered gene set gene names and annotations for the joint loci representative transcript assemblies (RTAs).

Table S4 Genes differentially expressed at each time point compared to control and expression trend. Fragments per kilobase of exon model per million fragments mapped (FPKM) values of genes or representative transcript assemblies differentially expressed 8-fold at time point comparisons 0 vs 24, 0 vs 36, 0 vs 48, and 0 vs 72 h in immature zygotic embryo explants of maize in inbred line A188 at 0, 24, 36, 48, and 72 h after placement on tissue culture initiation medium.

Table S5 Gene ontology describing genes differentially expressed compared to control. Gene ontology and enrichment summaries for genes that displayed an 8-fold expression change or greater at time point comparisons 0 vs 24, 0 vs 36, 0 vs 48, and 0 vs 72 h in immature zygotic embryo explants of maize in inbred line A188 at 0, 24, 36, 48, and 72 h after placement on tissue culture initiation medium.

Table S6 K-means clustering of genes expressed in early somatic embryogenesis. K-means clustering results of genes expressed in immature zygotic embryo explants of maize in inbred line A188 at 0, 24, 36, 48, and 72 h after placement on tissue culture initiation medium was performed allowing 6 clusters and gene annotations based on MapMan 5b filtered gene set gene names and annotations for the joint loci representative transcript assemblies (RTA).

Table S7 Gene onotology describing genes grouped by k-means clustering. Gene ontology enrichment analysis for genes in one of six clusters as determined by k-means analysis of transcripts detected in immature zygotic embryo explants of maize in inbred line A188 at 0, 24, 36, 48, and 72 h after placement on tissue culture initiation medium.

Table S8 Results of coexpression analysis between somatic embryogenesis-related genes. Coexpression analysis was done with transcripts detected in early somatic embryogenesis of the maize inbred line A188 at 0, 24, 36, 48, and 72 h after placement of explant tissues on tissue culture medium where select maize genes were determined as subjects such as maize genes with high sequence similarity to BABY BOOM1 (BBM1), LEAFY COTYLEDON2 (LEC2), and CLA-VATA1 (CLV1), and maize homologs for germin-like protein 2-1 (ZmGLP2-1), glutathione S-transferase 18, 19, and 22 (ZmGST 18, ZmGST 19, ZmGST22), LEAFY COTYLEDON1 (ZmLEC1), PINFORMED1 (ZmPIN1a), SOMATIC EMBRYO-GENESIS-LIKE KINASE1 and 3 (ZmSERK1 and ZmSERK3), WUSCHEL1 (ZmWUS1), and Wuschel-related homeobox domain 5 (ZmWOX5B) were used to determined corresponding maize genes as targets with a correlation co-efficient greater than and equal to 0.75.

Author Contributions

Conceived and designed the experiments: SAGDS SMK HFK. Performed the experiments: SAGDS CNH. Analyzed the data: SAGDS CNH. Contributed reagents/materials/analysis tools: SAGDS CNH HFK SMK CRB. Contributed to the writing of the manuscript: SAGDS HFK.

References

1. Borlaug NE (2000) Ending world hunger. The promise of biotechnology and the threat of antiscience zealotry. Plant Physiol 124: 487–490.
2. Delporte F, Jacquemin J-M, Masson P, Watillon B (2012) Insights into the regenerative property of plant cells and their receptivity to transgenesis: wheat as a research case study. Plant Signal Behav 7: 1608–1620.
3. Barampuram S, Zhang ZJ (2011) Recent advances in plant transformation. In: Birchler JA, editor. Plant Chromosome Engineering. pp. 1–35.
4. Gray DJ, Purohit A (1991) Somatic embryogenesis and development of synthetic seed technology. CRC Crit Rev Plant Sci 10: 33–61.
5. Murray SC, Eckhoff P, Wood L, Paterson AH (2013) A proposal to use gamete cycling in vitro to improve crops and livestock. Nat Biotechnol 31: 877–880.
6. Feher A, Pasternak TP, Dudits D (2003) Transition of somatic plant cells to an embryogenic state. Plant Cell Tissue Organ Cult 74: 201–228.
7. Yang XY, Zhang XL (2010) Regulation of somatic embryogenesis in higher plants. CRC Crit Rev Plant Sci 29: 36–57.
8. Zavattieri MA, Frederico AM, Lima M, Sabino R, Arnholdt-Schmitt B (2010) Induction of somatic embryogenesis as an example of stress-related plant reactions. Electron J Biotechnol 13: Available: http://www.ejbiotechnology.info/content/vol13/issue11/full/14/index.html Accessed 2011 January 12.
9. Karami O, Saidi A (2010) The molecular basis for stress-induced acquisition of somatic embryogenesis. Mol Biol Rep 37: 2493–2507.
10. Slater A, Scott NW, Fowler MR (2007) Plant tissue culture. In: Slater A, Scott NW, Fowler MR, editors. Plant biotechnology: The genetic manipulation of plants. pp. 37–53.
11. Targonska M, Hromada-Judycka A, Bolibok-Bragoszewska H, Rakoczy-Trojanowska M (2013) The specificity and genetic background of the rye (Secale cereale L.) tissue culture response. Plant Cell Rep 32: 1–9.
12. Armstrong CL, Romeroseverson J, Hodges TK (1992) Improved tissue-culture response of an elite maize inbred through backcross breeding, and identification

of chromosomal regions important for regeneration by RFLP analysis. Theor Appl Genet 84: 755–762.

13. Hodges TK, Kamo KK, Imbrie CW, Becwar MR (1986) Genotype specificity of somatic embryogenesis and regeneration in maize. Biotechnology (N Y) 4: 219–223.

14. Landi P, Chiappetta L, Salvi S, Frascaroli E, Lucchese C, et al. (2002) Responses and allelic frequency changes associated with recurrent selection for plant regeneration from callus cultures in maize. Maydica 47: 21–32.

15. Green CE, Phillips RL (1975) Plant regeneration from tissue cultures of maize. Crop Sci 15: 417–421.

16. Songstad DD, Armstrong CL, Petersen WL (1991) AgNO₃ increases type-II callus production from immature embryos of maize inbred B73 and its derivatives. Plant Cell Rep 9: 699–702.

17. Schnable PS, Ware D, Fulton RS, Stein JC, Wei FS, et al. (2009) The B73 maize genome: Complexity, diversity, and dynamics. Science 326: 1112–1115.

18. Hirsch CN, Foerster JM, Johnson JM, Sekhon RS, Muttoni G, et al. (2014) Insights into the maize pan-genome and pan-transcriptome. Plant Cell 26: 121–135.

19. Trapnell C, Williams BA, Pertea G, Mortazavi A, Kwan G, et al. (2010) Transcript assembly and quantification by RNA-Seq reveals unannotated transcripts and isoform switching during cell differentiation. Nat Biotechnol 28: 511–515. doi:510.1038/nbt.1621

20. Feher A (2008) The initiation phase of somatic embryogenesis: what we know and what we don't. Acta Biol Szeged 52: 53–56.

21. Zeng FC, Zhang XL, Cheng L, Hu LS, Zhu LF, et al. (2007) A draft gene regulatory network for cellular totipotency reprogramming during plant somatic embryogenesis. Genomics 90: 620–628.

22. Usadel B, Poree F, Nagel A, Lohse M, Czedik-Eysenberg A, et al. (2009) A guide to using MapMan to visualize and compare Omics data in plants: A case study in the crop species, Maize. Plant Cell Environ 32: 1211–1229.

23. McGonigle B, Keeler SJ, Lan SMC, Koeppe MK, O'Keefe DP (2000) A genomics approach to the comprehensive analysis of the glutathione S-transferase gene family in soybean and maize. Plant Physiol 124: 1105–1120.

24. Breen J, Bellgard M (2010) Germin-like proteins (GLPs) in cereal genomes: gene clustering and dynamic roles in plant defence. Funct Integr Genomics 10: 463–476.

25. Forestan C, Varotto S (2012) The role of PIN auxin efflux carriers in polar auxin transport and accumulation and their effect on shaping maize development. Mol Plant 5: 787–798.

26. Shen Y, Jiang Z, Yao XD, Zhang ZM, Lin HJ, et al. (2012) Genome expression profile analysis of the immature maize embryo during dedifferentiation. PloS ONE 7: e32237. doi:32210.31371/journal.pone.0032237

27. Huynh QK, Hironaka CM, Levine EB, Smith CE, Borgmeyer JR, et al. (1992) Anitfungal proteins from plants - purification, molecular cloning, and antifulgal properties of chitinases from maize seed. J Biol Chem 267: 6635–6640.

28. Chintamanani S, Hulbert SH, Johal GS, Balint-Kurti PJ (2010) Identification of a maize locus that modulates the hypersensitive defense response, using mutant-assisted gene identification and characterization. Genetics 184: 813–825.

29. Dejong AJ, Cordewener J, Loschiavo F, Terzi M, Vandekerckhove J, et al. (1992) A carrot somatic embryo mutant is rescued by chitinase. Plant Cell 4: 425–433.

30. Galland R, Blervacq A-S, Blassiau C, Smagghe B, Decottignies J-P, et al. (2007) Glutathione-S-transferase is detected during somatic embryogenesis in chicory. Plant Signal Behav 2: 343–348.

31. Hoenemann C, Ambold J, Hohe A (2012) Gene expression of a putative glutathione S-transferase is responsive to abiotic stress in embryogenic cell cultures of *Cyclamen persicum*. Electron J Biotechnol 15: Available: http://www.ejbiotechnology.info/index.php/ejbiotechnology/article/view/v15n11-19/1408 Accessed 2014 September 24.

32. Caliskan M, Turet M, Cuming AC (2004) Formation of wheat (*Triticum aestivum L.*) embryogenic callus involves peroxide-generating germin-like oxalate oxidase. Planta 219: 132–140.

33. Boutilier K, Offringa R, Sharma VK, Kieft H, Ouellet T, et al. (2002) Ectopic expression of BABY BOOM triggers a conversion from vegetative to embryonic growth. Plant Cell 14: 1737–1749.

34. Passarinho P, Ketelaar T, Xing MQ, van Arkel J, Maliepaard C, et al. (2008) BABY BOOM target genes provide diverse entry points into cell proliferation and cell growth pathways. Plant Mol Biol 68: 225–237.

35. Deng W, Luo K, Li Z, Yang Y (2009) A novel method for induction of plant regeneration via somatic embryogenesis. Plant Sci 177: 43–48.

36. Srinivasan C, Liu ZR, Heidmann I, Supena EDJ, Fukuoka H, et al. (2007) Heterologous expression of the BABY BOOM AP2/ERF transcription factor enhances the regeneration capacity of tobacco (*Nicotiana tabacum L.*). Planta 225: 341–351.

37. Heidmann I, de Lange B, Lambalk J, Angenent GC, Boutilier K (2011) Efficient sweet pepper transformation mediated by the BABY BOOM transcription factor. Plant Cell Rep 30: 1107–1115.

38. Ananiev EV, Wu C, Chamberlin MA, Svitashev S, Schwartz C, et al. (2009) Artificial chromosome formation in maize (*Zea mays L.*). Chromosoma 118: 157–177.

39. Sekhon RS, Lin HN, Childs KL, Hansey CN, Buell CR, et al. (2011) Genome-wide atlas of transcription during maize development. Plant J 66: 553–563.

40. Sekhon RS, Briskine R, Hirsch CN, Myers CL, Springer NM, et al. (2013) Maize gene atlas developed by RNA sequencing and comparative evaluation of transcriptomes based on RNA sequencing and microarrays. PLoS ONE 8: e61005. doi:61010.61371/journal.pone.0061005

41. Gaj MD, Zhang SB, Harada JJ, Lemaux PG (2005) Leafy cotyledon genes are essential for induction of somatic embryogenesis of Arabidopsis. Planta 222: 977–988.

42. Meinke DW, Franzmann LH, Nickle TC, Yeung EC (1994) LEAFY COTYLEDON mutants of Arabidopsis. Plant Cell 6: 1049–1064.

43. Braybrook SA, Harada JJ (2008) LECs go crazy in embryo development. Trends Plant Sci 13: 624–630.

44. Jia H, Suzuki M, McCarty DR (2014) Regulation of the seed to seedling developmental phase transition by the LAFL and VAL transcription factor networks. Wiley Interdiscip Rev Dev Biol 3: 135–145.

45. Lotan T, Ohto M, Yee KM, West MAL, Lo R, et al. (1998) Arabidopsis LEAFY COTYLEDON1 is sufficient to induce embryo development in vegetative cells. Cell 93: 1195–1205.

46. Ledwon A, Gaj MD (2011) LEAFY COTYLEDON1, FUSCA3 expression and auxin treatment in relation to somatic embryogenesis induction in Arabidopsis. Plant Growth Regul 65: 157–167.

47. Zhang SB, Wong L, Meng L, Lemaux PG (2002) Similarity of expression patterns of knotted1 and ZmLEC1 during somatic and zygotic embryogenesis in maize (*Zea mays L.*). Planta 215: 191–194.

48. Stone SL, Kwong LW, Yee KM, Pelletier J, Lepiniec L, et al. (2001) LEAFY COTYLEDON2 encodes a B3 domain transcription factor that induces embryo development. Proc Natl Acad Sci U S A 98: 11806–11811.

49. Stone SL, Braybrook SA, Paula SL, Kwong LW, Meuser J, et al. (2008) Arabidopsis LEAFY COTYLEDON2 induces maturation traits and auxin activity: Implications for somatic embryogenesis. Proc Natl Acad Sci U S A 105: 3151–3156.

50. Suzuki M, Wang HHY, McCarty DR (2007) Repression of the LEAFY COTYLEDON 1/B3 regulatory network in plant embryo development by VP1/ABSCISIC ACID INSENSITIVE 3-LIKE B3 genes. Plant Physiol 143: 902–911.

51. Petrasek J, Friml J (2009) Auxin transport routes in plant development. Development 136: 2675–2688.

52. Su YH, Zhao XY, Liu YB, Zhang CL, O'Neill SD, et al. (2009) Auxin-induced WUS expression is essential for embryonic stem cell renewal during somatic embryogenesis in Arabidopsis. Plant J 59: 448–460.

53. Weijers D, Sauer M, Meurette O, Friml J, Ljung K, et al. (2005) Maintenance of embryonic auxin distribution for apical-basal patterning by PIN-FORMED-dependent auxin transport in Arabidopsis. Plant Cell 17: 2517–2526.

54. Forestan C, Meda S, Varotto S (2010) ZmPIN1-mediated auxin transport is related to cellular differentiation during maize embryogenesis and endosperm development. Plant Physiol 152: 1373–1390.

55. Carraro N, Forestan C, Canova S, Traas J, Varotto S (2006) ZmPIN1a and ZmPIN1b encode two novel putative candidates for polar auxin transport and plant architecture determination of maize. Plant Physiol 142: 254–264.

56. Mayer KFX, Schoof H, Haecker A, Lenhard M, Jurgens G, et al. (1998) Role of WUSCHEL in regulating stem cell fate in the Arabidopsis shoot meristem. Cell 95: 805–815.

57. Leibfried A, To JPC, Busch W, Stehling S, Kehle A, et al. (2005) WUSCHEL controls meristem function by direct regulation of cytokinin-inducible response regulators. Nature 438: 1172–1175.

58. Liu XG, Kim YJ, Muller R, Yumul RE, Liu CY, et al. (2011) AGAMOUS terminates floral stem cell maintenance in Arabidopsis by directly repressing WUSCHEL through recruitment of polycomb group proteins. Plant Cell 23: 3654–3670.

59. Sang Y, Wu MF, Wagner D (2009) The stem cell-Chromatin connection. Semin Cell Dev Biol 20: 1143–1148.

60. Nardmann J, Werr W (2006) The shoot stem cell niche in angiosperms: Expression patterns of WUS orthologues in rice and maize imply major modifications in the course of mono- and dicot evolution. Mol Biol Evol 23: 2492–2504.

61. Gambino G, Minuto M, Boccacci P, Perrone I, Vallania R, et al. (2011) Characterization of expression dynamics of WOX homeodomain transcription factors during somatic embryogenesis in *Vitis vinifera*. J Exp Bot 62: 1089–1101.

62. Palovaara J, Hakman I (2008) Conifer WOX-related homeodomain transcription factors, developmental consideration and expression dynamic of WOX2 during *Picea abies* somatic embryogenesis. Plant Mol Biol 66: 533–549.

63. Clark SE, Williams RW, Meyerowitz EM (1997) The CLAVATA1 gene encodes a putative receptor kinase that controls shoot and floral meristem size in Arabidopsis. Cell 89: 575–585.

64. Schoof H, Lenhard M, Haecker A, Mayer KFX, Jurgens G, et al. (2000) The stem cell population of Arabidopsis shoot meristems is maintained by a regulatory loop between the CLAVATA and WUSCHEL genes. Cell 100: 635–644.

65. Heck GR, Perry SE, Nichols KW, Fernandez DE (1995) AGL15, a MADS domain protein expressed in developing embryos. Plant Cell 7: 1271–1282.

66. Lenhard M, Bohnert A, Jurgens G, Laux T (2001) Termination of stem cell maintenance in Arabidopsis floral meristems by interactions between WUSCHEL and AGAMOUS. Cell 105: 805–814.

67. Perry SE, Lehti MD, Fernandez DE (1999) The MADS-domain protein AGAMOUS-like 15 accumulates in embryonic tissues with diverse origins. Plant Physiol 120: 121–129.

68. Thakare D, Tang W, Hill K, Perry SE (2008) The MADS-domain transcriptional regulator AGAMOUS-LIKE15 promotes somatic embryo development in Arabidopsis and soybean. Plant Physiol 146: 1663–1672.

69. Zheng Y, Ren N, Wang H, Stromberg AJ, Perry SE (2009) Global identification of targets of the Arabidopsis MADS domain protein AGAMOUS-Like15. Plant Cell 21: 2563–2577.

70. Karlova R, Boeren S, Russinova E, Aker J, Vervoort J, et al. (2006) The Arabidopsis SOMATIC EMBRYOGENESIS RECEPTOR-LIKE KINASE1 protein complex includes BRASSINOSTEROID-INSENSITIVE1. Plant Cell 18: 626–638.

71. Frame B, Main M, Schick R, Wang K (2011) Genetic transformation using maize immature zygotic embryos. In: Thorpe TA, Yeung EC, editors. Plant embryo culture: Methods and protocols. pp. 327–341.

72. Chu CC, Wang CC, Sun CS, Chen H, Yin KC, et al. (1975) Establishment of an efficient medium for anther culture of rice through comparative experiments on nitrogen sources. Sci Sin 18: 659–668.

73. Langmead B, Trapnell C, Pop M, Salzberg SL (2009) Ultrafast and memory-efficient alignment of short DNA sequences to the human genome. Genome Biol 10: R25.

74. Trapnell C, Pachter L, Salzberg SL (2009) TopHat: Discovering splice junctions with RNA-Seq. Bioinformatics 25: 1105–1111.

75. Yi X, Du Z, Su Z (2013) PlantGSEA: A gene set enrichment analysis toolkit for plant community. Nucleic Acids Res 41: W98–W103.

76. Ledwon A, Gaj MD (2009) LEAFY COTYLEDON2 gene expression and auxin treatment in relation to embryogenic capacity of Arabidopsis somatic cells. Plant Cell Rep 28: 1677–1688.

77. Schmidt RJ, Veit B, Mandel MA, Mena M, Hake S, et al. (1993) Identification and molecular characterization of ZAG1, the maize homolog of the Arabidopsis floral homeotic gene AGAMOUS. Plant Cell 5: 729–737.

78. Shen B, Allen WB, Zheng P, Li C, Glassman K, et al. (2010) Expression of ZmLEC1 and ZmWRI1 increases seed oil production in maize. Plant Physiol 153: 980–987.

79. Mena M, Ambrose BA, Meeley RB, Briggs SP, Yanofsky MF, et al. (1996) Diversification of C-function activity in maize flower development. Science 274: 1537–1540.

80. Baudino S, Hansen S, Brettschneider R, Hecht VRG, Dresselhaus T, et al. (2001) Molecular characterisation of two novel maize LRR receptor-like kinases, which belong to the SERK gene family. Planta 213: 1–10.

81. Zhang S, Liu X, Lin Y, Xie G, Fu F, et al. (2011) Characterization of a ZmSERK gene and its relationship to somatic embryogenesis in a maize culture. Plant Cell Tissue Organ Cult 105: 29–37.

Permissions

List of Contributors

Blaire J. Steinwand, Shouling Xu, Joanna K. Polko, Stephanie M. Doctor, Mike Westafer and Joseph J. Kieber
Biology Department, University of North Carolina, Chapel Hill, North Carolina, United States of America

Ali Taheri, Isobel A. P. Parkin and Margaret Y. Gruber
Agriculture and Agri-Food Canada, Saskatoon Research Centre, Saskatoon, SK, Canada

Naghabushana K. Nayidu
Department of Biology, University of Saskatchewan, Saskatoon SK, Canada

Sateesh Kagale
Agriculture and Agri-Food Canada, Saskatoon Research Centre, Saskatoon, SK, Canada
National Research Council (NRC), Saskatoon SK, Canada, 4 POS Bio-Sciences, Saskatoon, SK, Canada

Andrew G. Sharpe
National Research Council (NRC), Saskatoon SK, Canada

Thushan S. Withana-Gamage
POS Bio-Sciences, Saskatoon, SK, Canada

Yong-Ju Huang, Aiming Qi and Bruce D. L. Fitt
School of Life and Medical Sciences, University of Hertfordshire, Hatfield, Hertfordshire, United Kingdom
Department of Plant Pathology and Microbiology, Rothamsted Research, Harpenden, Hertfordshire, United Kingdom

Graham J. King
Department of Plant Pathology and Microbiology, Rothamsted Research, Harpenden, Hertfordshire, United Kingdom
Southern Cross Plant Science, Southern Cross University, Lismore, Australia

Qingyao Lu, Lin Zhao, Dongmei Li, Diqiu Hao and Wenbin Li
Key Laboratory of Soybean Biology of Chinese Education Ministry (Key Laboratory of Biology and Genetics & Breeding for Soybean in Northeast China, Ministry of Agriculture), Northeast Agriculture University, Harbin, China

Yong Zhan
Agricultural Academy of Shi He Zi, Xinjiang Province, China

Danielle A. Orozco-Nunnelly, DurreShahwar Muhammad, Raquel Mezzich, Lon S. Kaufman and Katherine M. Warpeha
Molecular, Cell and Developmental Group, Department of Biological Sciences, Department of Biological Sciences, University of Illinois at Chicago (UIC), Chicago, Illinois, United States of America

Bao-Shiang Lee and Lasanthi Jayathilaka
Protein Research Laboratory, University of Illinois at Chicago (UIC), Chicago, Illinois, United States of America

Hung-Yi Wu and Erh-Min Lai
Institute of Plant and Microbial Biology, Academia Sinica, Taipei, Taiwan
Department of Plant Pathology and Microbiology, National Taiwan University, Taipei, Taiwan

Chao-Ying Chen
Department of Plant Pathology and Microbiology, National Taiwan University, Taipei, Taiwan

Rohan G. T. Lowe, Barbara J. Howlett, Bethany L. Clark and Angela P. Van de Wouw
School of Botany, The University of Melbourne, Parkville, Victoria, Australia

Andrew Cassin
ARC Centre of Excellence in Plant Cell Walls, School of Botany, The University of Melbourne, Parkville, Victoria, Australia

Jonathan Grandaubert and Thierry Rouxel
INRA-Bioger, UR1290, Thiverval-Grignon, France

Quélen L. Barcelos and Elaine A. Souza
Departamento de Biologia, Universidade Federal de Lavras, Lavras, Minas Gerais, Brazil

Joyce M. A. Pinto
Empresa Brasileira de Pesquisa Agropecuária (Embrapa), Sinop, Mato Grosso Brazil

Lisa J. Vaillancourt
Department of Plant Pathology, University of Kentucky, Lexington, Kentucky, United States of America

Ping Wang
School of Environmental Science and Public Health, Wenzhou Medical University, Wenzhou, China Key Laboratory of Coastal and Wetland Ecosystems (Xiamen University), Ministry of Education, Xiamen, China

Tun-Hua Wu
School of Information and Engineering, Wenzhou Medical University, Wenzhou, China,

Yong Zhang
State Key Laboratory of Marine Environmental Science (Xiamen University), College of the Environment and Ecology, Xiamen University, Xiamen, China

Wolfgang Goettel and Yong-Qiang (Charles) An
United States Department of Agriculture, Agricultural Research Service, Plant Genetics Research Unit, Donald Danforth Plant Science Center, Saint Louis, Missouri, United States of America

Zongrang Liu
United States Department of Agriculture, Agricultural Research Service, Appalachian Fruit Research Station, Kearneysville, West Virginia, United States of America

Jing Xia and Weixiong Zhang
Department of Computer Science and Engineering, Washington University, Saint Louis, Missouri, United States of America

Patrick X. Zhao
Plant Biology Division, The Samuel Roberts Noble Foundation, Ardmore, Oklahoma, United States of America

Ting Liao, De-Yi Yuan, Feng Zou, Chao Gao, Ya Yang, Lin Zhang and Xiao-Feng Tan
Key Laboratory of Cultivation and Protection for Non-Wood Forest Trees, Ministry of Education, The Key Lab of Non-Wood Forest Products of Forestry Ministry, Central South University of Forestry and Technology, Changsha, Hunan, China

Ken Harata and Yasuyuki Kubo
Laboratory of Plant Pathology, Graduate School of Life and Environmental Sciences, Kyoto Prefectural University, Kyoto, Japan

Ricardo Mir and José León
Instituto de Biología Molecular y Celular de Plantas, Consejo Superior de Investigaciones Científicas-Universidad Politécnica de Valencia, Valencia, Spain

Hongyan Meng and Bosi Xu
Biotechnology Research Institute, Chinese Academy of Agricultural Sciences, Beijing, People's Republic of China

Ling Jiang, Yunliu Fan and Chunyi Zhang
Biotechnology Research Institute, Chinese Academy of Agricultural Sciences, Beijing, People's Republic of China National Key Facility for Crop Gene Resources and Genetic Improvement (NFCRI), Beijing, People's Republic of China

Wenzhu Guo
Huazhong Agricultural University, Wuhan, People's Republic of China

Jinglai Li and Xiuqing Zhu
Beijing Institute of Pharmacology and Toxicology, Beijing, People's Republic of China

Xiaoquan Qi, Lixin Duan and Xianbin Meng
Institute of Botany, Chinese Academy of Sciences, Beijing, People's Republic of China

David G. Heckel
Department of Entomology, Max Planck Institute for Chemical Ecology, Jena, Germany

Francisco R. Badenes-Pere
Department of Entomology, Max Planck Institute for Chemical Ecology, Jena, Germany Instituto de Ciencias Agrarias, Consejo Superior de Investigaciones Cientı́ficas Madrid, Spain

Jonathan Gershenzon
Department of Biochemistry, Max Planck Institute for Chemical Ecology, Jena, Germany

Yuan Niu, Su-Hong Bu, Meng Li, Jian-Ying Feng, Jin Zhang, Sheng- Xian Yang, Medrine Mmayi Odinga, Shi-Ping Wei, Xiao-Feng Liu and Yuan-Ming Zhang
Section on Statistical Genomics, State Key Laboratory of Crop Genetics and Germplasm Enhancement, Department of Crop Genetics and Breeding, Nanjing Agricultural University, Nanjing, Jiangsu, China

Wen-Jie Zhang
Section on Statistical Genomics, State Key Laboratory of Crop Genetics and Germplasm Enhancement, Department of Crop Genetics and Breeding, Nanjing Agricultural University, Nanjing, Jiangsu, China Institute of Crop Research, Ningxia Academy of Agriculture and Forestry Sciences, Yinchuan, Ningxia, China

Stella A. G. D. Salvo, Shawn M. Kaeppler and Heidi F. Kaeppler
Department of Agronomy, University of Wisconsin, Madison, Wisconsin, United States of America

Candice N. Hirsch
Department of Agronomy and Plant Genetics, University of Minnesota, St. Paul, Minnesota, United States of America

C. Robin Buell
Department of Plant Biology, Michigan State University, East Lansing, Michigan, United States of America
DOE Great Lakes Bioenergy Research Center, Michigan State University, East Lansing, Michigan, United States of America

Index

www.ingramcontent.com/pod-product-compliance
Lightning Source LLC
Chambersburg PA
CBHW080412190526
45161CB00003B/212